Group Explicit Methods
for the Numerical Solution of
Partial Differential Equations

TOPICS IN COMPUTER MATHEMATICS

A series edited by David J. Evans, Loughborough University of Technology, UK

This book is part of a series. The publisher will accept continuation orders which may be cancelled at any time and which provide for automatic billing and shipping of each title in the series upon publication. Please write for details.

Group Explicit Methods for the Numerical Solution of Partial Differential Equations

By

DAVID J. EVANS

Loughborough University of Technology, UK

Routledge
Taylor & Francis Group

LONDON AND NEW YORK

First published 1997 by Overseas Publishers Association

Published 2019 by Routledge
2 Park Square, Milton Park, Abingdon, Oxon OX14 4RN
52 Vanderbilt Avenue, New York, NY 10017

Routledge is an imprint of the Taylor & Francis Group, an informa business

British Library Cataloging-in-Publication Data
Evans, David J.
   Group explicit methods for the numerical solution of
   partial differential equations. – (Topics in computer maths;
   v. 7)
   1. Differential equations, Partial – Numerical solutions
   2. Differential equations, Partial – Numerical solutions –
   Data processing
   I. Title
   515.3'53'0285

ISBN 13: 978-90-5699-019-0 (hbk)

Contents

Contents

Preface

The numerical solution of partial differential equations has been the subject of intense activity during the last 40 years or so, primarily due to advances in computer technology which in turn has led to improvements in the numerical methods that are used. Consequently, many intractable scientific and engineering problems that involve partial differential equations that were previously unsolved can now be resolved.

This book is primarily concerned with finite difference methods for solving partial differential equations as reflected by the contents of Chapters 1 and 2. However, the progress in computer technology still continues unabated and recently, with the introduction of parallel and vector computers, the emphasis on the early 'explicit methods' (that were limited in use) has become evident for they possess a quality which makes them amenable for use on parallel computers.

Hence, the introduction of a new concept 'the group explicit method' in Chapter 3 which is then applied to solve parabolic partial differential equations is explored. The method is then further developed as the Alternating Group Explicit (AGE) method and its analogy with the now famous Alternating Direction Implicit (ADI) method in Chapter 4. The AGE method is then extended to elliptic boundary value problems and hyperbolic equations in Chapters 5 and 6. The generality of the method is demonstrated further by its applicability to multi-dimensional problems which is an area that will shortly be widely explored by the new parallel computers. Finally in Chapter 8 the parallel implementation of the AGE method is developed for a shared memory parallel computer.

The author is greatly indebted to his many research students who have explored these methods with him, in particular, A. Abdul Rahman, M. S. Sahimi, W. S. Yousif, C. Li, A. Ahmad and A. Al-Wali. Final thanks must go to his secretary, Mrs. J. M. Poulton for swift and accurate typing of the manuscript.

1. Introduction

1.1. Introduction

The majority of the problems of physics and engineering fall naturally into one of three *physical* categories: *equilibrium problems*, *eigenvalue problems* and *propagation problems*. Equilibrium problems are problems of steady state in which the equilibrium configurations in a certain domain are to be determined. Eigenvalue problems may be thought of as extensions of equilibrium problems wherein critical values of certain parameters are to be determined in addition to the corresponding steady-state configurations. Propagation problems are initial value problems that have an unsteady state or transient nature (Ames, 1977). Normally, in most cases, the dependent variable to any of these problems is expressed in terms of several independent variables. Such problems inherently give rise to the need for partial derivatives in the description of their behaviour. The study of differential equations arising from these problems constitutes the field of *Partial Differential Equations*.

Mathematically a *partial differential equation* (henceforth abbreviated as p.d.e.) for a dependent variable $u(x, y, ...)$ is a relation of the form

$$F\left(x, y, ..., U, \frac{\partial U}{\partial x}, \frac{\partial U}{\partial y}, ..., \frac{\partial^2 U}{\partial x^2}, \frac{\partial^2 U}{\partial x \partial y}, \frac{\partial^2 U}{\partial y^2}, ...\right) = 0, \qquad (1.1.1)$$

where F is a given function of the independent variables $x, y, ...$ and the "unknown" function U and of a *finite* number of its partial derivatives. We call u a *solution* of (1.1.1) if after substitution of $U(x, y, ...)$ and its partial derivatives (1.1.1) is satisfied identically in $x, y, ...$ in some region Ω in the space of these independent variables. The independent variables $x, y, ...$ are real (unless if stated otherwise) and U and the derivatives of U occurring in (1.1.1) are continuous functions of $x, y, ...$ in the real domain Ω.

A p.d.e (1.1.1) is said to be of *order n* if the order of the highest partial derivatives involved is n. Equation (1.1.1) is said to be *linear* if F is linear in the unknown function and its derivatives and *quasi-linear* if F is linear in at least the highest order derivatives.

For example, the equation

$$\sqrt{x} \frac{\partial U}{\partial x} + U \frac{\partial U}{\partial y} = -U^2, \qquad (1.1.2)$$

1

is a first order quasi-linear p.d.e., and the equation

$$\frac{\partial^2 U}{\partial x^2} + \frac{\partial^2 U}{\partial y^2} - 32U = 0 \qquad (1.1.3)$$

is a second order linear p.d.e. Meanwhile the equation

$$\left(\frac{\partial^2 U}{\partial x^2}\right)\left(\frac{\partial^2 U}{\partial y^2}\right) - \left(\frac{\partial^2 U}{\partial x \partial y}\right)^2 = 0 \qquad (1.1.4)$$

is the example of a second order *non-linear* p.d.e. In equations (1.1.2–1.1.4) above, x and y are the independent variables and $U = U(x,y)$ is the dependent variable whose value is to be found.

A linear p.d.e. is said to be *homogeneous* if each term contains either the dependent variable or one of its derivatives. For example, equation

$$\frac{\partial U}{\partial t} - k\frac{\partial^2 U}{\partial x^2} = 0 \text{ (Heat Equation)} \qquad (1.1.5)$$

is homogeneous, whereas,

$$\frac{\partial U}{\partial t} - a\frac{\partial^2 U}{\partial x^2} = f(x,t), \quad a > 0, \qquad (1.1.6)$$

where $f(x,t)$ is a given function, is an *inhomogeneous* equation.

The problem of finding the solution to the p.d.e. is a very difficult problem as a general method of solution is not available except for certain special types of linear or quasi-linear equations. Furthermore, the *general solution* of the linear p.d.e. which contains *arbitrary functions* is of little use, since it has to be made to satisfy other conditions called *boundary conditions* which arise from the physical problem itself (see Section 1.3).

1.2. Classification of Partial Differential Equations

As the p.d.e. arises from different categories of physical phenomenon (e.g. steady viscous flow, resonance in electric circuits, propagation of heat, ...), this suggests that the governing equations are also quite different in nature. Normally, they are classified in terms of their mathematical form such as elliptic, hyperbolic or parabolic, or in terms of the type of problems to which they apply, i.e., the heat equation, the wave equation, ...

The most general second order linear p.d.e. in two independent variables is given by

$$L(U) \equiv A(x,y)\frac{\partial^2 U}{\partial x^2} + 2B(x,y)\frac{\partial^2 U}{\partial x \partial y} + C(x,y)\frac{\partial^2 U}{\partial y^2} + E\left(x, y, U, \frac{\partial U}{\partial x}, \frac{\partial U}{\partial y}\right) = 0. \quad (1.2.1)$$

It is called *elliptic, hyperbolic* or *parabolic* according to the determinant

$$\begin{vmatrix} B & A \\ C & B \end{vmatrix} \tag{1.2.2}$$

is negative, positive or zero, respectively. This classification depends in general on the region of the (x, y)-plane under consideration. Thus, it is possible for a p.d.e. to change its classification within different regions of the domain for which the problem is defined. The differential equation

$$y\frac{\partial^2 U}{\partial x^2} + x\frac{\partial^2 U}{\partial x\partial y} + y\frac{\partial^2 U}{\partial y^2} = F\left(x, y, U, \frac{\partial U}{\partial x}, \frac{\partial U}{\partial y}\right), \tag{1.2.3}$$

for instance, is elliptic for $|x| < 2|y|$, hyperbolic for $|x| > 2|y|$ and parabolic along $|x| = 2|y|$. This type of equation is said to be mixed type.

For a linear p.d.e. in more than two independent variables, i.e., for systems, and for a non-linear p.d.e., a similar but complicated classification can be carried out. In the case of three independent variables, the terms elliptic, parabolic and hyperbolic are replaced by their three-dimensional analogues such as ellipsoidal, etc.

The elliptic class are equilibrium problems and are usually described in terms of a closed region having boundary conditions prescribed at every point on the region's boundary. These are called the *boundary value problems*. The parabolic and hyperbolic problems are of propagation type and normally have a prescribed boundary condition on some part of the boundaries and an initial condition along the other part. It can also have open-ended regions into which the solution propagates. In mathematical parlance such problems are known as *initial (boundary) value problems*.

Another important aspect of the classification of a p.d.e. into hyperbolic, parabolic and elliptic is due to the characteristic equation. Here we can see that at every point of the x-y plane there are two directions in which the integration of the p.d.e. reduces to the integration of an equation involving total differentials only. In other words, the integration of an equation in certain directions is not complicated by the presence of partial derivatives in other directions (Smith, 1978).

Let the derivatives in equation (1.2.1) be denoted by

$$\frac{\partial U}{\partial x} = p; \quad \frac{\partial U}{\partial y} = q; \quad \frac{\partial^2 U}{\partial x^2} = r; \quad \frac{\partial^2 U}{\partial x\partial y} = s \quad \text{and} \quad \frac{\partial^2 U}{\partial y^2} = t.$$

Let C be a curve in the x-y plane on which the values of U and its derivatives satisfy equation (1.2.1). (C is not a curve on which initial values of U, p and q are given.) Therefore, the differentials of p and q in the directions tangential to C are given by

$$dp = \frac{\partial p}{\partial x}dx + \frac{\partial p}{\partial y}dy = rdx + sdy \tag{1.2.4}$$

and

$$dq = \frac{\partial q}{\partial x}dx + \frac{\partial q}{\partial y}dy = sdx + tdy, \tag{1.2.5}$$

where the p.d.e. (1.2.1) is written as

$$Ar + 2Bs + Ct + E = 0. \tag{1.2.6}$$

Elimination of r and t in (1.2.6) using (1.2.4) and (1.2.5) gives

$$A \frac{d}{dx}(dp - sdy) + 2Bs + C \frac{d}{dy}(dq - sdx) + E = 0,$$

i.e.,

$$s\left\{A\left(\frac{dy}{dx}\right)^2 - B\left(\frac{dy}{dx}\right) + C\right\} - \left\{A \frac{dp}{dx} \cdot \frac{dp}{dy} + C \frac{dq}{dx} + E \frac{dy}{dx}\right\} = 0. \tag{1.2.7}$$

Now choose dy/dx the tangent to C at a point $P(x, y)$ to satisfy

$$A\left(\frac{dy}{dx}\right)^2 - 2B\left(\frac{dy}{dx}\right) + C = 0. \tag{1.2.8}$$

Therefore (1.2.7) leads to

$$A \frac{dp}{dx} \cdot \frac{dp}{dy} + C \frac{dq}{dx} + E \frac{dy}{dx} = 0, \tag{1.2.9}$$

which gives the relationship between the total differentials dp and dq with respect to x and y.

This shows that at every point $P(x, y)$ of the solution domain there are two directions, given by the roots of equation (1.2.8), along which there is a relationship given by equation (1.2.9). The directions given by the roots of equation (1.2.8) are called the characteristic directions and the p.d.e. is said to be hyperbolic, parabolic or elliptic according to a similar determinant requirement as in (1.2.2).

1.3. Boundary Conditions

As mentioned earlier in the chapter, the solution of the p.d.e. has to be made to satisfy the boundary conditions which arise from the problem formulation. There are four main types of such conditions which arise frequently in the description of physical phenomena, these are:

1. *The First Boundary Value Problem (the Dirichlet Problem)*, where the solution u has to satisfy the given values

$$U\big|_s = \phi \tag{1.3.1}$$

on the boundary s. If $\phi \equiv 0$ the problem is called a *Homogeneous Dirichlet* problem.

2. *The Second Boundary Value Problem (the Neumann Problem)*, where the solution u has to satisfy the normal derivatives

$$\frac{\partial U}{\partial v}\bigg|_s = \Psi \tag{1.3.2}$$

on the boundary of that region, where Ψ is given.

3. *The Third Boundary Value Problem* (*Mixed or Robin's Problem*), where the solution *u* has to satisfy a combination of *u* and its derivatives, namely,

$$\left[\frac{\partial U}{\partial v} + hU \right]_s = \Psi \tag{1.3.3}$$

on the boundary *s*.

4. *The Fourth Boundary Value Problem* (*Periodic Boundary Problem*). In this case we seek the solution such that it satisfies the periodicity conditions, for example,

$$U|_x = U|_{x+\ell}, \quad \frac{\partial U}{\partial v}\bigg|_x = \frac{\partial U}{\partial v}\bigg|_{x+\ell}, \tag{1.3.4}$$

where ℓ is called the period.

The physical meaning of the first three boundary value problems can be illustrated by the problem of steady-state temperature distribution.

In the Dirichlet problem, the temperature is given on the boundary of a solid. In the Neumann problem, the loss or gain of heat through the boundary is given (it is proportional to $\partial U / \partial v$). In this problem in order to keep a steady-state distribution of temperature, the net flow of thermal energy passing through the boundary of a solid must be equal to zero, i.e.,

$$\int_s \Psi \, ds = 0. \tag{1.3.5}$$

The third-boundary problem deals with the heat exchange with the surrounding medium the temperature of which is Ψ/h, where *h* is the coefficient of thermal conductivity divided by the specific heat (Mikhlin, 1967).

1.4. Methods of Solution

Advancement in high-speed computers has greatly influenced the numerical methods used for the solution of p.d.e.'s. One measure of the growth is the upsurge of new methods for solving these problems. However these methods can be classified into two different main techniques, namely *finite difference* and *finite element* methods for solving p.d.e.'s.

Finite difference methods are still far and away the most widely used and understood for problems in p.d.e.'s. It is the only one that stands out as being universally applicable to both linear and non-linear problems. The main subclasses of finite difference methods are methods of *lines* and methods of *nets*.

The finite element method which initially arose in structural mechanics has expanded into other areas due to their promotion by Zienkiewicz (1971) and Oden (1972).

This book deals *mainly* with new finite difference methods when applied to solve parabolic partial differential equations and the extensions to elliptic and hyperbolic p.d.e.'s. However, for further generality, some brief references to finite element methods are also included.

1.5. Basic Matrix Algebra

In the numerical solution of p.d.e.'s by the finite difference method or finite element method, the differential system is replaced by a matrix system. In this section some useful properties of a matrix are outlined.

Notations 1.5.1.

A square matrix of order n

a_{ij} real number which is the element in the ith row and j^{th} column of the matrix A

A^{-1} inverse of A

A^T transpose of A

$|A|$ determinant of A

I unit matrix of order n

0 null matrix

$\rho(A)$ spectral radius of A

$\underline{x}$ column vector with element $x_i, i = 1, 2, \ldots, n$

$\underline{x}^T$ row vector with element $x_j, j = 1, 2, \ldots, n$

$\underline{\bar{x}}$ complex conjugate of $\underline{x}$

$\|A\|$ norm of A

$\|\underline{x}\|$ norm of vector $\underline{x}$

P permutation matrix which has entries of zeros and one only, with one non-zero entry in each row and column.

Definitions 1.5.2.

The $(n \times n)$ matrix A is said to be:

  (i) *non-singular if $|A| \neq 0$,*

 (ii) *diagonal if its only non-zero elements lie on the diagonal,*

(iii) *symmetric if $A = A^T$, i.e., $a_{ij} = a_{ji}, i, j = 1, 2, \ldots, n$,*

(iv) *orthogonal if $A^{-1} = A^T$,*

 (v) *null if $a_{ij} = 0$ for all i and j, $i, j = 1, 2, \ldots, n$,*

(vi) *diagonally dominant if*

$$|a_{ii}| \geq \sum_{\substack{j=1 \\ j \neq i}}^{n} |a_{ij}| \quad \text{for all } i,$$

(vii) *tridiagonal if $a_{ij} = 0$ for $|i - j| > 1$,*

(viii) *block diagonal if*

$$A = \begin{bmatrix} B_1 & & & \\ & B_2 & & 0 \\ & 0 & \ddots & \\ & & & B_s \end{bmatrix},$$

where each B_k $(k=1,2,...,s)$ is a square matrix of order $k_1,k_2,...,k_s$ with $k_1+k_2+\cdots+k_s=n$,

(ix) *upper triangular if* $a_{ij}=0$ *for* $i>j$,

(x) *lower triangular if* $a_{ij}=0$ *for* $j>i$,

(xi) *sparse if many of its elements are zero.*

For the matrix A whose elements a_{ij} which are not necessarily real numbers we denote A^H as the conjugate transpose of A. A is called *Hermitian* if $A^H=A$, i.e. if $\bar{a}_{j,i}=a_{i,j}$ for all i and j, $i,j=1,2,...,n$. The definition of a Hermitian matrix implies that the diagonal elements of the matrix are real. A real symmetric matrix is always Hermitian, but a Hermitian matrix is symmetric only if it is real.

If A is real and x is complex, then A is *positive definite* if

$$(\underline{x},A\underline{x})>0, \quad \text{for all } \underline{x}\neq 0.$$

(Note that the inner product $(\underline{x},y)$ of two complex vectors is $\sum_{i=1}^{n} x_i\bar{y}_i$ where $\bar{y}_i$ is the complex conjugate of y_i.) A is *non-negative* or *semi-positive definite* if $(\underline{x},A\underline{x})\geq 0$ for all $\underline{x}\neq 0$ with equality for at least one $\underline{x}\neq 0$.

A is a *band matrix* of bandwidth $w=p+q+1$ if $a_{ij}=0$ for $j>i+p$ or $i>j+q$. If $p=q=1$, then A is tridiagonal and a pentadiagonal matrix can be obtained when $p=q=2$.

Two matrices are called *commutative* if $AB=BA$. They then possess the same set of eigenvectors.

Theorem 1.1.

A real matrix is positive (non-negative) definite if and only if it is symmetric and all its eigenvalues are positive (non-negative, with at least one eigenvalue equal to zero).

This theorem is a very important one and sometimes used as a definition of positive (non-negative) definite.

The positive definite matrix can be written as $A=GJG^{-1}$ where J is a positive diagonal matrix and is called the *Jordan canonical form* of A. The matrix G can be taken to be an orthogonal matrix (Young, 1971). If $J^{1/2}$ denotes the diagonal matrix whose elements are the positive square roots of the elements of J, then $A^{1/2}=GJ^{1/2}G^{-1}$ is positive definite by Theorem 1.1. (It can be clearly seen that $A=(A^{1/2})^2=(GJ^{1/2}G^{-1})(GJ^{1/2}G^{-1})=GJG^{-1}$.)

Theorem 1.2.

A real symmetric matrix A of order n is positive (non-negative) definite if and only if it can be written in the form $A=P^TP$ where P is some non-singular (singular) matrix of the same order.

Proof. (i) Assume that $A=P^TP$ with P is some non-singular matrix, i.e. $|P|\neq 0$. Then for any vector $x\neq 0$

$$x^TAx=x^TP^TPx$$

$$=(Px)^T(Px)>0.$$

Thus, A is positive definite.

(ii) Let A be positive definite (and real). Since $A = A^{1/2}A^{1/2}$ and $A^{1/2}$ is symmetric therefore $A = (A^{1/2})^T A^{1/2}$. As $A^{1/2}$ is also positive definite $|A^{1/2}| \neq 0$. Thus, putting $P = A^{1/2}$ gives the required condition.

The proof for the case of A non-negative definite, follows in a similar fashion.

1.6. Vector and Matrix Norms

For the purpose of analysing the errors in later chapters, the approximate methods are usually associated with some vectors and matrices of which their magnitudes are measurable as non-negative scalars. Such a measuring concept is called a *norm*.

Definition 1.6.1.
Let the vector x be given by $x^T = [x_1, x_2, \ldots, x_n]$, the following scalars are defined as the $1, 2$ and ∞ norm of a vector $\underline{x}$:

$$\|x\|_1 = |x_1| + |x_2| + \cdots + |x_n|, \tag{1.6.1a}$$

$$\|x\|_2 = \left(\sum_{i=1}^n |x_i|^2 \right)^{1/2}, \tag{1.6.1b}$$

$$\|x\|_\infty = \sup_{1 \leq i \leq n} |x_i| \tag{1.6.1c}$$

($\|x\|_2$ *is often called the length of $\underline{x}$). In general L_p-norms are given by*

$$\|x\|_p = \left(\sum_{i=1}^n |x_i|^p \right)^{1/p}, \quad 1 \leq p \leq \infty. \tag{1.6.1d}$$

Since normally matrices and vectors appear simultaneously, it is convenient to introduce the norm of a matrix in such a way that it is compatible with a given vector norm.

Definition 1.6.2.
A matrix norm is said to be compatible with a vector norm $\|x\|$ if

$$\|Ax\| \leq \|A\| \, \|x\|,$$

for all non-zero x.

To construct the matrix norm compatible with the vector norm of (1.6.1a–1.6.1d), it is necessary that

$$\|A\| = \sup_{x \neq 0} \frac{\|Ax\|}{\|x\|} \tag{1.6.3}$$

(G. Dahlquist, 1974) which is equivalent to

$$\|A\| = \sup_{\|x\|=1} \|Ax\|. \tag{1.6.4}$$

Definition 1.6.5.
A matrix norm which is defined by (1.6.4) is said to be subordinate to the corresponding vector norm.

Definition 1.6.6.
Let A be a matrix of order n with eigenvalues $\lambda_i, 1 \leq i \leq n$, then the spectral radius $\rho(A)$ is given by

$$\rho(A) = \max_{1 \leq i \leq n} |\lambda_i|.$$

The matrix norm subordinate to $\| x \|_p$ is denoted by $\| A \|_p$ and these norms satisfy the relations:

$$\| A \|_1 = \max_{1 \leq j \leq n} \sum_{i=1}^{n} |a_{ij}| \quad \text{(maximum absolute column sum)}, \qquad (1.6.7)$$

$$\| A \|_2 = \sqrt{\rho(A^T A)}, \qquad (1.6.8)$$

$$\| A \|_\infty = \max_{1 \leq i \leq n} \sum_{j=1}^{n} |a_{ij}| \quad \text{(maximum absolute row sum)}. \qquad (1.6.9)$$

For the derivations of (1.6.7–1.6.9) see Varga (1962).

Theorem 1.3.
If x and y are vectors then:
  (i) $\| x \| > 0$ *for all* $x \neq 0$,
  (ii) $\| \alpha x \| = |\alpha| \, \| x \|$ *for any scalar* α,
  (iii) $\| x + y \| \leq \| x \| + \| y \|$.

The proof is obvious from the Definition 1.6.1.

Theorem 1.4.
If A and B are two matrices of order n, then:
  (i) $\| A \| > 0$ *if* $A \neq 0$,
  (ii) $\| kA \| = |k| \cdot \| A \|$ *for any scalar* k,
  (iii) $\| A + B \| \leq \| A \| + \| B \|$, *and*
  (iv) $\| AB \| \leq \| A \| \cdot \| B \|$.

For the proof see Varga (1962).

Theorem 1.5.
If A is a matrix of order n, then

$$\| A \| \geq \rho(A). \qquad (1.6.10)$$

Proof. If λ is any eigenvalue of A and $\underline{x}$ is an eigenvector associated with the eigenvalue λ, then $A\underline{x} = \lambda \underline{x}$.

Thus, from Theorems 1.3. and 1.4,

$$|\lambda| \cdot \| x \| = \| \lambda x \| = \| Ax \| \leq \| A \| \, \| x \|,$$

from which we conclude that $\| A \| \geq |\lambda|$ for all eigenvalues of A, which proves (1.6.10).

Theorem 1.6.
For any real symmetric matrix A of order n, $\| A \|_2 = \rho(A)$.

Proof. Since A is symmetric,

$$\| A \|_2^2 = \rho(A^T A) = \rho(A^2) = \rho^2(A),$$

and hence the result follows.

1.7. Eigenvalues and Eigenvectors of a Matrix

Definition 1.7.1.
If A is a square matrix of order n and if x is a non-zero vector such that $Ax = \lambda x$, where λ is some number, then x is said to be an eigenvector of A with corresponding eigenvalue λ.

Theorem 1.7.
If A is a square matrix of order n, any eigenvalue λ satisfies the nth degree polynomial equation $|A - \lambda I| = 0$; this equation is known as the characteristic equation of A.

Proof. We seek a scalar λ and non-zero vector x such that $Ax = \lambda x$ or $(A - \lambda I)x = 0$. Since this is a system of n simultaneous homogeneous equations in the n unknowns, $x_1, x_2, ..., x_n$ (not all are zero), therefore $A - \lambda I$ must be singular, or in other words $|A - \lambda I| = 0$.

Theorem 1.8 (Gerschgorin's first theorem).
The largest of the moduli of the eigenvalues of the square matrix A cannot exceed the largest sum of the moduli of the elements along any row or any column.

Proof. Let λ_i be an eigenvalue of A and x_i is the corresponding eigenvector with components $v_1, v_2, ..., v_n$. Then the equation $Ax_i = \lambda x_i$ is in detail given by

$$a_{1,1}v_1 + a_{1,2}v_2 + \cdots + a_{1,n}v_n = \lambda_i v_1$$

$$a_{2,1}v_1 + a_{2,2}v_2 + \cdots + a_{2,n}v_n = \lambda_i v_2$$

$$\vdots \qquad \vdots \qquad \qquad \vdots \qquad \vdots$$

$$a_{s,1}v_1 + a_{s,2}v_2 + \cdots + a_{s,n}v_n = \lambda_i v_s \qquad\qquad (1.7.2)$$

$$\vdots \qquad \vdots \qquad \qquad \vdots \qquad \vdots$$

$$a_{n,1}v_1 + a_{n,2}v_2 + \cdots + a_{n,n}v_n = \lambda_i v_n$$

Let v_s be the largest in modulus of $v_1, v_2, ..., v_n$. Select the sth equation and divide by v_s giving

$$\lambda_i = a_{s,1}\left(\frac{v_1}{v_s}\right) + a_{s,2}\left(\frac{v_2}{v_s}\right) + \cdots + a_{s,n}\left(\frac{v_n}{v_s}\right).$$

Since $|v_i/v_s| \leq 1$, $i = 1, 2, \ldots, n$, therefore,

$$|\lambda_i| \leq |a_{s,1}| + |a_{s,2}| + \cdots + |a_{s,n}|.$$

In particular this holds for $|\lambda_i| = \max |\lambda_s|$, $s = 1, 2, \ldots, n$.

As the eigenvalues of the transpose of A are the same as those of A, the proof is also true for any column (Smith, 1978).

Theorem 1.9 (Gerschgorin's circle theorem).
Let A have n eigenvalues $\lambda_i, i = 1, 2, \ldots, n$. Then each λ_i lies in the union of the n circles,

$$|\lambda - a_{i,i}| \leq \sum_{\substack{j=1 \\ j \neq i}}^{n} |a_{i,j}|.$$

Proof. By the previous proof, we have

$$\lambda_i = a_{s,1}\left(\frac{v_1}{v_s}\right) + a_{s,2}\left(\frac{v_2}{v_s}\right) + \cdots + a_{s,s} + \cdots + a_{s,n}\left(\frac{v_n}{v_s}\right).$$

Therefore

$$|\lambda_i - a_{s,s}| = \left| a_{s,1}\left(\frac{v_1}{v_s}\right) + a_{s,2}\left(\frac{v_2}{v_s}\right) + \cdots + a_{s,s-1}\left(\frac{v_{s-1}}{v_s}\right) + a_{s,s+1}\left(\frac{v_{s+1}}{v_s}\right) \right.$$

$$\left. + \cdots + a_{s,n}\left(\frac{v_n}{v_s}\right) \right| \leq |a_{s,1}| + |a_{s,2}| + \cdots + |a_{s,s-1}| + |a_{s,s+1}| + \cdots + |a_{s,n}|.$$

As λ_i is any eigenvalue, therefore,

$$|\lambda - a_{i,i}| \leq \sum_{\substack{j=1 \\ j \neq i}}^{n} |a_{i,j}|. \qquad (1.7.3)$$

Corollary 1.
If A is a square matrix of order n and

$$v = \max_{1 \leq i \leq n} \sum_{j=1}^{n} |a_{i,j}| = \|A\|_{\infty}, \qquad (1.7.4)$$

then $\rho(A) \leq v$.

Corollary 2.
If A is a square matrix of order n and

$$v' = \max_{1 \leq j \leq n} \sum_{i=1}^{n} |a_{i,j}| = \|A\|_{1}, \qquad (1.7.5)$$

then $\rho(A) \leq v'$.

These corollaries are the immediate result of Theorem 1.9.

Theorem 1.10.
If A is a symmetric, diagonally dominant matrix with positive diagonal elements, it is positive definite.

Proof. Since A is symmetric, the eigenvalues of A are real. As A is diagonally dominant, the application of Gerschgorin's Theorem shows that the eigenvalues are all positive. Therefore according to Theorem 1.1, A is positive definite.

Theorem 1.11.
Let λ be an eigenvalue of A with eigenvector x, with α, μ as scalar values. Then:
(i) *$\alpha\lambda$ is an eigenvalue of A with eigenvector x,*
(ii) *$\lambda - \mu$ is an eigenvalue of $A - \mu I$ with eigenvector x,*
(iii) *if A is non-singular, then $\lambda \neq 0$ and λ^{-1} is an eigenvalue of A^{-1} with eigenvector x*
   (Stewart, 1973).

1.8. Convergence of Sequences of Matrices

Definition 1.8.1.
The powers of a sequence involving A of order n is convergent to zero if the sequence of matrices $A, A^2, A^3, \ldots$, converges to the null matrix 0.

Theorem 1.12.

$$\lim_{r \to \infty} A^r = 0, \quad \text{if } \|A\| < 1, \tag{1.8.1}$$

Proof. $\|A^r\| = \|AA^{r-1}\| \le \|A\|\|A^{r-1}\|$

$$\le \|A\|^2\|A^{r-2}\|$$

.

.

.

$$\le \|A\|^r,$$

and so the result follows. This is only a sufficient condition but not a necessary one. The following theorem states the necessary and sufficient condition.

Theorem 1.13.

$$\lim_{r \to \infty} A^r = 0, \quad \text{if and only if } |\lambda_i| < 1,$$

for all eigenvalues λ_i $(i = 1, 2, \ldots, n)$ of A.

Proof. Consider the Jordan canonical form of A. A Jordan submatrix of A is of the form

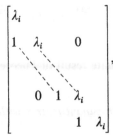

where λ_i is an eigenvalue of A. If this matrix is raised to the power r, then the result tends to the null matrix as $r \to \infty$ if and only if $|\lambda_i| < 1$.

Theorem 1.14.
Let

$$F(z) = c_0 + c_1 z + \cdots + c_k z^k + \cdots$$

be a power series with radius of convergence R (i.e. it is absolutely convergent for $|z| < R$). Then if A is a square matrix of order n with eigenvalues λ_i ($1 \leq i \leq n$) the matrix series

$$F(A) = c_0 I + c_1 A + c_2 A^2 + \cdots + c^k A^k + \cdots$$

will be convergent provided $|\lambda_i| < R$ for all i.

1.9. Eigenvalues of Some Common Matrices

The eigenvalues of the $(n \times n)$ matrix

$$A = \begin{bmatrix} a & b & & & \\ c & a & b & & 0 \\ & & \ddots & & \\ 0 & & c & a & b \\ & & & c & a \end{bmatrix}, \tag{1.9.1}$$

where b and c are both real and a is real or complex, are given by

$$\lambda_i = a + 2\sqrt{bc}\cos\left(\frac{i\pi}{n+1}\right), \quad i = 1, 2, \ldots, n \tag{1.9.2}$$

(Smith, 1978).

If A is an $(n \times n)$ cyclic tridiagonal matrix, i.e.,

$$A = \begin{bmatrix} a & b & & & c \\ c & a & b & & 0 \\ & & \ddots & & \\ 0 & & c & a & b \\ b & & & c & a \end{bmatrix}, \tag{1.9.3}$$

then the eigenvalues are given by

$$\lambda_i = a + 2\sqrt{bc}\cos\left(\frac{2i\pi}{n}\right), \quad i = 0, 1, \ldots, n-1. \tag{1.9.4}$$

1.10. The Solution of Systems of Linear Equations

Methods used to obtain the numerical solution of the linear system

$$A\underline{u} = \underline{b},\qquad(1.10.1)$$

that are best suited to the capability of modern electronic computers have been developed. The criteria to compare methods are measured by the number of computer operations needed to obtain a (sufficiently accurate) solution as well as the storage requirements which are measured by the size of computer memory needed to store the matrix A and other matrices occurring in the process of solution. Usually the methods used lie in two classes, the class of *direct methods* (or elimination methods) and the class of *iterative methods* (or indirect methods) which mainly depend upon the structure of the coefficient matrix A.

That is, if A is a large sparse matrix as in problems which arise in large order p.d.e.'s *iterative methods* are usually used, since these will not change the structure of the original matrix and therefore preserve sparsity. They are essentially based on generating a sequence of approximate solutions $\{\underline{u}^{(s)}\}$, $s = 0, 1, 2, \ldots$ for (1.10.1) and hope that this sequence approaches the solution $A^{-1}\underline{b}$ provided that the inverse exists. However, no arithmetic is associated with zero coefficients, so considerably fewer numbers have to be stored in the computer (the non-zero elements of the coefficient matrix) and hence minimise the amount of storage used. As a consequence they can be used to solve systems of equations that are too large for the use of direct methods.

Programming and data handling are also much simpler than for direct methods. Another advantage, not possessed by direct methods, is their frequent extension to the solution of sets of non-linear equations (Smith, 1978). The disadvantage of these methods are mainly the problem of selecting a good initial vector with which the iterative process may commence and also the accuracy of the final solution.

Direct methods, on the other hand, are used only when the coefficient matrix is dense as in statistical problems where the dimension is small. The advantages with these methods are that no initial vector is required and the accuracy of the final solution usually turns out to be satisfactory depending on the chosen word length of the machine. Direct methods cannot easily be used for large sparse unordered matrices because of the problem of fill-in which occurs during the elimination process and the storage requirements are prohibitive. However, if the matrix A is of regular shape, then special methods for storing the matrix can be devised in order to minimise the amount of storage used.

For non-linear problems, the most common procedure for solving these kinds of problems is to formulate them as the solution of an iterative sequence of linear (or rather linearised) equations to which solution methods for linear problems still apply.

In the next section we will describe some well known direct methods.

1.10.1. Direct Methods

Direct methods for solving linear systems are compared for efficiency on the basis of the number of arithmetical operations required. One of the methods used for obtaining the solution of a system of equations is given by *Cramer's rule*, which can be stated in the following theorem.

Theorem 1.15.
In the system (1.10.1), let $\det A \equiv |A| \neq 0$, *and let* A_i $(i = 1, 2, \ldots, m)$, *be the determinant of the matrix obtained from A by replacing the ith column of A by the column vector $\underline{b}$. Then, the solution of the system (1.10.1) is*

$$u_i = \frac{\det A_i}{\det A}, \quad i = 1, 2, \ldots, m. \tag{1.10.2}$$

The use of this method in actual numerical cases is generally grossly inefficient, because of the excessive labour involved in the evaluation of $(m + 1)$ determinants of order m, unless m is small. For example to solve (1.10.1) with order 10 (which is a trivially small problem by modern standards) would require some 68 million multiplications. The number of operations involved in this method is of order $(m!)$ if the system is of order m (Froberg, 1979).

A more efficient method of evaluating the determinants can reduce this to about 3000 multiplications, but even this is inefficient compared to Gaussian elimination, which requires about 380 multiplications (Gerald, 1980), this, as well as other methods are described as follows.

Systematic Elimination Methods
These methods are based ultimately on the process of the elimination of variables. The most widely used methods are:

(i) *Gaussian Elimination* which involves a finite number of transformations of a given rectangular system of equations into an upper triangular system which is a system which is much more easily solved. Precisely the number of the transformation is one less than the size of the given system. Thus, for the system (1.10.1) we have after $(m - 1)$ steps the final system (Ralston, 1965)

$$
\begin{bmatrix}
a_{1,1}^{(0)} & a_{1,2}^{(0)} & \cdots & & & a_{1,m}^{(0)} \\
& a_{2,2}^{(1)} & a_{2,3}^{(1)} & \cdots & & a_{2,m}^{(1)} \\
& & a_{3,3}^{(2)} & a_{3,4}^{(2)} & \cdots & a_{3,m}^{(2)} \\
& & & \ddots & & \vdots \\
& & & & a_{m-1,m-1}^{(m-2)} & a_{m-1,m}^{(m-2)} \\
& 0 & & & & a_{m,m}^{(m-1)}
\end{bmatrix}
\begin{bmatrix}
u_1 \\
u_2 \\
u_3 \\
\vdots \\
u_{m-1} \\
u_m
\end{bmatrix}
=
\begin{bmatrix}
b_1^{(0)} \\
b_2^{(1)} \\
b_3^{(2)} \\
\vdots \\
b_{m-1}^{(m-2)} \\
b_m^{(m-1)}
\end{bmatrix}, \tag{1.10.3}
$$

where

$$a_{i,j}^{(k)} = a_{i,j}^{(k-1)} - \frac{a_{i,k}^{(k-1)}}{a_{k,k}^{(k-1)}} a_{k,j}^{(k-1)}, \quad \begin{array}{l} k = 1, 2, \ldots, m-1, \\ i = k+1, \ldots, m, \\ j = k+1, \ldots, m, \end{array}$$

$$(1.10.4)$$

$$b_i^{(k)} = b_i^{(k-1)} - \frac{a_{i,k}^{(k-1)}}{a_{k,k}^{(k-1)}} b_k^{(k-1)}, \quad \begin{array}{l} a_{i,j}^{(0)} = a_{i,j}, \\ b_1^{(0)} = b_1. \end{array}$$

The solution for (1.10.3) is given by

$$u_m = \frac{b_m^{(m-1)}}{a_{m,m}^{(m-1)}},$$

$$u_i = \frac{1}{a_{i,i}^{(i-1)}} \left[b_i^{(i-1)} - \sum_{j=i+1}^{m} a_{i,j}^{(i-1)} u_j \right], \quad i = m-1, \ m-2, \ldots, 1. \quad (1.10.5)$$

Strategies for Avoiding a Small or Zero Pivot

If any of the diagonal elements of the matrix A in the system (1.10.1) becomes zero or small during the elimination process, i.e. $a_{k,k}^{(k-1)} = 0$ for any k, then the final upper triangular form will be unattainable, and hence the process will fail. To overcome this we locate a value of j in the range $(k+1 \leq j \leq m)$ such that $a_{j,k}^{(k-1)} \neq 0$ and largest in magnitude and we interchange the jth and k^{th} rows of $A^{(k)}$ and corresponding column in $b^{(k)}$ (i.e. we reorder the equations). The interchanges are referred to as *pivoting* which is normally employed to preserve arithmetic accuracy.

Definition 1.10.1.

Any of the diagonal elements in (1.10.3), i.e. $a_{k,k}^{(k-1)}$, $1 \leq k \leq m$, is termed the kth-pivot. If it is zero, then it is called a zero-pivot.

There are two basic well known pivoting schemes. They are mainly concerned with avoiding a zero pivot which may arise at any stage of the elimination process.

(1) Partial Pivoting

This strategy involves choosing an element of largest magnitude in the column of each reduced matrix as the pivot, elements of rows which have previously been pivotal being excluded from consideration. This way of pivoting can be easily illustrated in Figure (1.10.1) where x denotes a non-zero element.

$$A^{(2)} = \begin{bmatrix} x & x & x & x & x \\ 0 & x & x & x & x \\ 0 & 0 & \boxed{x} & x & x \\ 0 & 0 & \boxed{\otimes} & x & x \\ 0 & 0 & \boxed{x} & x & x \end{bmatrix}$$

Any element in the box can be taken as the pivot. If $\otimes$ is the largest magnitude then the 3rd and 4th row of $A^{(2)}$ have to be interchanged.

Figure 1.10.1

(2) Complete (Full) Pivoting

The pivot at each stage of the reduction is chosen as the element of largest magnitude in the submatrix of rows which have not been pivotal up to now, regardless of the position of the element in this matrix. This may require both row and column interchanges. Figure 1.10.2 illustrates the strategy.

$$A^{(2)} = \begin{bmatrix} x & x & x & x & x \\ 0 & x & x & x & x \\ 0 & 0 & x & x & x \\ 0 & 0 & x & x & \textcircled{x} \\ 0 & 0 & x & x & x \end{bmatrix}$$

Any element in the box can be taken as the pivot. If $\otimes$ is the element of largest magnitude then we interchange the 3rd and 4th rows of $A^{(2)}$, followed by the 3rd and 5th columns.

Figure 1.10.2

It is also necessary to interchange the elements of the solution vector to compensate for the column interchanges of $A^{(2)}$. Further, in both cases the elements of the right-hand side vector, $b^{(2)}$, are interchanged similarly to the rows of $A^{(2)}$.

In practice complete pivoting is time consuming in execution and is not frequently used. In addition, for large m and matrices with special pattern (e.g. band matrices) partial pivoting is less likely to completely destroy this pattern.

(ii) *Gauss–Jordan Scheme* An alternative to Gauss Elimination method (G.E.) is the Gauss Jordan scheme (G.J.) which leads to a diagonal matrix rather than triangular at the end of the process. In this method, the elements above the diagonal are made zero at the same time that zeros are created below the diagonal, and hence the solution can be obtained by dividing the components of the right-hand side vector, $\underline{b}$, by the corresponding diagonal elements, i.e., there is no need for the back substitution stage as in Gauss elimination.

However, G.E. can be shown to be superior to G.J. since it involves a smaller amount of work. For large m the G.J. method requires almost 50% more operations than the G.E. method (Ralston, 1965). It is shown by Fox (1964) that the amount of work involved in the G.E. and G.J. methods is as follows:

G.E.	*G.J.*	*Operations*
m	m	division
$\frac{1}{3}m^3 + m^2 - \frac{1}{3}m$	$\frac{1}{2}m^3 + m^2 - \frac{1}{2}m$	multiplication
$\frac{1}{3}m^3 + \frac{1}{2}m^2 - \frac{5}{6}m$	$\frac{1}{2}m^3 - \frac{1}{2}m$	addition

(iii) *Triangular, or LU, Decomposition* If we define the multipliers of the kth stage of the transformation for the G.E. method as

$$t_{i,k} = \frac{a_{i,k}}{a_{k,k}}, \quad \begin{array}{l} k = 1, 2, \ldots, m-1, \\ i = k+1, \ldots, m, \end{array} \tag{1.10.6}$$

where $t_{i,1} = a_{i,1}/a_{1,1}$, $i = 2, 3, \ldots, m$, is used to generate the zeros required in the first column. Then we define the $(m \times m)$ unit lower triangular matrices $M_1, M_2, \ldots, M_k$, $k = 1, 2, \ldots, m-1$, as follows (see Gault *et al.* 1974, Ralston, 1965)

$$
M_1 = \begin{bmatrix}
1 & & & & \\
-t_{2,1} & 1 & & & \\
-t_{3,1} & & 1 & 0 & \\
\vdots & & & \ddots & \\
& & 0 & & \\
-t_{m,1} & & & & 1
\end{bmatrix},
\quad
M_2 = \begin{bmatrix}
1 & & & & \\
0 & 1 & & & \\
0 & -t_{3,2} & & 0 & \\
0 & -t_{4,2} & & \ddots & \\
\vdots & & 0 & & \\
0 & -t_{m,2} & & & 1
\end{bmatrix}
\quad \cdots
$$

$$
M_k = \begin{bmatrix}
1 & & & & & \\
& \ddots & & & & \\
& & 1 & & 0 & \\
& 0 & -t_{k+1,k} & & & \\
& & -t_{k+2,k} & & & \\
& & \vdots & 0 & & 1 \\
& & -t_{m,k} & & &
\end{bmatrix}.
$$

Thus the matrix form (1.10.3) is equivalent to

$$MA\underline{u} = M\underline{b},\tag{1.10.7}$$

where

$$M = M_{m-1}M_{m-2}\cdots M_1,\tag{1.10.8}$$

$$
M_{m-1}M_{m-2}\cdots M_1 A =
\begin{bmatrix}
a_{1,1}^{(0)} & a_{1,2}^{(0)} & \cdots & & a_{1,m}^{(0)} \\
& a_{2,2}^{(1)} & a_{2,3}^{(1)} & \cdots & a_{2,m}^{(1)} \\
& & a_{3,3}^{(2)} & \cdots & a_{3,m}^{(2)} \\
& 0 & & \ddots & \vdots \\
& & & & a_{m,m}^{(m-1)}
\end{bmatrix}
\equiv
\begin{bmatrix}
v_{1,1} & v_{1,2} & \cdots & & v_{1,m} \\
& v_{2,2} & v_{2,3} & \cdots & v_{2,m} \\
& & & \ddots & \vdots \\
& 0 & & & \\
& & & & v_{m,m}^{(m-1)}
\end{bmatrix}
\equiv U.
$$

$$\tag{1.10.9}$$

Hence, we have

$$A = M_1^{-1}M_2^{-1}\cdots M_{m-1}^{-1}U.\tag{1.10.10}$$

But M_k^{-1}, $1 \leq k \leq m - 1$, is M_k itself with the signs of its off-diagonal elements reversed and the product $M_1^{-1} M_2^{-1}, \ldots, M_{m-1}^{-1}$ is given by

$$M_1^{-1} M_2^{-1} \ldots M_{m-1}^{-1} = \begin{bmatrix} 1 & & & & \\ t_{2,1} & 1 & & 0 & \\ t_{3,1} & t_{3,2} & 1 & & \\ \vdots & \vdots & & \ddots & \\ t_{m,1} & t_{m,2} & ---- & t_{m,m-1} & 1 \end{bmatrix} \equiv L. \qquad (1.10.11)$$

Thus,

$$A = LU, \qquad (1.10.12)$$

where L is a unit lower triangular matrix and U is an upper triangular matrix. The form (1.10.12) is termed *triangular or LU* decomposition.

Now we can write (1.10.1) as

$$LU\underline{u} = \underline{b}. \qquad (1.10.13)$$

The solution of (1.10.1) by this algorithm follows from (1.10.13) by introducing an auxiliary vector, $\underline{y}$ (say), such that the system (1.10.13) will be split into two triangular systems

$$L\underline{y} = \underline{b} \qquad (1.10.14)$$

and

$$U\underline{u} = \underline{y}. \qquad (1.10.15)$$

The two vectors $\underline{y}$ and $\underline{u}$ can be obtained easily from equations (1.10.14) and (1.10.15) by forward and backward substitution processes, respectively. The equations for the forward and backward substitutions can be summarised as follows:

$$y_1 = b_1,$$

$$y_i = b_i - \sum_{k=1}^{i-1} t_{i,k} y_k, \quad i = 2, 3, \ldots, m, \qquad (1.10.16)$$

and

$$u_m = \frac{y_m}{v_{m,m}},$$

$$u_j = \left(y_j - \sum_{k=j+1}^{m-1} v_{j,k} u_k \right) / v_{j,j}, \quad j = m-1, \ m-2, \ldots, 1, \qquad (1.10.17)$$

provided $v_{j,j} \neq 0$ ($i = 1, 2, \ldots, m$), i.e. U is non-singular.

Pivoting with the *LU* method is somewhat more complicated than with the G.E. method, because we do not usually handle the right-hand side vector simultaneously with our reduction of the matrix A (Gerald, 1978). This means we must keep a record of any row interchanges made during the formulation of L and U so that the elements of $\underline{b}$ can be similarly interchanged.

It can be shown that the amount of work in the LU method is the same as for G.E. method. But the disadvantage of Gauss's process in comparison with LU process is that a rounding error occurs when an element of a reduced matrix is computed and stored. This can be largely avoided in the LU process by the use of double-precision arithmetic in the calculation of the elements of L and U, and then the result may be rounded to single precision and recorded on the completion of each calculation. The removal of the need for computing and recording several intermediate matrices has localised what might otherwise be a significant source of error to a single step in the determination of each elements of L and U. The use of double-precision arithmetic in this step leads to a degree of accuracy comparable with that attained if the entire Gauss process were carried out with double-precision. This latter is an unattractive proposition, since it would require twice as much computer storage as the corresponding single precision solution (Gault *et al.*, 1974).

However, the LU decomposition may be applied when the following theorem is valid.

Theorem 1.10.2.
A non-singular matrix A may be decomposed into the product LU if and only if every leading principal submatrix of A is non-singular.

Corollary 1.10.1.
If L is a unit lower triangular matrix then the decomposition is unique.

Corollary 1.10.2.
If U is a unit upper triangular matrix then the decomposition is unique.

The LU decomposition is often called *Doolittle's method* if Corollary (1.10.1) is valid, whilst if Corollary (1.10.2) is valid it is called *Crout's method*.

The Decomposition of a Symmetric Matrix
If the matrix A in (1.10.1) is symmetric and satisfies the hypothesis of Theorem (1.16), then we can express A in the form

$$A = LDL^T, \tag{1.10.18}$$

where L is a unit lower triangular matrix and D is a diagonal matrix. This method will fail if any of the pivots are zero, since the use of row interchanges to avoid this destroys the symmetry of the system. But A is a real symmetric and positive definite matrix then a real lower triangular matrix can be found such that

$$A = LL^T, \tag{1.10.19}$$

so, if $L = (\ell_{i,j})$, where $\ell_{i,j} = 0$ for $i < j$, then,

$$\ell_{j,j} = \left[a_{j,j} - \sum_{k=1}^{j-1} \ell_{j,k}^2 \right]^{1/2}, \quad \text{if } i = j,$$

$$\ell_{i,j} = \frac{1}{\ell_{j,j}} \left[a_{j,i} - \sum_{k=1}^{j-1} \ell_{i,k} \ell_{j,k} \right], \quad j < i \le m, \ j = 1, 2, \ldots, m. \tag{1.10.20}$$

This is known as the Choleski decomposition (or the square root method), and it requires roughly half the storage space and the computational labour of *LU* decomposition. The main disadvantage being that it needs the calculation of the square roots, but this may be avoided by the decomposition (1.10.18). Thus, Choleski's method is an attractive proposition for problems involving symmetric positive definite matrices such as those occurring in discretised elliptic problems.

2. Numerical Solution of Partial Differential Equations by Finite Difference Methods

2.1. Introduction

In Chapter 1 some of the basic properties of partial differential equations were outlined. It is obvious from the accompanying discussion that the parabolic equations play a substantial role in many branches of science and engineering. Consequently much effort has been devoted to establishing efficient and accurate numerical methods to solve such equations. Many of these are now presented in Chapter 2.

2.2. Parabolic Equations in One Space Dimension

A linear parabolic partial differential equation takes the general form

$$\sigma(x,t)\frac{\partial U}{\partial t} = \frac{\partial}{\partial x}\left(a(x,t)\frac{\partial U}{\partial x}\right) + b(x,t)\frac{\partial U}{\partial x} + c(x,t)U + d(x,t), \qquad (2.2.1)$$

which is defined within some prescribed domain D of the (x,t) space. Within this domain, the functions $\sigma(x,t)$, $a(x,t)$ are strictly positive and $c(x,t)$ is non-negative.

Attention will now be focussed on developing finite difference methods for parabolic equations. A simplified form of (2.2.1), that is, the *diffusion (heat) equation with constant coefficients* $(\sigma(x,t) = a(x,t) = 1, b(x,t) = c(x,t) = 0)$ and $d(x,t) = 0$, i.e.,

$$\frac{\partial U}{\partial t} = \frac{\partial^2 U}{\partial x^2}, \quad 0 < x < 1,\ 0 \le t \le T, \qquad (2.2.2)$$

subject to the initial-boundary conditions

$$U(x,0) = f(x), \quad 0 < x < 1,$$
$$U(0,t) = g(t), \quad 0 \le t \le T, \qquad (2.2.2a)$$
$$U(1,t) = h(t), \quad 0 < t \le T,$$

will be used. For this purpose, a uniformly spaced network whose mesh points are $x_i = i\Delta x$, $t_j = j\Delta t$ for $i = 0, 1, 2, \ldots, m-1, m$ and $j = 0, 1, \ldots, n-1, n$ with $\Delta x = 1/m$ and $\Delta t = T/n$ is used.

2.3. Explicit Methods

If the space derivative is approximated by the second order central difference and the time derivative by the forward difference, then the resulting approximation is

$$\frac{1}{\Delta t}(u_{i,j+1}-u_{i,j})=\frac{1}{(\Delta x)^2}(u_{i+1,j}-2u_{i,j}+u_{i-1,j}), \tag{2.3.1}$$

where the truncation error terms have been omitted.

On solving this equation for $u_{i,j+1}$ we obtain the simplest explicit formula

$$u_{i,j+1}=ru_{i-1,j}+(1-2r)u_{i,j}+ru_{i+1,j}$$
$$=u_{i,j}+r(u_{i-1,j}-2u_{i,j}+u_{i+1,j}), \tag{2.3.1a}$$

where $r=\Delta t/(\Delta x)^2$ is the mesh ratio.

The *local truncation error* of (2.3.1) can be obtained as usual by Taylor's series expansion about the point (x_i, t_j). Hence, we have, if U is the exact solution

$$T=U_{i,j+1}-rU_{i-1,j}-(1-2r)U_{i,j}-rU_{i+1,j}, \tag{2.3.2}$$

which on expansion gives the result

$$T=\frac{1}{2}\left\{\Delta t\left(\frac{\partial^2 U}{\partial t^2}\right)_{i,j}-\frac{(\Delta x)^2}{6}\left(\frac{\partial^4 U}{\partial x^4}\right)_{i,j}\right\}+\frac{(\Delta x)^2}{6}\left(\frac{\partial^3 U}{\partial t^3}\right)_{i,j}-\frac{(\Delta x)^4}{360}\left(\frac{\partial^6 U}{\partial x^6}\right)_{i,j}+\cdots, \tag{2.3.3}$$

i.e.,

$$T=0([\Delta t])+0([\Delta x]^2).$$

We now study the stability of equation (2.3.1) by (a) the Maximum Principle, (b) the Energy Method, (c) Matrix Analysis and (d) Fourier Series.

(a) The Maximum Principle
From (2.3.1a) and (2.3.2), we have

$$U_{i,j+1}=rU_{i-1,j}+(1-2r)U_{i,j}+rU_{i+1,j}+0(\Delta t)+0([\Delta x]^2). \tag{2.3.4}$$

At the mesh point we have

$$\varepsilon_{i,j}=U_{i,j}-u_{i,j}. \tag{2.3.5}$$

Therefore, by subtracting equation (2.3.1a) from (2.3.4) leads to

$$\varepsilon_{i,j+1}=r\varepsilon_{i-1,j}+(1-2r)\varepsilon_{i,j}+r\varepsilon_{i+1,j}+0(\Delta t)+0([\Delta x]^2), \quad i=1,2,\ldots,m-1. \tag{2.3.6}$$

Since U agrees with u initially on the boundary then

$$\varepsilon_{i,0}=0, \qquad i=0,1,\ldots,m,$$

$$\varepsilon_{0,j}=\varepsilon_{m,j}=0, \quad j=0,1,\ldots,n.$$

We note that the sum of the coefficients of r in (2.3.6) is unity and if

$$r\leq\tfrac{1}{2}, \tag{2.3.7}$$

then they are all non-negative. Let E_j denote the maximum value of $|\varepsilon_{i,j}|$ along the jth time row. From (2.3.6), we find that if (2.3.7) holds, then

$$|\varepsilon_{i,j+1}| \leq r|\varepsilon_{i-1,j}| + (1-2r)|\varepsilon_{i,j}| + r|\varepsilon_{i+1,j}| + 0(\Delta t + [\Delta x]^2)$$
$$\leq rE_j + (1-2r)E_j + rE_j + M(\Delta t + [\Delta x]^2)$$
$$= E_j + M(\Delta t + [\Delta x]^2).$$

As this is true for all values of i, it is also true for $\max_i |\varepsilon_{i,j+1}|$. Hence

$$E_{j+1} \leq E_j + M(\Delta t + [\Delta x]^2),$$

i.e.,

$$E_{j+1} \leq E_{j-1} + 2M(\Delta t + [\Delta x]^2),$$

and so on, from which it follows that

$$E_j \leq E_0 + jM(\Delta t + [\Delta x]^2),$$
$$\leq nM(\Delta t + [\Delta x]^2),$$

i.e.,

$$E_j \leq N(\Delta t + [\Delta x]^2), \tag{2.3.8}$$

since $E_0 = 0$, $j \leq n$ and $N = nM$. However since $r = \Delta t/(\Delta x)^2$ then (2.3.8) can be expressed as $E_j = 0([\Delta x]^2)$.

The inequality (2.3.8) means that the finite difference analogue to the PDE *converges* to the solution of the differential equation as Δx and Δt tend uniformly to zero. The above bounded condition also implies stability and thus the stability condition is $r \leq 1/2$.

(b) The Energy Method
For convenience, we specify the boundary conditions in (2.2.2a) as

$$U(0,t) = U(1,t), \quad 0 \leq t \leq T. \tag{2.2.2b}$$

We first observe that if we multiply the differential equation (2.2.2) by U and integrate with respect to x, we obtain

$$\int_0^1 U \frac{\partial U}{\partial t} dx = \int_0^1 U \frac{\partial^2 U}{\partial x^2} dx,$$

which on integration by parts on the right-hand side gives

$$\frac{\partial}{\partial t} \int_0^1 U^2 dx = -\int_0^1 \left(\frac{\partial U}{\partial x}\right)^2 dx \leq 0.$$

We deduce from this, that

$$\int_0^1 U^2(x,t)dx \leq \int_0^1 U^2(x,0)dx = \int_0^1 f^2(x)dx,$$

and therefore the quantity $\int_0^1 U^2(x,t)dx$ remains bounded as $t \to \infty$.

As before, the perturbations at the mesh points satisfy the difference equation

$$\varepsilon_{i,j+1} - \varepsilon_{i,j} = r(\varepsilon_{i+1,j} + \varepsilon_{i-1,j} - 2\varepsilon_{i,j}), \qquad (2.3.9)$$

and the boundary conditions give

$$e_{0,j} = \varepsilon_{m,j} = 0, \quad \text{for all } j.$$

Equation (2.3.9) is multiplied by $(\varepsilon_{i,j+1} + \varepsilon_{i,j})$ and the result is summed over $i = 1, 2, \ldots, m-1$ to give

$$\|\varepsilon_{j+1}\|^2 - \|\varepsilon_j\|^2 = r \sum_{i=1}^{m-1} (\varepsilon_{i,j+1} + \varepsilon_{i,j})(\varepsilon_{i+1,j} + \varepsilon_{i-1,j} - 2\varepsilon_{i,j}), \qquad (2.3.10)$$

where $\|\varepsilon_{j+1}\|^2 = \sum_{i=1}^{m-1} (\varepsilon_{i,j})^2$. We shall now use the following identities

$$\sum_{i=1}^{m-1} e_i(\varepsilon_{i-1} - \varepsilon_i) = \sum_{i=1}^{m-1} \varepsilon_i(\varepsilon_{i+1} - e_i), \quad \text{if } \varepsilon_0 = e_m = 0, \qquad (2.3.11a)$$

and

$$\sum_{i=1}^{m-1} e_i(\varepsilon_{i+1} - \varepsilon_i) = \sum_{i=1}^{m-1} \varepsilon_{i+1}(e_i - e_{i+1}) - e_1\varepsilon_1, \quad \text{if } \varepsilon_0 = e_m = 0. \qquad (2.3.11b)$$

The right-hand side of (2.3.10) can be rearranged to read

$$r \sum_{i=1}^{m-1} e_i[(\varepsilon_{i+1,j} - \varepsilon_{i,j}) - (\varepsilon_{i,j} - \varepsilon_{i-1,j})], \qquad (2.3.12)$$

with $e_i = \varepsilon_{i,j+1} + \varepsilon_{i,j}$. Remembering that

$$\varepsilon_{0,j} = \varepsilon_{0,j+1} = \varepsilon_{m,j} = \varepsilon_{m,j+1} = 0,$$

then after substituting (2.3.11a) and (2.3.11b) into (2.3.12) we find that equation (2.3.10) becomes

$$\|\varepsilon_{j+1}\|^2 - \|\varepsilon_j\|^2 = -r\left[\sum_{i=1}^{m-1} \{(\varepsilon_{i+1,j} + \varepsilon_{i,j})^2 + (\varepsilon_{i+1,j+1} + \varepsilon_{i,j+1})(\varepsilon_{i+1,j} - \varepsilon_{i,j})\}\right]. \qquad (2.3.13)$$

If we define

$$E_j = \|\varepsilon_j\|^2 - \frac{1}{2r} \sum_{i=0}^{m-1} (\varepsilon_{i+1,j} - \varepsilon_{i,j})^2, \quad j = 0, 1, 2, \ldots,$$

it follows that

$$E_j \le \|\varepsilon_j\|^2,$$

and

$$E_{j+1} - E_j = \|\varepsilon_{j+1}\|^2 - \|\varepsilon_j\|^2 - \frac{1}{2r} \sum_{i=0}^{m-1} \{(\varepsilon_{i+1,j+1} - \varepsilon_{i,j+1})^2 - (\varepsilon_{i+1,j} - \varepsilon_{i,j})^2\}.$$

By using (2.3.12), we get

$$E_{j+1} - E_j = -\frac{1}{2r} \sum_{i=0}^{m-1} (\varepsilon_{i+1,j+1} - \varepsilon_{i,j+1} + \varepsilon_{i+1,j} - \varepsilon_{i,j})^2 \le 0.$$

We conclude therefore that E_j is a monotonic decreasing function of j. By summing the inequalities

$$(\varepsilon_{i+1,j} - \varepsilon_{i,j})^2 \leq (\varepsilon_{i+1,j} - \varepsilon_{i,j})^2 + (\varepsilon_{i+1,j} + \varepsilon_{i,j})^2 = 2(\varepsilon_{i+1,j})^2 + 2(\varepsilon_{i,j})^2,$$

for $i = 0, 1, \ldots, m - 1$, leads to

$$\sum_{i=0}^{m-1} (\varepsilon_{i+1,j} - \varepsilon_{i,j})^2 \leq 2 \sum_{i=0}^{m-1} [(\varepsilon_{i+1,j})^2 + (\varepsilon_{i,j})^2] = 4 \sum_{i=0}^{m-1} (\varepsilon_{i,j})^2 = 4\|\underline{\varepsilon}_j\|^2,$$

and so from the definition of E_j, we get

$$E_j \geq (1 - 2r)\|\underline{\varepsilon}_j\|^2.$$

Now if the condition

$$0 < r < \tfrac{1}{2}, \tag{2.3.14}$$

holds, then

$$\|\underline{\varepsilon}_j\|^2 \leq \frac{1}{(1 - 2r)} E_j,$$

and in view of the following inequalities

$$E_j \leq E_{j-1} \leq \ldots \leq E_0 \leq \|\underline{\varepsilon}_0\|^2,$$

we have

$$\|\underline{\varepsilon}_j\|^2 \leq \frac{1}{(1 - 2r)} \|\underline{\varepsilon}_0\|^2.$$

The vectors $\underline{\varepsilon}_j$, $j = 0, 1, \ldots$ are therefore bounded provided $0 \leq r \leq 1/2$ and stability of the difference equation results.

(c) Matrix Analysis

By taking account of the boundary conditions (2.2.2a), the explicit scheme (2.3.1a) can be written in matrix form as

$$\underline{u}_{j+1} = \Gamma \underline{u}_j, \tag{2.3.15}$$

where $\underline{u}_j = (u_{1,j}, u_{2,j}, \ldots, u_{m-1,j})^T$ and Γ is a square tridiagonal matrix termed the *amplification matrix* of order $(m - 1)$ given by

$$\Gamma = \begin{bmatrix} (1-2r) & r & & & & \\ r & (1-2r) & r & & & \\ & r & (1-2r) & r & & 0 \\ & & \ddots & \ddots & \ddots & \\ & 0 & & r & (1-2r) & r \\ & & & & r & (1-2r) \end{bmatrix}_{(m-1)\times(m-1)} = I + rT, \tag{2.3.16}$$

where

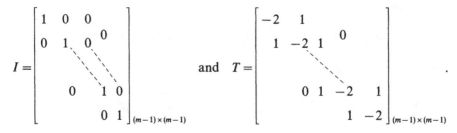

Since the matrix T is *tridiagonal* then the eigenvalues of T are given by

$$\eta_i = -4 \sin^2\left(\frac{i\pi}{2m}\right), \quad i = 1, 2, \ldots, m-1.$$

Hence, from (2.3.16) the eigenvalues of the *amplification matrix* Γ are $\mu_i = 1 + r\{-4 \sin^2(i\pi/2m)\}$ and the condition for stability is $|1 - 4r \sin^2(i\pi/2m)| \leq 1$. The only useful inequality is the lower which is $-1 \leq 1 - 4r \sin^2(i\pi/2m)$ giving

$$r \leq \frac{1}{2 \sin^2(i\pi/2m)}, \tag{2.3.17}$$

while the upper inequality produces the value $r > 0$.

This proves that the explicit scheme is stable for $r \leq 1/2$.

(d) Fourier Series Analysis
The substitution of the perturbation

$$\varepsilon_{i,j} = \xi^j \exp(i_c \beta i \Delta x), \tag{2.3.18}$$

into equation (2.3.1a) results in the stability equation

$$\gamma = 1 + r \left(\exp(-i_c \beta \Delta x) + \exp(i_c \beta \Delta x) - 2\right)$$

$$= 1 + r \left(2\cos(\beta \Delta x) - 2\right)$$

$$= 1 - 2r \left(1 - \cos(\beta \Delta x)\right)$$

$$= 1 - 4r \sin^2\left(\frac{\beta \Delta x}{2}\right). \tag{2.3.19}$$

The condition for stability, $|\gamma| \leq 1$, leads to

$$\left| 1 - 4r \sin^2\left(\frac{\beta \Delta x}{2}\right) \right| \leq 1.$$

By the same argument as above, we obtain conditional stability for $r \leq 1/2$.

2.4. Improving the Accuracy of Explicit Methods

In Richardson (1910) there is proposed the following explicit method to solve (2.2.2)

$$\frac{(u_{i,j+1} - u_{i,j-1})}{2\Delta t} = \frac{1}{(\Delta x)^2}(u_{i+1,j} - 2u_{i,j} + u_{i-1,j}). \tag{2.4.1}$$

By using the familiar Taylor series expansion about the point (x_i, t_j), the local truncation error of (2.4.1) is found to be

$$T = \left\{ \left(\frac{\partial^2 U}{\partial x^2} \right)_{i,j} - \left(\frac{\partial U}{\partial t} \right)_{i,j} \right\} - \frac{1}{6}(\Delta t)^2 \left(\frac{\partial^3 U}{\partial t^3} \right)_{i,j} + \frac{(\Delta x)^2}{12} \left(\frac{\partial^4 U}{\partial x^4} \right)_{i,j} + \cdots,$$

i.e.,

$$T = 0([\Delta t]^2) + 0([\Delta x]^2), \tag{2.4.2}$$

since the terms in brackets satisfy (2.2.2) at the mesh point (i, j). It is obvious that by comparing (2.4.2) and (2.3.3), equation (2.4.1), is of one order higher in time than equation (2.4.1). Unfortunately the difference equation (2.4.1) is now completely unstable for all choices of the mesh ratio r and hence the method is impractical. By applying the von Neumann criterion, the stability equation is

$$\gamma - \frac{1}{\gamma} = -8\lambda \sin^2 \left(\frac{\beta \Delta x}{2} \right), \tag{2.4.3}$$

or

$$\gamma^2 + 8\lambda \sin^2 \left(\frac{\beta \Delta x}{2} \right) \gamma - 1 = 0. \tag{2.4.4}$$

It is clear that the product of the roots of this equation is 1. Then if one root is less than 1, the other must be greater than 1. Also, the sum of the roots is $\gamma + (-1/\gamma)$. From (2.4.3) we see that unless $\beta = 0$, $\gamma - 1/\gamma < 0$ and either γ or $-1/\gamma$ is less than -1. Since γ must be assumed to be any real number, we conclude that the Richardson method is unconditionally unstable.

The accuracy of the explicit method (2.3.1) can be further improved and yet still maintain its stability by considering the truncation error (2.3.2). Let us assume that $U(x, t)$ possess continuous bounded derivatives up to order six in x and order three in t, i.e. $U \in C^{6,3}$. Hence, from the given diffusion equation $\partial U/\partial t = \partial^2 U/\partial x^2$, we get

$$\frac{\partial^2 U}{\partial t^2} = \frac{\partial^3 U}{\partial^2 x \partial t} = \frac{\partial^3 U}{\partial t \partial x^2} = \frac{\partial^4 U}{\partial x^4}, \tag{2.4.5}$$

and equation (2.3.2a) becomes

$$T = \frac{1}{2}\left[\Delta t - \frac{(\Delta x)^2}{6} \right] \left(\frac{\partial^4 U}{\partial x^4} \right)_{i,j} + \frac{(\Delta t)^2}{6} \left(\frac{\partial^4 U}{\partial t^4} \right)_{i,j} - \frac{(\Delta x)^4}{360} \left(\frac{\partial^6 U}{\partial x^6} \right)_{i,j} + \cdots.$$

Therefore, if Δt and Δx approach zero in such a way that

$$r = \frac{\Delta t}{(\Delta x)^2} = \frac{1}{6},$$

then the truncation error will be

$$T = 0([\Delta t]^2), \qquad (2.4.6a)$$

or

$$T = 0([\Delta x]^4), \qquad (2.4.6b)$$

and the bound (2.3.8) will be replaced by

$$E_j = \max |U_{i,j} - u_{i,j}| \le N_1([\Delta t]^2 + [\Delta x]^4). \qquad (2.4.7)$$

Thus the difference equation converges to the solution of the differential equation at *a rate much faster* than that exhibited by (2.3.8).

Another attempt to gain better accuracy is to expand $U_{i,j+1}$ by Taylor's series about the point (x_i, t_j), i.e.,

$$U_{i,j+1} = U_{i,j} + (\Delta t)\left(\frac{\partial U}{\partial t}\right)_{i,j} + \frac{(\Delta t)^2}{2}\left(\frac{\partial^2 U}{\partial t^2}\right)_{i,j} + \cdots$$

$$= U_{i,j} + (\Delta t)\left(\frac{\partial^2 U}{\partial x^2}\right)_{i,j} + \frac{(\Delta t)^2}{2}\left(\frac{\partial^4 U}{\partial x^4}\right)_{i,j} + \cdots, \qquad (2.4.8)$$

by virtue of (2.4.5). If we approximate $\partial^2 U/\partial x^2$ and $\partial^4 U/\partial x^4$ at the mesh points by second and fourth differences, respectively, i.e.,

$$\left(\frac{\partial^2 U}{\partial x^2}\right)_{i,j} \approx \frac{1}{(\Delta x)^2}[U_{i+1,j} - 2U_{i,j} + U_{i-1,j}],$$

and

$$\left(\frac{\partial^4 U}{\partial x^4}\right)_{i,j} \approx \frac{1}{(\Delta x)^4}\delta_x^4 U_{i,j}$$

$$\approx \frac{1}{(\Delta x)^4}[U_{i-2,j} - 4U_{i-1,j} + 6U_{i,j} - 4U_{i+1,j} + U_{i+2,j}],$$

then the difference analogue to (2.4.8) is

$$u_{i,j+1} = u_{i,j} + \frac{\Delta t}{(\Delta x)^2}\delta_x^2 u_{i,j} + \frac{1}{2}\frac{(\Delta t)^2}{(\Delta x)^4}\delta_x^4 u_{i,j}, \qquad (2.4.9)$$

with a local truncation error

$$T = 0([\Delta x]^2) + 0([\Delta t]^2). \qquad (2.4.10)$$

We also find that the stability requirement of (2.4.9) is the same as that of (2.3.1), i.e., $r \le 1/2$ and the boundedness condition (2.3.8) is replaced by

$$E_j = \max |U_{i,j} - u_{i,j}| \le N_2([\Delta x]^2 + [\Delta t]^2). \qquad (2.4.11)$$

Since $r = \Delta t/(\Delta x)^2$, (2.4.11) can also be written as $E_j = 0([\Delta x]^2)$ which is the same as before. We observe from (2.4.10) or (2.4.11) that although the error in the t-direction is one order higher, the error in the x-direction remains the same as that of the original explicit method. In fact, no overall improvement is obtained because the rate of convergence of the difference equation to the solution of the differential equation is of the same magnitude as before.

This can be rectified if we utilise the fact that

$$\left(\frac{\partial^2 U}{\partial x^2}\right)_{i,j} \approx \frac{1}{(\Delta x)^2} \delta_x^2 U_{i,j} - \frac{1}{12} (\Delta x)^2 \left(\frac{\partial^4 U}{\partial x^4}\right)_{i,j}$$

$$= \frac{1}{(\Delta x)^2} \delta_x^2 U_{i,j} - \frac{1}{12} (\Delta x)^{-2} \delta_x^4 U_{i,j}. \tag{2.4.12}$$

When (2.4.12) is used in (2.4.8) we obtain the approximation

$$u_{i,j+1} = u_{i,j} + \frac{\Delta t}{(\Delta x)^2} \delta_x^2 u_{i,j} + \frac{1}{(\Delta x)^4} \left[\frac{1}{2} (\Delta t)^2 - \frac{1}{12} (\Delta x)^2 \Delta t\right] \delta_x^4 u_{i,j} + T, \tag{2.4.13}$$

with a local truncation error

$$T = 0([\Delta x]^4) + 0([\Delta x]^2 \Delta t) + 0([\Delta t]^2), \tag{2.4.14}$$

and a stability ratio $r \le 2/3$. If the maximum analysis is now employed, it is found that

$$E_j \le N_3([\Delta x]^4 + [\Delta x]^2 \Delta t + [\Delta t]^2), \tag{2.4.15}$$

and for $r = \Delta t/(\Delta x)^2$, $E_j = 0((\Delta x)^4)$. Hence, we obtain improved accuracy and rate of convergence as well as a slightly better stability condition.

An adaptation of Richardson's equation (2.4.1) is the formula of Dufort and Frankel (1953) which is both *explicit and unconditionally stable*. As we shall see, however, these two desirable properties are achieved at the expense of consistency. Dufort and Frankel replaces the term $2u_{i,j}$ by the average quantity $(u_{i,j-1} + u_{i,j+1})$, thereby generating the *three-level formula*

$$\frac{(u_{i,j+1} - u_{i,j-1})}{2\Delta t} \approx \frac{(u_{i+1,j} - (u_{i,j-1} + u_{i,j+1}) + u_{i-1,j})}{(\Delta x)^2}, \tag{2.4.16}$$

which may be written as

$$(1 + 2r)u_{i,j+1} = 2r(u_{i-1,j} + u_{i+1,j}) + (1 - 2r)u_{i,j-1} + T, \tag{2.4.17}$$

where $r = \Delta t/(\Delta x)^2$ and T given later. For known boundary values with $m\Delta x = 1$, these equations when written in matrix form are

$$(1+2r)\begin{bmatrix} u_{1,j+1} \\ u_{2,j+1} \\ u_{3,j+1} \\ \vdots \\ \vdots \\ \vdots \\ u_{m-1,j+1} \end{bmatrix} = 2r\begin{bmatrix} 0 & 1 & & & & \\ 1 & 0 & 1 & & & \\ & 1 & 0 & 1 & 0 & \\ & & \ddots & & & \\ & & & & & 0 \\ & & & & 1 & 0 \end{bmatrix}\begin{bmatrix} u_{1,j} \\ u_{2,j} \\ u_{3,j} \\ \vdots \\ \vdots \\ \vdots \\ u_{m-1,j} \end{bmatrix} + (1-2r)\begin{bmatrix} u_{1,j-1} \\ u_{2,j-1} \\ u_{3,j-1} \\ \vdots \\ \vdots \\ \vdots \\ u_{m-1,j-1} \end{bmatrix} + 2r\begin{bmatrix} u_{0,j} \\ 0 \\ 0 \\ \vdots \\ 0 \\ u_m \end{bmatrix},$$

giving

$$\underline{u}_{j+1} = \frac{2r}{(1+2r)} A \underline{u}_j + \frac{(1-2r)}{(1+2r)} \underline{u}_{j-1} + \underline{c}_j, \tag{2.4.18}$$

where A is the matrix (of order $(m-1)$) as displayed and $\underline{c}_j$ is a vector of known values. If we put

$$v_j = (u_j, u_{j-1}),$$

then equation (2.4.18) can be written in the partitioned matrix form

$$\begin{bmatrix} u_{j+1} \\ \hline u_j \end{bmatrix} = \begin{bmatrix} \frac{2r}{(1+2r)}A & \frac{(1-2r)}{(1+2r)}I \\ \hline I & 0 \end{bmatrix} \begin{bmatrix} u_j \\ \hline u_{j-1} \end{bmatrix} + \begin{bmatrix} c_j \\ \hline \underline{0} \end{bmatrix},$$

i.e.,

$$\underline{v}_{j+1} = P\underline{v}_j + \underline{d}_j, \tag{2.4.19}$$

where I is the unit matrix of order $(m-1)$, P is the block matrix shown and $\underline{d}_j$, a column vector of known constants. The reduced two-level equations (2.4.19) will be stable when the eigenvalues of P are less than or equal to one in modulus. The matrix A has $(m-1)$ different eigenvalues and so it has $(m-1)$ linearly independent eigenvectors v_s, $s = 1, 2, \ldots, m-1$. Although I has $(m-1)$ eigenvalues, each equal to 1, it has $(m-1)$ linearly independent eigenvectors which may be taken as v_s, since $Iv_s = 1.v_s$. Hence, the eigenvalues μ of P are the eigenvalues of the matrix

$$\begin{bmatrix} \frac{2r\eta_k}{(1+2r)} & \frac{(1-2r)}{(1+2r)} \\ 1 & 0 \end{bmatrix},$$

where η_k is the kth eigenvalue of A which are given by

$$\eta_k = 2\cos\left(\frac{k\pi}{m}\right), \quad k = 1, 2, \ldots, m-1.$$

From

$$\det \begin{bmatrix} \left\{\frac{2r\eta_k}{(1+2r)} - \mu\right\} & \frac{(1-2r)}{(1+2r)} \\ 1 & -\mu \end{bmatrix} = 0,$$

we get the characteristic equation

$$\mu^2 - \frac{2r\mu_k}{(1+2r)}\mu - \frac{(1-2r)}{(1+2r)} = 0,$$

the roots of which are obtained from

$$\mu = \left\{ 2r \cos\left(\frac{k\pi}{m}\right) \pm \left(1 - 4r^2 \sin^2\left(\frac{k\pi}{m}\right)\right)^{1/2} \right\} \Big/ (1 + 2r). \qquad (2.4.20)$$

Case (i)
If $4r^2 \sin^2 (k\pi/m) \leq 1$, which corresponds to small values of Δt, then,

$$|\mu| < \frac{(2r + 1)}{(1 + 2r)} = 1.$$

Case (ii)
If $4r^2 \sin^2 (k\pi/m) > 1$, which corresponds to large time steps (Δt), then,

$$|\mu|^2 = \left\{ \left(2r \cos\left(\frac{k\pi}{m}\right)\right)^2 + 4r^2 \sin^2\left(\frac{k\pi}{m}\right) - 1 \right\} \Big/ (1 + 2r)^2,$$

$$= \frac{(4r^2 - 1)}{(4r^2 + 4r + 1)} < 1 \quad \text{since } \lambda > 0.$$

Therefore, the equations are unconditionally stable for all positive r.

By the usual Taylor series expansion, we find that the local truncation error of the Dufort–Frankel scheme is

$$T = \frac{1}{6} (\Delta t)^2 \left(\frac{\partial^3 U}{\partial t^3}\right)_{i,j} - \frac{1}{12} (\Delta x)^2 \left(\frac{\partial^4 U}{\partial x^4}\right)_{i,j} + \frac{(\Delta t)^2}{(\Delta x)^2} \left(\frac{\partial^2 U}{\partial t^2}\right)_{i,j} + 0\left(\frac{[\Delta t]^4}{[\Delta x]^2}\right). \qquad (2.4.21)$$

Hence

$$T = 0([\Delta t]^2) + 0([\Delta x]^2) + 0\left(\frac{[\Delta t]^4}{[\Delta x]^2}\right). \qquad (2.4.22)$$

For the difference approximation (2.4.16) to be consistent with the differential equation (2.2.2) requires that $\Delta t/\Delta x \to 0$ as $t \to 0$. This means that the Dufort–Frankel scheme (2.4.16) or (2.4.17) is *consistent* with the heat equation (2.2.2) if and only if $\Delta t \to 0$ *faster* than $\Delta x \to 0$. If $\Delta t/\Delta x$ is kept fixed say equal to β, then the third term of (2.4.21) would tend to the value $\beta^2 \partial^2 U/\partial t^2$, rather than to zero. Equation (2.4.16) or (2.4.17) would therefore be consistent with the hyperbolic equation

$$\frac{\partial U}{\partial t} - \frac{\partial^2 U}{\partial x^2} + \beta^2 \frac{\partial^2 U}{\partial t^2} = 0, \qquad (2.4.23)$$

and not with the parabolic equation (2.2.2).

2.5. Implicit Methods

The simplest implicit method is that which was first suggested by O'Brien *et al.* (1951). Upon approximating $\partial^2 U/\partial x^2$ of equation (2.2.2) on the $(j+1)^{th}$ row instead of the jth row, we obtain

$$\frac{(u_{i,j+1} - u_{i,j})}{\Delta t} = \frac{1}{(\Delta x)^2}(u_{i+1,j+1} - 2u_{i,j+1} + u_{i-1,j+1}) + T, \qquad (2.5.1)$$

or

$$-ru_{i-1,j+1} + (1+2r)u_{i,j+1} - ru_{i+1,j+1} = u_{i,j}, \qquad (2.5.2)$$

where T is given later. Crank and Nicolson (1947) used an average of the approximations on the jth and $(j+1)^{th}$ row. More generally, one can introduce a *weighting parameter* θ and replace equation (2.5.1) by

$$u_{i,j+1} - u_{i,j} = r\{\theta[u_{i-1,j+1} - 2u_{i,j+1} + u_{i+1,j+1}]$$
$$+ (1-\theta)[u_{i-1,j} - 2u_{i,j} + u_{i+1,j}]\}, \qquad (2.5.3)$$

with $0 \le \theta \le 1$. A single application of equation (2.5.3) equates a linear combination of three unknowns in the $(j+1)^{th}$ row to a linear combination of the three known values in the jth row and can be solved easily by the LU or Gaussian elimination algorithm. Thus, we obtain

$$-r\theta u_{i-1,j+1} + (1+2r\theta)u_{i,j+1} - r\theta u_{i+1,j+1}$$
$$= r(1-\theta)u_{i-1,j} + [1-2r(1-\theta)]u_{i,j} + r(1-\theta)u_{i+1,j}. \qquad (2.5.4)$$

If $\theta = 1$, equation (2.5.4) becomes the fully implicit (O'Brien *et al.* 1951) form of (2.5.2). If $\theta = 1/2$ we obtain the Crank–Nicolson formula. On the other hand, if $\theta = 0$ the explicit relation (2.3.1a) is recovered.

We now examine the stability of (2.5.4.) by the matrix method. We assume that the boundary values are zero and the initial values are known. With $u_{0,j} = u_{m,j} = 0$, equation (2.5.4), can be written in the matrix form as

$$\begin{bmatrix} (1+2r\theta) & -r\theta \\ -r\theta & (1+2r\theta) & -r\theta & & 0 \\ & & \ddots & \ddots & \ddots \\ 0 & & -r\theta & (1+2r\theta) & -r\theta \\ & & & -r\theta & (1+2r\theta) \end{bmatrix} \begin{bmatrix} u_1 \\ u_2 \\ \vdots \\ \vdots \\ u_{m-2} \\ u_{m-1} \end{bmatrix}_{j+1}$$

$$= \begin{bmatrix} (1+2r(1-\theta)) & r(1-\theta) & & & \\ r(1-\theta) & (1-2r(1-\theta)) & r(1-\theta) & & 0 \\ & & & & \\ 0 & r(1-\theta) & (1-2r(1-\theta)) & r(1-\theta) \\ & & r(1-\theta) & (1-2r(1-\theta)) \end{bmatrix} \begin{bmatrix} u_1 \\ u_2 \\ \vdots \\ u_{m-2} \\ u_{m-1} \end{bmatrix}_j ,$$

i.e.,

$$(I - r\theta T)\underline{u}_{j+1} = (I + r(1-\theta)T)\underline{u}_j, \tag{2.5.5}$$

where I is the identity matrix of order $(m-1)$ and T is as given in (2.3.16). From (2.5.5), we have

$$\underline{u}_{j+1} = \Gamma\underline{u}_j, \tag{2.5.6}$$

where

$$\Gamma = (I - r\theta T)^{-1}(I + r(1-\theta)T), \tag{2.5.7}$$

the amplification matrix. As before, the eigenvalues of T are given by

$$\eta_i = -4\sin^2\left(\frac{i\pi}{2m}\right), \quad i = 1, 2, \dots, m-1.$$

Hence, from equation (2.5.7) the eigenvalues of Γ are

$$\mu_i = \frac{(1 - 4r(1-\theta)\sin^2(i\pi/2m))}{(1 + 4r\theta\sin^2(i\pi/2m))}, \tag{2.5.8}$$

and the condition for stability is

$$-1 \le \frac{1 - 4r(1-\theta)\sin^2(i\pi/2m)}{1 + 4r\theta\sin^2(i\pi/2m)} \le 1. \tag{2.5.9}$$

The upper inequality is automatically satisfied for $r > 0$. The lower inequality gives

$$(8r\theta - 4r)\sin^2\left(\frac{i\pi}{2m}\right) \ge -2.$$

Hence $2r(1 - 2\theta) \le 1$. We conclude that the equation (2.5.4) is conditionally stable for

$$r \le \frac{1}{2(1 - 2\theta)}, \quad \text{when } 0 \le \theta \le \frac{1}{2}, \tag{2.5.10}$$

and is unconditionally stable for all

$$r > 0, \quad \text{when } 1/2 \le \theta \le 1. \tag{2.5.11}$$

To analyse the local truncation error of the general method, we have from (2.5.4)

$$T_{2.5.3} = -r\theta(U_{i-1,j+1} + U_{i+1,j+1}) - r(1-\theta)(U_{i-1,j} + U_{i+1,j})$$
$$+ (1 + 2r\theta)U_{i,j+1} - (1 - 2r(1-\theta))U_{i,j}. \tag{2.5.12}$$

On expanding the terms $U_{i,j+1}, U_{i-1,j}, U_{i+1,j}, U_{i-1,j+1}$ and $U_{i+1,j+1}$ about the mesh points (x_i, t_j) by Taylor's series and on recognising that,

$$\left(\frac{\partial U}{\partial t}\right)_{i,j} = \left(\frac{\partial^2 U}{\partial x^2}\right)_{i,j},$$

$$\left(\frac{\partial^2 U}{\partial t^2}\right)_{i,j} = \left(\frac{\partial^3 U}{\partial x^2 \partial t}\right)_{i,j} = \left(\frac{\partial^4 U}{\partial x^4}\right)_{i,j},$$

$$\left(\frac{\partial^3 U}{\partial t^3}\right)_{i,j} = \left(\frac{\partial^4 U}{\partial x^2 \partial t^2}\right)_{i,j} = \left(\frac{\partial^5 U}{\partial x^4 \partial t}\right)_{i,j}$$

(2.5.13)

and

$$\left(\frac{\partial^4 U}{\partial t^4}\right)_{i,j} = \left(\frac{\partial^5 U}{\partial x^2 \partial t^3}\right)_{i,j},$$

we get

$$T_{2.5.3} = \left(\frac{(\Delta t)^2}{2} - r\theta(\Delta x)^2 \Delta t - \frac{r}{12}(\Delta x)^4\right)\left(\frac{\partial^2 U}{\partial t^2}\right)_{i,j}$$

$$+ \left(\frac{(\Delta t)^2}{6} - \frac{r\theta}{2}(\Delta x)^2 (\Delta t)^2 - \frac{r\theta}{12}(\Delta x)^4 \Delta t\right)\left(\frac{\partial^3 U}{\partial t^3}\right)_{i,j}$$

$$+ \left(\frac{(\Delta t)^4}{24} - \frac{1}{6}r\theta(\Delta x)^2 (\Delta t)^3\right)\left(\frac{\partial^4 U}{\partial t^4}\right)_{i,j} + \frac{(\Delta t)^5}{120}\left(\frac{\partial^5 U}{\partial t^5}\right)_{i,j} + \cdots.$$

(2.5.14)

But $r = \Delta t/(\Delta x)^2$. Therefore, we obtain

$$T_{2.5.3} = \left(\frac{(\Delta t)^2}{2} - (\Delta t)^2 \theta - \frac{\Delta t}{12}(\Delta x)^2\right)\left(\frac{\partial^2 U}{\partial t^2}\right)_{i,j} + \left(\frac{(\Delta t)^3}{6} - \frac{(\Delta t)^3}{2}\theta - \frac{(\Delta t)^2}{12}(\Delta x)^2 \theta\right)$$

$$\left(\frac{\partial^3 U}{\partial t^3}\right)_{i,j} + \left(\frac{(\Delta t)^4}{24} - \frac{1}{6}(\Delta t)^4 \theta\right)\left(\frac{\partial^4 U}{\partial t^4}\right)_{i,j} + \frac{(\Delta t)^5}{120}\left(\frac{\partial^5 U}{\partial t^5}\right)_{i,j} + \cdots,$$

(2.5.15)

i.e.,

$$T_{2.5.3} = \Delta t \left\{ \left[\Delta t(1/2 - \theta) - \frac{(\Delta x)^2}{12}\right]\left(\frac{\partial^2 U}{\partial t^2}\right)_{i,j} + \left[\frac{(\Delta t)^2}{2}(1/3 - \theta) - \frac{(\Delta t)}{12}(\Delta x)^2 \theta\right] \right.$$

$$\left. \left(\frac{\partial^3 U}{\partial t^3}\right)_{i,j} + \frac{(\Delta t)^3}{6}(1/4 - \theta)\left(\frac{\partial^4 U}{\partial t^4}\right)_{i,j} + \frac{(\Delta t)^4}{120}\left(\frac{\partial^5 U}{\partial t^5}\right)_{i,j} + \cdots \right\} = 0.$$

Hence, an expression for the local truncation error is

$$T_{2.5.3} = \left[\Delta t\left(\frac{1}{2} - \theta\right) - \frac{(\Delta x)^2}{12}\right]\left(\frac{\partial^2 U}{\partial t^2}\right)_{i,j} + \left[\frac{(\Delta t)^2}{2}\left(\frac{1}{3} - \theta\right) - \frac{(\Delta t)}{12}(\Delta x)^2 \theta\right]\left(\frac{\partial^3 U}{\partial t^3}\right)_{i,j}$$

$$+ \frac{(\Delta t)^3}{6}\left(\frac{1}{4} - \theta\right)\left(\frac{\partial^4 U}{\partial t^4}\right)_{i,j} + \frac{(\Delta t)^4}{120}\left(\frac{\partial^5 U}{\partial t^5}\right)_{i,j} + \cdots.$$

(2.5.16)

We find that for $\theta = 0$ (the explicit scheme (2.3.1))

$$T_1 = 0(\Delta t) + 0([\Delta x]^2), \qquad (2.5.17)$$

for $\theta = 1/2$ (the Crank–Nicolson scheme)

$$T_2 = 0([\Delta x]^2) + 0([\Delta t]^2), \qquad (2.5.18)$$

and for $\theta = 1$ (the fully implicit scheme (2.5.4))

$$T_3 = 0(\Delta t) + 0([\Delta x]^2). \qquad (2.5.19)$$

2.6. Parabolic Problems with Derivative Boundary Conditions

Boundary conditions which are expressed in terms of derivatives occur very frequently in practice. To solve (2.2.2), the boundary conditions (2.2.2a) are replaced by

$$\frac{\partial U}{\partial x} = h_1(U - v_1) \qquad \text{at } x = 0, \ 0 < t \leq T,$$

$$\frac{\partial U}{\partial x} = -h_2(U - v_2) \quad \text{at } x = 1, \ 0 < t \leq T, \qquad (2.6.1)$$

where h_1, h_2, v_1 and v_2 are constants with $h_1 \geq 0$, $h_2 \geq 0$. The boundary conditions may be approximated in a number of ways. For example, they may be approximated by central differences and an explicit scheme is used to approximate the differential equation. Alternatively, they can be replaced by forward differences at $x = 0$ and backward differences at $x = 1$, and an explicit scheme is employed for the solution.

When the boundary conditions are approximated by the central difference equations

$$\frac{(u_{1,j} - u_{-1,j})}{2\Delta x} = h_1(u_{0,j} - v_1),$$

$$\frac{(u_{m+1,j} - u_{m-1,j})}{2\Delta x} = -h_2(u_{m,j} - v_2),$$

with $m\Delta x = 1$, and the differential equation by the Crank–Nicolson scheme (2.5.3) with $\theta = 1/2$, i.e.,

$$-\frac{r}{2}u_{i-1,j+1} + (1+r)u_{i,j+1} - \frac{r}{2}u_{i+1,j+1} = \frac{r}{2}u_{i-1,j} + (1-r)u_{i,j} + \frac{r}{2}u_{i+1,j}, \quad (2.6.2)$$

the elimination of $u_{-1,j}$ and $u_{m+1,j}$ leads to the equation

$$A\underline{u}_{j+1} = B\underline{u}_j + \underline{b}_j, \qquad (2.6.3)$$

where

$$\underline{u}_j = (u_{0,j}, u_{1,j}, \ldots, u_{m,j})^T,$$

$$A = I - \tfrac{1}{2}rQ, \tag{2.6.4}$$

and

$$B = I + \tfrac{1}{2}rQ, \tag{2.6.5}$$

where Q is a matrix of order $(m+1)$ given by

$$Q = \begin{bmatrix} -2(1+\Delta x h_1) & 2 & & & 0 \\ 1 & -2 & 1 & & \\ & 1 & -2 & 1 & \\ 0 & & 1 & -2 & 1 \\ & & & 2 & -2(1+\Delta x h_2) \end{bmatrix}, \tag{2.6.6}$$

and

$$\underline{b}_j^T = (2rv_1 \Delta x h_1, 0, \ldots, 0, 2rv_2 \Delta x h_2)^T.$$

For the analysis of stability we write (2.6.3) as

$$\underline{u}_{j+1} = A^{-1}B\underline{u}_j + A^{-1}\underline{b}_j,$$

i.e.,

$$\underline{u}_{j+1} = (I - \tfrac{1}{2}rQ)^{-1}(I + \tfrac{1}{2}rQ)\underline{u}_j + \hat{\underline{b}}_j, \tag{2.6.7}$$

where

$$\hat{\underline{b}}_j = (I - \tfrac{1}{2}rQ)^{-1}\underline{b}_j,$$

and it is assumed that $\det(I - (1/2)rQ) \neq 0$.

We note that Q is unsymmetric. Therefore, we introduce the diagonal matrix

$$D = \begin{bmatrix} \sqrt{2} & & & \\ & 1 & 0 & \\ & 0 & \ddots & \\ & & & 1 \\ & & & & \sqrt{2} \end{bmatrix}, \tag{2.6.8}$$

of order $(m+1)$ such that Q is similar to the symmetric matrix

$$\tilde{Q} = D^{-1}QD. \tag{2.6.9}$$

Then

$$(\tilde{A}^{-1}B) = D^{-1}(I - \tfrac{1}{2}rQ)^{-1}(I + \tfrac{1}{2}rQ)D$$

$$= [D^{-1}(I - \tfrac{1}{2}rQ)^{-1}D][D^{-1}(I + \tfrac{1}{2}rQ)D]$$

$$= [D^{-1}(I - \tfrac{1}{2}rQ)D]^{-1}[D^{-1}(I + \tfrac{1}{2}rQ)D]$$

$$= [I - \tfrac{1}{2}r\tilde{Q}][I + \tfrac{1}{2}r\tilde{Q}]. \tag{2.6.10}$$

However, the matrices $(I - (1/2)r\tilde{Q})$ and $(I + (1/2)r\tilde{Q})$ are symmetric and commute, and so $(\tilde{A}^{-1}B)$ is symmetric. Therefore, $A^{-1}B$ is similar to the symmetric matrix $(\tilde{A}^{-1}B)$ and

$$\rho(\tilde{A}^{-1}B) = \rho(\tilde{A}^{-1}B) \leq 1 \tag{2.6.11}$$

is a necessary and sufficient condition for stability, where ρ denotes the spectral radius.

The eigenvalues η_j $(j = 0, 1, \ldots, m)$ of $A^{-1}B$ are given by

$$\eta_j = \frac{(1 + \frac{1}{2}r\mu_j)}{(1 - \frac{1}{2}r\mu_j)}, \tag{2.6.12}$$

where μ_j are the eigenvalues of the matrix Q. Since η and ρ are related by

$$\rho(A^{-1}B) = \max_j |\eta_j|,$$

the condition for stability (2.6.11) together with (2.6.12) gives

$$\mu_j \leq 0 \quad \text{for all } j.$$

By means of the Gerschgorin Theorem, it is easily seen that μ_j lies on the negative line for all j. Hence, equation (2.6.3) is unconditionally stable.

2.7. Improving the Accuracy of Implicit Methods

(a) Reduction of the Local Truncation Error

The derivatives of $U(x,t)$ can be expressed exactly in terms of the *infinite series* of forward, backward or central differences. For example, in the previous analysis, we used

$$\frac{\partial^2 U}{\partial x^2} = \frac{1}{(\Delta x)^2}\left(\delta_x^2 U - \frac{1}{12}\delta_x^4 U + \frac{1}{90}\delta_x^6 U + \cdots\right). \tag{2.7.1}$$

When the diffusion equation (2.2.2) is approximated at the point $(x_i, t_{j+1/2})$ by

$$\frac{1}{\Delta t}(u_{i,j+1} - u_{i,j}) = \frac{1}{2}\left\{\left(\frac{\partial^2 U}{\partial x^2}\right)_{i,j+1} + \left(\frac{\partial^2 U}{\partial x^2}\right)_{i,j}\right\}$$

$$= \frac{1}{2(\Delta x)^2}\left(\delta_x^2 - \frac{1}{12}\delta_x^4 + \frac{1}{90}\delta_x^6 + \cdots\right)(u_{i,j+1} + u_{i,j}), \tag{2.7.2}$$

then the terms involving δ_x^4 can be eliminated by operating on both sides with $(1 + 1/12\,\delta_x^2)$. This gives us

$$(1 + \tfrac{1}{12}\delta_x^2)(u_{i,j+1} - u_{i,j}) = \tfrac{1}{2}r(\delta_x^2 u_{i,j+1} + \delta_x^2 u_{i,j}), \tag{2.7.3}$$

where we have neglected the terms of order δ_x^6.

Equation (2.7.3) can be rearranged as

$$[1 + (\tfrac{1}{12} - \tfrac{1}{2}r)\delta_x^2]u_{i,j+1} = [1 + (\tfrac{1}{12} + \tfrac{1}{2}r)\delta_x^2]u_{i,j},$$

which on expanding leads to the *Douglas equation* (Douglas, 1961)

$$(1 - 6r)u_{i-1,j+1} + (10 + 12r)u_{i,j+1} + (1 - 6r)u_{i+1,j+1}$$

$$= (1 + 6r)u_{i-1,j} + (10 - 12r)u_{i,j} + (1 + 6r)u_{i+1,j}. \qquad (2.7.4)$$

We note that the Douglas equation (2.7.4) can be considered as a special case of the weighted equation (2.5.3) by putting $\theta = 1/2 - 1/12r$.

By substituting this value of θ into (2.5.8) we find that

$$\mu_i = \frac{(1 - \tfrac{1}{3}(6r + 1)\sin^2(i\pi/2m))}{(1 + \tfrac{1}{3}(6r + 1)\sin^2(i\pi/2m))}, \quad i = 1, 2, \dots, m - 1,$$

and it is clear that $|\mu_i| \le 1$ for all values of $\lambda > 0$. Hence, the Douglas equation has unrestricted stability. Furthermore, with the same value of θ, the expression for the truncation error, i.e., equation (2.5.16) becomes

$$T_{2.7.4} = \frac{1}{12}\left[-(\Delta t)^2 + \frac{1}{12}(\Delta x)^4 \right]\left(\frac{\partial^3 U}{\partial t^3}\right)_j^i$$

$$+ \frac{1}{72}[\Delta t]^2\left((\Delta x)^2 - 3(\Delta t)\right)\left(\frac{\partial^4 U}{\partial t^4}\right)_j^i + \frac{(\Delta t)^2}{120}\left(\frac{\partial^5 U}{\partial t^5}\right)_j^i + \cdots,$$

i.e.,

$$T_{2.7.4} = 0([\Delta t]^2) + 0([\Delta x]^4) \text{ as opposed to } T = 0([\Delta t]^2) + 0([\Delta x]^2),$$

for the Crank–Nicolson method. We note that the resulting tridiagonal system of equations for the Douglas scheme can be solved by the Gauss elimination algorithm and requires the same amount of arithmetic as the Crank–Nicolson method.

(b) Use of Three-Level Difference Equations
In the construction of difference equations of high accuracy, or, occasionally, to improve stability, one often uses more time levels than the minimum number required by the given differential equations. As an example, a fully implicit two-level approximation to the heat equation (2.2.2) is from (2.5.1)

$$\frac{(u_{i,j+1} - u_{i,j})}{\Delta t} = \frac{(u_{i+1,j+1} - 2u_{i,j+1} + u_{i-1,j+1})}{(\Delta x)^2},$$

and from (2.5.19) this has a local truncation error

$$T = 0(\Delta t) + 0([\Delta x]^2).$$

When the analogue for $\partial U/\partial t$ is replaced by $(3/2(u_{i,j+1}) - 2u_{i,j} + 1/2(u_{i,j-1}))/\Delta t$, we obtain a three level equation given by

$$\frac{1}{\Delta t}\left(\frac{3}{2}u_{i,j+1} - 2u_{i,j} + \frac{1}{2}u_{i,j-1}\right) = \frac{1}{(\Delta x)^2}(u_{i+1,j+1} - 2u_{i,j+1} + u_{i-1,j+1}), \quad (2.7.5)$$

which has a local truncation error of the order

$$T_{2.7.5} = 0([\Delta t]^2) + 0([\Delta x]^2). \tag{2.7.6}$$

Now initial data are required at $t = 0$ and $t = \Delta t$ to start the calculation. The data at $t = 0$ are given whilst the data at $t = \Delta t$ are calculated from a two-level difference formula or by using a power series expansion. The data calculated at $t = \Delta t$ should be of an accuracy comparable with that of the three-level scheme.

Equation (2.7.5) can be written as

$$- 2r u_{i-1,j+1} + (3 + 4r) u_{i,j+1} - 2r u_{i+1,j+1} = 4u_{i,j} - u_{i,j-1}.$$

In matrix form, this becomes

$$\underline{u}_{j+1} = 4A^{-1}\underline{u}_j - A^{-1}\underline{u}_{j-1} + A^{-1}\underline{c}_{j+1}, \tag{2.7.7}$$

where $\underline{u}_{j+1}$ is a vector of known boundary values and

$$A = \begin{bmatrix} (3+4r) & -2r & & & 0 \\ -2r & (3+4r) & -2r & & \\ & \ddots & \ddots & \ddots & \\ & & & & -2r \\ 0 & & & -2r & (3+4r) \end{bmatrix} \quad \text{of order } (m-1).$$

Equation (2.7.7) can be written as

$$\left[\begin{array}{c} u_{j+1} \\ \hline u_j \end{array}\right] = \left[\begin{array}{c|c} 4A^{-1} & -A^{-1} \\ \hline I & 0 \end{array}\right]\left[\begin{array}{c} u_j \\ \hline u_{j-1} \end{array}\right] + \left[\begin{array}{c} A^{-1}c_{j+1} \\ \hline \underline{0} \end{array}\right],$$

i.e., $\underline{v}_{j+1} = P\underline{v}_j + \underline{c}'$. As before, the eigenvalues μ of P are the eigenvalues of

$$\begin{bmatrix} \dfrac{4}{\eta_k} & -\dfrac{1}{\eta_k} \\ 1 & 0 \end{bmatrix},$$

where η_k is the kth eigenvalue of A given by

$$\eta_k = 3 + 8r \sin^2\left(\frac{k\pi}{2m}\right), \quad k = 1, 2, \dots, m-1.$$

Since $\det(P - \mu I) = 0$, we have

$$\mu^2 - \left(\frac{4}{\eta_k}\right)\mu + \left(\frac{1}{\eta_k}\right) = 0$$

then

$$\mu = [2 \pm \sqrt{(4 - \eta_k)}]/\eta_k$$

$$= \left[2 \pm \sqrt{\left(1 - 8r\sin^2\left(\frac{k\pi}{2m}\right)\right)}\right] \Big/ \left(3 + 8r\sin^2\left(\frac{k\pi}{2m}\right)\right).$$

When the roots are real, $|\mu| < (2 + \sqrt{1 - \delta})/(3 + \delta)$, $\delta > 0$. Hence $|\mu| < 1$.

When the roots are complex, $|\mu| = 1/\sqrt{\eta_k} < 1$. Therefore, the approximation (2.7.7) is unconditionally stable.

We note that the truncation error of (2.7.5) is of the same order as the Crank–Nicolson method and is often used when the initial data is discontinuous or varies rapidly with x since it dampens the short-wavelength components (components with $\beta\Delta x \approx \pi$) more rapidly.

The Crank–Nicolson approximation should be used when the initial data and its derivatives are continuous. This is because the coefficient of $(\Delta t)^2$ in the truncation error is smaller. A three-level variation of the Douglas equation is

$$\frac{1}{12(\Delta t)}\left\{\frac{3}{2}(u_{i+1,j+1} - u_{i+1,j}) - \frac{1}{2}(u_{i+1,j} - u_{i+1,j-1})\right\}$$

$$+ \frac{5}{6(\Delta t)}\left\{\frac{3}{2}(u_{i,j+1} - u_{i,j}) - \frac{1}{2}(u_{i,j} - u_{i,j-1})\right\}$$

$$+ \frac{1}{12(\Delta t)}\left\{\frac{3}{2}(u_{i-1,j+1} - u_{i-1,j}) - \frac{1}{2}(u_{i-1,j} - u_{i-1,j-1})\right\}$$

$$= \frac{1}{(\Delta x)^2}(u_{i+1,j+1} - 2u_{i,j+1} + u_{i-1,j+1}), \qquad (2.7.8)$$

and like the Douglas equation, is stable and has a truncation error

$$T_{2.7.8} = 0([\Delta t]^2) + 0([\Delta x]^4).$$

Let us now consider another method which is second-order accurate in both space and time. If we replace $\partial^2 U/\partial x^2$ by the average of the second differences at t_{j-1}, t_j and t_{j+1} and $\partial U/\partial t$ by a central first difference we obtain

$$\frac{(u_{i,j+1} - u_{i,j-1})}{2(\Delta t)} = \frac{1}{3}\frac{\delta_x^2}{(\Delta x)^2}(u_{i,j+1} + u_{i,j} + u_{i,j-1}), \qquad (2.7.9)$$

where $\delta_x^2 u_{i,j} = u_{i+1,j} - 2u_{i,j} + u_{i-1,j}$ and $T = 0([\Delta t]^2) + 0([\Delta x]^2)$. The method (2.7.9) can now be improved to attain a higher order accuracy. As

$$\left(\frac{\partial^2 U}{\partial x^2}\right)_{i,j} = \frac{1}{3}\frac{\delta_x^2}{(\Delta x)^2}(U_{i,j+1} + U_{i,j} + U_{i,j-1}) - \frac{(\Delta x)^2}{12}\left(\frac{\partial^4 U}{\partial x^4}\right)_{i,j} + 0([\Delta t]^2) + 0([\Delta x]^4),$$

$$(2.7.10)$$

then

$$\left(\frac{\partial^2 U}{\partial x^2}\right)_{i,j} = \frac{1}{3}\frac{\delta_x^2}{(\Delta x)^2}(U_{i,j+1}+U_{i,j}+U_{i,j-1}) - \frac{(\Delta x)^2}{12}\left(\frac{\partial^2 U}{\partial t^2}\right)_{i,j} + 0([\Delta t]^2) + 0([\Delta x]^4)$$

$$= \frac{1}{3}\frac{\delta_x^2}{(\Delta x)^2}(U_{i,j+1}+U_{i,j}+U_{i,j-1}) - \frac{(\Delta x)^2}{12}\frac{\delta_t^2}{(\Delta t)^2}U_{i,j} + 0([\Delta t]^2) + 0([\Delta x]^4).$$

(2.7.11)

Thus we obtain the difference equation

$$\frac{(u_{i,j+1}-u_{i,j-1})}{2(\Delta t)} = \frac{1}{3}\frac{\delta_x^2}{(\Delta x)^2}(u_{i,j+1}+u_{i,j}+u_{i,j-1}) - \frac{(\Delta x)^2}{12}\frac{\delta_t^2}{(\Delta t)^2}U_{i,j}, \qquad (2.7.12)$$

with a local truncation error

$$T_{2.7.12} = 0([\Delta t]^2) + 0([\Delta x]^4).$$

2.8. Asymmetric Finite Difference Schemes

We now discuss a class of *asymmetric finite difference equations* of order $0(\Delta x)$ introduced by Saul'yev (1964) of the form

$$(1+\alpha r)u_{i,j+1} - \alpha r u_{i-1,j+1} = (1-\alpha)r u_{i-1,j} + [1+(\alpha-2)r]u_{i,j} + r u_{i+1,j}, \quad (2.8.1)$$

and

$$(1+\alpha r)u_{i,j+1} - \alpha r u_{i+1,j+1} = r u_{i-1,j} + [1+(\alpha-2)r]u_{i,j} + (1-\alpha)r u_{i+1,j}, \quad (2.8.2)$$

where $0 \le \alpha \le 1$. The computational molecules of these equations are given by Figure 2.8.1 below.

For $\alpha = 0$, both formulae reduce to the classical explicit form of (2.3.1a). When $\alpha = 1$, we obtain

$$(1+r)u_{i,j+1} - r u_{i-1,j+1} = (1-r)u_{i,j} + r u_{i+1,j}, \qquad (2.8.3)$$

and

$$(1+r)u_{i,j+1} - r u_{i+1,j+1} = r u_{i-1,j} + (1-r)u_{i,j}, \qquad (2.8.4)$$

whose computational molecules are given in Figure 2.8.2.

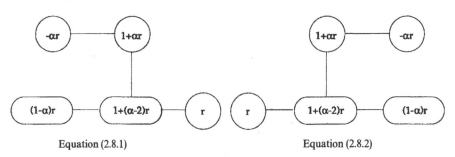

Equation (2.8.1)   Equation (2.8.2)

Figure 2.8.1

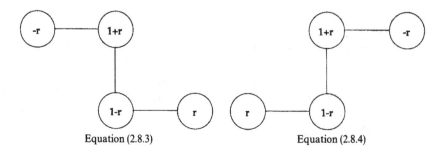

Equation (2.8.3)   Equation (2.8.4)

Figure 2.8.2

For an initial-boundary value problem, equations (2.8.3) and (2.8.4) are both implicit and explicit if the computations proceed in the positive left to right (LR) and negative right to left (RL) directions, respectively. Hence, equations (2.8.1) and (2.8.2) may be considered as generalisations of the classical explicit equation, i.e., semi-explicit.

The stability of the asymmetric formulae (2.8.1) and (2.8.2) for the case $\alpha = 1$ is now investigated by using the von Neumann criterion. The equation governing the stability of equation (2.8.1) is found to be

$$-\alpha r \xi^{j+1} \exp(i_c \beta(i-1)\Delta x) + (1+\alpha r)\xi^{j+1}\exp(i_c \beta i \Delta x)$$

$$= (1-\alpha)r\xi^j \exp(i_c \beta(i-1)\Delta x) + (1+(\alpha-2)r)\xi^j \exp(i_c \beta i \Delta x)$$

$$+ r\xi^j \exp(i_c \beta(i+1)\Delta x),$$

which reduces to

$$\gamma = \frac{(1-\alpha)r \exp(-i_c \beta \Delta x) + r \exp(i_c \beta \Delta x) + [1+(\alpha-2)r]}{(1+\alpha r) - \alpha r \exp(-i_c \beta \Delta x)},$$

where $\gamma = \xi^{j+1}/\xi^j$. For $\alpha = 1$, we get

$$\gamma = \frac{1 - r[1 - \exp(i_c \beta \Delta x)]}{1 + r[1 - \exp(-i_c \beta \Delta x)]},$$

i.e.,

$$\gamma = \frac{(1 + 2r^2 \cos(\beta \Delta x) - 2r^2 \cos^2(\beta \Delta x)) + i_c r \sin(\beta \Delta x)[2r - 2r \cos(\beta \Delta x)]}{1 + 2r(1 - \cos(\beta \Delta x)) + 2r^2(1 - \cos(\beta \Delta x))},$$

therefore

$$|\gamma| = \frac{\{1 + 8r^2 \sin^2(\beta \Delta x/2) - 16r^2 \sin^4(\beta \Delta x/2) + 16r^4 \sin^4(\beta \Delta x/2)\}^{1/2}}{[1 + 4r \sin^2(\beta \Delta x/2) + 4r^2 \sin^2(\beta \partial x/2)]}$$

$$= \frac{\{[1 + 4r^2 \sin^2(\beta \Delta x/2)]^2 - [4r \sin^2(\beta \Delta x/2)]^2\}^{1/2}}{\{1 + 4r \sin^2(\beta \Delta x/2) + 4r^2 \sin^2(\beta \Delta x/2)\}}$$

$$= \frac{[1+4r^2\sin^2(\beta\Delta x/2)+4r\sin^2(\beta\Delta x/2)]^{1/2}[1+4r^2\sin^2(\beta\Delta x/2)-4r\sin(\beta\Delta x/2)]^{1/2}}{[1+4r\sin^2(\beta\Delta x/2)+4r^2\sin^2(\beta\Delta x/2)]}$$

$$= \frac{[1+4r\sin^2(\beta\Delta x/2)-4r^2\sin^2(\beta\Delta x/2)]^{1/2}}{[1+4r\sin^2(\beta\Delta x/2)+4r^2\sin^2(\beta\Delta x/2)]^{1/2}}.$$

Therefore

$$|\gamma| \le 1 \quad \text{for all } r > 0.$$

In the same manner, it can also be shown that $|\gamma| \le 1$ for equation (2.8.2) with $\alpha = 1$. We conclude that both asymmetric equations (2.8.3) and (2.8.4) are always stable. More generally, it was mentioned by A. F. Filipov (Saul'yev, 1964) that in the case of the Cauchy problem, a necessary and sufficient condition for the stability of the equations (2.8.1) and (2.8.2) is

$$\gamma \le \frac{1}{2(1-\alpha)}, \quad 0 \le \alpha \le 1. \tag{2.8.5}$$

Due to their low accuracy, the asymmetric formulae (2.8.1) and (2.8.2) are not highly recommended for use in the numerical integration of the heat equation. In the method of Saul'yev, however, different equations are used on *alternate time steps*, that is,

$$u_{i,j+1} = \frac{1}{(1+\alpha r)} \{ \alpha u_{i-1,j+1} + (1-\alpha)r u_{i-1,j} + [1+(\alpha-2)r]u_{i,j} + r u_{i+1,j} \} \tag{2.8.6}$$

and

$$u_{i,j+2} = \frac{1}{(1+\alpha r)} \{ \alpha u_{i+1,j+2} + r u_{i-1,j+1} + [1+(\alpha-2)r]u_{i,j+1} + (1-\alpha)r u_{i+1,j+1} \}. \tag{2.8.7}$$

The first system of equations (2.8.6) can be solved in the left–right (LR) order $i = 1, 2, \ldots, m-1$ and the second system of equations (2.8.7) takes the right–left (RL) order $i = m-1, m-2, \ldots, 2, 1$ in turn for $j = 0, 2, 4, \ldots$. In particular, for $\alpha = 1$, equations (2.8.6) and (2.8.7) reduce to the equations

$$u_{i,j+1} - u_{i,j} = r(u_{i+1,j} - u_{i,j} - u_{i,j+1} + u_{i-1,j+1}) \tag{2.8.8}$$

and

$$u_{i,j+2} - u_{i,j+1} = r(u_{i+1,j+2} - u_{i,j+2} - u_{i,j+1} + u_{i-1,j+1}). \tag{2.8.9}$$

A stability analysis and an estimate of the accuracy can be obtained by eliminating the $u_{i,j+1}$ from the two equations. From the equations (2.8.8) and (2.8.9), we have, respectively

$$(1+r)u_{i,j+1} - r u_{i-1,j+1} = (1-r)u_{i,j} + r u_{i+1,j} \tag{2.8.9a}$$

and

$$(1-r)u_{i,j+1} - r u_{i-1,j+1} = (1+r)u_{i,j} + r u_{i+1,j+2}. \tag{2.8.9b}$$

If we add the two equations, we get

$$u_{i,j+1} = \tfrac{1}{2}\{(1-r)u_{i,j} + (1+r)u_{i,j+2} + r(u_{i+1,j} - u_{i+1,j+2})\}$$

and similarly

$$u_{i-1,j+1} = \tfrac{1}{2}\{(1-r)u_{i-1,j} + (1+r)u_{i-1,j+2} + r(u_{i,j} - u_{i,j+2})\}.$$

By substituting these equations into equation (2.8.9) and after some manipulation of terms, leads to the *implicit scheme*

$$u_{i,j+2} - (r+r^2)\delta_x^2 u_{i,j+2} = u_{i,j} + (r-r^2)\delta_x^2 u_{i,j}. \qquad (2.8.10)$$

If the stability analysis is carried out by means of the von Neumann criterion, we arrive at the overall stability polynomial

$$\gamma^2\{1 - (r+r^2)[\exp(i_c\beta\Delta x) + \exp(-i_c\beta\Delta x) - 2]\}$$

$$= 1 + (r-r^2)[\exp(i_c\beta\Delta x) + \exp(-i_c\beta\Delta x) - 2].$$

Therefore

$$\gamma^2 = \frac{\{1 + (r-r^2)[\exp(i_c\beta\Delta x) + \exp(-i_c\beta\Delta x) - 2]\}}{\{1 - (r+r^2)[\exp(i_c\beta\Delta x) + \exp(-i_c\beta\Delta x) - 2]\}},$$

which upon simplification, becomes

$$\gamma^2 = \frac{1 + 2(r-r^2)[\cos(\beta\Delta x) - 1]}{1 - 2(r+r^2)[\cos(\beta\Delta x) - 1]}.$$

Hence

$$|\gamma| = \frac{\{[1 - r(1 - \cos(\beta\Delta x))]^2 + r^2\sin^2(\beta\Delta x)\}^{1/2}}{\{[1 + r(1 - \cos(\beta\Delta x))]^2 + r^2\sin^2(\beta\Delta x)\}^{1/2}}.$$

It is clear that since $1 - \cos(\beta(\Delta x) \geq 0$, $|\gamma| \leq 1$ which implies that *Saul'yev's alternating method* has unrestricted stability.

Let us now consider equation (2.8.10), i.e.,

$$u_{i,j+2} - (r+r^2)\delta_x^2 u_{i,j+2} = u_{i,j} + (r-r^2)\delta_x^2 u_{i,j}.$$

If the terms r^2 were absent then this equation would be the Crank–Nicolson scheme whose truncation error is $0([\Delta t]^2) + 0([\Delta x]^2)$. With the presence of the r^2 terms, however, their contribution to $(u_{i,j+2} - u_{i,j})/2\Delta t$, is

$$0\left([r\Delta x]^2 \frac{\partial^3 u}{\partial x^2 \partial t}\right) = 0\left(\frac{[\Delta t]^2}{[\Delta x]^2} \frac{\partial^3 u}{\partial x^2 \partial t}\right).$$

For any fixed r, the truncation error is $0([\Delta t]) = 0([\Delta x]^2)$. If Δt and Δx are regarded as being independent, then, just as we have seen for the Dufort–Frankel scheme, the Saul'yev scheme will be consistent with the heat flow equation only if $\Delta t/\Delta x \to 0$ as the net is refined.

2.9. Parabolic Equations in Several Space Dimensions

We now consider finite-difference methods of the solution of the equation

$$\frac{\partial U}{\partial t} = \mathcal{L}(U),$$ (2.9.1)

where

$$\mathcal{L} = \sum_{i=1}^{n} \frac{\partial}{\partial x_i} \left(a_i(x_1, x_2, \ldots, x_n, t) \frac{\partial}{\partial x_i} \right) + b_i(x_1, x_2, \ldots, x_n, t) \frac{\partial}{\partial x_i} - c(x_1, x_2, \ldots, x_n, t),$$ (2.9.2)

with values of $a_1, a_2, \ldots, a_n$ strictly positive and the value of c non-negative.

There are two categories of finite-difference methods to be considered for the parabolic equations with several space variables. Firstly, the generalisation of the standard methods previously discussed for one-dimensional problems and secondly, splitting methods which have no single space variable analogue.

For simplicity, we consider the two-dimensional heat flow equation in the rectangular region bounded by $R \times [0 < t \le T]$ where R is a closed connected region in the x-y plane, with continuous boundary ∂R given by $R = [0 \le x, y \le 1]$, namely

$$\frac{\partial U}{\partial t} = \frac{\partial^2 U}{\partial x^2} + \frac{\partial^2 U}{\partial y^2}.$$ (2.9.3)

Appropriate initial and boundary data are provided as

$$U(x, y, 0) = f(x, y), \quad t = 0,$$
$$U(x, y, t) = g(x, y, t), \quad (x, y, t) \in \partial R \times [0 < t \le T],$$ (2.9.4)

where f and g are given for prescribed values of (x, y, t).

In the same manner as the one-dimensional case, the region R is covered by a rectilinear grid with sides, parallel to the axes, with Δx and Δy the grid spacings in the direction x and y and Δt in the time direction, respectively. Unless otherwise stated the space intervals are taken such that $\Delta x = \Delta y$, throughout the discussion. The grid points (x, y, t) are denoted by (i, j, k) where $x = i\Delta x$, $y = j\Delta y$, and $t = k\Delta t$.

The point $i = 0$, $j = 0$, $k = 0$ is the origin (Fig. 2.9.1). The exact and approximate values at the grid point (i, j, k) are denoted by $U_{i,j,k}$ and $u_{i,j,k}$ respectively.

(a) Explicit Methods
To extend the standard explicit methods to a second space dimension, we write equation (2.9.3) as

$$\frac{\partial U}{\partial t} = \mathcal{L} U,$$ (2.9.5)

where

$$\mathcal{L} \equiv \frac{\partial^2}{\partial x^2} + \frac{\partial^2}{\partial y^2} \equiv D_1^2 + D_2^2.$$

DAVID J. EVANS

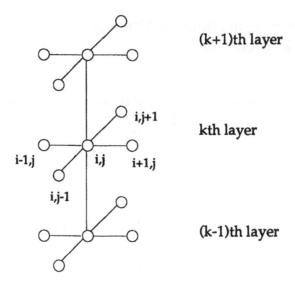

(k+1)th layer

i,j+1

kth layer

i-1,j   i,j   i+1,j

i,j-1

(k-1)th layer

Figure 2.9.1

From Taylor's expansion, it is possible to write $U(i\Delta x, j\Delta y, (k+1)\Delta t)$ in terms of $u(i\Delta x, j\Delta y, k\Delta t)$, i.e.,

$$U(i\Delta x, j\Delta y, (k+1)\Delta t) = u_{i,j,k+1} = \left(1 + \Delta t \frac{\partial}{\partial t} + \frac{\Delta t^2}{2} \frac{\partial^2 u}{\partial t^2} + \cdots\right) u_{i,j,k}$$

$$= \exp\left(\Delta t \frac{\partial}{\partial t}\right) u_{i,j,k}$$

$$= \exp(\Delta t \mathcal{L}) u_{i,j,k}$$

$$= \exp(\Delta t D_1^2) \exp(\Delta t D_2^2) u_{i,j,k}, \qquad (2.9.6)$$

provided D_1^2 and D_2^2 commute, where

$$D_1^2 = \frac{1}{(\Delta x)^2} \left(\delta_x^2 - \frac{1}{12}\delta_x^4 + \frac{1}{90}\delta_x^6 \cdots\right), \qquad (2.9.7)$$

and

$$D_2^2 = \frac{1}{(\Delta y)^2} \left(\delta_y^2 - \frac{1}{12}\delta_y^4 + \frac{1}{90}\delta_y^6 \cdots\right). \qquad (2.9.8)$$

Elimination of D_1^2 and D_2^2 leads to the formula

$$u_{i,j,k+1} = [1 + r\delta_x^2 + \tfrac{1}{2}r(r - \tfrac{1}{6})\delta_x^4 \cdots][1 + r\delta_y^2 + \tfrac{1}{2}r(r - \tfrac{1}{6})\delta_y^4 \cdots] u_{i,j,k}, \qquad (2.9.9)$$

where $r = \Delta t/(\Delta s)^2$ with $\Delta x = \Delta y = \Delta s$. Various explicit formulae can be obtained from equation (2.9.9) of which one is the well-known equation

$$u_{i,j,k+1} = [1 + r(\delta_x^2 + \delta_y^2)] u_{i,j,k}, \qquad (2.9.10)$$

which is the standard explicit formulae involving five points at the time level $t = k\Delta t$.

Equation (2.9.10) can be written as

$$u_{i,j,k+1} = (1-4r)u_{i,j,k} + r(u_{i-1,j,k} + u_{i+1,j,k} + u_{i,j-1,k} + u_{i,j+1,k}). \quad (2.9.11)$$

If we define the difference between the exact solution of differential and difference equations at the mesh point $(i\Delta x, j\Delta y, k\Delta t)$ as

$$e_{i,j,k} = U_{i,j,k} - u_{i,j,k}, \quad (2.9.12)$$

then from (2.9.11) we obtain

$$e_{i,j,k+1} = (1-4r)e_{i,j,k} + r(e_{i-1,j,k} + e_{i+1,j,k} + e_{i,j-1,k} + e_{i,j+1,k})$$

$$+ \frac{\Delta s^2}{2}\left(\frac{\partial^2 u}{\partial t^2}\right)_{i,j,k} - \Delta t \frac{\Delta s^2}{12}\left(\frac{\partial^4 u}{\partial x^4} + \frac{\partial^4 u}{\partial y^4}\right)_{i,j,k} + \cdots. \quad (2.9.13)$$

This equation shows that the Local Truncation Error (LTE) of formula (2.9.11) is $0((\Delta t)^2 + (\Delta t)(\Delta x)^2)$ has an order of accuracy of $0(\Delta t + (\Delta x)^2)$. Using the maximum principle for the stability analysis similar to section 2.3a for the one-dimensional case, it is simple to deduce that stability is obtained for $0 < r \le 1/4$.

Another simple explicit formula which expands the stability condition up to $r \le 1/2$ is the form

$$u_{i,j,k+1} = (1 + r\delta_x^2)(1 + r\delta_y^2)u_{i,j,k}, \quad (2.9.14)$$

which has the LTE of $0((\Delta t)^2 + (\Delta t)(\Delta x)^2)$ or order of accuracy of $0(\Delta t + (\Delta x)^2)$.

For $r = 1/6$, this formulae has a greater order of accuracy, i.e., $0((\Delta t)^2 + (\Delta x)^4)$, i.e.,

$$u_{i,j,k+1} = (1 + \tfrac{1}{6}\delta_x^2)(1 + \tfrac{1}{6}\delta_y^2)u_{i,j,k}, \quad (2.9.15)$$

but it is of limited use since the step forward in time is so small, i.e., $\Delta t = (1/6)\Delta x^2$.

In order to obtain the unconditionally stable explicit difference schemes, Larkin (1963), introduced a class of explicit formulae of four basic different types, namely

$$(1 + w + c^2)v_{i,j,k+1} = v_{i-1,j,k+1} + v_{i+1,j,k} - (1 - w + c^2)v_{i,j,k}$$

$$+ c^2[v_{i,j-1,k+1} + v_{i,j+1,k}], \quad (2.9.16)$$

$$(1 + w + c^2)z_{i,j,k+1} = z_{i-1,j,k} + z_{i+1,j,k+1} - (1 - w + c^2)z_{i,j,k}$$

$$+ c^2[z_{i,j+1,k+1} + z_{i,j-1,k}] \quad (2.9.17)$$

$$(1 + w + c^2)v_{i,j,k+1} = v_{i-1,j,k+1} + v_{i+1,j,k} - (1 - w + c^2)v_{i,j,k}$$

$$+ c^2[v_{i,j-1,k} + v_{i,j+1,k+1}], \quad (2.9.18)$$

and

$$(1 + w + c^2)z_{i,j,k+1} = z_{i-1,j,k} + z_{i+1,j,k+1} - (1 - w + c^2)z_{i,j,k}$$

$$+ c^2[z_{i,j-1,k+1} + z_{i,j+1,k}], \quad (2.9.19)$$

where $c = \Delta x / \Delta y$ and $w = (\Delta x)^2 / \Delta t = 1/r$. Dependent variables v and z are given diagrammatically by Figure 2.9.2 (a–d).

Besides the individual schemes, combination of these schemes also offers good approximation such as

    (i) the use of (2.9.16) and (2.9.17) alternately to (2.9.18) alternately with (2.9.19),

    (ii) the use of (2.9.16) and (2.9.17) at each time level and average the results as

$$u_{i,j+1} = \frac{v_{i,j+1} + z_{i,j+1}}{2}. \tag{2.9.20}$$

Then a similar strategy is applied to (2.9.18) and (2.9.19).

The truncation error (LTE) for equations (2.9.16) and (2.9.18) is given by

$$T_{2.9.16} = \left(\frac{\Delta t}{\Delta x}\right)\frac{\partial^2 v}{\partial x \partial t} + \frac{\Delta t}{2}\left(\frac{\partial^2 v}{\partial t^2} - \frac{\partial^3 v}{\partial t \partial x^2}\right) + \frac{(\Delta t)^2}{2\Delta x}\frac{\partial^3 v}{\partial x \partial t^2} + \cdots, \tag{2.9.21}$$

whereas equations (2.9.17) and (2.9.19) have truncation errors of

$$T_{2.9.17} = -\left(\frac{\Delta t}{\Delta x}\right)\frac{\partial^2 z}{\partial x \partial t} + \frac{\Delta t}{2}\left(\frac{\partial^2 z}{\partial t^2} - \frac{\partial^3 z}{\partial t \partial z^2}\right) + \frac{(\Delta t)^2}{2\Delta x}\frac{\partial^3 z}{\partial x \partial t^2} + \cdots. \tag{2.9.22}$$

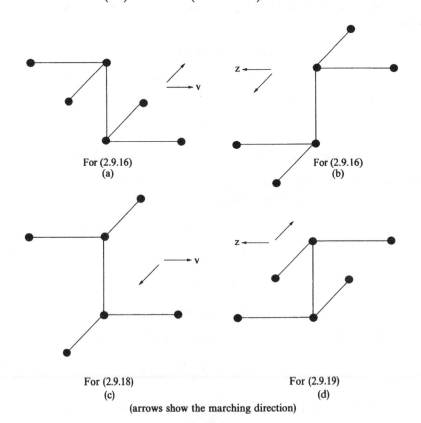

For (2.9.16)　　　　　　　　　　　　　　For (2.9.16)
(a)　　　　　　　　　　　　　　　　　　　(b)

For (2.9.18)　　　　　　　　　　　　　　For (2.9.19)
(c)　　　　　　　　　　　　　　　　　　　(d)

(arrows show the marching direction)

Figure 2.9.2

Consistency of the schemes with the equation (2.9.3) requires that $\Delta t/\Delta x \to 0$ and $\Delta x \to 0$. By diminishing Δx and Δt such that w is constant, the leading term of equation (2.9.21) and (2.9.22) would be of first degree in Δx with all other terms being of higher order. Because the leading term of (2.9.21) and (2.9.22) have opposite signs, averaging (2.9.20) will effectively produce a truncation error of second degree. The alternating use of the equations (2.9.16) and (2.9.17) or (2.9.18) and (2.9.19) is termed the *Alternating Direction Explicit* (ADE) method or semi-explicit method.

Another explicit scheme, i.e., the Dufort–Frankel scheme for two space dimensions can be written as

$$(\nabla_t - (1/2)\nabla_t^2)u_{i,j,k+1} = r(\delta_x^2 + \delta_y^2 - 2\delta_t^2)u_{i,j,k} \tag{2.9.23}$$

is an unconditionally stable scheme also. Here the terms $-(1/2)\nabla_t^2$ and $2r\delta_t^2$ have been added to both sides in order to obtain stability. The truncation error is given by

$$\frac{1}{\Delta t}T_{i,j,k} = 0\left((\Delta t)^2 + (\Delta s)^2 + \left(\frac{\Delta t}{\Delta s}\right)^2\right), \tag{2.9.24}$$

where $\Delta x = \Delta y = \Delta s$, which tends to zero if $\Delta t/\Delta x \to 0$ as $\Delta t \to 0$, $\Delta s \to 0$. In order to apply this formulae, $u_{i,j,k}$ are normally taken from an accurate two level difference scheme such as (2.9.15).

(b) Implicit Difference Schemes

The general implicit difference scheme for two dimensions is given by

$$[1 + \{\sigma - (1 - \gamma_1)r\}\delta_x^2][1 + \{\sigma - (1 - \gamma_1)r\}\delta_y^2]u_{i,j,k+1}$$

$$= [1 + (\sigma + r\gamma_1)\delta_x^2][1 + (\sigma + r\gamma_1)\delta_y^2]u_{i,j,k} + (1 - 2\gamma_1)r^2\delta_x^2\delta_y^2 u_{i,j,k}, \tag{2.9.25}$$

for arbitrary constants γ_1 and σ. The stability condition will be satisfied if

$$\sigma < \tfrac{1}{4} \quad \text{and} \quad \gamma_1 \leq \tfrac{1}{2}, \tag{2.9.26}$$

and the truncation error is given by

$$\frac{1}{(\Delta t)}T_{i,j,k} = \left[\left(\sigma - \frac{1}{12}\right)(\Delta s)^2 + \left(\gamma_1 - \frac{1}{2}\right)\Delta t\right]\left(\frac{\partial^4 U}{\partial x^4}i,j,k + \frac{\partial^4 U}{\partial y^4}i,j,k\right)$$

$$- (1 - 2\gamma_1)\Delta t\frac{\partial^4 U}{\partial x^2\partial y^2}i,j,k + 0((\Delta t)^2 + (\Delta s)^4). \tag{2.9.27}$$

From (2.9.25) and (2.9.27), we find that the values:

(i) $\gamma_1 = 0$ and $\sigma = 0$ give an implicit method of order $(\Delta t + (\Delta s)^2)$ namely

$$(1 - r\delta_x^2)(1 - r\delta_y^2)u_{i,j,k+1} = u_{i,j,k}, \tag{2.9.28}$$

and by neglecting the term $r^2\delta_x^2\delta_y^2 u_{i,j,k+1}$ it gives the classical fully implicit scheme (Fig. 2.9.3 (a))

$$u_{i,j,k+1} + r(-u_{i-1,j,k+1} + 2u_{i,j,k+1} - u_{i+1,j,k+1})$$

$$+ r(-u_{i,j-1,k+1} + 2u_{i,j,k+1} - u_{i,j+1,k+1}) = u_{i,j,k}, \tag{2.9.29}$$

for $i, j = 1, 2, \ldots, N$ and $N\Delta x = N\Delta y = 1$. The compact form of (2.9.29) is

$$(I + rA)\underline{u}_{i+1} = \underline{u}_k, \quad k \geq 0, \qquad (2.9.30)$$

where A is a $N^2 \times N^2$ block tridiagonal matrix given by

$$A = \begin{bmatrix} D & -I_N & & & \\ -I_N & D & -I_N & & 0 \\ & & \ddots & & \\ 0 & & -I_N & D & -I_N \\ & & & -I_N & D \end{bmatrix}, \quad D = 4I_N - (L_N + L_N^T), \qquad (2.9.31)$$

with

$$L_N = \begin{bmatrix} 0 & & & & \\ 1 & 0 & & 0 & \\ & \ddots & \ddots & & \\ 0 & & 1 & 0 & \\ & & & 1 & 0 \end{bmatrix}, \qquad (2.9.32)$$

hence at each step, we are required to solve a very large system of linear equations which in general is the drawback of the implicit method.

(ii) $\gamma_1 = 1/2, \sigma = 0$, gives the Crank–Nicolson method (Fig. 2.9.3 (b)),

$$(1 - (1/2)r\delta_x^2)(1 - (1/2)r\delta_y^2)u_{i,j,k+1} = (1 + (1/2)r\delta_x^2)(1 + (1/2)r\delta_y^2)u_{i,j,k}, \qquad (2.9.33)$$

of order $((\Delta t)^2 + (\Delta s)^2)$.

(iii) $\gamma_1 = 1/2$ and $\sigma = 1/12$, gives the Mitchell–Fairweather formula (Fig. 2.9.3(b)),

$$[1 + (1/12 - (1/2)r)\delta_x^2][1 + (1/12 - (1/2)r)\delta_y^2]u_{i,j,k+1}$$
$$= [1 + (1/12 + (1/2)r)\delta_x^2][1 + (1/12 + (1/2)r)\delta_y^2]u_{i,j,k}, \qquad (2.9.34)$$

of order $((\Delta t)^2 + (\Delta s)^4)$.

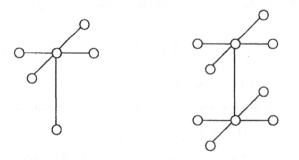

For equation (2.9.29)   For equations (2.9.33) and (2.9.34)
(a)   (b)

Figure 2.9.3

Even though the implicit methods are mostly unconditionally stable, however, due to the large system of linear equations to be solved, they are computationally expensive. We will discuss below alternative ways to manipulate the implicit methods in order to introduce computationally inexpensive explicit type schemes.

(c) Splitting Methods
In the numerical treatment of parabolic differential equations, splitting is referred to as a method of breaking down a process into a series of simple processes. Amongst these are the *Alternating Direction Implicit (ADI)*, the *Locally One Dimensional (LOD)*, and the *Hopscotch methods*.

Alternating Direction Implicit Methods
The ADI method which was first introduced by Peaceman and Rachford (1955) is a two-step process involving the solution of sets of tridiagonal equations along lines parallel to the x and y axes at the first and second time steps, respectively.

In the Peaceman–Rachford ADI method, the first step in advancing from t_k to $t_k + \Delta t/2$, implicit differences are used for $\partial^2 u/\partial x^2$ and explicit differences are used for $\partial^2 u/\partial y^2$. In the second step in advancing from $t_k + \Delta t/2$ to t_{k+1}, a reverse procedure is used i.e., explicit differences for $\partial^2 x/\partial x^2$ and implicit differences for $\partial^2 x/\partial y^2$. Accordingly, the difference approximation to (2.9.3) can be expressed as:

(i) $\quad \dfrac{u_{i,j,k+1/2} - u_{i,j,k}}{\Delta t/2} = \dfrac{1}{(\Delta x)^2} \delta_x^2 u_{i,j,k+1/2} + \dfrac{1}{(\Delta y)^2} \delta_y^2 u_{i,j,k},$

$$(2.9.35)$$

(ii) $\quad \dfrac{u_{i,j,k+1} - u_{i,j,k+1/2}}{\Delta t/2} = \dfrac{1}{(\Delta x)^2} \delta_x^2 u_{i,j,k+1/2} + \dfrac{1}{(\Delta y)^2} \delta_y^2 u_{i,j,k+1},$

which may also be written as (Fig. 2.9.4 (a)),

(i) $\quad (1 - (r/2)\delta_x^2) u_{i,j,k+1/2} = (1 + (r/2)\delta_y^2) u_{i,j,k},$

$$(2.9.36)$$

(ii) $\quad (1 - (r/2)\delta_y^2) u_{i,j,k+1} = (1 + (r/2)\delta_x^2) u_{i,j,k+1/2}.$

The equation (2.9.36) can also be obtained directly from the splitting of equation (2.9.33). The method is computationally feasible as it only requires the solution of sets of tridiagonal equations, and that each scheme, used on its own is conditionally stable. However, if they are used *alternately*, then the overall scheme is unconditionally stable.

The higher-order accuracy formulae of Mitchell–Fairweather (2.9.34) can be shown as in Figure 2.9.4 (a):

(i) $\quad [1 + (\tfrac{1}{12} - \tfrac{1}{2}r)\delta_x^2] u_{i,j,k+1/2} = [1 + (\tfrac{1}{12} + \tfrac{1}{2}r)\delta_y^2] u_{i,j,k},$ $\qquad$ (2.9.37)

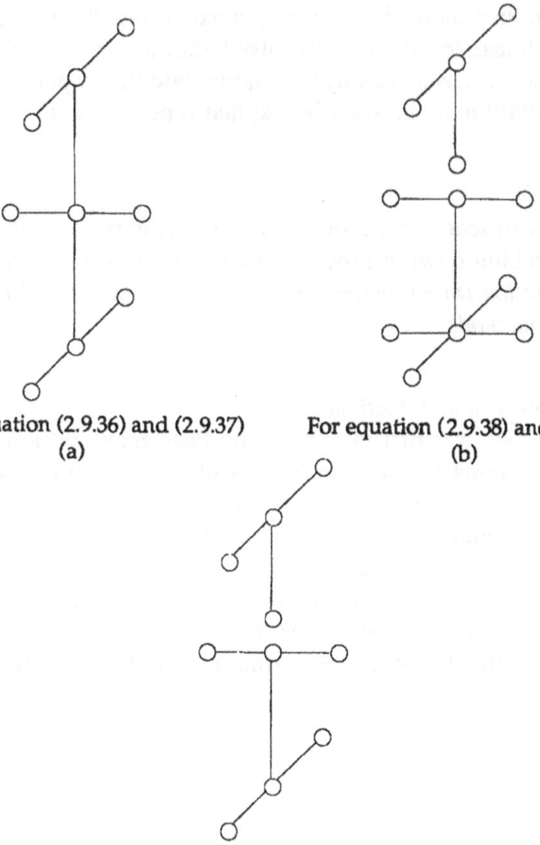

For equation (2.9.36) and (2.9.37)   For equation (2.9.38) and (2.9.39)
              (a)   (b)

For equation (2.9.42)
        (c)

Figure 2.9.4

(ii) $[1 + (\tfrac{1}{12} - \tfrac{1}{2}r)\delta_y^2]u_{i,j,k+1} = [1 + (\tfrac{1}{12} + \tfrac{1}{2}r)\delta_x^2]u_{i,j,k+1/2}$.

The implicit Crank–Nicolson formulae can be split in an alternative manner as suggested by D'Yakonov (1963), to give:

(i) $(1 - \tfrac{1}{2}r\delta_x^2)u_{i,j,k+1/2} = (1 + \tfrac{1}{2}r\delta_x^2)(1 + \tfrac{1}{2}r\delta_y^2)u_{i,j,k}$,

$$(2.9.38)$$

(ii) $(1 - \tfrac{1}{2}r\delta_y^2)u_{i,j,k+1} = u_{i,j,k+1/2}$,

and the corresponding higher-order accurate formulae to give:

(i) $[1 + (\tfrac{1}{12} - \tfrac{1}{2}r)\delta_x^2]u_{i,j,k+1/2} = [1 + (\tfrac{1}{12} + \tfrac{1}{2}r)\delta_x^2][1 + (\tfrac{1}{12} + \tfrac{1}{2}r)\delta_y^2]u_{i,j,k}$, $(2.9.39)$

(ii) $[1 + (\frac{1}{12} - \frac{1}{2}r)\delta_y^2]u_{i,j,k+1} = u_{i,j,k+1/2}$

as in Figure 2.9.4 (b).

The Peaceman–Rachford method (2.9.36) in matrix form can be written as:

(i) $(I + rH)\underline{u}_{k+1/2} = (I - rV)\underline{u}_k,$

$$(2.9.40)$$

(ii) $(I + rV)\underline{u}_{k+1} = (I - rH)\underline{u}_{k+1/2},$

where

$$H = 2I - (L + L^T),$$

$$(2.9.41)$$

$$V = 2I - (B + B^T),$$

$$L = \begin{bmatrix} L_N & & & \\ & L_N & 0 & \\ & & \ddots & \\ 0 & & & \\ & & & L_N \end{bmatrix}, \quad \text{and} \quad B = \begin{bmatrix} 0 & & & \\ I_N & 0 & & 0 \\ & I_N & \ddots & \\ & & \ddots & \\ 0 & & I_N & 0 \end{bmatrix}, \quad (2.9.42)$$

with L_N given in (2.9.32). Douglas and Rachford (1956) proposed an alternative to the Peaceman–Rachford method as in Figure 2.9.4 (c), i.e.,

(i) $(1 - r\delta_x^2)u_{i,j,k+1/2} = (1 + r\delta_y^2)u_{i,j,k},$

$$(2.9.43)$$

(ii) $(1 - r\delta_y^2)u_{i,j,k+1} = u_{i,j,k+1/2} - r\delta_x^2 u_{i,j,k},$

which in matrix form can be written as,

(i) $(I + rH)\underline{u}_{k+1/2} = (I - rV)\underline{u}_k,$

$$(2.9.44)$$

(ii) $(I + rV)\underline{u}_{k+1} = \underline{u}_{k+1/2} + rV\underline{u}_k,$

respectively. This method is also unconditionally stable and computationally feasible involving the solution of systems of tridiagonal difference equations along horizontal lines for $\underline{u}_{k+1/2}$ and then along vertical lines for $\underline{u}_{k+1}$.

It is important to note that $\underline{u}_{k+1/2}$ has no physical significance and it is only the first estimate or intermediate value (some refer to this as $\underline{u}_{k+1}^*$). It is not necessarily an approximation to the solution at any value of time. As a result, particularly with the high accuracy methods, the boundary values at the intermediate level must be obtained, if possible, in terms of the boundary values at $t = k\Delta t$ and $t = (k + 1)\Delta t$. The methods of coping with this problem are given in Mitchell A. R. and Griffiths D. F. (1980).

Locally One-Dimensional Methods
If we consider the equation (2.9.3) as the pair of equations

$$\frac{1}{2}\frac{\partial U}{\partial t} = \frac{\partial^2 U}{\partial x^2} \tag{2.9.45}$$

and

$$\frac{1}{2}\frac{\partial U}{\partial t} = \frac{\partial^2 U}{\partial y^2}, \tag{2.9.46}$$

then the simplest explicit formulae can be obtained as

$$u_{i,j,k+1/2} = (1 + r\delta_y^2)u_{i,j,k} \tag{2.9.47}$$

and

$$u_{i,j,k+1} = (1 + r\delta_x^2)u_{i,j,k+1/2}, \tag{2.9.48}$$

with the order of accuracy given by $0(\Delta t + (\Delta s)^2)$. The strategy of splitting a two-dimensional problem into a one-dimensional problem is called Locally One Dimensional (LOD) method, and has been developed extensively by D'Yakonov (1963) and Samarskii (1964). The equations (2.9.47) and (2.9.48) form the explicit scheme of LOD and is described in Figure 2.9.5. It is important to note that the elimination of $u_{i,j,k+1/2}$ from both equations leads to

$$u_{i,j,k+1} = (1 + r\delta_x^2)(1 + r\delta_y^2)u_{i,j,k},$$

which is the explicit equation (2.9.14).

Mitchell (1969) has exploited some interesting connections between the Peaceman–Rachford and LOD methods.

Hopscotch Methods
Another class of splitting method was introduced by Saul'yev (1964), and developed by Gourlay (1970) is called the Hopscotch method. To briefly discuss this class of

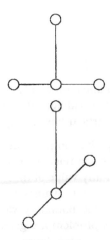

Figure 2.9.5

method, we restrict ourselves to the linear parabolic equation of two dimension, i.e.,

$$\frac{\partial U}{\partial t} = \mathcal{L}(U) + g(x, y, t), \tag{2.9.49}$$

where $\mathcal{L}$ is the second-order linear, elliptic differential operator in the space variables x and y.

A general splitting formula for the equation (2.9.49) is given by Danaee (1980),

$$u_{i,j,k+1} - \Delta t(\theta_{i,j,k+1} \mathcal{L}_{\Delta x}^{(1)} + \eta_{i,j,k+1} \mathcal{L}_{\Delta x}^{(2)}) u_{i,j,k+1}$$

$$= u_{i,j,k} + \delta t(\theta_{i,j,k} \mathcal{L}_{\Delta x}^{(1)} + \eta_{i,j,k} \mathcal{L}_{\Delta x}^{(2)}) u_{i,j,k}$$

$$+ \Delta t(\theta_{i,j,k+1} g_{i,j,k+1}^{(1)} + \eta_{i,j,k+1} g_{i,j,k+1}^{(2)})$$

$$+ \Delta t(\theta_{i,j,k} g_{i,j,k}^{(1)} + \eta_{i,j,k} g_{i,j,k}^{(2)}), \tag{2.9.50}$$

with the restrictions

$$\theta_{i,j,k+1} + \theta_{i,j,k} = 1,$$

$$\eta_{i,j,k+1} + \eta_{i,j,k} = 1, \tag{2.9.51}$$

where

$$g_{i,j,k}^{(1)} + g_{i,j,k}^{(2)} = g_{i,j,k},$$

and

$$\mathcal{L}_{\Delta x}^{(1)} + \mathcal{L}_{\Delta x}^{(2)} = \mathcal{L}_{\Delta x}.$$

The *odd–even* Hopscotch method can be obtained by defining

$$\theta_{i,j,k} = \eta_{i,j,k} = \begin{cases} 1 & \text{if } i+j+k \text{ even,} \\ 0 & \text{if } i+j+k \text{ odd,} \end{cases}$$

which from (2.9.50) gives the explicit formula

$$u_{i,j,k+1} = u_{i,j,k} + \Delta t(\mathcal{L}_{\Delta x}^{(1)} + \mathcal{L}_{\Delta x}^{(2)}) u_{i,j,k} + \Delta t(g_{i,j,k}^{(1)} + g_{i,j,k}^{(2)}), \tag{2.9.52}$$

alternately with the implicit formula

$$u_{i,j,k+1} - \Delta t(\mathcal{L}_{\Delta x}^{(1)} + g_{i,j,k+1}^{(1)}) u_{i,j,k+1} - \Delta t(\mathcal{L}_{\Delta x}^{(2)} + g_{i,j,k+1}^{(2)}) = u_{i,j,k}. \tag{2.9.53}$$

Similar to the *odd–even* Hopscotch, the *line-Hopscotch* method in the y-direction can be obtained if we define

$$\theta_{i,j,k} = \eta_{i,j,k} = \begin{cases} 1 & \text{if } i+k \text{ odd,} \\ 0 & \text{if } i+k \text{ even,} \end{cases} \tag{2.9.54}$$

and in the x-direction if we define,

$$\theta_{i,j,k} = \eta_{i,j,k} = \begin{cases} 1 & \text{if } j+k \text{ odd,} \\ 0 & \text{if } j+k \text{ even.} \end{cases} \tag{2.9.55}$$

The general splitting formulae, can also be used to derive the standard formulae like Crank–Nicolson equation if we take $\theta_{i,j,k} = \eta_{i,j,k} = \theta_{i,j,k+1} = \eta_{i,j,k+1} = 1/2$ for all i,j

and k. If $\theta_{i,j,k+1} = \eta_{i,j,k+1} = 0$ and $\theta_{i,j,k} = \eta_{i,j,k} = 1$ for all i, j, k we obtain the explicit scheme. If

$$\theta_{i,j,k} = 1/2(1 + (-1)^k) = \begin{cases} 1 & \text{if } k \text{ is even,} \\ 0 & \text{if } k \text{ is odd,} \end{cases}$$

$$\eta_{i,j,k} = 1 - \theta_{i,j,k} \quad \text{for all } i, j, k,$$

then we get the Peaceman–Rachford method with a time step of $2\Delta t$.

2.10. Alternative Methods to Solve the Diffusion Equation

Some of the well-known finite difference methods that are available for the treatment of parabolic equations are summarised in Table 2.10.1. The truncation error and the stability requirement of each method are also included.

2.11. Elliptic Partial Differential Equations – The Model Problem

(I) Two Dimensional Case

Let us consider the Dirichlet problem for the Laplace equation which requires to determine the solution $u(x, y)$ satisfying (2.11.1) inside a closed region R and is determined on the boundary ∂R by the boundary condition (2.11.2), i.e.,

$$\frac{\partial^2 U}{\partial x^2} + \frac{\partial^2 U}{\partial y^2} = 0, \quad (x, y) \in R, \tag{2.11.1}$$

$$U(x, y) = g(x, y), \quad (x, y) \in \partial R. \tag{2.11.2}$$

The region R under consideration is covered, by a rectilinear net with mesh spacing h in the X and Y direction and mesh points (x_i, y_j), where $x_i = ih, y_j = jh, (i, j = 0, 1, \ldots, m + 1)$.

This problem is a special case of the general two-dimensional self adjoint elliptic equation

$$\frac{\partial}{\partial x}\left(A\frac{\partial U}{\partial x}\right) + \frac{\partial}{\partial y}\left(B\frac{\partial U}{\partial y}\right) + FU = G,$$

where A, B, F and G are functions of x, y the independent variables.

Substituting the finite difference approximations for the derivatives in (2.11.1) the following five-point formula is obtained

$$-u_{i+1,j} - u_{i-1,j} + 4u_{i,j} - u_{i,j+1} - u_{i,j-1} = 0, \tag{2.11.3}$$

$$u_{i,j} = g_{i,j} \equiv g(ih, jh), \quad i = 0, m + 1 \text{ for } j = 1, 2, \ldots, m,$$

$$j = 0, m + 1 \text{ for } i = 1, 2, \ldots, m, \tag{2.11.4}$$

Table 2.10.1 Finite-difference Approximations to $\partial U/\partial t = \partial^2 U/\partial x^2$.

Finite Difference Form	Explicit/Implicit	Stability Condition	Truncation Error	Computational Molecule
1. $u_{i,j+1} = u_{ij} + r(u_{i+1,j} - 2u_{ij} + u_{i-1,j})$ (Classical explicit method)	Explicit	$r \le 1/2$	$T = 0(\Delta t) + 0([\Delta x]^2)$	
2. $\frac{r}{2}u_{i+1,j+1} - (1+r)u_{i,j+1} + \frac{r}{2}u_{i-1,j+1}$ $= -\frac{r}{2}u_{i+1,j} - (1-r)u_{ij} - \frac{r}{2}u_{i-1,j}$ (Crank–Nicolson method)	With $\theta = 1/2$ Implicit	Always stable	$T = 0([\Delta t]^2) + 0([\Delta x]^2)$	
3. $-ru_{i+1,j+1} + (1+2r)u_{i,j} - ru_{i-1,j+1} = u_{ij}$ (Fully implicit method)	With $\theta = 1$ Implicit	Always stable	$T = 0(\Delta t) + 0([\Delta x]^2)$	
4. $u_{i,j+1} = u_{ij} + \frac{1}{6}(u_{i+1,j} - 2u_{ij} + u_{i-1,j})$ (Special explicit method)	With $r = 1/6$ Explicit	Stable	$T = 0([\Delta t]^2) = 0([\Delta x]^4)$	
5. $r\theta u_{i+1,j+1} - (1+2r\theta)u_{i,j+1} + r\theta u_{i-1,j+1}$ $= -r(1-\theta)u_{i+1,j} - (1-2r(1-\theta))u_{ij}$ $- r(1-\theta)u_{i-1,j}$ (Weighted formula $(0 \le \theta \le 1)$)	Implicit	$r \le \dfrac{1}{(2-4\theta)}$ for $0 \le \theta \le 1$ Always stable for $1/2 \le \theta \le 1$	$T = \begin{cases} 0((\Delta t)^2) + 0((\Delta x)^2) \\ \quad \text{for } \theta = 1/2 \\ 0(\Delta t) + 0((\Delta x)^2) \\ \quad \text{for } \theta \ne 1/2 \end{cases}$	

Table 2.10.1 (continued ...)

Finite Difference Form	Explicit/Implicit	Stability Condition	Truncation Error	Computational Molecule
6. As in 5. with $\theta = 1/2 - \dfrac{1}{12r}$ (Douglas Method)	Implicit	Always stable	$T = 0([\Delta t]^2) + 0([\Delta x]^4)$	
7. $u_{i,j+1} = u_{i,j-1} + 2r(u_{i+1,j} - 2u_{ij} + u_{i-1,j})$ (Richardson Method)	Explicit, 3-Level	Always unstable	$T = 0([\Delta t]^2) + 0([\Delta x]^2)$	
8. $u_{i,j+1} = \dfrac{1}{(1+2r)}\{2r(u_{i+1,j} + u_{i-1,j})$ $+ (1-2r)u_{i,j-1}\}$ (Dufort–Frankel Method)	Explicit, 3-Level	Always stable	$T = 0[(\Delta t)^2] + 0([\Delta x]^2)$ $+ 0\left(\left[\dfrac{\Delta t}{\Delta x}\right]^2\right)$ with $\dfrac{\Delta t}{\Delta x} \to 0$, as Δt, $\Delta x \to 0$	
9. $ru_{i+1,j+1} - (1/2)(3+4r)u_{i,j+1}$ $+ ru_{i-1,j+1} + 1/2u_{i,j-1}$	Implicit, 3-Level	Always stable	$T = 0((\Delta t)^2) + 0([\Delta x]^2)$	
10. $ru_{i+1,j+1} - (1+\theta+2r)u_{i,j+1} + ru_{i-1,j}$ $= -(1+2\theta)u_{ij} + u_{i,j-1}$ (where $\theta \geq 0$)	Implicit, 3-Level	Always stable	$T = \begin{cases} 0([\Delta t]^2) + 0([\Delta x]^2) \\ \quad \text{for } \theta = 1/2 \\ 0(\Delta t) + 0([\Delta x]^2) \end{cases}$	

11. As in 10. with $\theta = \dfrac{1}{2} + \dfrac{1}{12r}$ — Implicit, 3-Level — Always stable — $T = 0([\Delta t]^2) = 0([\Delta x]^4)$

12. $\left(\dfrac{1}{12} - \dfrac{r}{2}\right)u_{i+1,j+1} + \left(\dfrac{5}{6} + r\right)u_{i,j+1}$
$+ \left(\dfrac{1}{12} - \dfrac{r}{2}\right)u_{i-1,j+1}$
$= \left(\dfrac{1}{12} + \dfrac{r}{2}\right)u_{i+1,j} + \left(\dfrac{5}{6} - r\right)u_{ij} + \left(\dfrac{1}{12} + \dfrac{r}{2}\right)u_{i-1,j}$

(*Douglas Method*) — Implicit — Always stable — $T = 0([\Delta t]^2) + 0([\Delta x]^4)$

13. $\left(\dfrac{1}{8} - r\right)u_{i+1,j+1} + \left(\dfrac{5}{4} + 2r\right)u_{i,j+1}$
$+ \left(\dfrac{1}{8} - r\right)u_{i-1,j+1}$
$= \dfrac{1}{6}u_{i+1,j} + \dfrac{5}{3}u_{ij} + \dfrac{1}{6}u_{i-1,j} - \dfrac{1}{24}u_{i+1,j-1}$
$- \dfrac{5}{12}u_{i,j-1} - \dfrac{1}{24}u_{i-1,j-1}$

(*Variation of Douglas Equation*) — Implicit, 3-Level — Always stable — $T = 0([\Delta t]^2) + 0([\Delta x]^4)$

14. (a) $u_{i,j+1} - u_{ij} = r(u_{i+1,j} - u_{ij} - u_{i,j+1} + u_{i-1,j+1})$

(b) $u_{i,j+2} - u_{i,j+1} = r(u_{i+1,j+2} - u_{i,j+2} - u_{i,j+1} + u_{i-1,j+1})$

(*Saul'yev Alternating Method*) — Semi-implicit — Always stable — $T = 0(\Delta t) = 0([\Delta x]^2)$ for fixed r

The solution $u_{i,j}$ at the point (ih, jh) can be obtained by solving the linear system (2.11.3), which can be represented by the computational *molecule* shown in Figure 2.11.1 which when applied at each mesh point yields the system

$$A\underline{u} = \underline{b}.$$ (2.11.5)

Now we order the m^2 internal points column-wise (Fig. 2.11.2), the coefficient matrix A of the obtained linear system (2.11.5) is a real symmetric, square, sparse matrix

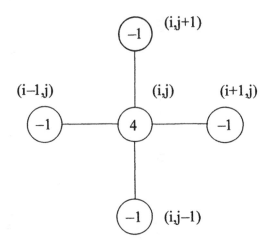

Figure 2.11.1 Five-Point Computational Molecule for the Laplace Operator.

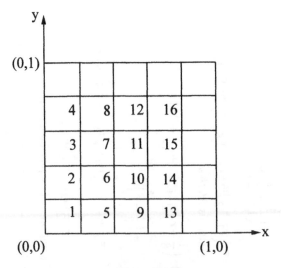

Figure 2.11.2

which has the block tridiagonal form

$$A = \begin{bmatrix} D_1 & -I & & & & \\ -I & D_2 & -I & & & \\ & -I & D_3 & -I & & \\ & & \ddots & \ddots & \ddots & \\ & & & & -I & \\ & & & & -I & D_m \end{bmatrix} \qquad (2.11.6a)$$

where I is the unit matrix of order m and the D_i $(1 \le i \le m)$ also of order m, are given by

$$D_i = \begin{bmatrix} 4 & -1 & & & & \\ -1 & 4 & -1 & & & \\ & -1 & 4 & -1 & & \\ & & \ddots & \ddots & \ddots & \\ & & & & & -1 \\ & & & & -1 & 4 \end{bmatrix}, \quad 1 \le i \le m. \qquad (2.11.6b)$$

In addition to the column-wise ordering of the interior mesh points, there exists many different other methods of ordering. In general, for a given set of mesh points $(x_0 + p_i h, y_0 + q_i h)$, $i = 1, 2, \ldots, m$, some of the methods can be given as:

(i) *Natural ordering*   i.e. row or column-wise: a point $(x_0 + ph, y_0 + qh)$ occurs before $(x_0 + p'h, y_0 + q'h)$, if $q < q'$ or if $q = q'$ and $p < p'$.

(ii) *Diagonal ordering*   a point $(x_0 + ph, y_0 + qh)$ occurs before $(x_0 + p'h, y_0 + q'h)$ if $p + q < p' + q'$.

(iii) *Red-black ordering*   all points $(x_0 + ph, y_0 + qh)$ with $(p + q)$ even (red points) occur before those with $(p + q)$ odd (black points).

For a problem arising from the solution of a five-point difference equation on a square mesh, the above kinds of ordering all lead to consistently ordered matrices.

Now, we demonstrate the procedure of deriving a formula which is more accurate than the five-point formula, i.e., the *nine-point* formula where the order of the local truncation error is increased.

Consider the two-dimensional Laplacian operator $\nabla^2 (\nabla^2 \equiv \partial/\partial x^2 + \partial/\partial y^2)$. We define D_x and D_y so ∇^2 can be written as $\nabla^2 = D_x^2 + D_y^2$.

If we assume that $h_x = h_y = h$, then by Taylor's expansion

$$U(x + h, y) = e^{hD_x} U(x, y),$$

$$U(x, y + h) = e^{hD_y} U(x, y), \qquad (2.11.7)$$

DAVID J. EVANS

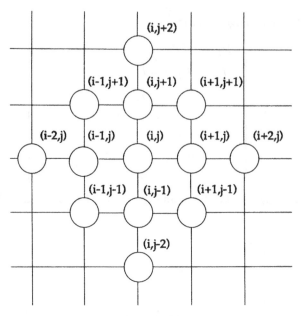

Figure 2.11.3

hence we may write on the basis of the above result

$$u_{i\pm1,j} = e^{\pm hD_x}u_{i,j},$$

$$u_{i,j\pm1} = e^{\pm hD_y}u_{i,j}, \qquad (2.11.8a)$$

$$u_{i\pm1,j\pm1} = e^{\pm h(D_x+D_y)}u_{i,j},$$

$$u_{i\pm2,j} = e^{\pm 2hD_x}u_{i,j},$$

$$u_{i,j\pm2} = e^{\pm 2hD_y}u_{i,j}, \qquad (2.11.8b)$$

where $u_{i,j}$ represents values of the numerical approximation to $U_{i,j}$. Then we define S_1, S_2 and S_3 as follows:

$$S_1 = u_{i+1,j} + u_{i,j+1} + u_{i-1,j} + u_{i,j-1},$$

$$S_2 = u_{i+1,j+1} + u_{i-1,j+1} + u_{i-1,j-1} + u_{i+1,j-1}, \qquad (2.11.9)$$

$$S_3 = u_{i+2,j} + u_{i,j+2} + u_{i-2,j} + u_{i,j-2}.$$

Hence from Figure 2.11.3, i.e.,

$$S_1 = (e^{hD_x} + e^{hD_y} + e^{-hD_x} + e^{-hD_y})u_{i,j}$$

$$= [2\cosh(hD_x) + 2\cosh(hD_y)]u_{i,j}$$

$$= \left[4 + h^2(D_x^2 + D_y^2) + \frac{h^4}{12}(D_x^4 + D_y^4) + \frac{h^6}{360}(D_x^6 + D_y^6) + \cdots\right]u_{i,j}$$

$$= \left[4 + h^2\nabla^2 + \frac{h^4}{12}(\nabla^4 - 2D_x^2D_y^2) + \frac{h^6}{360}(\nabla^6 - 3\nabla^2D_x^4D_y^4) + \cdots\right]u_{i,j}, \qquad (2.11.10)$$

$$S_2 = (e^{h(D_x+D_y)} + e^{h(-D_x+D_y)} + e^{h(-D_x-D_y)} + e^{h(D_x-D_y)})u_{i,j}$$

$$= 4[\cosh(hD_x)\cosh(hD_y)]u_{i,j}$$

$$= \left[4 + 2h^2(D_x^2+D_y^2) + \frac{h^4}{6}(D_x^4 + 6D_x^2D_y^2 + D_y^4) \right.$$

$$\left. + \frac{h^6}{180}(D_x^6 + 15D_x^4D_y^2 + 15D_x^2D_y^4 + D_y^6) + \cdots \right]u_{i,j}$$

$$= \left[4 + 2h^2\nabla^2 + \frac{h^4}{6}(\nabla^4 + 4D_x^2D_y^2) + \frac{h^6}{180}(\nabla^6 - 12\nabla^2D_x^4D_y^4) + \cdots \right]u_{i,j}$$

$$(2.11.11)$$

and

$$S_3 = (e^{2hD_x} + e^{2hD_y} + e^{-2hD_x} + e^{-2hD_y})u_{i,j}$$

$$= \left[4 + 4h^2\nabla^2 + \frac{3}{4}h^4(\nabla^4 - 2D_x^2 + D_y^2) + \frac{8h^6}{45}(\nabla^6 - 3\nabla^2D_x^4D_y^4) + \cdots \right]u_{i,j}.$$

$$(2.11.12)$$

By eliminating the term $D_x^2D_y^2$ between S_1 and S_2, we obtain the following nine-point formula

$$\nabla^2 u_{i,j} = \frac{1}{6h^2}(4S_1 + S_2 - 20u_{i,j}) - \frac{1}{12}h^2\nabla^4 u_{i,j}$$

$$- \frac{1}{180}h^4\left(\frac{1}{2}\nabla^6 u_{i,j} + \nabla^2 u_{i,j}D_x^4D_y^4 + 0(h^6) \right) \qquad (2.11.13)$$

for the Laplace equation $\nabla^2 u = 0$. Hence $\nabla^4 u = (\nabla^2 u)^2 = 0$ and $\nabla^6 u = 0$. So the nine-point formula for the Laplace operator is given by

$$20u_{i,j} - 4S_1 - S_2 = 0, \qquad (2.11.14)$$

with a truncation error of order h^6.

The linear system (2.11.15) comprising the equation (2.11.14) can be represented by the computational molecule shown in Figure 2.11.4.

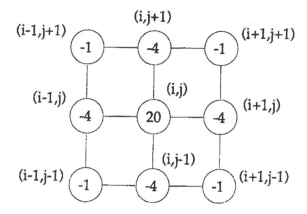

Figure 2.11.4

Other combinations between S_1, S_2 and S_3 for the Laplace equation may yield different finite difference formulae, for example, $S_3 - 16S_1$ gives the molecule shown in Figure 2.11.5.

This molecule suffers from the difficulty that its accuracy is difficult to match near the boundary ∂R. Moreover, its use of points at a distance $2h$ from $u_{i,j}$ is a mild disadvantage in the solution of the linear equations causing a matrix with a wider bandwidth. It is a more serious disadvantage for nodes near the boundary, because the greater size of the molecule causes many more interior points of the net to become irregular, Forsythe and Wasow (1960). Techniques for dealing with such points (see Mitchell and Griffiths (1980)).

The *Biharmonic equation* which is a more complicated partial differential equation of elliptic type has the form (for the model problem)

$$\nabla^4 U \equiv \nabla^2 \left(\frac{\partial^2}{\partial x^2} + \frac{\partial^2}{\partial y^2} \right) U = 0, \tag{2.11.15}$$

which can be written as,

$$\frac{\partial^4 U}{\partial x^4} + 2 \frac{\partial^4 U}{\partial x^2 \partial y^2} + \frac{\partial^4 U}{\partial y^4} = 0, \quad (x, y) \in R. \tag{2.11.16}$$

The appropriate boundary conditions usually take one of two forms

(i) $\qquad \left. \begin{array}{l} U = g_1(x, y) \\[2mm] \dfrac{\partial U}{\partial n} = g_2(x, y) \end{array} \right\} \quad \text{for } (x, y) \in \partial R \qquad (2.11.17)$

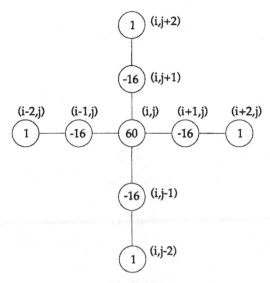

Figure 2.11.5

or

(ii)
$$\left.\begin{array}{l} U = g_1(x, y) \\ \dfrac{\partial^2 U}{\partial n^2} = g_2(x, y) \end{array}\right\} \quad \text{for } (x, y) \in \partial R,$$
(2.11.18)

where $\partial/\partial n$ is the derivative in the direction of the outward normal to the boundary ∂R.

If we take R to be the unit square $0 \leq x, y \leq 1$, from equation (2.11.16) by using difference replacements, assuming that $h_x = h_y = h$, we get

$$\delta_x^4 u_{i,j} + 2\delta_x^2 \delta_y^2 u_{i,j} + \delta_y^4 u_{i,j} = 0,$$
(2.11.19)

where $\delta_x^4 = \delta_x^2(\delta_x^2)$ and $\delta_y^4 = \delta_y^2(\delta_y^2)$.

By a further manipulation equation (2.11.19) can be written as

$$20u_{i,j} - 8S_1 + 2S_2 + S_1 = 0,$$
(2.11.20)

with a truncation error of order h^2.

The system (2.11.20) can be represented by the molecule shown in Figure 2.11.6.

In this 13-point molecule the presence of the terms $u_{i\pm2,j}$ and $u_{i,j\pm2}$ means that special modifications have to be made when it is applied at nodes distance h from the boundary. Techniques for dealing with such points are discussed in Mitchell and Griffiths (1980).

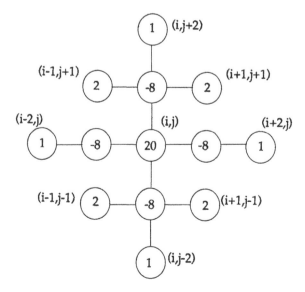

Figure 2.11.6 13-Point Computational Molecule for Biharmonic Operator.

(II) Three Dimensional Model Problem

We consider the Laplace equation in three space dimensions:

$$\frac{\partial^2 U}{\partial x^2} + \frac{\partial^2 U}{\partial y^2} + \frac{\partial^2 U}{\partial z^2} = 0, \quad (x, y) \in R, \tag{2.11.21}$$

subject to the Dirichlet boundary condition

$$U(x, y, z) = g(x, y, z), \quad (x, y, z) \in \partial R, \tag{2.11.22}$$

where R is the unit cube $0 \le x, y, z \le 1$ and ∂R its boundary.

The volumetric region under consideration R is covered by an equally spaced three-dimensional net with mesh spacing h in the x, y and z directions. The mesh points are defined by (x_i, y_j, z_k) where $x_i = ih$, $y_j = jh$ and $z_k = kh$, $(0 \le i, j, k \le m + 1)$, where $(m + 1)h = 1$.

Discrete approximations to the derivatives in (2.11.21), leads to the following seven-point finite difference equation

$$6u_{i,j,k} - u_{i+1,j,k} - u_{i-1,j,k} - u_{i,j+1,k} - u_{i,j-1,k} - u_{i,j,k+1} - u_{i,j,k-1} = 0, \tag{2.11.23}$$

$$u_{i,j,k} = g_{i,j,k}, \quad i = 0, N \text{ for } 1 \le j, k \le m$$

$$j = 0, N \text{ for } 1 \le i, k \le m \tag{2.11.24}$$

$$k = 0, N \text{ for } 1 \le i, j \le m.$$

Equation (2.11.23) is generally represented by the molecule shown in Figure 2.11.7.

By ordering the m^3 internal mesh points with increasing values of i, then for j and then for k (Fig. 2.11.8)

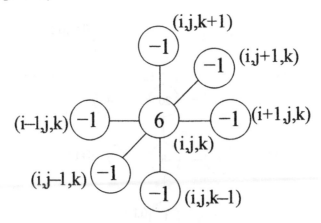

Figure 2.11.7

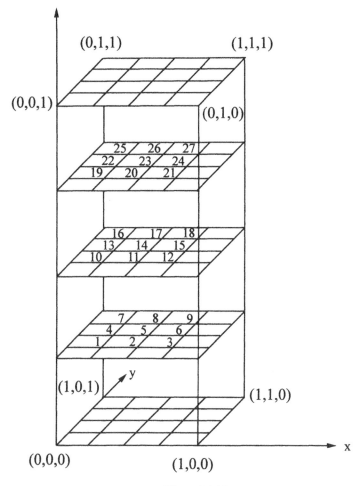

Figure 2.11.8

and applying the seven-point equation at each mesh point we get the system (2.11.5). The coefficient matrix A of the obtained system is a real, symmetric, square, seven-diagonal, sparse matrix of order m^3 and of the general form

$$
A = \begin{bmatrix}
A_1 & -I & & & & \\
-I & A_2 & -I & & 0 & \\
& -I & A_3 & -I & & \\
& & & \ddots & \ddots & \\
& 0 & & & \ddots & -I \\
& & & & -I & A_m
\end{bmatrix}, \qquad (2.11.25a)
$$

where I is the unit matrix of order m^2 and $A_i, 1 \leq i \leq m$, are matrices of order m^2

given by

$$A_i = \begin{bmatrix} D_{j+1} & -J & & & & \\ -J & D_{j+2} & -J & & 0 & \\ & -J & \ddots & \ddots & & \\ & & \ddots & \ddots & \ddots & \\ & 0 & & \ddots & \ddots & -J \\ & & & & -J & D_{j+m} \end{bmatrix}, \quad \begin{array}{l} j = (i-1)m, \\ 1 \leq i \leq m, \end{array} \qquad (2.11.25b)$$

where J is now the unit matrix of order m and D_r, $1 \leq r \leq m^2$, are matrices of order m given by

$$D_r = \begin{bmatrix} 6 & -1 & & & & \\ -1 & 6 & -1 & & 0 & \\ & -1 & 6 & -1 & & \\ & & \ddots & \ddots & \ddots & \\ & 0 & & \ddots & \ddots & -1 \\ & & & & -1 & 6 \end{bmatrix}, \quad 1 \leq r \leq m^2. \qquad (2.11.25c)$$

The solution vector $\underline{u}$ and the right hand side vector $\underline{b}$ of (2.11.5) are $(m^3 \times 1)$ column vectors.

Thus we have seen that the application of the finite difference and finite element methods for solving p.d.e.'s yield a system of linear simultaneous equations which can be represented in matrix notation as

$$A\underline{u} = \underline{b}, \qquad (2.11.26)$$

where A is a coefficient matrix of order $(m \times m)$, $(m \equiv N - 1)$, the order of the matrix A equals the number of interior mesh points, $\underline{b}$ is a column vector containing known sources and boundary terms, $\underline{u}$ is an unknown column vector.

The remainder of this chapter deals with well-known methods of solving the system (2.11.26).

2.12. Iterative Methods

In Chapter 1 some basic definitions were given for the iterative methods, in which a sequence of approximate solution vectors $\{\underline{u}^{(k)}\}$ are derived to solve the non-singular system

$$A\underline{u} = \underline{b}, \qquad (2.12.1)$$

such that $\underline{u}^{(k)} \rightarrow A^{-1}\underline{b}$ as $k \rightarrow \infty$, were given.

If we define a sequence of functions (iteration functions), $F_0(A, \underline{b})$, $F_1(A, \underline{b}, \underline{u}^{(0)})$, ...,
$F_k(A, \underline{b}, \underline{u}^{(0)}, \underline{u}^{(1)}, ..., \underline{u}^{(k-1)})$, where

$$\underline{u}^{(0)} = F_0(A, \underline{b}),$$

$$\underline{u}^{(k+1)} = F_{k+1}(A, \underline{b}, \underline{u}^{(0)}, \underline{u}^{(1)}, ..., \underline{u}^{(k)}), \qquad (2.12.2)$$

then, the iterative method is:

Stationary: If for some integer $\ell > 0$, F_k is independent of k for all $k \geq \ell$.
Otherwise, it is *non-stationary*.
Linear: If for each k, F_k is a linear function of $\underline{u}^{(0)}, \underline{u}^{(1)}, ...,$ and $\underline{u}^{(k-1)}$. Otherwise,
it is *non-linear*.
Consistent: If at any stage of the iterative process we obtain a solution to (2.12.1),
then all subsequent iterants remain unchanged.

The most general linear stationary iteration has the form

$$\underline{u}^{(k+1)} = G\underline{u}^{(k)} + \underline{d}, \qquad (2.12.3)$$

where G, *the iteration matrix*, is a matrix depending upon A and $\underline{b}$, and $\underline{d}$ is a vector
to be defined later. The exact solution of (2.12.1) i.e., $A^{-1}\underline{b}$, at a fixed point of (2.12.3)
can be reproduced, and we have

$$A^{-1}\underline{b} = GA^{-1}\underline{b} + \underline{d}, \qquad (2.12.4)$$

hence

$$\underline{d} = (I - G)A^{-1}\underline{b}. \qquad (2.12.5)$$

This is called the *consistency condition*. It can be shown (see Young, 1971), that if the
consistency condition applies then for some k

$$\underline{u}^{(k+1)} = G\underline{u}^{(k)} + \underline{d} = G\underline{u} + \underline{d} = \underline{u}.$$

That is, once the solution is obtained, the iterative process makes no further
modification of successive iterates.

We consider two classes of stationary iterative methods. These are (I) *The simultaneous methods* and (II) *The Successive methods*. The main difference between these
two classes is the ordering. In (I), the order in which the solution vectors are
calculated does not affect the values of successive iterates at a particular vector, in
other words, all elements of the approximate solution are modified at the same time,
i.e., the $(k+1)^{st}$ iterate is a function of the kth (and earlier iterates) only. In (II), the
approximate solution values are modified one after the other, using the latest
available values of the iterate, so that some elements of the $(k+1)^{st}$ iterate are
functions of the $(k+1)^{st}$ and earlier iterates. In this case, the ordering of the mesh
points is significant. Successive methods are themselves of two types, *point methods*
in which each component of the iterate is modified by an explicit calculation and
block methods in which blocks of equations are modified successively, the blocks
themselves being solved simultaneously by a direct method.

Basic Iterative Methods

Let us consider, without loss of generality that the $(m \times m)$ non-singular coefficient matrix A of the system (2.12.1) can be expressed as

$$A = Q - S, \qquad (2.12.6)$$

where Q and S are also $(m \times m)$ matrices and Q is non-singular. This expression then represents a *splitting* of the matrix A. Equation (2.12.1) then becomes

$$Q\underline{u} = S\underline{u} + \underline{b}. \qquad (2.12.7)$$

Different splittings of the matrices Q and S will clearly give different iterative methods. The main iterative methods are given by:

(i) *The Jacobi Method or the method of Simultaneous Displacement*   This is one of the well known methods in class (I). In this method we assume, without loss of generality, that $Q = D$ and $S = (E + F)$, where D is the main diagonal elements of A, E and F are strictly lower and upper $(m \times m)$ triangular matrices respectively. Hence equation (2.12.7) can be written as

$$D\underline{u} = (E + F)\underline{u} + \underline{b}. \qquad (2.12.8)$$

By the assumption that A is a non-singular matrix, thus D^{-1} exists and we can replace the system (2.12.8) by the equivalent system

$$\underline{u} = D^{-1}(E + F)\underline{u} + D^{-1}\underline{b}. \qquad (2.12.9)$$

The Jacobi iterative method is defined by

$$\underline{u}^{(k+1)} = B\underline{u}^{(k)} + \underline{g}, \qquad (2.12.10)$$

where B is the *Jacobi iterative matrix* associated with the matrix A and is given by

$$B = D^{-1}(E + F) \qquad (2.12.11)$$

and $\underline{g} = D^{-1}\underline{b}$. In this method the components of the vector $\underline{u}^{(k)}$ must be saved while computing the components of $\underline{u}^{(k+1)}$.

(ii) *The Gauss–Siedel Method (The GS Method)*   This is also known as the Successive Displacement method. It is based on the immediate use of the improved calculated values of $u_i^{(k+1)}$ instead of $u_i^{(k)}$ in the calculation of $u_{i+1}^{(k+1)}$, $1 < i < m$. By setting $Q = D - E$ and $S = F$, the matrices, D, E and F as defined before, equation (2.12.7) becomes

$$(D - E)\underline{u} = F\underline{u} + \underline{b}. \qquad (2.12.12)$$

Then, the GS iterative method is defined by

$$D\underline{u}^{(k+1)} = E\underline{u}^{(k+1)} + F\underline{u}^{(k)} + \underline{b}. \qquad (2.12.13)$$

By multiplying both sides by D^{-1}, the following equation is obtained

$$\underline{u}^{(k+1)} = L\underline{u}^{(k+1)} + R\underline{u}^{(k)} + \underline{g}, \qquad (2.12.14)$$

where $L = D^{-1}E$, $R = D^{-1}F$ and $\underline{g} = D^{-1}\underline{b}$. Here L and R are strictly lower and strictly upper triangular matrices. Equation (2.12.14) can be written as

$$(I - L)\underline{u}^{(k+1)} = R\underline{u}^{(k)} + \underline{g}. \qquad (2.12.15)$$

Since L is strictly lower triangular matrix then $\det(I-L)=1$, hence $(I-L)$ is a non-singular matrix, therefore $(I-L)^{-1}$ exists and the GS iterative method will take the form

$$\underline{u}^{(k+1)} = \mathscr{L}\,\underline{u}^{(k)} + \underline{t},\tag{2.12.16}$$

where $\mathscr{L}$ is the GS *iterative matrix* and is given by

$$\mathscr{L} = (I-L)^{-1}R\tag{2.12.17}$$

and

$$\underline{t} = (I-L)^{-1}\underline{g}.\tag{2.12.18}$$

The computational advantage of this method is that it does not require the simultaneous storage of the two approximations $u_i^{(k+1)}$ and $u_i^{(k)}$ in the course of the computations as does the Jacobi iterative method.

The next two methods are related to the Jacobi method and GS method respectively.

(iii) *The Simultaneous Over-relaxation Method (JOR Method)* This method is a modification to the Jacobi method. If we assume that $\tilde{\underline{u}}^{(k+1)}$ is the vector obtained from the Jacobi method, then from (2.12.10)

$$\tilde{\underline{u}}^{(k+1)} = B\underline{u}^{(k)} + \underline{g},\tag{2.12.19}$$

and by choosing a real parameter ω, the actual vector $\underline{u}^{(k+1)}$ of this iteration method is determined from

$$\underline{u}^{(k+1)} = \omega\tilde{\underline{u}}^{(k+1)} + (1-\omega)\underline{u}^{(k)}.\tag{2.12.20}$$

Elimination of $\tilde{\underline{u}}^{(k+1)}$ between equations (2.12.19) and (2.12.20) leads to

$$\underline{u}^{(k+1)} = B_\omega\underline{u}^{(k)} + \omega\underline{g},\tag{2.12.21}$$

where B_ω is the iteration matrix of the JOR method and is given by

$$B_\omega = [\omega D^{-1}(E+F) + (1-\omega)I].\tag{2.12.22}$$

The real parameter ω is called the *relaxation factor*.

If $\omega = 1$, we have the Jacobi method. If $\omega > 1$ (< 1) then we are in a sense carrying out the operation of "over-relaxation" (under-relaxation) at each of the nodal points. Both, the Jacobi and JOR methods are clearly independent of the order in which the equations are solved.

(iv) *The Successive Over-relaxation Method (SOR Method)* Similarly related to the GS method is the SOR method. This method is the same as the JOR method except that one uses the recent values of $u_i^{(k+1)}$ whenever possible. Hence, from equation (2.12.14), the SOR method is defined as

$$\underline{u}^{(k+1)} = \omega(L\underline{u}^{(k+1)} + R\underline{u}^{(k)} + \underline{g}) + (1-\omega)\underline{u}^{(k)}.\tag{2.12.23}$$

Equation (2.12.23) can be written in the form

$$(I - \omega L)\underline{u}^{(k+1)} = [\omega R + (1-\omega)I]\underline{u}^{(k)} + \omega\underline{g}.\tag{2.12.24}$$

However $(I - \omega L)$ is non-singular for any choice of ω, since det $(I - \omega L) = 1$. So we can solve (2.12.24) for $\underline{u}^{(k+1)}$ obtaining

$$\underline{u}^{(k+1)} = \mathscr{L}_\omega \underline{u}^{(k)} + (I - \omega L)^{-1}\omega \underline{g}, \tag{2.12.25}$$

where $\mathscr{L}_\omega$ is *the SOR iteration matrix* and is given by

$$\mathscr{L}_\omega = (I - \omega L)^{-1}(\omega R + (1-\omega)I). \tag{2.12.26}$$

For $\omega = 1$, we have the GS method. Again $\omega > 1$ (<1) corresponds to the cases of over-relaxation (under-relaxation) as discussed before. The GS method and the SOR method both depend upon the order in which the equations are solved.

(v) *The Symmetric SOR Method (The SSOR Method)* This method involves two half iterations using the SOR method. The first half iteration is the ordinary SOR method while the second half iteration is the SOR method using the nodal points in reverse order. Hence, we can define the SSOR iterative method by

$$\underline{u}^{(k+1/2)} = \mathscr{L}_\omega \underline{u}^{(k)} + (I - \omega L)^{-1}\omega \underline{g}, \tag{2.12.27a}$$

$$\underline{u}^{(k+1)} = \mathscr{K}_\omega \underline{u}^{(k+1/2)} + (I - \omega R)^{-1}\omega \underline{g}, \tag{2.12.27b}$$

where $\underline{u}^{(k+1/2)}$ is an intermediate approximation to the solution, $\mathscr{L}_\omega$ is given in (2.12.26) and $\mathscr{K}_\omega$ is given by

$$\mathscr{K}_\omega = (I - \omega R)^{-1}[\omega L + (1-\omega)I]. \tag{2.12.28}$$

By eliminating $\underline{u}^{(k+1/2)}$ between equations (2.12.27a) and (2.12.27b), we have

$$\underline{u}^{(k+1)} = \mathscr{H}_\omega \underline{u}^{(k)} + \underline{d}_\omega, \tag{2.12.29}$$

where $\mathscr{H}_\omega$ is the SSOR iteration matrix and is given by

$$\mathscr{H}_\omega = \mathscr{K}_\omega \mathscr{L}_\omega = (I - \omega R)^{-1}(I - \omega L)^{-1}[\omega L + (1-\omega)I][\omega R + (1-\omega)I] \tag{2.12.30}$$

and

$$\underline{d}_\omega = \omega(2-\omega)(I - \omega R)^{-1}(I - \omega L)^{-1}\underline{g}. \tag{2.12.31}$$

In conclusion, it is clear that the iterative methods described above can all be presented in the form

$$\underline{u}^{(k+1)} = G\underline{u}^{(k)} + \underline{r}, \tag{2.12.32}$$

according to the following table.

Table 2.12.1

Method	The Iteration Matrix G	The Vector $\underline{r}$
Jacobi	$D^{-1}(E+F) = L+R$	$D^{-1}\underline{b}$
GS	$(I-L)^{-1}R$	$(I-L)^{-1}D^{-1}\underline{b}$
JOR	$\omega D^{-1}(E+F) + (1-\omega)I$ $= \omega(L+R) + (1-\omega)I$	$\omega D^{-1}\underline{b}$
SOR	$(I-\omega L)^{-1}[\omega R + (1-\omega)I]$	$(I-\omega L)^{-1}\omega D^{-1}\underline{b}$
SSOR	$I - \omega(2-\omega)(1-\omega R)^{-1}(I-\omega L)^{-1}D^{-1}A$	$\omega(2-\omega)(I-\omega R)^{-1}(I-\omega L)^{-1}D^{-1}\underline{b}$

where $G, \underline{r}, D, E, F, L$ and R are as defined earlier in this section.

Non-linear Over-relaxation Method (NLOR)

When the finite difference approximations are directly applied to a nonlinear elliptic partial differential equation, a system of non-linear algebraic equations is obtained. Several methods have been proposed to solve these systems (see Ames, 1965) in which the solution is obtained by an outer iteration (Newton's method) which linearizes the equations, followed by some iterative technique, for example, the SOR method. This process is repeated until convergence is obtained, thus constructing a cascade of outer iterations alternated with a large sequence of inner linear iterations.

Alternative to the above strategy, a direct and simple method which is particularly well adapted for solving algebraic systems associated with non-linear elliptic equations is due to Leiberstein (1968)–a method called *non-linear over-relaxation* (NLOR).

Consider a system of m equations each having continuous first derivatives

$$\underline{f}(\underline{u}) = f_p(u_1, u_2, \ldots, u_m) = 0, \quad p = 1, 2, \ldots, m. \tag{2.12.33}$$

The basic idea in this NLOR method is to introduce a relaxation factor ω and solve the iterative sequence of equations given by

$$u_1^{(k+1)} = u_1^{(k)} - \omega \frac{f_1(u_1^{(k)}, u_2^{(k)}, \ldots, u_m^{(k)})}{f_{11}(u_1^{(k)}, u_2^{(k)}, \ldots, u_m^{(k)})},$$

$$u_2^{(k+1)} = u_2^{(k)} - \omega \frac{f_2(u_1^{(k+1)}, u_2^{(k)}, \ldots, u_m^{(k)})}{f_{22}(u_1^{(k+1)}, u_2^{(k)}, \ldots, u_m^{(k)})}, \tag{2.12.34}$$

$$\cdot$$
$$\cdot$$
$$\cdot$$

$$u_m^{(k+1)} = u_m^{(k)} - \omega \frac{f_m(u_1^{(k+1)}, u_2^{(k+1)}, \ldots, u_m^{(k)})}{f_{mm}(u_1^{(k+1)}, u_2^{(k+1)}, \ldots, u_m^{(k)})},$$

where $f_{pq} = \partial f_p / \partial u_q$.

This method has a feature of the Gauss–Seidel method in that it uses corrected intermediate results immediately upon them becoming available. In addition if the f_p are linear functions of the x_j values, this method reduces to the SOR method.

The convergence criteria for the NLOR method can be shown to be the same as that for the SOR method with the coefficient matrix A replaced by the Jacobian matrix of the equations (2.12.33), $J(f_{pq}^{(k)})$, where

$$J(f_{pq}^{(k)}) = \left[\frac{\partial f_{pq}^{(k)}(u)}{\partial x_q} \right], \quad p, q = 1, 2, \ldots, m. \tag{2.12.35}$$

This is accomplished by use of the Taylor series so that the vector form for $\underline{u}^{(k+1)} - \underline{u}^{(k)}$ is

$$\underline{u}^{(k+1)} - \underline{u}^{(k)} = -\omega [D_k^{-1} L_k \underline{e}^{(k+1)} + (I + D_k^{-1} U_k) \underline{e}^{(k)}], \tag{2.12.36}$$

where $\underline{e}^{(k)} = \underline{u}^{(k)} - \underline{u}$ is the error vector and L_k, D_k and U_k are the lower triangular, diagonal and upper triangular components matrices of the Jacobian matrix respectively, i.e.,

$$J(f_{pq}^{(k)}) = L_k + D_k + U_k. \tag{2.12.37}$$

For convergence, the Jacobian matrix (2.12.35), at each stage of the iteration, must have the same properties required for A in the SOR method, e.g., Property (A) (see Ames, 1965).

2.13. The Basic Convergence Criterion

Definition 2.13.1.
The iterative method (2.12.32) converges if

$$\lim_{k \to \infty} u_i^{(k)} = u_i, \quad \text{for all } i \tag{2.13.1}$$

and for all starting vectors $\underline{u}^{(0)}$ and u_i is the exact solution.

Theorem 2.13.1.
An iterative method which can be expressed in the form of equation (2.12.32) converges if and only if $\rho(G) < 1$.

Proof. We define the error vector after k iterations to be

$$\underline{e}^{(k)} = \underline{u}^{(k)} - \underline{u}, \tag{2.13.2}$$

where $\underline{u}$ is the unique vector solution of (2.12.1) and assume that the iterative method is consistent, i.e.,

$$\underline{u} = G\underline{u} + \underline{r}, \tag{2.13.3}$$

then from (2.12.32) and (2.13.3) we obtain

$$\underline{e}^{(k+1)} = G\underline{e}^{(k)}, \tag{2.13.4}$$

and hence

$$\underline{e}^{(k)} = G\underline{e}^{(k-1)} = G^2\underline{e}^{(k-2)} = \cdots = G^k\underline{e}^{(0)}, \tag{2.13.5}$$

where $\underline{e}^{(0)}$ is the error vector associated with the initial vector $\underline{u}^{(0)}$. We require the conditions under which $\underline{e}^{(k)} \to 0$ and $k \to \infty$. From equation (2.13.5) this can happen if and only if $G^k \to (0)$ (the null matrix) as $k \to \infty$. By Theorem (1.13) this will be true if and only if $\rho(G) \leq 1$, which completes the proof.

Corollary 2.13.1.
A sufficient condition for convergence of (2.12.32) is merely that

$$\|G\| < 1, \tag{2.13.6}$$

since $\rho(G) \leq \|G\|$.

It is not a necessary condition because it may happen in some cases that $\|G\| > 1$ but $\rho(G) < 1$ which guarantees the convergence of the iteration process according to Theorem (2.13.1). On the other hand as confirmed by Theorem (2.13.1) the convergence of (2.12.32) is totally independent of the choice of the initial vector $\underline{u}^{(0)}$, as long as the matrix A in (2.12.1) is non-singular, whilst it is dependent on $\underline{u}^{(0)}$ if A is singular.

2.14. Rate of Convergence

In practical computation, even if an iterative method converges, it may converge too slowly to be of practical value. Hence, it is essential to evaluate the effectiveness of an iterative method. To carry out this we should consider both the work required per iteration and the number of iterations required for convergence to a specified accuracy. For the latter and normally in practice, the usual approach is to iterate until the norm of the error vector $\underline{e}^{(k)}$ is reduced to less than some predetermined factor, say ε, of the norm of the initial vector $\underline{e}^{(0)}$. From (2.13.5) we have

$$\|\underline{e}^{(k)}\| = \|G^k\|\cdot\|\underline{e}^{(0)}\|. \tag{2.14.1}$$

Then, if $\underline{e}^{(0)} \neq \underline{0}$

$$\|\underline{e}^{(k)}\|/\|\underline{e}^{(0)}\| \leq \|G^k\|. \tag{2.14.2}$$

We require

$$\|\underline{e}^{(k)}\| \leq \varepsilon\|\underline{e}^{(0)}\|, \tag{2.14.3}$$

where $\|\cdot\|$ denotes $\|\cdot\|_2$. By Theorem (2.13.1) we know that $\|G^k\| \to 0$ as $k \to \infty$ if and only if $\rho(G) < 1$. Hence, equation (2.14.3) can be satisfied by choosing k sufficiently large that

$$\|G^k\| \leq \varepsilon. \tag{2.14.4}$$

If k is large enough so that $\|G^k\| < 1$, it follows that (2.14.4) is equivalent to

$$k \geq -\log\varepsilon/((-1/k)\log\|G^k\|), \tag{2.14.5}$$

and from this inequality we can obtain a lower bound for the number of iterations for the iterative methods (2.12.32).

Definition 2.14.1.
For any convergent iterative method of the form (2.12.32), the quantity

$$R_k(G) = \frac{-\log\|G^k\|}{k} \tag{2.14.6}$$

is the average rate of convergence after k iterations.

Corollary 2.14.1.
If $R_k(G_1) < R_k(G_2)$, then G_2 is iteratively faster for k iterations than G_1.

Definition 2.14.2.
The asymptotic average rate of convergence is defined by

$$R(G) = \lim_{k \to \infty} R_k(G) = -\log\rho(G). \tag{2.14.7}$$

The latter equality holds, since

$$\rho(G) = \lim_{k \to \infty}(\|G^k\|)^{1/k}, \tag{2.14.8}$$

which is a result proved by Young (1971). We shall refer to $R(G)$ as the *rate of convergence*.

It is normal for iterative processes to converge slowly in substantially large problems corresponding to $\rho(G)$ being only slightly less than 1 and a rate of convergence approaching 0.

Now, to obtain a crude estimate of the number of iterations, k, upon replacing $\|G^k\|$ by $[\rho(G)]^k$ in equation (2.14.5) we see that $\varepsilon \approx [\rho(G)]^k$ and hence

$$k \approx \frac{-\log \varepsilon}{-\log \rho(G)} = \frac{-\log \varepsilon}{R(G)}. \tag{2.14.9}$$

However, the value of k obtained from (2.14.9) could be very much lower when compared with the number required, in which $\|G^k\|$ will behave like $k\rho(G)^{k-1}$, rather than $\rho(G)^k$ (see Young, 1971). In this case, the smallest value of k is such that

$$k[\rho(G)]^{k-1} \leq \varepsilon \tag{2.14.10}$$

estimates the number of iterations required in practice.

2.15. Convergence Theorems for Basic Iterative Methods

The basic criterion, regarding convergence, as shown in Section 2.13 is that the relevant iteration matrix G must have its spectral radius less than 1. Unfortunately, this test is difficult to apply in practice for large matrices, since the determination of the largest eigenvalues of G might well involve more computation than the actual solution of the equations. However, in this section we will state some convergence theorems for the basic iterative methods which were considered in Section 2.12.

Theorem 2.15.1.
Let A be a strictly or irreducibly diagonally dominant $(m \times m)$ complex matrix. Then, both the associated point Jacobi and the point GS matrices are convergent and the Jacobi iterative method and GS iterative method for the matrix problem $A\underline{u} = b$ are convergent for any initial approximation vector $\underline{u}^{(0)}$.

Proof. (see Varga, 1962).

Theorem 2.15.2. (*The Stein–Rosenberg comparison theorem*, 1948)
If the system $A\underline{u} = \underline{b}$ has a Jacobi iteration matrix $B = (L + R)$ which contains no negative elements, and if $\mathcal{L}$ is the GS iteration matrix, then one and only one of the following mutually exclusive relations holds:

    (a) $\rho(B) = \rho(\mathcal{L}) = 0$,

    (b) $0 < \rho(\mathcal{L}) < \rho(B) < 1$,

    (c) $\rho(B) = \rho(\mathcal{L}) = 1$,

or

    (d) $1 < \rho(B) < \rho(\mathcal{L})$.

Proof. (see Varga, 1962).

The contents of this theorem is that the Jacobi iteration matrix and the GS iteration matrix are either both convergent or both divergent together. Furthermore, if the matrix B is non-negative and $0 < \rho(B) < 1$, then the GS iterative method is asymptotically faster than the Jacobi iterative method. Hence

$$R(\mathcal{L}) > R(B). \tag{2.15.1}$$

Notice the importance of the two conditions, B contains no negative elements and $0 < \rho(B) < 1$, for the results given above, since sometimes it happens that the Jacobi iterative method might converge and the GS iterative method diverge. We can illustrate this by the following example (Fox, 1964).

The matrix

$$A = \begin{bmatrix} 1 & 0 & 1 \\ -1 & 1 & 0 \\ 1 & 2 & -3 \end{bmatrix},$$

gives rise to the iteration matrices

$$B = \begin{bmatrix} 0 & 0 & 1 \\ -1 & 0 & 0 \\ -1/3 & -2/3 & 0 \end{bmatrix} \text{ and } \mathcal{L} = \begin{bmatrix} 0 & 0 & 1 \\ 0 & 0 & 1 \\ 0 & 0 & 1 \end{bmatrix}.$$

If λ_i and ξ_i are the eigenvalues of the matrix B and $\mathcal{L}$ respectively, for $i = 1, 2, 3$, then

$$\lambda_1 = 0.748, \quad \lambda_{2,3} = -0.374 \pm 0.868i$$

$$\xi_1 = 1, \quad \xi_2 = 0, \quad \xi_3 = 0,$$

so that the Jacobi process converges very slowly and the GS process diverges.

But in general the GS method is superior to the Jacobi method and its superiority is given by the following theorem:

Theorem 2.15.3.
If A is a symmetric, positive definite matrix, then its GS iteration matrix $\mathcal{L} = (D + E)^{-1}F$ has spectral radius less than unity (i.e., in this case the GS iterative method always converges).

Proof. (see Lieberstein 1968, Fox, 1964 and Gault et al., 1974).

In Lieberstein (1968), there is given a counter example which verifies the invalidity of Theorem (2.15.3) for the Jacobi process, i.e. although the matrix A is symmetric and positive definite, its iteration matrix $D^{-1}(E + F)$ may have eigenvalue(s) greater than 1 in modulus.

Theorem 2.15.4.
If the Jacobi method converges, then the JOR method converges for $0 < \omega \leq 1$.

Proof. (see Young, 1971).

Theorem 2.15.5.
Let A be an irreducible matrix with weak diagonal dominance. Then:

    (a) *The Jacobi method converges and the JOR method converges for* $0 < \omega \le 1$.
    (b) *The GS method converges, and the SOR method converges for* $0 < \omega \le 1$.

Proof. (see Young, 1971).

So far, the Jacobi and GS iterative methods were compared in terms of their spectral radii and we saw how the matrix property of being irreducible with weak diagonal dominance was a sufficient condition for the convergence of the Jacobi and GS iterative methods. Whereas $\rho(G) < 1$ is both a necessary and sufficient condition that G be a convergent matrix. Next we give different necessary and sufficient conditions for the convergence of the SOR and GS iterative methods. We commence with the following theorem of Kahan (1958).

Theorem 2.15.6.
If $\mathscr{L}_\omega$ is the SOR iteration matrix then

$$\rho(\mathscr{L}_\omega) \ge |\omega - 1|, \tag{2.15.2}$$

for all real ω. Moreover, if the SOR method converges, then,

$$0 < \omega < 2. \tag{2.15.3}$$

Proof. If η_i are the eigenvalues of $\mathscr{L}_\omega$ we have

$$\det(\mathscr{L}_\omega) = \prod_{i=1}^{m} \eta_i. \tag{2.15.4}$$

Hence by (2.12.26) we have

$$\det(\mathscr{L}_\omega) = \det[(I - \omega L)^{-1}(\omega R + (1 - \omega)I)]$$
$$= \det(I - \omega L)^{-1}\det[\omega R + (1 - \omega)I].$$

But $(I - \omega L)$ is a lower triangular matrix with diagonal elements equal to unity and $\omega R + (1 - \omega)I$ is an upper triangular matrix with diagonal elements $1 - \omega$. Hence

$$\det(\mathscr{L}_\omega) = 1 \cdot (1 - \omega)^m. \tag{2.15.5}$$

Also, from (2.15.4) and (2.15.5) we have

$$\prod_{i=1}^{m} \eta_i = (1 - \omega)^m. \tag{2.15.6}$$

But

$$\rho(\mathscr{L}_\omega) = \max_i |\eta_i|, \tag{2.15.7}$$

hence

$$\rho(\mathscr{L}_\omega) \ge [|1 - \omega|^m]^{1/m},$$

which implies

$$\rho(\mathscr{L}_\omega) \ge |1 - \omega|.$$

Now, if the SOR method converges, then from Theorem 2.13.1

$$\rho(\mathcal{L}_\omega) < 1,$$

$$\Rightarrow |1 - \omega| < 1,$$

and this implies $0 < \omega < 2$.

Theorem 2.15.7.
Let A be a symmetric matrix with positive diagonal elements. Then, the SOR method converges if and only if A is positive definite and $0 < \omega < 2$.

Proof. (see Young, 1971 or Varga, 1962).

Corollary 2.15.1.
Let A be a symmetric matrix with positive diagonal elements. Then, the GS method converges if and only if A is positive definite.

Now the following theorem is an extension of the result of Theorem (2.15.2). (The theorem and the corollaries which follow are given in Young (1971)).

Theorem 2.15.8.
If A is an L-matrix and if $0 < \omega \le 1$, and $B = (L + R)$ then

(a) $\rho(B) < 1$, *if and only if $\rho(\mathcal{L}_\omega) < 1$,*
(b) $\rho(B) < 1$, *(and $\rho(\mathcal{L}_\omega) < 1$) if and only if A is an M-matrix, if $\rho(B) < 1$, then,*

$$\rho(\mathcal{L}_\omega) \le (1 - \omega + \omega\rho(B)), \tag{2.15.8}$$

(c) *if $\rho(B) \ge 1$ and $\rho(\mathcal{L}_\omega) \ge 1$, then*

$$\rho(\mathcal{L}_\omega) \ge (1 - \omega + \omega\rho(B)) \ge 1. \tag{2.15.9}$$

Corollary 2.15.2.
If A is an L-matrix, then

(a) $\rho(B) < 1$, *if and only if $\rho(\mathcal{L}) < 1$,*
(b) $\rho(B) < 1$ *(and $\rho(\mathcal{L}) < 1$) if and only if A is an M-matrix, if $\rho(B) < 1$, then,*

$$\rho(\mathcal{L}) < \rho(B), \tag{2.15.10}$$

(c) *if $\rho(B) \ge 1$ and $\rho(\mathcal{L}) \ge 1$, then,*

$$\rho(\mathcal{L}) \ge \rho(B). \tag{2.15.11}$$

Corollary 2.15.3.
If A is a Stieltjes matrix, then the Jacobi iterative method converges and the JOR iterative method converges for $0 < \omega \le 1$. (See definition for the L-matrix, M-matrix and Stieltjes matrix), (Young 1971, Varga 1962).

For the SSOR iterative method, the eigenvalues of its iteration matrix $\mathcal{H}_\omega$, are real and non-negative and the method is convergent for $0 < \omega < 2$. In other words, if $\rho(\mathcal{H}_\omega) < 1$, then $0 < \omega < 2$.

In practice the SSOR iterative method is not as efficient as the normal SOR method. However it is superior in problems with limited application, (Habetler and Wachspress, 1961) especially when combined with Chebyshev acceleration.

2.16. Determination of the Optimum Relaxation Factor and Comparison of Rates of Convergence

The determination of a suitable value for the relaxation factor ω of the SOR method is of paramount importance, and in particular the optimum value of ω, denoted by ω_b, which minimises the spectral radius of the SOR iteration matrix and thereby maximise the rate of convergence of the method. For an arbitrary set of linear equations, no formula exists for the determination of ω_b, though a simple but time consuming procedure for estimating ω_b is to run the problem on a computer for a range of values of ω to obtain some idea of the value which gives the most rapid convergence. However it can be calculated for many of the difference equations approximating second order partial differential equations because their matrices are of a special type which possess Property (A), and the significance of this was first revealed by Young (1954). He proved that when a matrix possesses Property (A) then it can be transformed into what he termed a consistently ordered matrix. Under this condition the eigenvalues of the SOR iteration matrix $\mathcal{L}_\omega$ associated with A are related to the eigenvalues μ of the corresponding Jacobi iteration matrix B of A by the equation

$$\frac{(\lambda + \omega - 1)^2}{\lambda} = \omega^2 \mu^2, \qquad (2.16.1)$$

from this equation it can be seen that

$$\lambda \pm \omega \mu \lambda^{1/2} + \omega - 1 = 0. \qquad (2.16.2)$$

If we assume that A is symmetric, then, the eigenvalues of B are real and occur in pairs $\pm \mu$. Let $\bar{\mu}$ denote the largest eigenvalue of B, it can be shown that ω_b defined by

$$\omega_b^2 \bar{\mu}^2 = 4(\omega_b - 1), \quad 1 \le \omega_b \le 2, \qquad (2.16.3a)$$

$$\omega_b = \frac{2}{1 + \sqrt{1 - \bar{\mu}^2}}, \qquad (2.16.3b)$$

is the value of ω which minimises $\rho(\mathcal{L}_\omega)$, i.e. $\omega \ne \omega_b$, then

$$\rho(\mathcal{L}_\omega) > \rho(\mathcal{L}_{\omega_b}). \qquad (2.16.4)$$

Using this optimum value of ω, it can also be shown that

$$\rho(\mathcal{L}_{\omega_b}) = \frac{1 - \sqrt{1 - \bar{\mu}^2}}{1 + \sqrt{1 - \bar{\mu}^2}} = \omega_b - 1. \qquad (2.16.5)$$

Moreover, for any ω in the range $0 < \omega < 2$, we have

$$\rho(\mathcal{L}_\omega) = \begin{cases} \left[\dfrac{\omega \bar{\mu} + (\omega^2 \bar{\mu}^2 - 4(\omega - 1))^{1/2}}{2} \right]^2 & \text{if } 0 < \omega \le \omega_b, \\[4mm] \omega - 1 & \text{if } \omega_b \le \omega < 2. \end{cases} \qquad (2.16.6)$$

For the Gauss–Seidel iterative method ($\omega = 1$), equation (2.16.2) gives

$$\rho(\mathscr{L}) = [\rho(B)]^2 = (\bar{\mu})^2. \tag{2.16.7}$$

Using (2.16.7) it is clear from (2.16.3b) that

$$\omega_b = \frac{2}{1 + \sqrt{1 - \rho(\mathscr{L})}}. \tag{2.16.8}$$

The estimation of ω_b depends on whether $\rho(B)$ or $\rho(\mathscr{L})$ can be easily estimated. Methods have been suggested by many authors (Carré, 1961, Varga, 1962, etc.,) one of which is the power method that can be described as follows:

Assuming the matrix of the finite difference equations is consistently ordered and has Property (A), calculate the sequence of approximations $\underline{u}^{(1)}, \underline{u}^{(2)}, \dots, \underline{u}^{(k)}$, to the solution of the system of equations $A\underline{u} = \underline{b}$ by the Gauss–Seidel method and then we have

$$\rho(\mathscr{L}) = \lim_{k \to \infty} \frac{\|\underline{d}^{(k)}\|}{\|\underline{d}^{(k-1)}\|}, \tag{2.16.9}$$

where $\underline{d}^{(k)}$ is defined as

$$\underline{d}^{(k)} = \underline{u}^{(k)} - \underline{u}^{(k-1)}, \tag{2.16.10}$$

or equivalently

$$\|\underline{d}^{(k)}\| = \left\{ \sum_{j=1}^{n} (u_j^{(k)} - u_j^{(k-1)})^2 \right\}^{1/2}. \tag{2.16.11}$$

Hence, using the power method we can determine an approximate value of $\rho(\mathscr{L})$, which in turn can be substituted into equation (2.16.8) to give an estimate of the optimum acceleration factor, ω_b.

For Poisson's equation over a rectangle of sides ph and qh, with Dirichlet boundary conditions, it can be shown (see Smith, 1978), that $\rho(B)$ for the five point difference approximation, using a square mesh of side h, is

$$\rho(B) = \frac{1}{2}\left(\cos\frac{\pi}{p} + \cos\frac{\pi}{q}\right). \tag{2.16.12}$$

Hence for the model problem of Laplace's equation in the unit square with m^2 internal mesh points

$$\rho(B) = \bar{\mu} = \cos\frac{\pi}{m+1} = \cos\pi h. \tag{2.16.13}$$

In this case, if we choose $h = 1/20$ then $\bar{\mu} \approx 0.9877$, it is interesting to see the behaviour of $\rho(\mathscr{L}_\omega)$ as a function of ω. It can be shown that as ω increases from 0 to 1, $\rho(\mathscr{L}_\omega)$ decreases very slowly from 1 to $\bar{\mu}^2 \approx 0.9755$. As ω increases further $\rho(\mathscr{L}_\omega)$ decreases slightly more rapidly until ω gets close to $\omega_b \approx 1.72945$ at which point the decrease is very rapid such that, as $\omega \to \omega_b -$, the slope of $\rho(\mathscr{L}_\omega)$ approaches $-\infty$ (see Young, 1971).

The value of $\rho(\mathscr{L}_\omega)$ for $\omega = \omega_b$ is $\omega_b - 1 \cong 0.72945$. As ω increases further, $\rho(\mathscr{L}_\omega)$ increases linearly to a value of unity when $\omega = 2$. This is best seen by plotting $\rho(\mathscr{L}_\omega)$ as a function of ω for $0 \leq \omega \leq 2$, as shown in Figure 2.16.1.

In a small number of rather specialised cases $\rho(B)$ can be calculated quite easily, for example if A is an $(m \times m)$ tridiagonal matrix whose Jacobi matrix is

$$B = \begin{bmatrix} 0 & b & & & & \\ a & 0 & b & & 0 & \\ & a & 0 & b & & \\ & & a & 0 & b & \\ & 0 & & & & b \\ & & & & a & 0 \end{bmatrix}$$

it is known that

$$\rho(B) = 2\sqrt{ab}\,\cos\frac{\pi}{(m+1)}. \qquad (2.16.14)$$

Direct comparisons of the Jacobi, Gauss–Seidel and SOR iterative methods can be made when the matrix A has Property (A) and is consistently ordered. From equation (2.16.7) and by using the definition of the rate of convergence given in (2.16.7), we obtain

$$R(\mathscr{L}) = 2R(B). \qquad (2.16.15)$$

Thus, the rate of convergence of the GS method is twice that of the Jacobi method. By use of equation (2.16.5) it can be verified that, asymptotically as $\rho(B) \to 0$

$$R(\mathscr{L}_{\omega_b}) \approx 2\sqrt{R(\mathscr{L})}. \qquad (2.16.16)$$

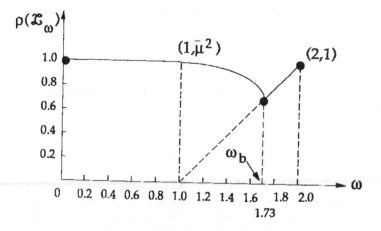

Figure 2.16.1

For the model problem of the Laplace equation in the unit square with m^2 internal mesh points (Varga, 1962) we have the following results as $m \to \infty$ or $h \to 0$

(i) $\quad R(B) \approx \dfrac{\pi^2}{2(m+1)^2} \approx \dfrac{1}{2}\pi^2 h^2,$

(ii) $\quad R(\mathscr{L}) \approx \dfrac{\pi^2}{(m+1)^2} \approx \pi^2 h^2,$ \hfill (2.16.17)

(iii) $\quad R(\mathscr{L}_{\omega_b}) \approx \dfrac{2\pi}{(m+1)} \approx 2\pi h,$

therefore

$$\frac{R(\mathscr{L}_{\omega_b})}{R(B)} \approx \frac{4(m+1)}{\pi} \approx \frac{4}{\pi h}, \quad \text{as } m \to \infty, \quad (h \to 0),$$ \hfill (2.16.18)

where $h = 1/(m+1)$.

Thus, for m large, the SOR iterative method with optimum relaxation factor is an order of magnitude faster than the Jacobi iterative method for the model problem, and

$$\frac{R(\mathscr{L}_{\omega_b})}{R(\mathscr{L})} \approx \frac{2(m+1)}{\pi} \approx \frac{2}{\pi h}, \quad h \to 0.$$ \hfill (2.16.19)

This shows the superiority of the SOR iterative method over the GS iterative method by an order of magnitude of approximately h^{-1}.

2.17. Related Point Iterative Methods

In this section, we will briefly discuss some other iterative methods for solving the system

$$A\underline{u} = \underline{b},$$ \hfill (2.17.1)

which have been developed and successfully applied.

2.17.1. Simultaneous Displacement Method (Gradient Method)

The Jacobi iterative method described in Section 2.12 can be extrapolated to give a faster rate of convergence when the coefficient matrix A (of order m) in (2.17.1) is real and positive definite. We assume, without loss of generality, that all the diagonal elements of A are unity. Hence, the Jacobi iterative method

$$D\underline{u}^{(k+1)} = (E+F)\underline{u}^{(k)} + \underline{b}$$ \hfill (2.17.2)

can be written as

$$\underline{u}^{(k+1)} = H\underline{u}^{(k)} + \underline{b} \quad (\text{since } D^{-1} = I),$$ \hfill (2.17.3)

where D, E, and F as defined in Section 2.11 and by assuming that $H = D^{-1}(E+F)$.

Alternatively equation (2.17.3) can be written as

$$\underline{u}^{(k+1)} = (I - A)\underline{u}^{(k)} + \underline{b},$$

$$= \underline{u}^{(k)} + \underline{q}^{(k)}, \qquad (2.17.4)$$

which shows that the change in the solution vector is equal to the kth residual vector $\underline{q}^{(k)} = \underline{b} - A\underline{u}^{(k)}$. Equation (2.17.4) can then be extrapolated by multiplying $\underline{q}^{(k)}$ by a positive constant α, to be determined later, giving rise to the iterative method

$$\underline{u}^{(k+1)} = (I - \alpha A)\underline{u}^{(k)} + \alpha\underline{b}, \qquad (2.17.5)$$

so that the error vector after $k+1$ iterations $\underline{e}^{(k+1)} = \underline{u}^{(k+1)} - A^{-1}\underline{b}$ satisfies the recurrence relationship

$$\underline{e}^{(k+1)} = (I - \alpha A)\underline{e}^{(k)}, \qquad (2.17.6)$$

which is a linear stationary iteration with iteration matrix $G = I - \alpha A$. Hence, for convergence, we require

$$\rho(I - \alpha A) < 1. \qquad (2.17.7)$$

If we denote the eigenvalues of A by λ_i, $1 \le i \le m$, then the condition (2.17.7) can be written as

$$|1 - \alpha\lambda_i| < 1, \quad 1 \le i \le m, \qquad (2.17.8)$$

which gives the permissible range of values of α for convergence of (2.17.5) to be

$$0 < \alpha < \frac{2}{\max\limits_i \lambda_i}. \qquad (2.17.9)$$

However, to obtain the fastest possible convergence, we must choose α so that

$$\max\limits_i |1 - \alpha\lambda_i| = \min. \qquad (2.17.10)$$

Very often we know that the λ_i lie on the real axis in an interval $a \le \lambda_i \le b$ for all i $(0 < a < b < \infty)$. For every α, the function $|1 - \alpha\lambda|$ takes its maximum at one of the end points $\lambda = a$, or $\lambda = b$. So the best choice of α is that one for which

$$\max(|1 - \alpha a|, |1 - \alpha b|) \text{ is smallest}.$$

Clearly, it is the one for which

$$1 - \alpha a = -(1 - \alpha b),$$

i.e.,

$$\alpha = \frac{2}{a+b}. \qquad (2.17.11)$$

With this choice of α we have, for all i,

$$|1 - \alpha\lambda_i| \le \frac{b-a}{b+a} = \frac{P-1}{P+1} < 1, \qquad (2.17.12)$$

where $P = b/a$ is the *P-condition number* of A, defined by Todd (1950), as the ratio of the maximum eigenvalue to the minimum eigenvalue of a positive definite matrix.

2.17.2. Richardson's Method

Richardson (1910), presented a method for solving the linear system (2.17.1) when A is a positive definite matrix. In this method the relaxation factor α of equation (2.17.5) may depend upon the iteration number k, so that $\alpha = \alpha_k$, giving rise to the iterative method

$$\underline{u}^{(k+1)} = (I - \alpha_k A)\underline{u}^{(k)} + \alpha_k \underline{b}. \tag{2.17.13}$$

The iteration matrix $(I - \alpha_k A)$ changes at each iteration and so the rate of convergence of the method cannot easily be found. By (2.17.6) we see that

$$\underline{e}^{(k+1)} = (I - \alpha_k A)\underline{e}^{(k)}.$$

We have, upon successive application, the expression

$$\underline{e}^{(k+1)} = \left[\prod_{i=0}^{k}(I - \alpha_i A)\right]\underline{e}^{(0)}$$

$$= Q_{k+1}(A)\underline{e}^{(0)} \quad \text{(say)}, \tag{2.17.14}$$

where

$$Q_{k+1} = \prod_{i=0}^{k}(I - \alpha_i A) \tag{2.17.15}$$

is a polynomial of degree $(k+1)$ in A such that $Q_{k+1}(0) = 1$ and $Q_{k+1}(1/\alpha_i) = 0$.

This is a non-stationary process, since the error operator changes for each iteration.

If the eigenvalues of A are λ_i, $1 \leq i \leq m$ as before, with corresponding eigenvectors $\underline{v}_i$, then the eigenvalues and the eigenvectors of $Q_{k+1}(A)$ are $Q_{k+1}(\lambda_i)$ and $\underline{v}_i$ respectively. Now represent $\underline{e}^{(0)}$ in the basis of the eigenvectors $\underline{v}_i$ of A then

$$\underline{e}^{(0)} = \sum_{i=1}^{m}\gamma_i\underline{v}_i, \tag{2.17.16}$$

the γ_i being arbitrary, hence, from equation (2.17.14)

$$\underline{e}^{(k+1)} = \sum_{i=1}^{m}\gamma_i Q_{k+1}(\lambda_i)\underline{v}_i. \tag{2.17.17}$$

For the error $\underline{e}^{(k+1)}$ to be small, $Q_{k+1}(x)$ must be small for x throughout the interval $0 < a \leq x \leq b < \infty$. The precise problem may be formulated in the following manner:

Find $Q_{k+1}(x)$ defined for $a \leq x \leq b$ with $Q_{k+1}(0) = 1$, such that

$$\max_{a \leq x \leq b}|Q_{k+1}(x)| = \min. \tag{2.17.18}$$

Such a polynomial is defined by

$$Q_{k+1}(x) = \frac{T_{k+1}[(b + a - 2x)/(b - a)]}{T_{k+1}[(b + a)/(b - a)]}, \tag{2.17.19}$$

where T_k is a Chebyshev polynomial of degree k, defined by

$$T_k(x) = \cos [k \cos^{-1} x]$$

$$= 1/2[(x + \sqrt{x^2 - 1})^k + (x - \sqrt{x^2 - 1})^k] \qquad (2.17.20)$$

adjusted to the interval $[-1, 1]$.

Hence, $Q_{k+1}(x)$ is a Chebyshev polynomial of degree $(k+1)$ over the interval $[a, b]$, and scaled so that $Q_{k+1}(0) = 1$. Equation (2.17.19) can be written as

$$Q_{k+1} = \frac{T_{k+1}(y)}{T_{k+1}(y_0)}, \qquad (2.17.21)$$

where

$$y = \frac{b + a - 2x}{b - a} \quad \text{and} \quad y_0 = \frac{b + a}{b - a} = \frac{P + 1}{P - 1} > 1$$

(y_0 is the value of y when $x = 0$). Thus $y = 1$ for $x = a$ and $y = -1$ for $x = b$.

Now, since the maximum absolute value of the numerator of equation (2.17.21) is unity, equation (2.17.19) can be written as

$$\max_{a \leq x \leq b} |Q_{k+1}(x)| = \frac{1}{T_{k+1}(y_0)} = [T_{k+1}(y_0)]^{-1}, \qquad (2.17.22)$$

and

$$T_{k+1}(y) = \cos[(k+1)\theta], \qquad (2.17.23)$$

where $\cos \theta = y$ and θ is a complex number. Hence

$$2T_{k+1}(y) = e^{i(k+1)\theta} + e^{-i(k+1)\theta}$$

$$= (\cos \theta + i \sin \theta)^{k+1} + (\cos \theta - i \sin \theta)^{k+1},$$

since $\cos \theta = y$, and θ is a complex number. Hence

$$2T_{k+1}(y) = e^{i(k+1)\theta} + e^{-i(k+1)\theta}$$

$$= (\cos \theta + i \sin \theta)^{k+1} + (\cos \theta - i \sin \theta)^{k+1},$$

since $\cos \theta = y$ thus $i \sin \theta = \sqrt{y^2 - 1}$

$$2T_{k+1}(y) = (y + \sqrt{y^2 - 1})^{k+1} + (y - \sqrt{y^2 - 1})^{k+1}, \quad y \geq 1.$$

From (2.17.22) as $k \to \infty$,

$$\max_{a \leq x \leq c} |Q_{k+1}(x)| = \frac{2}{(y_0 + \sqrt{y_0^2 - 1})^{k+1} + (y_0 - \sqrt{y_0^2 - 1})^{k+1}} \leq \frac{2}{(y_0 + \sqrt{y_0^2 - 1})^{k+1}}$$

$$= 2(y_0 - \sqrt{y_0^2 - 1})^{k+1} \qquad (2.17.24)$$

and it follows that the eigenvalues $|Q_{k+1}(\lambda_i)|$ of $Q_{k+1}(A)$ are uniformly bounded by $2(y_0 - \sqrt{y_0^2 - 1})^{k+1}$ as $k \to \infty$ and the asymptotic bound to the average rate of convergence is given by $\log (y_0 + \sqrt{y_0^2 - 1})^{k+1}$.

For substantially large problems we have

$$y_0 \approx 1 + \frac{2}{P}, \tag{2.17.25}$$

which gives the rate of convergence of Richardson's method

$$R \approx \frac{2}{\sqrt{P}}. \tag{2.17.26}$$

(See Forsythe and Wasow, 1960).

Now, the relaxation factors α_i are the reciprocals of the zeros of the Chebyshev polynomial $Q_{k+1}(x)$, or equivalently, the reciprocal of the zeros of $T_{k+1}(y)$, where the zeros of $T_{k+1}(y)$ are given by

$$y_i^{(k+1)} = \cos\left[\frac{(2i-1)\pi}{2k}\right], \quad 1 \le i \le k, \tag{2.17.27}$$

since, $y = (b + a - 2x)/(b - a)$, then $x = (b + a - y(b - a))/2$ \hfill (2.17.28)

and

$$\alpha_i = \frac{2}{b + a - y^{(k+1)}(b - a)}, \quad 1 \le i \le k, \tag{2.17.29}$$

which defines the iteration method (2.17.13) for a given a and b.

This choice would ensure the fastest convergence in the absence of rounding errors, but in practice the iteration is very sensitive to rounding errors when a/b is very small.

2.17.3. Second Order Methods

Consider the following linear stationary iteration of second degree, known as the second order Richardson's iterative method

$$\underline{u}^{(k+1)} = \underline{u}^{(k)} + \alpha(\underline{b} - A\underline{u}^{(k)}) + \beta(\underline{u}^{(k)} - \underline{u}^{(k-1)}), \tag{2.17.30}$$

where the parameters α and β remain constant throughout the iteration and are chosen to provide maximum convergence for this method. The error vector for (2.17.30) satisfies

$$\underline{e}^{(k+1)} = [(1 + \beta)I - \alpha A]\underline{e}^{(k)} - \beta\underline{e}^{(k-1)}, \tag{2.17.31}$$

thus, for each error mode we have

$$e_i^{(k+1)} = [1 + \beta - \alpha\lambda_i]e_i^{(k)} - \beta e_i^{(k-1)}, \tag{2.17.32}$$

where λ_i are the eigenvalues of A. Let γ_i be the eigenvalues of the matrix associated with the iterative process then

$$\underline{e}_i^{(k+1)} = \gamma_i \underline{e}_i^{(k)} = \gamma_i^2 \underline{e}_i^{(k-1)}, \tag{2.17.33}$$

so from (2.17.32) and (2.17.33) we obtain

$$e_i^{(k+1)} = \left[1 + \beta - \alpha\lambda_i - \frac{\beta}{\gamma_i}\right] e_i^{(k)} \tag{2.17.34}$$

and the combination of (2.17.33) and (2.17.34) leads to

$$\gamma_i^2 - (1 + \beta - \alpha\gamma_i)\gamma_i + \beta = 0 \tag{2.17.35}$$

which yields

$$\gamma_i = \frac{\delta_i \pm \sqrt{\delta_i^2 - 4\beta}}{2} \tag{2.17.36}$$

where $\delta_i = 1 + \beta - \alpha\lambda_i$.

Choosing α and β so that $(\delta_i^2 - 4\beta) < 0$ for all λ_i (which means that γ_i will be complex) all $|\gamma_i|$ will be identical. Then, the choices

$$1 + \beta - \alpha a = 2\sqrt{\beta}, \tag{2.17.37a}$$

$$1 + \beta - \alpha b = -2\sqrt{\beta} \tag{2.17.37b}$$

leads to

$$\alpha = \left[\frac{2}{\sqrt{a} + \sqrt{b}}\right]^2 \text{ and } \beta = \left[\frac{\sqrt{b} - \sqrt{a}}{\sqrt{b} + \sqrt{a}}\right]^2. \tag{2.17.38}$$

(See Frankel, 1950). The spectral radius of the method is $\sqrt{\beta}$ and the rate of convergence of the second order Richardson method is

$$R \approx \frac{2}{\sqrt{b/a}} = \frac{2}{\sqrt{P}}. \tag{2.17.39}$$

Note that in this iteration process we need to store two vector iterants for each iteration and this storage requirement could be severe for large systems of equations.

Another second order method (non-stationary), is the 'Chebyshev second order method' defined by

$$u^{(k+1)} = u^{(k)} + \alpha_k(b - Au^{(k)}) + \beta_k(u^{(k)} - u^{(k-1)}) \tag{2.17.40}$$

where α_k and β_k are defined as follows:

$$\alpha_k = \frac{4T_k(b+a)/(b-a)}{(b-a)T_{k+1}(b+a)/(b-a)} \text{ and } \beta_k = \frac{T_{k-1}(b+a)/(b-a)}{T_{k+1}(b+a)/(b-a)}. \tag{2.17.41}$$

The error vector is given by

$$e^{(k+1)} = \frac{T^{k+1}(b+a-2x)/(b-a)}{T_{k+1}(b+a)/(b-a)} e^{(0)} \tag{2.17.42}$$

which leads to identical results for the rate of convergence to those obtained using Richardson's method.

Since α_k and β_k are less than unity, round off errors do not appear and the method is preferred to Richardson's method with the only inconvenient item being the estimation of the extreme eigenvalues a and b.

2.17.4. Semi-iterative Methods

Consider the completely consistent linear stationary iterative method

$$\underline{u}^{(k+1)} = G\underline{u}^{(k)} + \underline{r}, \tag{2.17.43}$$

where $\underline{r} = (I - G)A^{-1}\underline{b}$ and $(I - G)$ is non-singular.

Given the sequence $\underline{u}^{(0)}, \underline{u}^{(1)}, \underline{u}^{(2)}, \ldots$, we can often develop another sequence $\underline{y}^{(1)}$, $\underline{y}^{(2)}, \ldots$ by the theory of summability so that either the new sequence converges when the old one does not, or else the new one converges faster than the old one if the old one converges. Assume that there is given a set of coefficients $\alpha_{k,\ell}$, $\ell = 0, 1, 2, \ldots, k$ such that

$$\sum_{\ell=0}^{k} \alpha_{k,\ell} = 1, \quad k = 0, 1, 2, \ldots, \tag{2.17.44}$$

so the new sequence can be considered as a linear combination of the old. Then the relation

$$\underline{y}^{(k)} = \sum_{\ell=0}^{k} \alpha_{k,\ell}\, \underline{u}^{(\ell)}, \quad k = 0, 1, 2, \ldots, \tag{2.17.45}$$

defines a *semi-iterative method* with respect to the iterative method of equation (2.17.43). If $\alpha_{k,\ell} = 0$ for all $\ell < k$ and $\alpha_{k,k} = 1$, the semi-iterative method reduces to the original method.

Our aim here is to determine $\alpha_{k,\ell}$ such that the rate of convergence of (2.17.45) is greater than that of (2.17.43). Such a determination is described in Frank (1960).

Let the error vector of the kth iterant be

$$\underline{e}^{(k)} = \underline{u}^{(k)} - \underline{u} \tag{2.17.46a}$$

and

$$\underline{\tilde{e}}^{(k)} = \underline{y}^{(k)} - \underline{u}, \tag{2.17.46b}$$

where $\underline{u}$ is the unique vector solution of (2.17.1), hence from equation (2.17.45)

$$\begin{aligned}
\underline{\tilde{e}}^{(k)} &= \sum_{\ell=0}^{k} \alpha_{k,\ell}\, \underline{u}^{(\ell)} - \underline{u} \\
&= \sum_{\ell=0}^{k} \alpha_{k,\ell}\, \underline{u}^{(\ell)} - \underline{u} \sum_{\ell=0}^{k} \alpha_{k,\ell} \\
&= \sum_{\ell=0}^{k} \alpha_{k,\ell}\, \underline{e}^{(\ell)}.
\end{aligned} \tag{2.17.47}$$

But from equation (2.13.5)

$$\underline{e}^{(\ell)} = G^{\ell}\underline{e}^{(0)},$$

hence

$$\underline{\tilde{e}}^{(\ell)} = \left(\sum_{\ell=0}^{k} \alpha_{k,\ell}\, G^{\ell} \right)\underline{e}^{(0)}. \tag{2.17.48}$$

If we now define a polynomial

$$P_k(\underline{u}) = \left(\sum_{\ell=0}^{k} \alpha_{k,\ell} \right)\underline{u}^{\ell} \tag{2.17.49}$$

then, equation (2.17.48) can be expressed as

$$\tilde{\underline{e}}^{(\ell)} = P_k(G)\underline{e}^{(0)} = P_k(G)\tilde{\underline{e}}^{(0)}. \qquad (2.17.50)$$

Suppose that the iteration matrix G of the iterative system (2.17.36) has real eigenvalues μ and they lie in the range

$$a \le \mu \le b < 1, \qquad (2.17.51)$$

where a and b are real numbers, then, from equation (2.17.43) we have

$$\|\tilde{\underline{e}}^{(\ell)}\| \le \|P_k(G)\| \, \|\underline{e}^{(0)}\| \qquad (2.17.52)$$

and since $\rho(P_k(G)) = \max_{a \le \mu \le b} |P_k(\mu)|$, it follows that the problem becomes that of minimising

$$\max_{a \le \mu \le b} |P_k(\mu)|, \qquad (2.17.53)$$

such that, $P_k(1) = 1$, for all k.

Let us introduce as before a new variable, $\gamma = \gamma(\mu)$, in which for $a \le \mu \le b$ we have, $-1 \le \gamma \le 1$ (i.e., $\gamma(a) = -1$, $\gamma(b) = 1$). So this will define the transformation

$$\gamma(\mu) = \frac{2\mu - (b+a)}{b-a} \qquad (2.17.54a)$$

and

$$\mu = \frac{1}{2[(b-a)\gamma + (b+a)]}, \qquad (2.17.54b)$$

so

$$\gamma(1) = \frac{2 - (b+a)}{b-a} = z \text{ (say)},$$

then,

$$z > 1.$$

Let

$$Q_k(\gamma)_i = P_k\left(\frac{(b-a)\gamma_i + (b+a)}{2}\right), \qquad (2.17.55)$$

hence,

$$\max_{a \le \mu \le c} |P_k(\mu)| = \max_{-1 \le \gamma \le 1} |Q_k(\gamma)|, \qquad (2.17.56)$$

then the problem is to find a polynomial $Q_k(\gamma)$ such that $Q_k(z) = P_k(1) = 1$ and

$$\max_{-1 \le \gamma \le 1} |Q_k(\gamma)| = \min. \qquad (2.17.57)$$

The answer to this problem is, as before, supplied by the work of Markoff (1916) and is given by

$$Q_k(\gamma) = \frac{T_k(\gamma)}{T_k(z)}, \qquad (2.17.58)$$

where

$$T_k(u) = \begin{cases} \cos(k.\cos^{-1}u), & |u| \leq 1, \\ \cosh(k\cosh^{-1}u), & |u| > 1. \end{cases} \tag{2.17.59}$$

Moreover,

$$\max_{-1 \leq \gamma \leq 1} |Q_k(\gamma)| = \frac{1}{T_k(z)} = \frac{1}{T_k[(2-(b+a))/(b-a)]}, \tag{2.17.60}$$

therefore we have

$$P_k(\mu) = Q_k\left[\frac{2\mu-(b+a)}{b-a}\right]$$

$$= \frac{T_k[(2\mu-(b+a))/(b-a)]}{T_k[(2-(b+a))/(b-a)]}. \tag{2.17.61}$$

Now, since $z > 1$ and $T_k(z) > 1$, then the method is convergent even though the basic method (2.17.43) may not converge.

Since $T_0(x) = 1$, $T_1(x) = x$ and $T_{k+1}(x) = 2xT_k(x) - T_{k-1}(x)$ for $k \geq 1$. So, $T_2(x) = 2x^2 - 1$, and we have

$$P_0(\mu) = 1, \tag{2.17.62}$$

and $\alpha_{0,0} = 1$,

$$P_1(\mu) = \frac{2\mu-(b+a)}{2-(b+a)}, \tag{2.17.63}$$

and

$$\alpha_{1,0} = -\frac{b+a}{2-(b+a)}, \quad \alpha_{1,1} = \frac{2}{2-(b+a)};$$

$$P_2(\mu) = \frac{[2((2\mu-(b+a))/(b-a))^2 - 1]}{[2((2-(b+a))/(b-a))^2 - 1]}$$

$$= \alpha_{2,2}\mu^2 + \alpha_{2,1}\mu + \alpha_{2,0}, \tag{2.17.64}$$

and

$$\alpha_{2,0} = \frac{(a+b)^2 + 4ab}{(a+b)^2 + 8(1-a-b) + 4ab},$$

$$\alpha_{2,1} = \frac{-8(a+b)}{(a+b)^2 + 8(1-a-b) + 4ab},$$

$$\alpha_{2,2} = \frac{8}{(a+b)^2 + 8(1-a-b) + 4ab}.$$

The $\{\alpha_{k,\ell}\}$ can be determined in this way for any value of k. However, it is more convenient to develop a relation between $\underline{y}^{(k+1)}$, $\underline{y}^{(k)}$, and $\underline{y}^{(k-1)}$ and determine $\underline{y}^{(k)}$

directly without findng $\underline{u}^{(k)}$ first (see Young, 1971). We state the final result

$$\underline{y}^{(k+1)} = 2\left[\frac{2}{b-a}G - \left(\frac{b+a}{b-a}\right)\right]\frac{T_k(z)}{T_{k+1}(z)}\underline{y}^{(k)} - \left(\frac{T_{k-1}(z)}{T_{k+1}(z)}\right)\underline{y}^{(k-1)} + \frac{4}{b-a}\frac{T_k(z)}{T_{k+1}(z)}\underline{r},$$

(2.17.65)

for $k \geq 1$.

It can be seen that the application of a semi-iterative method is similar to SOR with a variable relaxation factor. The additional storage of a vector ($\underline{y}^{(k)}$ and $\underline{y}^{(k-1)}$) could be an important consideration limiting the utility of this method in the solution of large problems.

2.18. Block Iterative Methods

In our previous discussion of iterative methods for solving a system of linear equations

$$\sum_{j=1}^{m} a_{i,j}u_j = b_i \quad (i = 1, 2, ..., m),$$

(2.18.1)

or

$$A\underline{u} = \underline{b},$$

we dealt with point iterative methods, that is, at each step the approximate solution is modified at a single point of the domain. An extension of these methods are *the block (group) iterative methods* in which several unknowns are connected together in the iteration formula in such a way that a linear system must be solved before any one of them can be determined.

In the system (2.18.1), assume that the equations and the unknowns u_i are partitioned into ℓ groups such that u_i, $i = 1, 2, ..., m_1$, constitute the first group; $i = m_1 + 1$, $m_1 + 2, ..., m_2$, constitute the second group and in general, u_i, $m_{s-1} < i \leq m_s$, constitute the sth group and $m_\ell = m$. The following definition is due to Young (1971).

Definition 2.18.1.
An ordered grouping π of $W = \{1, 2, ..., m\}$ is a subdivision of W into disjoint subsets $G_1, G_2, ..., G_\ell$ such that $G_1 \cup G_2 \cup \cdots \cup G_\ell = W$. Two ordered groupings π and π' defined by $G_1, G_2, ..., G_\ell$ and $G'_1, G'_2, ..., G'_\ell$, respectively are identical of $\ell = \ell'$ and if $G_1 = G'_1, G_2 = G'_2, ..., G\ell = G'_\ell$.

Evidently, this partitioning π imposes a partitioning of the matrix A into blocks (groups) of the form

$$A = \begin{bmatrix} A_{1,1} & A_{1,2}\text{-----}A_{1,\ell} \\ A_{2,1} & A_{2,2}\text{-----}A_{2,\ell} \\ \vdots & \vdots \qquad \vdots \\ A_{\ell,1} & A_{\ell,2}\text{-----}A_{\ell,\ell} \end{bmatrix},$$

(2.18.2)

where the diagonal blocks $A_{i,i}$, $1 \le i \le \ell$ are square, non-singular matrices. From this partitioning of the matrix A, we define the matrices

$$D = \begin{bmatrix} A_{1,1} & & & \\ & A_{2,2} & & 0 \\ & & \ddots & \\ 0 & & & A_{\ell,\ell} \end{bmatrix}, \quad E = -\begin{bmatrix} 0 & & & \\ A_{2,1} & 0 & & 0 \\ \vdots & & \ddots & \\ A_{\ell,1} & \cdots\cdots & A_{\ell,\ell-1} & 0 \end{bmatrix},$$

and

$$F = -\begin{bmatrix} 0 & A_{1,2} & \cdots\cdots & A_{1,\ell} \\ & 0 & & \vdots \\ & & \ddots & A_{\ell-1,\ell} \\ 0 & & & 0 \end{bmatrix}, \tag{2.18.3}$$

where D is a block diagonal matrix and the matrices E and F are strictly lower and upper block triangular matrices, respectively, and

$$A = D - E - F. \tag{2.18.4}$$

Now, for the column vector $\underline{u}$ in equation (2.18.1) we define column vectors $\underline{U}_1, \underline{U}_2, \ldots, \underline{U}_\ell$ where $\underline{U}_s$ is formed from $\underline{u}$ by deleting all elements of $\underline{u}$ except those corresponding to group s. Similarly, define column vectors $\underline{B}_1, \underline{B}_2, \ldots, \underline{B}_\ell$ for the given vector $\underline{b}$, we also define the submatrices $\underline{A}_{i,j}$, for $i, j = 1, 2, \ldots, \ell$, such that each $A_{i,j}$ are formed from the matrix A by deleting all rows except those corresponding to G_i and all columns except those corresponding to G_j. The system (2.18.1) evidently can be written now in the equivalent form

$$\sum_{j=1}^{\ell} A_{i,j} \underline{U}_j = \underline{B}_i, \quad i = 1, 2, \ldots, \ell. \tag{2.18.5}$$

We shall assume from now on that all submatrices $A_{i,i}$ are non-singular. Then, the *block Jacobi iterative method* is defined by

$$A_{i,i} \underline{U}_i^{(k+1)} = - \sum_{\substack{j=1 \\ j \ne i}}^{\ell} A_{i,j} \underline{U}_j^{(k)} + B_i, \quad 1 \le i \le \ell, \tag{2.18.6}$$

or equivalently,

$$\underline{U}_i^{(k+1)} = - \sum_{\substack{j=1 \\ j \ne i}}^{\ell} B_{i,j} \underline{U}_j^{(k)} + \underline{V}_i, \quad 1 \le i \le \ell, \tag{2.18.7}$$

where

$$B_{i,j} = \begin{cases} -A_{i,i}^{-1} A_{i,j}, & \text{if } i \ne j, \\ 0, & \text{if } i = j \end{cases} \tag{2.18.8}$$

and

$$\underline{V}_i = A_{i,i}^{-1} \underline{B}_i, \quad 1 \le i \le \ell.$$

We may write (2.18.7) in the matrix form

$$\underline{u}^{(k+1)} = B^{(\pi)}\underline{u}^{(k)} + \underline{v}^{(\pi)}, \qquad (2.18.9)$$

where

$$B^{(\pi)} = (D^{(\pi)})^{-1}(E^{(\pi)} + F^{(\pi)}), \quad \text{``block Jacobi matrix''}, \qquad (2.18.10a)$$

and

$$\underline{v}^{(\pi)} = (D^{(\pi)})^{-1}\underline{b}, \qquad (2.18.10b)$$

$$D^{(\pi)} = \text{diag}_{(\pi)} A,$$

$E^{(\pi)}$ and $F^{(\pi)}$ are again strictly lower and upper triangular matrices.

For the *block Gauss–Seidel iterative method* we have

$$A_{i,i}\underline{U}_i^{(k+1)} = -\sum_{j=1}^{i-1} A_{i,j}\underline{U}_j^{(k+1)} - \sum_{j=i+1}^{\ell} A_{i,j}\underline{U}_j^{(k)} + \underline{B}_i, \quad 1 \le i \le \ell, \qquad (2.18.11)$$

or equivalently,

$$\underline{U}_i^{(k+1)} = -\sum_{j=1}^{i-1} B_{i,j}\underline{U}_j^{(k+1)} + \sum_{j=i+1}^{\ell} B_{i,j}\underline{U}_j^{(k)} + \underline{V}_i, \quad 1 \le i \le \ell, \qquad (2.18.12)$$

where $B_{i,j}$ and $\underline{V}_i$ are as given in (2.18.8). This can also be written in the matrix form

$$\underline{U}^{(k+1)} = \mathscr{L}^{(\pi)}\underline{U}^{(k)} + (D^{(\pi)} - E^{(\pi)})^{-1}B, \qquad (2.18.13)$$

where

$$\mathscr{L}^{(\pi)} = (D^{(\pi)} - E^{(\pi)})^{-1}F^{(\pi)}, \quad \text{``block GS matrix''}. \qquad (2.18.14)$$

The matrix form of the block JOR (BJOR) and the block SOR (BSOR) iterative methods are given by (2.18.15) and (2.18.16), respectively,

$$\underline{U}^{(k+1)} + B_\omega^{(\pi)}\underline{U}^{(k)} + \underline{V}^{(\pi)}, \qquad (2.18.15)$$

$$\underline{U}^{(k+1)} = \mathscr{L}_\omega^{(\pi)}\underline{U}^{(k)} + \omega(I - \omega L^{(\pi)})^{-1}\underline{V}^{(\pi)}, \qquad (2.18.16)$$

where

$$B_\omega^{(\pi)} = \omega B^{(\pi)} + (1 - \omega)I, \qquad (2.18.17a)$$

$$\underline{V}^{(\pi)} = \omega(D^{(\pi)})^{-1}\underline{B}, \qquad (2.18.17b)$$

$$\mathscr{L}_\omega^{(\pi)} = (I - \omega L^{(\pi)})^{-1}(\omega R^{(\pi)} + (1 - \omega)I), \qquad (2.18.18a)$$

$$L^{(\pi)} = (D^{(\pi)})^{-1}E^{(\pi)}, \qquad (2.18.18b)$$

and

$$R^{(\pi)} = (D^{(\pi)})^{-1}F^{(\pi)}. \qquad (2.18.18c)$$

$B_\omega^{(\pi)}$ and $\mathscr{L}_\omega^{(\pi)}$ are the BJOR and BSOR iteration matrices, respectively.

For the convergence of the methods and subsequent analysis we follow the generalisations of Young's definition of Property A, consistently ordered matrices, and ordering vectors. Some theorems concerning the convergence criteria are presented.

First, we define the $(\ell \times \ell)$ matrix $Z = (z_{i,j})$ by

$$z_{i,j} = 1 \quad \text{if } A_{i,j} \ne 0$$

and

$$z_{i,j} = 0 \quad \text{if } A_{i,j} = 0, \tag{2.18.19}$$

where the matrix A and an ordered grouping π, with ℓ groups, are given.

Definition 2.18.2.
The matrix has Property $A^{(\pi)}$ if Z has Property A.

Definition 2.18.3.
The matrix A is a π-consistently ordered matrix if Z is consistently ordered.

Arms, Gates and Zondek (1956) gave the following definition of Property $A^{(\pi)}$ as a generalisation of Young's definition (1954).

Definition 2.18.4.
The matrix A has Property $A^{(\pi)}$ for a given partition π if there exist two disjoint subsets S_1 and S_2 of W, the set of the first ℓ integers, such that $S_1 \cup S_2 = W$, and if $A_{i,j} \neq 0$, then either $i = 1$, or $i \in S_1$ and $j \in S_2$ or $i \in S_2$ and $j \in S_1$.

As with Property $A^{(\pi)}$, the following definition is a generalisation of Young's definition of an "ordering vector".

Definition 2.18.5.
An ordering ℓ-tuple for A will be an ℓ-tuple $\underline{v}^{(\pi)} = (v_1^\pi, v_2^\pi, \ldots, v_\ell^\pi)$, where each v_s^π is an integer, such that, if $A_{i,j} \neq 0$, and $i \neq j$, then $|v_i^\pi - v_j^\pi| = 1$.

Theorem 2.18.1.
A matrix A has Property $A^{(\pi)}$ if and only if there exists an ordering ℓ-tuple for A.

The proof is given in Young (1971).

Theorem 2.18.2.
If A is symmetric matrix and $D^{(\pi)}$ is positive definite, then $\rho(\mathcal{L}_\omega^{(\pi)}) < 1$ if and only if A is positive definite and $0 < \omega < 2$.

Theorem 2.18.3.
If A has Property $A^{(\pi)}$ and is consistently ordered, with $0 < \omega < 2$, and if λ is a non-zero eigenvalue of $\mathcal{L}_\omega^{(\pi)}$, and if μ satisfies

$$(\lambda + \omega - 1)^2 = \lambda \omega^2 \mu^2, \tag{2.18.20}$$

then μ is an eigenvalue of $B^{(\pi)}$. Conversely if μ is an eigenvalue of $B^{(\pi)}$ and if λ satisfies (2.12.20), then λ is an eigenvalue of $\mathcal{L}_\omega^{(\pi)}$.

Equation (2.18.20) leads to an explicit formula for the optimal relaxation factor ω_b in terms of $\bar{\mu}$ (the spectral radius of $(B^{(\pi)})$ where the matrix A is symmetric, positive definite and has Property $A^{(\pi)}$. The value of ω_b is optimal in the sense that the spectral radius $\bar{\lambda}$ of $\mathcal{L}_\omega^{(\pi)}$ is minimal so that the convergence rate is greatest. The relations are given by

$$\omega_b = \frac{2}{1 + \sqrt{1 - \bar{\mu}^2}} \tag{2.18.21}$$

and

$$\bar{\lambda} = \rho(\mathcal{L}_\omega^{(\pi)}) = \omega_b - 1. \tag{2.18.22}$$

Further we have

$$\rho(\mathscr{L}_{\omega_b}^{(\pi)}) < \rho(\mathscr{L}_{\omega}^{(\pi)}), \quad \omega \neq \omega_b, \tag{2.18.23}$$

and, asymptotically, as $\bar{\mu} \to 1$, we obtain the relation

$$R(\mathscr{L}_{\omega_b}^{(\pi)}) \approx 2\sqrt{R(\mathscr{L}^{(\pi)})} = 2\sqrt{2R(B^{(\pi)})}. \tag{2.18.24}$$

To illustrate the foregoing, we consider blocks each consisting of all points on a column (or row), or on two columns (or rows) on our model problem. Methods based on the use of such blocks are called *line iterative methods* and *two-line iterative methods*, respectively, they are special cases of a wider class of methods known as the k-line iterative methods. But we shall only consider the first two types. For convenience, let us assume that our model problem has an even number of columns, so that they are an exact number of two-line blocks. We have ℓ line blocks and $\ell/2$ two-line blocks.

The Line Iterative Method
In this method the points are ordered as in Figure 2.11.2 and we consider all the mesh points of a particular line as a block, then for the five-point difference approximation, the resulting coefficient matrix A is block tridiagonal with each diagonal submatrix as a tridiagonal matrix. Hence

$$A = \begin{bmatrix} \beta_1 & \gamma_1 & & & & \\ \alpha_2 & \beta_2 & \gamma_2 & & 0 & \\ & \ddots & \ddots & \ddots & & \\ & & & & \gamma_{\ell-1} & \\ 0 & & & & \alpha_\ell & \beta_\ell \end{bmatrix}, \tag{2.18.25}$$

where

$$\beta_r = \begin{bmatrix} 4 & -1 & & & \\ -1 & 4 & -1 & & 0 \\ & \ddots & \ddots & \ddots & \\ & & & & -1 \\ 0 & & & -1 & 4 \end{bmatrix}, \quad 1 \le r \le \ell$$

and

$$\alpha_{r+1} = \gamma_r = \begin{bmatrix} -1 & & & \\ & -1 & & 0 \\ & & \ddots & \\ 0 & & & -1 \end{bmatrix}, \quad 1 \le r \le \ell-1. \tag{2.18.26}$$

The system (2.18.25) can be written in this case as

$$
\begin{bmatrix}
\beta_1 & \gamma_1 & & & \\
\alpha_2 & \beta_2 & \gamma_2 & & 0 \\
& & \ddots & \ddots & \\
& & & & \gamma_{\ell-1} \\
0 & & & \alpha_\ell & \beta_\ell
\end{bmatrix}
\begin{bmatrix}
\underline{U}_1 \\
\underline{U}_2 \\
\vdots \\
\vdots \\
\underline{U}_\ell
\end{bmatrix}
=
\begin{bmatrix}
\underline{C}_1 \\
\underline{C}_2 \\
\vdots \\
\vdots \\
\underline{C}_\ell
\end{bmatrix}.
\tag{2.18.27}
$$

It is clear that we must solve sub-systems of the form

$$
\alpha_r \underline{U}_{r-1} + \beta_r \underline{U}_r + \gamma_r \underline{U}_{r+1} = \underline{C}_r, \quad 1 \le r \le m,
\tag{2.18.28}
$$

for $\beta_r \underline{U}_r$, i.e. solve

$$
\beta_r \underline{U}_r = \underline{t}_r \equiv (\underline{C}_r - \alpha_r \underline{U}_{r-1} - \gamma_r \underline{U}_{r+1}),
\tag{2.18.29}
$$

for $\underline{U}_r$, with $\alpha_r, \beta_r, \gamma_r, 1 \le r \le \ell$, as in (2.18.26) with $\alpha_1 = \gamma_\ell = 0$ and $\underline{U}_r$ is the column vector of values on the rth line. The solution of equation (2.18.28) by the block SOR iterative method is obtained by solving

$$
\alpha_r \underline{U}_{r-1}^{(k+1)} + \beta_r \underline{\hat{U}}_r^{(k+1)} + \gamma_r \underline{U}_{r+1}^{(k)} = \underline{C}_r,
\tag{2.18.30}
$$

for $\beta_r \underline{\hat{U}}_r^{(k+1)}$, i.e.,

$$
\beta_r \underline{\hat{U}}_r^{(k+1)} = \underline{t}_r \equiv (\underline{C}_r - \alpha_r \underline{U}_{r-1}^{(k+1)} - \gamma_r \underline{U}_{r+1}^{(k)}),
\tag{2.18.31}
$$

for $\underline{\hat{U}}_r^{(k+1)}$, and extrapolating

$$
\underline{U}_r^{(k+1)} = \omega(\underline{\hat{U}}_r^{(k+1)} - \underline{U}_r^{(k)}) + \underline{U}_r^{(k)}
\tag{2.18.32}
$$

to obtain the final solution. This particular block iterative method is often called the *successive line over-relaxation method* (SLOR).

Efficient algorithms exist for the solution of equation (2.18.31), where the matrices β_r have the general form

$$
\beta_r =
\begin{bmatrix}
b_1 & c_1 & & & \\
a_2 & b_2 & c_2 & & 0 \\
& & \ddots & \ddots & \\
& & & & c_{m-1} \\
0 & & & a_m & b_m
\end{bmatrix}, \quad 1 \le r \le \ell
\tag{2.18.33}
$$

A well known algorithm for the problem (based on Gaussian elimination) is as follows:

Compute the initial values

$$
c_1^* = \frac{c_1}{b_1},
$$

$$
t_1^* = \frac{t_1}{b_1}, \quad b_1 \ne 0.
$$

Compute recursively,

$$c_i^* = \frac{c_i}{b_i - a_i c_{i-1}^*}, \quad i = 2, 3, \ldots, m,$$

$$t_i^* = \frac{t_i - a_i t_{i-1}^*}{b_i - a_i c_{i-1}^*}, \quad i = 2, 3, \ldots, m, \tag{2.18.34a}$$

which transforms the matrix equation to

$$\begin{bmatrix} 1 & c_1^* & & & & \\ & 1 & c_2^* & & 0 & \\ & & \ddots & \ddots & & \\ & 0 & & 1 & c_{m-1}^* \\ & & & & 1 \end{bmatrix} \begin{bmatrix} u_1 \\ u_2 \\ \vdots \\ \vdots \\ u_m \end{bmatrix} = \begin{bmatrix} t_1^* \\ t_2^* \\ \vdots \\ \vdots \\ t_m^* \end{bmatrix}. \tag{2.18.34b}$$

The components u_i of the solution vector $\underline{u}$ can then be computed by the back substitution process

$$u_m = t_m^*,$$

$$u_i = t_i^* - c_i^* u_{i+1}, \quad i = m-1, \, m-2, \ldots, 1. \tag{2.18.34c}$$

The Two-line Iterative Method
In this method we consider the block as consisting of the mesh points lying on two grid lines and the points may be re-ordered among the two lines as shown in Figure 2.18.1.

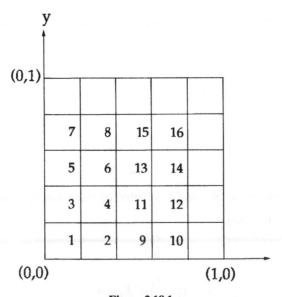

Figure 2.18.1

Then, for the five-point difference approximation, the resulting coefficient matrix A has a block structure similar to that shown in (2.18.25) but we will have $\ell/2$ blocks rather than ℓ blocks, and each block submatrix β_r, $1 \leq r \leq \ell/2$, is a quindiagonal matrix of the form, i.e.,

$$\beta_r = \begin{bmatrix} 4 & -1 & -1 & & & & & \\ -1 & 4 & 0 & -1 & & & & \\ -1 & 0 & 4 & -1 & -1 & & & \\ & -1 & -1 & 4 & 0 & -1 & & \\ & & & & & & & \\ & & & & -1 & -1 & 4 & 0 & -1 \\ & & & & & -1 & 0 & 4 & -1 \\ & & & & & & -1 & -1 & 4 \end{bmatrix}, \quad 1 \leq r \leq \ell/2,$$

$$\alpha_r = \begin{bmatrix} 0 & -1 & & & \\ & 0 & 0 & & \\ & & & -1 & \\ & & & & \\ & & & & \\ & & & 0 & -1 \\ & & & & 0 \end{bmatrix}, \quad 2 \leq r \leq \ell/2$$

and

$$\gamma_r = \begin{bmatrix} 0 & & & \\ -1 & 0 & & \\ & 0 & & \\ & & -1 & \\ & & & \\ & & & 0 & 0 \\ & & & -1 & 0 \end{bmatrix}, \quad 1 \leq r \leq \ell/2 - 1. \quad (2.18.35)$$

Similar to the line iterative method we can write (2.18.5) in the form of (2.18.27) with ℓ replaced by $\ell/2$ and by following the same steps, equation (2.18.31) can be solved for $\hat{U}_r^{(k+1)}, 1 \le r \le \ell/2$, with β_r a quindiagonal matrix, then by extrapolating we obtain the final solution. This iterative method is known as the *successive two-line overrelaxation method* (S.2.LOR).

For solving equation (2.18.31), where the matrices β_r have the general form

$$\beta_r = \begin{bmatrix} c_1 & d_1 & e_1 \\ b_2 & c_2 & d_2 & e_3 \\ a_3 & b_3 & c_3 & d_3 & & e_3 \\ & a_4 & b_4 & c_4 & & d_4 & & e_4 \\ & & & & & & & & & 0 \\ & & & & & & & & & \\ & 0 & & a_{m-2} & b_{m-2} & c_{m-2} & d_{m-2} & e_{m-2} \\ & & & & a_{m-1} & b_{m-1} & c_{m-1} & d_{m-1} \\ & & & & & a_m & b_m & c_m \end{bmatrix}, \quad 1 \le r \le \ell/2, \quad (2.18.36)$$

we can use the following algorithm.

Compute the initial values

$$d_1^* = \frac{d_1}{c_1}, \quad d_m = 0, \quad b_1 = 0 \quad \text{and} \quad c_1 \ne 0,$$

$$e_1^* = \frac{e_1}{d_1}, \quad e_m = e_{m+1} = 0,$$

$$c_2^* = c_2 - b_2 d_1^*,$$

$$d_2^* = \frac{d_2 - b_2 e_1^*}{c_2^*},$$

$$e_2^* = \frac{e_2}{c_2^*}, \quad c_2^* \ne 0, \quad\quad\quad\quad\quad\quad (2.18.37a)$$

then compute recursively,

$$b_i^* = b_i - a_i d_{i-1}^*,$$

$$c_i^* = c_i - a_i e_{i-2}^* - b_i^* d_{i-1}^*, \quad i = 3, 4, \ldots, m,$$

$$d_i^* = \frac{d_i - b_i^* e_{i-1}^*}{c_i^*}$$

and

$$e_i^* = \frac{e_i}{c_i^*}, \quad c_i^* \ne 0, \quad i = 3, 4, \ldots, m. \quad\quad\quad (2.18.37b)$$

Compute

$$t_1^* = \frac{t_i}{c_i},$$

$$t_2^* = \frac{t_2 - b_2 t_1^*}{c_2^*},$$

$$t_i^* = \frac{t_i - a_i t_{i-2}^* - b_i t_{i-1}^*}{c_i^*}, \quad i = 3, 4, \ldots, m. \tag{2.18.37c}$$

The components u_i of the solution vector $\underline{u}$ can then be computed by the back substitution

$$u_m = t_m^*,$$

$$u_{m-1} = t_{m-1}^* - d_{m-1}^* u_m,$$

$$u_i = t_1^* - d_i^* u_{i+1} - e_i^* u_{i+2}, \quad i = m-2, \ m-3, \ldots, 1. \tag{2.18.37d}$$

It can be shown that by going from point to block iterative methods we can obtain improved rates of convergence. For the SLOR method, Cuthill and Varga (1959) have shown that it is possible to normalise the equation in such a way that the SOR and SLOR both require exactly the same arithmetic computations per mesh point. Also, it has been shown that this normalised method is stable with respect to the growth of rounding errors. Varga (1962), shows for the model problem

$$\rho(B^{(line)}) = \frac{\cos \pi h}{2 - \cos \pi h}. \tag{2.18.38}$$

From this we can easily calculate that

$$R(B^{(line)}) \approx \pi^2 h^2$$

and

$$R(\mathscr{L}^{(line)}) \approx 2\pi^2 h^2, \quad h \to 0. \tag{2.18.39}$$

Hence by comparing these results with (2.16.13), we note that the line Jacobi method converges at the same rate as the point GS method.

Consequently, for the SLOR and S2LOR methods we have

$$R(\mathscr{L}_{\omega_b}^{(line)}) \approx 2\sqrt{2\pi h}, \quad h \to 0, \tag{2.18.40}$$

$$R(\mathscr{L}_{\omega_b}^{(line)}) \approx 4\pi h, \quad h \to 0. \tag{2.18.41}$$

Hence

$$\frac{R(\mathscr{L}_{\omega_b}^{(line)})}{R(\mathscr{L}_{\omega_b})} \approx \frac{2\sqrt{2\pi h}}{2\pi h} \approx \sqrt{2} \tag{2.18.42a}$$

and

$$\frac{R(\mathscr{L}_{\omega_b}^{(2\ line)})}{R(\mathscr{L}_{\omega_b})} \approx \frac{4\pi h}{2\sqrt{2\pi h}} \approx \sqrt{2}, \tag{2.18.42b}$$

which shows that the rate of convergence of the S.2.LOR method is approximately twice that of the SOR method. Parter (1961) showed that the k-line SOR method with optimum ω converges approximately $\sqrt{2k}$ as fast as the SOR method. However, one must solve more complicated systems of equations of wider bandwidth at each iteration. Parter (1986) gave a formula for evaluating the spectral radius p of the k-line Jacobi scheme which is given by

$$\rho(B^{(k\ line)}) \approx 1 - k\pi^2 h^2. \tag{2.18.43}$$

Alternative block forms have been proposed by Evans and Benson (1971), i.e., the peripheral block and Evans and Yousif (1984), i.e., the 4 point explicit block.

2.19. Alternating Direction Implicit Methods

Alternating Direction Implicit (ADI) methods are somewhat similar to the line Jacobi iterative methods but with alternating directions. There are two basic ADI methods due to Peaceman and Rachford (1955) and Douglas and Rachford (1956). In this section we will consider the former, which is known as the Peaceman–Rachford (PR) method for solving the system

$$A\underline{u} = \underline{b}, \tag{2.19.1}$$

where the $(m \times m)$ matrix A is non-singular and can be represented as the sum of three $(m \times m)$ matrices

$$A = H + V + \Sigma. \tag{2.19.2}$$

We can make the following assertions about the matrix A:

   (a) A is an irreducible Stieltjes matrix, i.e., A is a real, symmetric, and positive definite irreducible matrix with non-positive off-diagonal entries.
   (b) H and V are real, symmetric, diagonally dominant matrices with positive diagonal entries and non-positive off-diagonal entries.
   (c) Σ is a non-negative diagonal matrix.

In a typical situation H and V would be tridiagonal or could be made so by a permutation of the rows and corresponding columns.

By using (2.19.2) we can write the matrix equation (2.19.1) as a pair of matrix equations

$$(H + (1/2)\Sigma + rI)\underline{u} = \underline{b} - (V + (1/2)\Sigma - rI)\underline{u},$$

$$(V + (1/2)\Sigma + rI)\underline{u} = \underline{b} - (H + (1/2)\Sigma - rI)\underline{u}, \tag{2.19.3}$$

for any positive scalar r. If we let

$$H_1 = H + (1/2)\Sigma, \quad V_1 = V + 1/2\Sigma, \tag{2.19.4}$$

then the *Peaceman–Rachford Alternating Direction Implicit Method* is defined by

$$(H_1 + r_{k+1}I)\underline{u}^{(k+1/2)} = \underline{b} - (V_1 - r_{k+1}I)\underline{u}^{(k)},$$

$$(V_1 + r_{k+1}I)\underline{u}^{(k+1)} = \underline{b} - (H_1 - r_{k+1}I)\underline{u}^{(k+1/2)}, \tag{2.19.5}$$

where the r_k's are positive *acceleration parameters* chosen to make the process converge rapidly, and $\underline{u}^{(0)}$ is an arbitrary initial vector approximation to the unique solution of (2.19.1).

Since both H_1 and V_1 are, by suitable rearrangement of their rows and corresponding columns, tridiagonal matrices, the iterative method (2.19.5) can be carried out directly using the algorithm given in section 2.18. The vector $\underline{u}^{(k+1/2)}$ is treated as an auxiliary vector which is discarded as soon as it has been used in the calculation of $\underline{u}^{(k+1)}$.

The two equations of (2.19.5) are now combined to give

$$\underline{u}^{(k+1)} = T_{r_{k+1}} \underline{u}^{(k)} + \underline{g}_{r_{k+1}}, \quad k \geq 0, \tag{2.19.6}$$

where

$$T_r = (V_1 + rI)^{-1}(H_1 - rI)(H_1 + rI)^{-1}(V_1 - rI), \tag{2.19.7a}$$

$$g_r = (V_1 + rI)^{-1} \{I - (H_1 - rI)(H_1 + rI)^{-1}\} \underline{b}. \tag{2.19.7b}$$

It will be noticed that (2.19.6) is of the same form as (2.12.3).

We consider the case, where all the constants r_j are equal to the fixed constant $r > 0$. The following convergence theorem will be stated here without proof (see Varga (1962)).

Theorem 2.19.1.
Let H_1 and V_1 be $m \times m$ symmetric non-negative definite matrices, where at least one of the matrices H_1 and V_1 is positive definite. Then, for any $r > 0$, the PR matrix T_r of (2.19.7a) is convergent.

For the Laplace model problem with $h = 1/(m+1)$, if the vector $\underline{\alpha}^{(p,q)}$ is defined so that its component for the ith column and j^{th} row of the mesh is

$$\alpha_{i,j}^{(p,q)} = \beta_{p,q} \sin\left(\frac{p\pi i}{m+1}\right) \cdot \sin\left(\frac{q\pi i}{m+1}\right), \quad 1 \leq i,j \leq m, \quad 1 \leq p,q \leq m, \tag{2.19.8}$$

it follows that

$$H\underline{\alpha}^{(p,q)} = 4\sin^2\left(\frac{p\pi}{2(m+1)}\right)\underline{\alpha}^{(p,q)}, \quad \text{for all } 1 \leq p, 1 \leq m, \tag{2.19.9}$$

$$V\underline{\alpha}^{(p,q)} = 4\sin^2\left(\frac{q\pi}{2(m+1)}\right)\underline{\alpha}^{(p,q)}.$$

Since for this problem $\Sigma = 0$, hence from (2.19.9) and (2.19.7a) we have

$$T_r\underline{\alpha}^{(p,q)} = \left[\frac{r - 4\sin^2(q\pi/2(m+1))}{r + 4\sin^2(q\pi/2(m+1))}\right]\left[\frac{r - 4\sin^2(p\pi/2(m+1))}{r + 4\sin^2(p\pi/2(m+1))}\right]\underline{\alpha}^{(p,q)}, \quad 1 \leq p,q \leq m. \tag{2.19.10}$$

We therefore conclude that

$$\rho(T_r) = \left\{ \max_{1 \leq q \leq m} \left|\frac{r - 4\sin^2(q\pi/2(m+1))}{r + 4\sin^2(q\pi/2(m+1))}\right| \right\}^2. \tag{2.19.11}$$

To minimise this as a function of r, we consider the simple function

$$g(x; r) = \frac{r - x}{r + x}, \quad r > 0, \tag{2.19.12}$$

where $0 < x_1 \le x \le x_2$. The derivative of $g(x; r)$ with respect to x is negative for all $x \ge 0$, so that

$$\max_{x_1 \le x \le x_2} |g(x; r)| = \max \left\{ \left| \frac{r - x_1}{r + x_1} \right|, \left| \frac{r - x_2}{r + x_2} \right| \right\}. \tag{2.19.13}$$

From this, it is easily shown that

$$\max_{x_1 \le x \le x_2} |g(x; r)| = \begin{cases} \dfrac{x_2 - r}{x_2 + r}, & 0 < r \le \sqrt{x_1 x_2}, \\[2mm] \dfrac{r - x_1}{r + x_1}, & r \ge \sqrt{x_1 x_2}, \end{cases} \tag{2.19.14}$$

hence

$$\min_{r > 0} \left\{ \max_{x_1 \le x \le x_2} |g(x; r)| \right\} = g(x_1; \sqrt{x_1 x_2}) = \frac{\sqrt{x_2} - \sqrt{x_1}}{\sqrt{x_2} + \sqrt{x_1}}. \tag{2.19.15}$$

Thus, by setting $x_1 = 4 \sin^2 (\pi/2(m + 1))$ and $x_2 = 4 \cos^2 (\pi/2(m + 1))$, we have the value $r = \sqrt{x_1 x_2}$ is optimum in the sense that the spectral radius $\rho(T_r)$ is minimised. It can further be shown that the spectral radius (and consequently, the rate of convergence) of this method is asymptotically the same as that of the point SOR with optimum relaxation parameter for all $h > 0$ for the model problem.

A substantial improvement in the convergence of the Peaceman–Rachford method for solving Laplace's equation in a square can be obtained by the use of a sequence of iteration parameters r_{k+1} (see equation (2.19.5)). In this case the parameters are used successively in a cyclic order.

At least two methods have been given for choosing the iteration parameters, which though not optimal, nevertheless are nearly as good as the optimal parameters in many cases. The first set of parameters given by Peaceman and Rachford (1955) are

$$r_i^{(p)} = \bar{b}(\bar{a}/\bar{b})^{(2i-1)/2n}, \quad i = 1, 2, \ldots, n, \tag{2.19.16}$$

and the second set are the Wachspress parameters (Wachspress, 1957), which are given by

$$r_i^{(w)} = \bar{b}(\bar{a}/\bar{b})^{(i-1)/(n-1)}, \quad n \ge 2, \quad i = 1, 2, \ldots, n, \tag{2.19.17}$$

where, $\bar{a} = \min(a, \alpha)$, $\bar{b} = \max(b, \beta)$, with a and b, α and β being the least and greatest eigenvalues of H_1 and V_1 of equations (2.19.5), respectively.

2.20. Hyperbolic Equations–Finite Difference Methods

It has been shown that finite difference methods have been widely used with great success in parabolic and elliptic problems. However, use of these methods for hyperbolic equations are in a sense quite restricted.

Alternative methods segment the solution domain by the use of families of character-istics from which efficient methods of solution can be derived. In other words, these characteristics play the role of natural boundaries in the solution domain. We also note that the existence of discontinuities in the initial values and initial derivatives leads to discontinuities across the characteristics along its entire length from the point of discontinuity.

Finite difference methods, on the other hand, do not accommodate these possibilities. The central feature of these methods is the replacement or approximation of derivatives at a grid point of the finite difference network, say the rectangular (square) grid by some difference quotients over a small interval. This process of approximation is extended to all other grid points in the network over the area of integration of interest without taking into consideration such features as the role of the character-istics. Further difficulties arise if these characteristics are not straight lines or are dependent on the solution.

We shall attempt to derive a *generalised two-time level finite difference approximation* to the first order hyperbolic or *convection equation* in one space dimension of the form

$$-a\frac{\partial U}{\partial x} = \frac{\partial U}{\partial t}, \tag{2.20.1}$$

where $a > 0$ and is constant. We shall also examine the stability and truncation errors of these approximations.

The space derivative $\partial U/\partial x$ may be approximated by

$$\frac{\partial U}{\partial x} \approx \frac{(u_{i+1/2} - u_{i-1/2})}{\Delta x}. \tag{2.20.2}$$

$u_{i+1/2}$ may be chosen in a number of ways. The simplest option is the arithmetic averaging given by

$$u_{i+1/2} = (u_i + u_{i+1})/2. \tag{2.20.3}$$

Another option which is more commonly used is

$$u_{i+1/2} = u_i. \tag{2.20.4}$$

A "distance-weighting" parameter w can therefore be used to characterise these options as follows:

backward-in-distance weighting: $w = 1$,
centred-in-distance weighting: $w = 1/2$,
forward-in-distance weighting: $w = 0$,

and clearly equations (2.20.3) and (2.20.4) can be combined into the general equation

$$u_{i+1/2} = wu_i + (1 - w)u_{i+1}. \tag{2.20.5}$$

Similarly, we have

$$u_{i-1/2} = wu_{i-1} + (1 - w)u_i. \tag{2.20.6}$$

We can use a similar **parametric** approach to the differentials, for example, the *explicit*, *implicit* and *centred-in-time* difference equations, where the spatial derivatives are evaluated, respectively, at times t_j, t_{j+1} or $t_{j+1/2}$. If we use θ as the "time-weighting" parameter and prescribe the values:

Implicit:
(backward-in-time) $\theta = 1$,
centred-in-time: $\theta = 1/2$,

explicit:
(forward-in-time) $\theta = 0$,

then equation (2.20.1) can be replaced by the finite difference analogue

$$-\frac{a}{\Delta x}[\theta(u_{i+1/2, j+1} - u_{i-1/2, j+1}) + (1 - \theta)(u_{i+1/2, j} - u_{i-1/2, j})] = \frac{u_{i, j+1} - u_{ij}}{\Delta t}, \quad (2.20.7)$$

where we have employed the usual forward difference approximation for $\partial U / \partial t$.

By substituting equations (2.20.5) and (2.20.6) into equation (2.20.7) we arrive at the final, generalised difference analogue for the first order hyperbolic equation (2.20.1) as

$$\begin{aligned} -r[\theta\{(1-w)u_{i+1, j+1} + (2w-1)u_{i, j+1} - wu_{i-1, j+1}\} \\ + (1-\theta)\{(1-w)u_{i+1, j} + (2w-1)u_{ij} - wu_{i-1, j}\}] = u_{i, j+1} - u_{ij}, \end{aligned} \quad (2.20.8)$$

where $r = a\Delta t / \Delta x$ *is the mesh ratio.* Most of the well-known standard methods may be obtained from formula (2.20.8). For example, by putting $\theta = 0$ and $w = (1 + r)/2 \le 1$ gives us

$$u_{i, j+1} = (1/2r)(1+r)u_{i-1, j} + (1-r^2)u_{ij} - (1/2r)(1-r)u_{i+1, j}, \quad (2.20.9)$$

which is called the *Lax–Wendroff explicit formula* (Richtmyer and Morton, 1967). This is probably the most well known method for first-order hyperbolic equations and as we shall see later is second-order accurate. For $\theta = 1/2$ and $w = 1/2$, we get the implicit centre-in-distance and centred-in-time formula

$$\frac{r}{4}u_{i-1, j+1} - u_{i, j+1} - \frac{r}{4}u_{i+1, j+1} = -\frac{r}{4}u_{i-1, j} - u_{i, j} + \frac{r}{4}u_{i+1, j}, \quad (2.20.10)$$

which is also second-order accurate in both space and time. Equation (2.20.10) is of the *Crank–Nicolson type* that is frequently encountered in *parabolic* problems and can be expressed as

$$[1 + \tfrac{1}{4}(\Delta_x + \nabla_x)]u_{i, j+1} = [1 - \tfrac{1}{4}r(\Delta_x + \nabla x)]u_{ij}, \quad (2.20.11)$$

or

$$-\frac{a}{4\Delta x}(\Delta_x + \nabla_x)(u_{i, j+1} + u_{ij}) = \frac{u_{i, j+1} - u_{ij}}{\Delta t}, \quad (2.20.12)$$

where

$$\Delta_x u_{i, j} = u_{i+1, j} - u_{ij}$$

and

$$\nabla_x u_{i, j} = u_{i, j} - u_{i-1, j}.$$

Other schemes of interest are displayed in Table 2.20.1 below together with their computational molecules and stability conditions and truncation errors, both of which will be derived in the next two sections.

Table 2.20.1

θ/w	Backward-in-distance w=1	Centred-in-distance w=1/2	Forward-in-distance w=0
Backward-in-time $\theta=1$ (implicit)	Always stable. $T=0(\Delta x)+0([\Delta t]^2)$	Always stable. $T=0([\Delta x]^2)+0(\Delta t)$	Stable if $\lambda \geq 1$. $T=0(\Delta x)+0(\Delta t)$
Centred-in-time $\theta=1/2$	Always stable. $T=0(\Delta x)+0([\Delta t]^2)$	Always (neutrally) stable. $T=0([\Delta x]^2)+0([\Delta t]^2)$	Always unstable. $T=0(\Delta x)+0([\Delta t]^2)$
Forward-in-time $\theta=1$ (explicit)	Stable if $\lambda \leq 1$. $T=0(\Delta x)+0(\Delta t)$	Always unstable. $T=0([\Delta x]^2)+0(\Delta t)$	Always unstable. $T=0(\Delta x)+0(\Delta t)$

2.21. Stability Analysis of the Generalised Finite Difference Methods

An analysis using *the von Neumann criterion* will be carried out to investigate the stability of (2.20.8). We shall assume that if u_{ij} is a solution to the difference equation at the point (x_i, t_j) then its perturbation $u_{ij} + \varepsilon_{ij}$ also satisfies the difference equation and we examine the possible growth of ε_{ij}. Specifically, if u_{ij} satisfies equation (2.20.8), then

$$-r[\theta\{(1-w)(u_{i+1,j+1}+\varepsilon_{i+1,j+1})+(2w-1)(u_{i,j+1}+\varepsilon_{i,j+1})$$
$$-w(u_{i-1,j+1}+\varepsilon_{i-1,j+1})\}+(1-\theta)\{(1-w)(u_{i+1,j}+\varepsilon_{i+1,j})$$
$$+(2w-1)(u_{ij}+\varepsilon_{ij})-w(u_{i-1,j}+\varepsilon_{i-1,j})\}]$$
$$= (u_{i,j+1}+\varepsilon_{i,j+1})-u_{ij}+\varepsilon_{ij}). \qquad (2.21.1)$$

On subtracting equation (2.20.8) from (2.21.1) we get

$$-r[\theta\{(1-w)\varepsilon_{i+1,j+1}+(2w-1)\varepsilon_{i,j+1}-w\varepsilon_{i-1,j+1}\}$$
$$+(1-\theta)\{(1-w)\varepsilon_{i+1,j}+(2w-1)\varepsilon_{ij}-w\varepsilon_{i-1,j}\}]=\varepsilon_{i,j+1}-\varepsilon_{i,j}. \qquad (2.21.2)$$

We see that the error equation (2.21.2) has exactly the same form as the original difference equation (2.20.8). It will generally be true that a difference equation and its error equation will be identical when the difference equation is linear and homogeneous.

The von Neumann stability analysis consists of expanding the error ε_{ij} in a Fourier series of the form

$$\varepsilon_{ij} = \sum_{\beta} \xi_{\beta}^j \exp(i_c \beta x_i), \qquad (2.21.3)$$

where $i_c = \sqrt{-1}$. This is followed by substituting the series into the error equation and solving for the amplification factor $\gamma_\beta = \xi_\beta^{j+1}/\xi_\beta^j$ for each component. For stability, the modulus of *the amplification factor* must be less than or equal to one for all the components. The analysis is somewhat simplified by omitting the subscript β and by taking $x_i = i\Delta x$. Equation (2.21.3) will then take the form

$$\varepsilon_{ij} = \xi^j \exp(i_c \beta i \Delta x) \qquad (2.21.4)$$

and substituting this directly into the difference equation and cancelling the resulting common factor, $\xi^j \exp(i_c \beta i \Delta x)$, leads to an equation that must be satisfied by the parameters $\gamma, \beta, \Delta x$ and Δt.

We now apply the criterion to equation (2.20.8). By substituting the error (2.21.4) into (2.20.8) and cancelling the common factor $\exp(i_c \beta i \Delta x)$, we obtain

$$-r[\theta\xi^{j+1}\{(1-w)\exp(i_c\beta\Delta x)+(2w-1)-w\exp(-i_c\beta\Delta x)\}+(1-\theta)\xi^j$$
$$\{(1-w)\exp(i_c\beta\Delta x)+(2w-1)-w\exp(-i_c\beta\Delta x)\}]=\xi^{j+1}-\xi^j.$$

By using the identity

$$\exp(\pm i_c\beta\Delta x) = \cos(\beta\Delta x) \pm i_c\sin(\beta\Delta x),$$

we get

$$-r[\theta\xi^{j+1} + (1-\theta)\xi^j][(2w-1)\{1-\cos(\beta\Delta x)\} + i_c\sin(\beta\Delta x)] = \xi^{j+1} - \xi^j.$$

Hence

$$\frac{\xi^{j+1}}{\xi^j} = \gamma$$

$$= \frac{1 - r(1-\theta)(2w-1)\{1-\cos(\beta\Delta x)\} - i_c\lambda(1-\theta)\sin(\beta\Delta x)}{1 + r\theta(2w-1)\{1-\cos(\beta\Delta x)\} + i_c\lambda\theta\sin(\beta\Delta x)}, \qquad (2.21.5)$$

the amplification factor which is complex. Clearly, by taking the square of the modulus of r we obtain

$$|\gamma|^2 = \frac{[1 - r(1-\theta)(2w-1)\{1-\cos(\beta\Delta x)\}]^2 + r^2(1-\theta)^2\sin^2(\beta\Delta x)}{[1 + r\theta(2w-1)\{1-\cos(\beta\Delta x)\}]^2 + r^2\theta^2(\beta\Delta x)}. \qquad (2.21.6)$$

We now proceed to analyse, separately, the stability of: (a) the *Lax–Wendroff equation*, (b) *centred-in-distance equations*, (c) *backward-in-distance equations* and (d) *forward-in-distance equations*.

(a) *Stability of the Lax–Wendroff Equation.* With $w = (1+r)/2$ and $\theta = 0$, we have from (2.21.6)

$$|\gamma|^2 = [1 - r^2(1-\cos(\beta\Delta x))]^2 + r^2\sin^2(\beta\Delta x)$$

$$= \left[1 - 2r^2\sin^2\left(\frac{\beta\Delta x}{2}\right)\right]^2 + r^2\sin^2(\beta\Delta x)$$

$$= 1 - 4r^2\sin^2\left(\frac{\beta\Delta x}{2}\right) + 4r^4\sin^4\left(\frac{\beta\Delta x}{2}\right) + r^2\sin^2(\beta\Delta x).$$

Since $\sin(\beta\Delta x) = 2\sin(\beta\Delta x/2)\cos(\beta\Delta x/2)$ we have

$$r^2\sin^2(\beta\Delta x) = 4r^2\sin^2\left(\frac{\beta\Delta x}{2}\right)\cos^2\left(\frac{\beta\Delta x}{2}\right)$$

$$= 4r^2\sin^2\left(\frac{\beta\Delta x}{2}\right) - 4r^2\sin^4\left(\frac{\beta\Delta x}{2}\right).$$

Hence

$$|\gamma| = \left[1 - 4r^2(1-r^2)\sin^4\frac{\beta\Delta x}{2}\right]^{1/2}, \qquad (2.21.7)$$

and for stability

$$|\gamma| \le 1 \text{ implies } 0 \le 4r^2(1-r^2) \quad \text{giving } 0 < r \le 1.$$

Therefore the Lax–Wendroff method is conditionally stable for $0 < r \le 1$.

(b) *Stability of the centred-in-distance equations*   For $w = 1/2$, equation (2.21.6) reduces to

$$|\gamma|^2 = \frac{1 + r^2(1-\theta)^2 \sin^2(\beta\Delta x)}{1 + r^2\theta^2 \sin^2(\beta\Delta x)}. \tag{2.21.8}$$

The stability requirement $|\gamma|^2 \leq 1$ leads to

$$1 + r^2(1-\theta)^2 \sin^2(\beta\Delta x) \leq 1 + r^2\theta^2 \sin^2(\beta\Delta x),$$

which simplifies to

$$(1-\theta)^2 \leq \theta^2. \tag{2.21.9}$$

For $\theta = 1/2$, $|\gamma|$ is exactly one; for $\theta > 1/2$, inequality (2.21.9) is always satisfied; for $\theta < 1/2$, inequality (2.21.9) is never satisfied.

We conclude that the centred-in-distance, centred-in-time difference equation is neutrally stable, the centred-in-distance, backward-in-time equation is unconditionally stable while centred-in-distance forward-in-time equation is always unstable.

(c) *Stability of the backward-in-distance equations*   For $w = 1$, equation (2.21.6) reduces to

$$|\gamma|^2 = \frac{[1 - r(1-\theta)\{1 - \cos(\beta\Delta x)\}]^2 + r^2(1-\theta)^2 \sin^2(\beta\Delta x)}{[1 + r\theta\{1 - \cos(\beta\Delta x)\}]^2 + r^2\theta^2 \sin^2(\beta\Delta x)}. \tag{2.21.10}$$

It is immediately evident that for $\theta = 1$, the numerator takes on exactly the value of one while the denominator is larger than one. Therefore, for this particular case, $|\gamma| \leq 1$ and satisfies the stability requirement.

For the more general case, $|\gamma|^2 \leq 1$, the numerator of equation (2.21.10) must be less than or equal to the denominator. On expanding and after some manipulation, this leads to

$$-2r\{1 - \cos(\beta\Delta x)\} + r^2(1-2\theta)\{1 - \cos(\beta\Delta x)\}^2 + r^2(1-2\theta)\sin^2(\beta\Delta x) \leq 0,$$

which reduces to

$$-2r\{1 - \cos(\beta\Delta x)\}[1 - r(1-2\theta)] \leq 0.$$

Since $r > 0$ and $\cos(\beta\Delta x) \leq 1$, then we must have

$$1 - r(1-2\theta) \geq 0, \tag{2.21.11}$$

which is automatically satisfied for $\theta \geq 1/2$ and for all $r > 0$. However, when $\theta < 1/2$ we now have a restriction on r for stability. That is, the equations are stable for $r \leq (1/(1-2\theta))$ and they are unstable otherwise for $r > (1/(1-2\theta))$. In particular, for $\theta = 0$, the stability requirement is $r \leq 1$. We conclude that both the backward-in-distance, backward-in-time and the backward-in-distance, centred-in-time formulae are always stable while the backward-in-distance, forward-in-time equation is conditionally stable for $r \leq 1$.

(d) *Stability of the forward-in-distance equations* For $w = 0$ equation (2.21.6) becomes

$$|\gamma|^2 = \frac{[1 + r(1 - \theta)\{1 - \cos(\beta\Delta x)\}]^2 + r^2(1 - \theta)^2 \sin^2(\beta\Delta x)}{[1 - r\theta\{1 - \cos(\beta\Delta x)\}]^2 + r^2\theta^2 \sin^2(\beta\Delta x)}. \qquad (2.21.12)$$

Again, the condition for stability is that the numerator be less than the denominator. This leads to

$$2r\{1 - \cos(\beta\Delta x)\} + r^2(1 - 2\theta)\{1 - \cos(\beta\Delta x)\}^2 + r^2(1 - 2\theta)\sin^2(\beta\Delta x) \leq 0,$$

which reduces to

$$2r\{1 - \cos(\beta\Delta x)\}[1 + r(1 - 2\theta)] \leq 0.$$

Stability then requires that

$$1 + r(1 - 2\theta) \leq 0, \qquad (2.21.13)$$

which is never satisfied for any $\theta \leq 1/2$.

When $\theta > 1/2$, inequality (2.21.13) leads to the following restriction on r

$$r \geq \frac{1}{2\theta - 1}.$$

In particular, for $\theta = 1$ (corresponding to the forward-in-distance, backward-in-time equation), the stability requirement is $r \geq 1$ or

$$a\frac{\Delta t}{\Delta x} \geq 1. \qquad (2.21.14)$$

In practice, it is difficult to satisfy the above stability condition at all mesh points as it entails excessively large time steps. We therefore conclude that not one of the forward-in-distance equations is useful, either because of their unconditional instability (when $\theta = 0$ or $\theta = 1/2$), or because of the difficult stability restriction (when $\theta > 1/2$).

2.22. Truncation Error Analysis of the Generalised Finite Difference Methods

Let us consider the differential equation at the point $(i\Delta x, j\Delta t)$,

$$\mathcal{L}(U_{ij}) = 0, \qquad (2.22.1)$$

where U is the *exact solution of the differential equation* at that point. For example, for equation (2.20.1) we have

$$\mathcal{L}(U_{ij}) = \left(-a\frac{\partial U}{\partial x} - \frac{\partial U}{\partial t}\right)_{ij} = 0. \qquad (2.22.2)$$

The derivatives in this equation may be replaced exactly at the point $(i\Delta x, j\Delta t)$ by an appropriate infinite (difference) series. To derive a finite-difference equation approximating the differential equation, however, each series is truncated after a certain number of terms. Let us denote this approximating difference equation by

$$F(u_{ij}) = 0, \tag{2.22.3}$$

where u is the *exact solution of the difference equation* at the point $(i\Delta x, j\Delta t)$.

The amount by which the exact solution U of the partial differential equation does not satisfy the differential equation at the point $(i\Delta x, j\Delta t)$ is called the *local truncation error* T_{ij}. Clearly

$$T_{ij} = F(U_{ij}). \tag{2.22.4}$$

By means of a Taylor series expansion we are able to find the principal part of the local truncation error and hence deduce the order of accuracy of the difference equation.

To study the accuracy of the generalised finite difference methods, we use equation (2.20.8),

$$-\frac{a}{\Delta x}[\theta\{(1-w)u_{i+1,j+1} + (2w-1)u_{i,j-1} - wu_{i-1,j+1}\}$$

$$+ (1-\theta)\{(1-w)u_{i+1,j} + (2w-1)\, u_{ij} - wu_{i-1,j}\}] - \frac{u_{i,j+1} - u_{ij}}{\Delta t} = F(u_{ij}). \tag{2.22.5}$$

We note that the expression

$$(1-w)u_{i+1} + (2w-1)u_i - wu_{i-1},$$

can also be written as

$$(\tfrac{1}{2})\,(u_{i+1} - u_{i-1}) + (\tfrac{1}{2} - w)\,(u_{i+1} - 2u_i + u_{i-1}). \tag{2.22.6}$$

Hence, using equations (2.22.4–2.22.6) we can now write an expression for the local truncation error as

$$T = -a\left[\theta\left\{\frac{U_{i+1,j+1} - U_{i-1,j+1}}{2\Delta x} + \Delta x(1/2 - w)\frac{U_{i+1,j+1} - 2U_{i,j+1} + U_{i-1,j+1}}{(\Delta x)^2}\right\}\right.$$

$$\left. + \frac{U_{i+1,j} - U_{i-1,j}}{2\Delta x} + \Delta x(1/2 - w)\frac{U_{i+1,j} - 2U_{i,j} + U_{i-1,j}}{(\Delta x)^2}\right] - \frac{U_{i,j+1} - U_{ij}}{\Delta t}. \tag{2.22.7}$$

From finite difference approximations to $\partial U/\partial x$ and $\partial^2 U/\partial x^2$, we have

$$\frac{U_{i+1} - U_{i-1}}{2\Delta x} = \left(\frac{\partial U}{\partial x}\right)_i + 0([\Delta x]^2),$$

and

$$\frac{U_{i+1} - 2U_i + U_{i-1}}{(\Delta x)^2} = \left(\frac{\partial^2 U}{\partial x^2}\right)_i + 0([\Delta x]^2).$$

Substituting these into equation (2.22.7) gives us

$$T = -a\left[\theta\left(\frac{\partial U}{\partial x}\right)_{i,j+1} + \theta\Delta x(1/2 - w)\left(\frac{\partial^2 U}{\partial x^2}\right)_{i,j+1} + (1-\theta)\left(\frac{\partial U}{\partial x}\right)_{i,j}\right.$$

$$\left. + (1-\theta)\Delta x(1/2 - w)\left(\frac{\partial^2 U}{\partial x^2}\right)_{i,j}\right] - \frac{U_{i,j+1} - U_{i,j}}{\Delta t} + 0([\Delta x]^2). \tag{2.22.8}$$

We can now apply Taylor's series expansions on $U_{i,j+1}$ and U_{ij} about the point $(i\Delta x, (j+\theta)\Delta t)$ using the obvious relations

$$j + 1 = (j + \theta) + (1 - \theta) \tag{2.22.9a}$$

and

$$j = (j + \theta) - \theta, \tag{2.22.9b}$$

to give

$$T = -a\Delta x(1/2 - w)\left(\frac{\partial^2 U}{\partial x^2}\right)_{i,j+\theta} + (\theta - 1/2)\Delta t a^2 \left(\frac{\partial^2 U}{\partial x^2}\right)_{i,j+\theta} + 0([\Delta x]^2) + 0([\Delta t]^2), \tag{2.22.13}$$

i.e.,

$$T = a\Delta x[(w - 1/2) + r(\theta - 1/2)]\left(\frac{\partial^2 U}{\partial x^2}\right)_{i,j+\theta} + 0([\Delta x]^2) + 0([\Delta t]^2).$$

The principal part of the local truncation error is therefore,

$$a\Delta x[(w - 1/2) + r(\theta - 1/2)]\left(\frac{\partial^2 U}{\partial x^2}\right)_{i,j+\theta}.$$

For the Lax–Wendroff formula (2.20.9), we find that with $\theta = 0$ and $w = 1/(2(1 + r))$, $T = 0([\Delta x]^2) + 0([\Delta t]^2)$ confirming that the method is second-order accurate in both space and time. The local truncation errors of the other formulae are described in Table 2.20.1.

2.23. Other Approximations for the Convection Equation

Other methods can be developed in much the same way as for the generalised schemes to approximate the convection equation (2.20.1).

(i) *Leapfrog method*   If we discretise $\partial U/\partial x$ and $\partial U/\partial t$ by their second order centred analogues as in equation (2.20.2) then we are led to the *three-time level formula*

$$\frac{u_{i,j+1} - u_{i,j-1}}{2\Delta t} = -a\frac{(u_{i+1,j} - u_{i-1,j})}{2\Delta x}$$

or for $r = a\Delta t/\Delta x$,

$$u_{i,j+1} = u_{i,j-1} - r(u_{i+1,j} - u_{i-1,j}), \tag{2.23.1}$$

which is known as the explicit leapfrog scheme. We shall now establish the stability condition of this scheme. By substituting the error

$$\varepsilon_{ij} = \xi^j \exp(i_c\beta i\Delta x)$$

into equation (2.23.1) results in

$$\xi^{j+1} \exp(i_c\beta i\Delta x) = \xi^{j-1} \exp(i_c\beta i\Delta x) - r\xi^j(\exp(i_c\beta(i + 1)\Delta x) - \exp(i_c\beta(i - 1)\Delta x)),$$

or

$$\gamma = \gamma^{-1} - r(\exp(i_c\beta\Delta x) - \exp(-i_c\beta\Delta x)). \tag{2.23.2}$$

But we know from the identity

$$\exp(i_c\beta\Delta x) - \exp(-i_c\beta\Delta x) = -2i_c\sin(\beta\Delta x).$$

Hence equation (2.23.2) becomes

$$\gamma^2 + (2ri_c\,\sin(\beta\Delta x))\gamma - 1 = 0.$$

The roots of this equation (the amplification factors) are

$$\gamma = i_c r\sin(\beta\Delta x) \pm \sqrt{1 - r^2\sin2(\beta\Delta x)}.$$

If $r\sin(\beta\Delta x) > 1$, then the absolute value of one of these roots exceeds unity; if $r\sin(\beta\Delta x) \le 1$ then both roots satisfy $|\gamma| = 1$. Thus, assuming that the maximum value for $\sin(\beta\Delta x)$ must be considered possible, then the stability condition is $r \le 1$.

From equation (2.23.1) we observe that

$$\frac{1}{\Delta t}(u_{i,j+1} - u_{i,j-1}) = \frac{1}{\Delta x}a(u_{i+1,j} - u_{i-1,j}).$$

An expression for the local truncation error is

$$T = f(U_{ij})$$

$$= \frac{1}{\Delta t}(U_{i,j+1} - U_{i,j-1}) - \frac{1}{\Delta x}a(U_{i+1,j} - U_{i-1,j}).$$

On applying Taylor's theorem on this expression at the point $(i\Delta x, j\Delta t)$ we find that

$$T = \frac{1}{\Delta t}\left(2\Delta t\left(\frac{\partial U}{\partial t}\right)_{ij} + 0([\Delta t]^3)\right) - \frac{1}{\Delta x}a\left(2\Delta x\frac{\partial U}{\partial x}\right)_{ij} + 0([\Delta x]^3)\right)$$

$$= 2\left(\frac{\partial U}{\partial t}\right)_{ij} - 2a\left(\frac{\partial U}{\partial x}\right)_{ij} + 0([\Delta x]^2) + 0([\Delta t]^2)$$

$$= 2\left\{\left(\frac{\partial U}{\partial t}\right)_{ij} - a\left(\frac{\partial U}{\partial x}\right)_{ij}\right\} + 0([\Delta x]^2) + 0([\Delta t]^2).$$

Hence $T = 0([\Delta x]^2) + 0([\Delta t]^2)$ for the leapfrog scheme.

(ii) *Wendroff implicit method*   The algorithm due to Wendroff may be obtained by representing $\partial U/\partial x$ at the $(j + 1/2)$ level as the average of the derivatives at j and $j+1$ and by representing $\partial U/\partial t$ at the $(i + 1/2)$ level as the average of the derivatives at i and $i+1$. Thus

$$\left(\frac{\partial U}{\partial t}\right)_{i+1/2,j+1/2} \approx 1/2\left(\frac{u_{i+1,j+1} - u_{i+1,j}}{\Delta t} + \frac{u_{i,j+1} - u_{ij}}{\Delta t}\right)$$

and

$$\left(\frac{\partial U}{\partial x}\right)_{i+1/2,j+1/2} \approx 1/2\left(\frac{u_{i+1,j+1}-u_{i,j+1}}{\Delta x}+\frac{u_{i+1,j}-u_{ij}}{\Delta x}\right)$$

The Wendroff algorithm then assumes the form

$$(1+r)u_{i+1,j+1}+(1-r)u_{i,j+1}=(1+r)u_{ij}+(1-r)u_{i+1,j}. \tag{2.23.3}$$

It cannot be used for pure initial-value problems, that is, conditions on $t=0$ only because it would give an inifinite number of simultaneous equations. If, however, initial values are known on the x-axis, $x \geq 0$, and boundary values on the t-axis, $t \geq 0$, the equation can be used explicitly by writing it as

$$u_{i+1,j+1}=u_{ij}+\frac{(1-r)}{(1+r)}\{u_{i+1,j}-u_{i,j+1}\}.$$

To analyse its stability, we again insert the error $\varepsilon_{ij}=\xi^j\exp(i_c\beta i\Delta x)$ into (2.23.3). This provides

$$(1+r)\xi^{j+1}\exp(i_c\beta(i+1)\Delta x)+(1-r)\xi^{j+1}\exp(i_c\beta i\Delta x)$$
$$=(1+r)\xi^j\exp(i_c\beta i\Delta x)+(1-r)\xi^j\exp(i_c\beta(i+1)\Delta x),$$

i.e.,

$$\gamma=\frac{(1+r)+(1-r)\exp(i_c\beta\Delta x)}{(1-r)+(1+r)\exp(i_c\beta\Delta x)},$$

the amplification factor.

This can be written as

$$\gamma=\frac{[(1+r)+(1-r)\cos(\beta\Delta x)]+i_c(1-r)\sin(\beta\Delta x)}{[(1-r)+(1+r)\cos(\beta\Delta x)]+i_c(1+r)\sin(\beta\Delta x)}.$$

Hence

$$|\gamma|^2=\frac{(1+r)^2+(1-r)^2+2(1-r^2)\cos(\beta\Delta x)}{(1+r)^2+(1-r)^2+2(1-r^2)\cos(\beta\Delta x)}$$

$$=1.$$

Therefore, the Wendroff scheme is unconditionally stable for all values of r.

Finally, it can be easily derived that

$$T=0([\Delta x]^2)+0([\Delta t]^2),$$

showing that the Wendroff scheme is second-order accurate in both space and time.

2.24. Finite Difference Approximations for Second Order Hyperbolic Equations

A natural extension to the first-order hyperbolic equation is to employ finite difference procedures to second-order equations. In the process of developing these methods, we will, of course, bear in mind, the limitations imposed by the characteristics.
(a) *Explicit Methods*   Let us consider the simplest of the second-order hyperbolic equations called the Wave Equation given by

$$\frac{\partial^2 U}{\partial x^2} = \frac{\partial^2 U}{\partial t^2} \tag{2.24.1}$$

and the Cauchy condition

$$U(x,0) = f(x), \quad \frac{\partial U}{\partial t}(x,0) = g(x). \tag{2.24.2}$$

As before, we take a rectangular net with constant space and time intervals given by Δx and Δt, respectively, and we write $u_{ij} = u(i\Delta x, j\Delta t), -\infty < i < \infty, 0 \le j \le \infty$. Both second partial derivatives are approximated by central difference expressions whose truncation error is $0([\Delta x]^2)$. Thus equation (2.24.1) is approximated by the explicit formula

$$u_{i,j+1} = r^2(u_{i-1,j} + u_{i+1,j}) + 2(1 - r^2)u_{ij} - u_{i,j-1}, \tag{2.24.3}$$

where $r = \Delta t/\Delta x$. The first initial condition of (2.24.2) specifies $u_{i,0}$ on the line $t = 0$. We can use the second condition to find values on the line $t = \Delta t$ by employing a 'false' boundary and the second-order central difference formula

$$\frac{\partial U}{\partial t}\bigg|_{i,0} = \frac{U_{i,1} - U_{i,-1}}{2\Delta t} + 0([\Delta t]^2). \tag{2.24.4}$$

Writing $g(i\Delta x) = g_i$, we have the approximation

$$u_{i,1} - u_{i,-1} = 2\Delta t g_i. \tag{2.24.5}$$

From equation (2.24.3) with $j = 0$, we have

$$u_{i,1} = r^2(u_{i-1,0} + u_{i+1,0}) + 2(1 - r^2)u_{i,0} - u_{i,-1}.$$

Upon replacing $u_{i,-1}$ with its value, from equation (2.24.5), and solving for $u_{i,1}$ we find

$$u_{i,1} = (\tfrac{1}{2}) r^2(f_{i-1} + f_{i+1}) + (1 - r^2)f_i + \Delta t g_i. \tag{2.24.6}$$

The computational molecule for equation (2.24.3) is shown in **Figure 2.24.1**.

(b) *Implicit Methods*   With the expectation of gaining stability advantages, we shall now attempt to derive implicit methods for the second-order hyperbolic equations. For the wave equation (2.24.1) the simplest implicit system is obtained by approximating

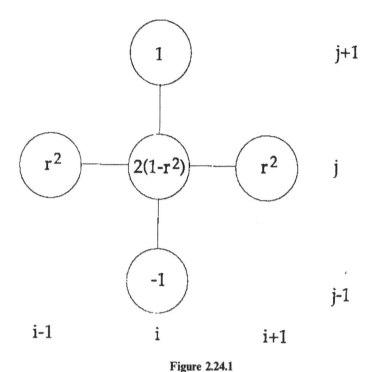

Figure 2.24.1

$\partial^2 U/\partial t^2$ as before, by a second central difference centred at (i,j) while $\partial^2 U/\partial x^2$ is approximated by the average of two central differences, one centred at $(i, j+1)$ and the other at $(i, j-1)$. Thus, one simple implicit approximation takes the form

$$u_{i,j+1} - 2u_{ij} + u_{i,j-1} = \frac{r^2}{2}\{(u_{i+1,j+1} - 2u_{i,j+1} + u_{i-1,j+1})$$

$$+ (u_{i+1,j-1} - 2u_{i,j-1} + u_{i-1,j-1})\}. \qquad (2.24.7)$$

The implicit nature of this formula is obvious by rewriting the expression to be solved on the $(j+1)^{th}$ line in terms of the values on the two preceding lines. Thus one finds the equation

$$-r^2 u_{i+1,j+1} + 2(1 + r^2)u_{i,j+1} - r^2 u_{i-1,j+1}$$

$$= 4u_{ij} + r^2 u_{i+1,j-1} - 2(1 + r^2)u_{i,j-1} + r^2 u_{i-1,j-1}. \qquad (2.24.8)$$

Another implicit method discussed by Richtmyer and Morton (1967) to approximate (2.24.1) is given by

$$-\frac{1}{4}r^2 u_{i-1,j+1} + \left(1 + \frac{r^2}{2}\right)u_{i,j+1} - \frac{r^2}{4}u_{i+1,j+1} = \frac{1}{2}r^2 u_{i-1,j} + (2 - r^2)u_{ij}$$

$$+ \frac{r^2}{2}u_{i+1,j} + \frac{r^2}{4}u_{i-1,j-1} + \left(-1 - \frac{r^2}{2}\right)u_{i,j-1} + \frac{r^2}{4}u_{i+1,j-1}. \qquad (2.24.9)$$

By assuming that there are m mesh values to be determined, then writing equations (2.24.8) and (2.24.9) for each $i, i = 1, 2, \ldots, m$ and inserting the discretised boundary conditions, *the tridiagonal nature* of the system becomes clear. Thus *the direct algorithm* may be applied to find a non-iterative solution.

Equations (2.24.3), (2.24.4) and (2.24.5) are special cases of a *general three-level implicit form* obtained by approximating, $(\partial^2 U/\partial t^2)_{k,j}$ by $1/(\Delta t)^2 \, \delta_t^2 u_{ij}$ and approximating $(\partial^2 U/\partial x^2)_{k,j}$ with

$$\frac{1}{(\Delta x)^2} [\alpha \delta_x^2 u_{i,j+1} + (1 - 2\alpha)\delta_x^2 u_{ij} + \alpha \delta_x^2 u_{i,j-1}].$$

Hence

$$\frac{1}{(\Delta t)^2} \delta_t^2 u_{ij} = \frac{1}{(\Delta x)^2} [\alpha \delta_x^2 u_{i,j+1} + (1 - 2\alpha)\delta_x^2 u_{ij} + \alpha \delta_x^2 u_{i,j-1}], \qquad (2.24.10)$$

where α is a weighting factor and δ^2 is the operator defined by

$$\delta_x^2 u_{ij} = u_{i+1,j} - 2u_{ij} + u_{i-1,j}.$$

Note that $\alpha = 0$ gives the explicit method (2.24.3), $\alpha = 1/2$ gives the implicit method (2.24.18) and for $\alpha = 1/4$ we obtain the implicit equation (2.24.19).

2.25. Stability Analysis of the General Three-Level Formula

To determine the conditions of stability of the general three level implicit method, we perform a Fourier series analysis on equation (2.24.10) by inserting the error

$$\varepsilon_{ij} = \xi^j \exp(i_c \beta i \Delta x).$$

This leads to

$$\frac{1}{(\Delta t)^2} \delta_t^2 \, \varepsilon_{ij} = \frac{1}{(\Delta x)^2} [\alpha \delta_x^2 \, \varepsilon_{i,j+1} + (1 - 2\alpha)\delta_x^2 \, \varepsilon_{ij} + \alpha \delta_x^2 \, \varepsilon_{i,j-1}],$$

or

$$\xi^{j-1} \exp(i_c \beta \Delta x)(1 - 2\gamma + \gamma^2) = r^2 [\alpha \xi^{j-1} \exp(i_c \beta \Delta x)\{\exp(i_c \beta \Delta x)$$
$$+ \exp(i_c \beta \Delta x) - 2\} + (1 - 2\alpha)\xi^j \exp(i_c \beta \Delta x)$$
$$\{\exp(i_c \beta \Delta x) + \exp(i_c \beta \Delta x) - 2\}$$
$$+ \alpha \xi^{j-1} \exp(i_c \beta \Delta x)\{\exp(i_c \beta \Delta x) + \exp(i_c \beta \Delta x) - 2\}].$$

After the cancellation and grouping of terms, we get

$$\gamma^2 - 2\gamma + 1 = r^2 \exp(i_c \beta \Delta x) + \exp(i_c \beta \Delta x) - 2)(\alpha \gamma^2 + (1 - 2\alpha)\gamma + \alpha)$$
$$= r^2 [2(\cos(\beta \Delta x) - 1)\{\alpha \gamma^2 + (1 - 2\alpha)\gamma + \alpha\}]. \qquad (2.25.1)$$

But

$$\cos(\beta \Delta x) - 1 = -2\sin^2\left(\frac{\beta \Delta x}{2}\right).$$

Hence (2.25.1) reduces to the quadratic

$$\left(1 + 4\alpha r^2 \sin^2\left(\frac{\beta\Delta x}{2}\right)\right)\gamma^2 + 2\left(2(1 - 2\alpha)\gamma^2 \sin^2\left(\frac{\beta\Delta x}{2}\right) - 1\right)\gamma$$

$$+ \left(1 + 4\alpha\gamma^2 \sin^2\left(\frac{\beta\Delta x}{2}\right)\right) = 0. \quad (2.25.2)$$

The roots of this quadratic are given by

$$\gamma = \frac{-[2(1 - 2\alpha)\gamma^2 \sin^2(\beta\Delta x/2) - 1] \pm 2r\sin(\beta\Delta x/2)\sqrt{r^2 \sin^2(\beta\Delta x/2)[1 - 4\alpha] - 1}}{(1 + 4\alpha r^2 \sin^2(\beta\Delta x/2))}.$$

$$(2.25.3)$$

We now discuss the stability requirement of the explicit formula (2.24.3). Putting $\alpha = 0$ into equations (2.25.2) and (2.25.3) we obtain

$$\gamma^2 - 2\left(1 - 2r^2 \sin^2\left(\frac{\beta\Delta x}{2}\right)\right)\gamma + 1 = 0 \quad (2.25.4)$$

and

$$\gamma = -\left[2r^2 \sin^2\left(\frac{\beta\Delta x}{2}\right) - 1\right] \pm 2r\sin\left(\frac{\beta\Delta x}{2}\right)\sqrt{r^2 \sin^2\left(\frac{\beta\Delta x}{2}\right) - 1},$$

or

$$\gamma = \left(1 - 2r^2 \sin^2\left(\frac{\beta\Delta x}{2}\right)\right) \pm \sqrt{4r^4 \sin^4\left(\frac{\beta\Delta x}{2}\right) - 4r^2 \operatorname{sn}^2\left(\frac{\beta\Delta x}{2}\right)}. \quad (2.25.5)$$

Letting

$$A = 1 - 2r^2 \sin^2\left(\frac{\beta\Delta x}{2}\right), \quad (2.25.6)$$

equation (2.25.4) becomes

$$\gamma^2 - 2A\gamma + 1 = 0, \quad (2.25.7)$$

and using (2.25.5), the values of γ are

$$\gamma_1 = A + \sqrt{A^2 - 1} \quad \text{and} \quad \gamma_2 = A - \sqrt{A^2 - 1}.$$

As r, β and Δx are real, then by (2.25.6), $A \leq 1$. When $A < -1$, $|\gamma_2| > 1$ giving instability.

When

$$-1 \leq A \leq 1, \quad A^2 \leq 1, \quad \gamma_1 = A + i_c\sqrt{1 - A^2} \quad \text{and} \quad \gamma^2 = A - i_c\sqrt{1 - A^2}.$$

It is clear that $|\gamma_1| = |\gamma_2| = 1$ proving that the explicit method is stable for $-1 \leq A \leq 1$. From equation (2.25.6), we then have

$$-1 \leq 1 - 2r^2 \sin^2\left(\frac{\beta\Delta x}{2}\right) \leq 1.$$

The only useful inequality is $-1 \leq 1 - 2r^2 \sin^2(\beta \Delta x/2)$ giving $r \leq 1$. Hence (2.24.3) is conditionally stable for $r \leq 1$.

For $\alpha > 0$, the stability equation is, from (2.25.2),

$$\gamma^2 - 2B\gamma + 1 = 0, \tag{2.25.8}$$

where $B = 1 - [2A'/(1 + 4A'\alpha)]$ and $A' = r^2 \sin^2(\beta \Delta x/2)$. We note that (2.25.8) is of the same form as (2.25.7) implying that $|\gamma_1| \leq 1$ and $|\gamma_2| \leq 1$, if and only if $-1 \leq B \leq 1$. Thus, $-1 \leq 1 - 2A'/(1 + A'\alpha) \leq 1$, or,

$$1 \geq A'/(1 + 4A'\alpha) \geq 0.$$

Since A' and α are non-negative, the right inequality is trivial. The left inequality yields the two inequalities

$$A' \leq \frac{1}{1 - 4\alpha}, \quad \alpha \leq \frac{1}{4} \quad \text{and} \quad A' > \frac{1}{1 - 4\alpha}, \quad \alpha > \frac{1}{4}.$$

The second inequality is trivial since $A' = \lambda^2 \sin^2(\beta \Delta x/2) \geq 0$ always. If we allow $\sin^2(\beta \Delta x/2)$ to take on its largest possible value, we obtain from the first inequality

$$r \leq 1/\sqrt{1 - 4\alpha}, \quad \alpha < \tfrac{1}{4}.$$

For stability r must satisfy the above condition. If $\alpha \geq 1/4$, stability is obtained for all values of r.

The implicit methods above clearly need some extra boundary conditions, for example along two lines parallel to the x-axis, since otherwise we have more unknowns than equations along the new line. We can then use a matrix method for analysing stability which will automatically include the effects of the boundaries.

The application of the matrix method can, perhaps, be better illustrated by first considering a general two-level finite difference scheme approximating a given first-order hyperbolic differential equation with which the initial and boundary values are specified for example at $x = x_0$ and $x = x_m$. In particular, if the boundary values are zero then the tridiagonal system of equations generated by the finite difference approximation can be written in the matrix form

$$A\underline{u}_{j+1} = B\underline{u}_j, \tag{2.25.9}$$

where A and B are square matrices of order $(m-1)$ and $\underline{u}_j$ is a column vector consisting of the u-values along the j-line, that is $\underline{u}_j = (u_{1,j}, u_{2,j}, \ldots, u_{m-1,j})^T$. The equation governing stability is (2.25.9); with other than zero boundary conditions a vector will be added to equation (2.25.9) which can, at most, depend upon i, i.e.,

$$A\underline{u}_{j+1} = B\underline{u}_j + \underline{c}.$$

The non-singular nature of A allows us to rewrite equation (2.15.9) as

$$\underline{u}_{j+1} = P\underline{u}_j, \quad P = A^{-1}B. \tag{2.25.10}$$

Upon repeated application of equation (2.25.10) leads to

$$\underline{u}_{j+1} = P\underline{u}_j = P^2\underline{u}_{j-1} + \cdots + P^j\underline{u}_1 = P^{j+1}\underline{u}_0,$$

where $\underline{u}_0$ is the vector of initial values. Now suppose we introduce errors at every mesh point along $t = 0$ and start the computation with the vector of values $\underline{u}_0^*$ instead of $\underline{u}_0$,

$$\underline{u}_1^* = P\underline{u}_0^*, \quad \underline{u}_2^* = P\underline{u}_0^*, \ldots, \underline{u}_j^* = P^j\underline{u}_0^*,$$

where we assume that no further errors are introduced.

If we define the error vector $\underline{\varepsilon}$ by

$$\underline{\varepsilon} = \underline{u} - \underline{u}^*,$$

then $\underline{\varepsilon}_j = \underline{u}_j - \underline{u}_j^* = P^j(\underline{u}_0 - \underline{u}_0^*) = P^j\underline{\varepsilon}_0$.

The finite difference scheme will be stable when $\underline{\varepsilon}_j$ remains bounded as j increases indefinitely. This can always be investigated by expressing the initial error vector in terms of the eigenvectors of P. We assume that the matrix P has $(m-1)$ linearly independent eigenvectors $\underline{v}_s$, which will always be so if the eigenvalues μ_s of P are all distinct or if P is real and symmetric (Hermitian). Then these eigenvectors can be used as a basis for our $(m-1)$-dimensional vector space and the $\underline{\varepsilon}_0$ with its $(m-1)$ components, can be expressed uniquely as a linear combination of them, namely,

$$\underline{\varepsilon}_0 = \sum_{s=1}^{m-1} c_s \underline{v}_s, \quad \text{where the } c_s, s = 1, \ldots, m-1$$

are known scalars.

The errors along the time level $t = \Delta t$, resulting from the initial perturbations $\underline{\varepsilon}_0$ will be given by

$$\underline{\varepsilon}_1 = P\underline{\varepsilon}_0 = P\sum_{s=1}^{m-1} c_s \underline{v}_s = \sum_{s=1}^{m-1} c_s P\underline{v}_s.$$

But $P\underline{v}_s = \mu_s \underline{v}_s$ by the definition of an eigenvalue. Therefore

$$\underline{\varepsilon}_1 = \sum_{s=1}^{m-1} c_s \mu_s \underline{v}_s.$$

Similarly

$$\underline{\varepsilon}_j = \sum_{s=1}^{m-1} c_s \mu_s^j \underline{v}_s.$$

This shows that the errors will not increase exponentially with j provided the eigenvalue with largest modulus (the spectral radius of P) has a modulus less than or equal to unity. We call P the *amplification matrix*.

Before we proceed to establish stability, we state the following theorem which is useful for the analysis of three or more time level difference equations.

Theorem 2.25.1.
Stability of Three or More Time Level Difference Equations

If the matrix P can be written as

$$
P = \begin{bmatrix}
P_{1,1} & P_{1,2}\text{------} & P_{1,m} \\
P_{2,1} & P_{2,2}\text{-----} & P_{2,m} \\
\vdots & \vdots & \vdots \\
P_{m,1} & P_{m,1}\text{-----} & P_{m,m}
\end{bmatrix},
$$

where each $P_{i,j}$ is an $n \times n$ matrix, and all the $P_{i,j}$ have a common set of n linearly independent eigenvectors, then the eigenvalues of P are given by the eigenvalues of the matrices

$$
\begin{bmatrix}
\eta^{(k)}_{1,1} & \eta^{(k)}_{1,2}\text{------} & \eta^{(k)}_{1,m} \\
\eta^{(k)}_{2,1} & \eta^{(k)}_{2,2}\text{-----} & \eta^{(k)}_{2,m} \\
\vdots & \vdots & \vdots \\
\eta^{(k)}_{m,1} & \eta^{(k)}_{m,1}\text{-----} & \eta^{(k)}_{m,m}
\end{bmatrix}, \quad k = 1,\ldots,n,
$$

where $\eta^{(k)}_{i,j}$ is the kth eigenvalue of $P_{i,j}$ corresponding to the kth eigenvector v_k common to all the P_{ij}'s.

Proof. Let $\underline{v}_k$ be an eigenvector common to all the submatrices P_{ij}, $i,j = 1,2,\ldots,m$ and denote the corresponding eigenvalues of $P_{1,1},\ldots,P_{2,1},\ldots$ by $\eta^{(k)}_{1,1}, \eta^{(k)}_{2,1},\ldots$, respectively. For simplicity, consider $i,j = 1,2$ and denote $\underline{v}_k$ by $\underline{v}$, $\eta^{(k)}_{i,j}$ by $\eta_{i,j}$. Then

$$
P_{1,1}\,\underline{v} = \eta_{1,1}\,\underline{v}, \quad P_{1,2}\,\underline{v} = \eta_{1,2}\,\underline{v},
$$
$$
P_{2,1}\,\underline{v} = \eta_{2,1}\,\underline{v}, \quad P_{2,2}\,\underline{v} = \eta_{2,2}\,\underline{v}.
$$

We multiply these equations, respectively, by the non-zero constants α_1 and α_2 and write them as

$$
\begin{bmatrix}
P_{1,1} & P_{1,2} \\
P_{2,1} & P_{2,2}
\end{bmatrix}
\begin{bmatrix}
\alpha_1 \underline{v} \\
\alpha_2 \underline{v}
\end{bmatrix}
=
\begin{bmatrix}
(\eta_{1,1}\alpha_1 + \eta_{1,2}\alpha_2)\underline{v} \\
(\eta_{2,1}\alpha_1 + \eta_{2,2}\alpha_2)\underline{v}
\end{bmatrix}. \tag{2.25.11}
$$

Let us assume that

$$
P = \begin{bmatrix}
P_{1,1} & P_{1,2} \\
P_{2,1} & P_{2,2}
\end{bmatrix}
$$

has an eigenvalue μ corresponding to the eigenvector

$$
\begin{bmatrix}
\alpha_1 \underline{v} \\
\alpha_2 \underline{v}
\end{bmatrix}
$$

so that

$$
\begin{bmatrix}
P_{1,1} & P_{1,2} \\
P_{2,1} & P_{2,2}
\end{bmatrix}
\begin{bmatrix}
\alpha_1 \underline{v} \\
\alpha_2 \underline{v}
\end{bmatrix}
= \mu
\begin{bmatrix}
\alpha_1 \underline{v} \\
\alpha_2 \underline{u}
\end{bmatrix}. \tag{2.25.12}
$$

By the right-hand sides of equations (2.25.11) and (2.25.12)

$$(\eta_{1,1} - \mu)\alpha_1 + \eta_{1,2}\alpha_2 = 0$$

and

$$\eta_{2,1}\alpha_1 + (\eta_{2,2} - \mu)\alpha_2 = 0.$$

These two equations will have a non-trivial solution for α_1 and α_2 if and only if

$$\det \begin{bmatrix} (\eta_{1,1} - \mu) & \eta_{1,2} \\ \eta_{2,1} & (\eta_{2,2} - \mu) \end{bmatrix} = 0,$$

i.e., if and only if μ is an eigenvalue of the matrix

$$\begin{bmatrix} \eta_{1,1} & \eta_{1,2} \\ \eta_{2,1} & \eta_{2,2} \end{bmatrix}.$$

We are now in a position to investigate the stability conditions that equation (2.24.16) has to fulfil. In matrix form, (2.24.16) may be written as

$$A\underline{u}_{j+1} = B\underline{u}_j + C\underline{u}_{j-1} + \underline{b}_j, \tag{2.25.13}$$

where

$$A = (1 + 2\alpha r^2)I - \alpha r^2 E,$$

$$B = 2(1 - (1 - 2\alpha)r^2)I + (1 - 2\alpha)r^2 E \tag{2.25.14}$$

and

$$C = (-1 - 2\alpha r^2)I + \alpha r^2 E.$$

E is the matrix with unity values along each diagonal immediately above and below the main diagonal and zeros elsewhere, and $\underline{b}_j$ is a column vector of known constants (boundary values). From (2.15.13) we have,

$$\underline{u}_{j+1} = A^{-1}B\underline{u}_j + A^{-1}C\underline{u}_{j-1} + A^{-1}\underline{b}_j. \tag{2.25.15}$$

Therefore, a perturbation ε_0 of the initial values will satisfy

$$\underline{\varepsilon}_{j+1} = A^{-1}B\underline{\varepsilon}_j + A^{-1}C\,\underline{\varepsilon}_{j-1}.$$

Hence

$$\begin{bmatrix} \underline{\varepsilon}_{j+1} \\ \underline{\varepsilon}_j \end{bmatrix} = \begin{bmatrix} A^{-1}B & A^{-1}C \\ I & 0 \end{bmatrix} \begin{bmatrix} \underline{\varepsilon}_j \\ \underline{\varepsilon}_{j-1} \end{bmatrix},$$

i.e., $\underline{v}_{j+1} = P\underline{v}_j$. The matrices $A, B,$ and C have the same system of linearly independent eigenvectors as E. So have the matrices $A^{-1}B$ and $A^{-1}C$.

Therefore, applying Theorem 2.25.1, the eigenvalues μ of P are given by

$$\det \begin{bmatrix} (a_k^{-1}b_k - \mu) & a_k^{-1}c_k \\ 1 & -\mu \end{bmatrix} = 0, \quad k = 1, 2, \dots, (m-1), \tag{2.25.16}$$

where a_k, b_k and c_k are the eigenvalues of A, B, C, respectively. We note from (2.25.14) that each of the matrices A, B and C is of a common tridiagonal form and

$$a_k = 1 + 4\alpha r^2 \sin^2\left(\frac{k\pi}{2m}\right),$$

$$b_k = 2 - 4(1 - 2\alpha)r^2 \sin^2\left(\frac{k\pi}{2m}\right), \quad k = 1, 2, \ldots, m-1,$$

$$c_k = -1 - 4\alpha r^2 \sin^2\left(\frac{k\pi}{2m}\right).$$

Using equation (2.25.16), we have

$$a_k \mu^2 - b_k \mu - c_k = 0,$$

or

$$\left(1 + 4\alpha r^2 \sin^2\left(\frac{k\pi}{2m}\right)\right)\mu^2 + 2\left(2(1 - 2\alpha)r^2 \sin^2\left(\frac{k\pi}{2m}\right) - 1\right)\mu$$

$$+ \left(1 + 4\alpha r^2 \sin^2\left(\frac{k\pi}{2m}\right)\right) = 0, \quad (2.25.17)$$

which is exactly of the form (2.25.2). To avoid repetition of the mathematics involved to solve (2.25.17), we conclude that $\mu = 1$ if $\alpha \geq 1/4$ and for all values of $r = \Delta t/\Delta x$. Hence the implicit equation (2.24.16) is *always stable*. We observe that the stability criterion has not been changed by the incorporation of the boundary values, and in general different types of boundary conditions have only negligible effect.

2.26. Truncation Error Analysis of the General Three-Level Formula

As before, by virtue of equation (2.24.16), we have

$$F(u_{i,j}) = \delta_t^2 u_{ij} - r^2[\alpha\delta_x^2 u_{i,j+1} + (1 - 2\alpha)\delta_x^2 u_{ij} + \alpha\delta_x^2 u_{i,j-1}],$$

$$F(U_{i,j}) = \delta_t^2 U_{ij} - r^2[\alpha\delta_x^2 U_{i,j+1} + (1 - 2\alpha)\delta_x^2 U_{ij} + \alpha\delta_x^2 U_{i,j-1}]$$

and

$$T = F(U_{ij}).$$

If we expand $U_{i-1,j-1}, U_{i-1,j}, U_{i-1,j+1}, U_{i,j-1}, U_{i,j+1}, U_{i+1,j-1}, U_{i+1,j}$ and $U_{i+1,j+1}$ about the point (x_i, t_j), we find that

(a) $\quad \delta_t^2 U_{ij} = (\Delta t)^2\left(\frac{\partial^2 U}{\partial t^2}\right)_{i,j} + \frac{2(\Delta t)^4}{4!}\left(\frac{\partial^4 U}{\partial t^4}\right)_{i,j} + \frac{2(\Delta t)^6}{6!}\left(\frac{\partial^6 U}{\partial t^6}\right)_{i,j} + \cdots$

(b) $\quad \delta_x^2 U_{i,j+1} = (\Delta x)^2\left(\frac{\partial^2 U}{\partial x^2}\right)_{i,j} + (\Delta x)^2(\Delta t)\left(\frac{\partial^3 U}{\partial x^2\partial t}\right)_{i,j} + \frac{(\Delta x)^4}{12}\left(\frac{\partial^4 U}{\partial x^4}\right)_{i,j}$

$$+ (1/2)(\Delta x)^2(\Delta t)^2\left(\frac{\partial^4 U}{\partial x^2\partial t^2}\right)_{i,j} + \cdots$$

(c) $\qquad \delta_x^2 U_{i,j} = (\Delta x)^2 \left(\dfrac{\partial^2 U}{\partial x^2}\right)_{i,j} + \dfrac{1}{12}(\Delta x)^4 \left(\dfrac{\partial^4 U}{\partial x^4}\right)_{i,j} + \cdots$

(d) $\qquad \delta_x^2 U_{i,j-1} = (\Delta x)^2 \left(\dfrac{\partial^2 U}{\partial x^2}\right)_{i,j} - (\Delta x)^2 (\Delta t) \left(\dfrac{\partial^3 U}{\partial x^2 \partial t}\right)_{i,j} + \dfrac{1}{12}(\Delta x)^4 \left(\dfrac{\partial^4 U}{\partial x^4}\right)_{i,j}$

$$+ (1/2)(\Delta x)^2 (\Delta t)^2 \left(\dfrac{\partial^4 U}{\partial x^2 \partial t^2}\right)_{i,j} + \cdots .$$

Hence

$$T = (\Delta t)^2 \left(\dfrac{\partial^2 U}{\partial t^2}\right)_{i,j} + \dfrac{(\Delta t)^4}{12}\left(\dfrac{\partial^4 U}{\partial x^4}\right)_{i,j} + \cdots$$

$$- r^2 \left\{ (\Delta x)^2 \left(\dfrac{\partial^2 U}{\partial x^2}\right)_{i,j} + \dfrac{(\Delta x)^4}{12}\left(\dfrac{\partial^4 U}{\partial x^4}\right)_{i,j} + \alpha(\Delta x)^2 (\Delta t)^2 \left(\dfrac{\partial^4 U}{\partial x^2 \partial t^2}\right)_{i,j} + \cdots \right\}$$

$$= (\Delta t)^2 \left(\dfrac{\partial^2 U}{\partial t^2}\right)_{i,j} + \dfrac{(\Delta t)^4}{12}\left(\dfrac{\partial^4 U}{\partial x^4}\right)_{i,j} - (\Delta t)^2 \left(\dfrac{\partial^2 U}{\partial x^2}\right)_{i,j}$$

$$- \dfrac{1}{12}(\Delta t)^2 (\Delta x)^2 \left(\dfrac{\partial^4 U}{\partial x^4}\right)_{i,j} - \alpha(\Delta t)^4 \left(\dfrac{\partial^4 U}{\partial x^2 \partial t^2}\right)_{i,j} + \cdots$$

$$= (\Delta t)^2 \left\{ \left(\dfrac{\partial^2 U}{\partial t^2}\right)_{i,j} - \left(\dfrac{\partial^2 U}{\partial x^2}\right)_{i,j} \right\} - \dfrac{1}{12}(\Delta t)^2 (\Delta x)^2 \left(\dfrac{\partial^4 U}{\partial x^4}\right)_{i,j}$$

$$- \alpha(\Delta t)^4 \left(\dfrac{\partial^4 U}{\partial x^2 \partial t^2}\right)_{i,j} + \dfrac{(\Delta t)^2}{12}\left(\dfrac{\partial^4 U}{\partial x^4}\right)_{i,j} + \cdots .$$

From the given differential equation, $(\partial^2 U/\partial t^2)_{i,j} - (\partial^2 U/\partial x^2)_{i,j} = 0$ and $(\partial^4 U/\partial t^4)_{i,j} = (\partial^4 U/\partial x^2 \partial t^2)_{i,j} = (\partial^4 U/\partial x^4)_{i,j}$. Therefore

$$T = -(\Delta t)^2 \left\{ \dfrac{(\Delta x)^2}{12}(1 + (12\alpha - 1)r^2)\left(\dfrac{\partial^4 U}{\partial x^4}\right)_{i,j} + \cdots \right\},$$

from which the principal part is $(\Delta t)^2/12 (\Delta x)^2 (1 + (12\alpha - 1)r^2 (\partial^4 U/\partial x^4)_{i,j}$. We deduce that for $\alpha = 0$, $1/4$ and $1/2$

$$T = 0([\Delta x]^2) + 0([\Delta t]^2).$$

In fact, for the explicit formula (with $\alpha = 0$) the truncation error is

$$T = -\dfrac{(\Delta t)^2 (\Delta x)^2}{12}[(1 - r^2)\left(\dfrac{\partial^4 U}{\partial x^4}\right)_{i,j} + \dfrac{1}{30}(\Delta x)^2 (1 - r^4)\left(\dfrac{\partial^6 U}{\partial x^6}\right)_{i,j} + \cdots].$$

It vanishes completely when $\lambda = 1$, and so the difference formula

$$u_{i,j+1} = u_{i+1,j} + u_{i-1,j} - u_{i,j-1}$$

is *an exact difference representation* of wave equation.

2.27. Other Approximations for the Wave Equation

Mitchell (1969) uses the following method for deriving implicit approximations to the wave equation (2.24.1) that are accurate to fourth-order differences. If $U_{i,\,j+1}$ and $U_{i,\,j-1}$ are expanded about the point (x_i, t_j) by Taylor's series, we get

$$U_{i,\,j+1} - 2U_{i,\,j} + U_{i,\,j-1} = (\Delta t)^2 \left(\frac{\partial^2 U}{\partial t^2}\right)_{i,j} + \frac{1}{12}(\Delta t)^4 \left(\frac{\partial^4 U}{\partial t^4}\right)_{i,j} + \cdots.$$

If U is a solution of the wave equation, then

$$\left(\frac{\partial^4 U}{\partial x^4}\right)_{i,j} = \left(\frac{\partial^4 U}{\partial t^4}\right)_{i,j}, \left(\frac{\partial^6 U}{\partial x^6}\right)_{i,j} = \left(\frac{\partial^6 U}{\partial t^6}\right)_{i,j}, \cdots.$$

Hence

$$U_{i,\,j+1} - 2U_{ij} + U_{i,\,j-1} = (\Delta t)^2 \left(\frac{\partial^2 U}{\partial x^2}\right)_{i,j} + \frac{1}{12}(\Delta t)^4 \left(\frac{\partial^4 U}{\partial x^4}\right)_{i,j} + \cdots. \qquad (2.27.1)$$

From the central difference formula

$$\frac{\partial^2 U}{\partial x^2} = \frac{1}{(\Delta x)^2}\left(\delta_x^2 - \frac{1}{12}\delta_x^4 + \frac{1}{90}\delta_x^6 + \cdots\right) U,$$

it follows that for fourth-order differences

$$\frac{\partial^2 U}{\partial x^2} = \frac{1}{(\Delta x)^2}(\delta_x^2 - \frac{1}{12}\delta_x^4) U$$

and

$$\frac{\partial^4 U}{\partial x^4} = \frac{\partial^2}{\partial x^2}\left(\frac{\partial^2 U}{\partial x^2}\right) = \frac{1}{(\Delta x)^2}\delta_x^4 U.$$

By substituting these approximations into equation (2.27.1), we find that to this order of accuracy

$$U_{i,\,j+1} - 2U_{ij} + U_{i,\,j-1} = r^2\{1 + \tfrac{1}{12}(r^2 - 1)\delta_x^2\} \delta_x^2 U_{i,j}, \qquad (2.27.2)$$

where $r = \Delta t/\Delta x$. Now if we operate on both sides of this equation with $\{1 + 1/12(r^2 - 1)\delta_x^2\}^{1/2}$ and expand each operator upto terms in δ_x^4 by the binomial expansion, we arrive at the following implicit difference approximation

$$u_{i,\,j+1} - 2u_{ij} + u_{i,\,j-1} = \tfrac{1}{24}(r^2 - 1)(\delta_x^2 u_{i,\,j+1} + \delta_x^2 u_{i,\,j-1})$$

$$+ \tfrac{1}{12}(11r^2 + 1)\delta_x^2 u_{ij} + \tfrac{1}{192}(r^2 - 1)(9r^2 - 1)\delta_x^4 u_{ij}$$

$$- \tfrac{1}{384}(r^2 - 1)^2(\delta_x^4 u_{i,\,j+1} + \delta_x^4 u_{i,\,j-1}). \qquad (2.27.3)$$

Similarly, if both sides of (2.27.2) are operated on by $(1 + 1/12(r^2 - 1)\delta_x^2)^{-1}$, the corresponding difference equation is

$$u_{i,j+1} - 2u_{ij} + u_{i,j-1} = \tfrac{1}{12}(r^2 - 1)(\delta_x^2 u_{i,j+1} + \delta_x^2 u_{i,j-1}) + \tfrac{1}{6}(5r^2 + 1)\delta_x^2 u_{ij}$$

$$+ \tfrac{1}{144}(r^2 - 1)^2\{2\delta_x^4 u_{ij} - (\delta_x^4 u_{i,j+1} + \delta_x^4 u_{i,j-1})\}. \quad (2.27.4)$$

These high-order difference approximations would be difficult to implement in practice because of the problems associated with the boundary conditions.

Von Neumann (cf. O'Brien *et al.* 1951) introduced the difference equation

$$\frac{\delta_t^2 u_{i,j}}{(\Delta t)^2} = \frac{a^2 \delta_x^2 u_{i,j}}{(\Delta x^2)} + \omega\left[\frac{1}{(\Delta x)^2(\Delta t)^2}\delta_t^2\delta_x^2 u_{i,j}\right] \quad (2.27.5)$$

to solve the wave equation (2.24.1). Except for $\omega = 1$, (2.27.5) is an implicit equation whose solution at each time step is obtained by solving a tridiagonal system of linear equations. For $\omega = 0$, the equation reduces to the classical explicit method (2.24.3). Von Neumann proved that (2.27.5) is unconditionally stable if $\omega > 1/4$ and conditionally stable if $\omega \leq 1/4$, the stability condition in the latter case being $r \leq 1/(1 - 4\omega)^{1/2}$.

This result was generalised to quasi-linear hyperbolic equations of the form

$$\frac{\partial^2 U}{\partial t^2} = a(x, t)\frac{\partial^2 U}{\partial x^2} + F\left(x, t, U, \frac{\partial U}{\partial x}, \frac{\partial U}{\partial t}\right) \quad (2.27.6)$$

and extended to *certain linear multi-dimensional* systems. However, the linear equations that arise are no longer tridiagonal. To overcome this problem, Lees (1962) proposed modifications to equation (2.27.5) by applying *an alternating direction procedure* and employing *the energy method*, he showed that the modified schemes are unconditionally stable if $\omega > 1/4$.

3. Group Explicit Methods for Parabolic Equations

3.1. Introduction

In the determination of attempting to attain more accurate numerical solutions to the exact solution of a problem without upgrading the order of the approximation the strategy of combining different numerical algorithms can be used. These can give truncation errors of different signs which when combined may then cancel producing higher order accuracy results. To illustrate the different algorithms which can be utilised we derive a generalised approximation, so that the differences in the sign of the truncation errors can be clearly observed.

It is well known that in general an explicit type difference method is the simplest to use. As simplicity is normally related to the computational cost, it is worth trying to retain the explicit type of approximation as much as possible. This study, as will be shown later, results in a new class of Group Explicit methods. The derivation of the new class of methods is based on the simplest heat-conduction problem

$$\frac{\partial U}{\partial t} = \frac{\partial^2 U}{\partial x^2}, \quad 0 \leq x \leq 1, \; t \geq 0. \tag{3.1.1}$$

This equation is chosen for the sake of simplicity in the following discussion. However the basic principles can be carried over to a more general class of problems as will be shown in later chapters.

3.2. A Generalised Two Time Level Finite Difference Approximation

We approximate equation (3.1.1) at the point $(i\Delta x, (j + 1/2)\Delta t)$ by the finite-difference approximation

$$\frac{u_{i,j+1} - u_{i,j}}{\Delta t} = \frac{1}{(\Delta x)^2} \{ \theta_1 \delta_x u_{i+1/2,j+1} - \theta_2 \delta_x u_{i-1/2,j+1} + \theta_1' \delta_x u_{i+1/2,j}$$

$$- \theta_2' \delta_x u_{i-1/2,j} \}, \tag{3.2.1}$$

131

with the compulsory conditions on the parameters $\theta_i, \theta_i', i = 1, 2$ given by

$$0 \le \theta_i, \quad \theta_i' \le 1, \qquad i = 1, 2, \tag{3.2.2}$$

$$\sum_{i=1}^{2} (\theta_i + \theta_i') = 2, \tag{3.2.3}$$

$$-\theta_1 + \theta_2 - \theta_1' + \theta_2' = 0, \tag{3.2.4}$$

and the optional conditions:

$$-\theta_1 + \theta_2 + \theta_1' - \theta_2' = 0, \tag{3.2.5}$$

$$-\theta_1 - \theta_2 + \theta_1' + \theta_2' = 0. \tag{3.2.6}$$

The need for the compulsory and optional conditions will be given later. However, in general, all well known two-time level formulae fulfill the compulsory conditions, while the optional conditions determine the order of accuracy and the consistency of the formulae derived.

Examples of how known standard formulae are derived from (3.2.1–3.2.6) are as follows:

(a) If all of the conditions (3.2.2–3.2.6) are fulfilled $\theta_1 = \theta_2 = \theta_1' = \theta_2' = 1/2$, then (3.2.1) yields

$$\frac{u_{i,j+1} - u_{i,j}}{\Delta t} = \frac{1}{2(\Delta x)^2} \{(u_{i+1,j+1} - 2u_{i,j+1} + u_{i-1,j+1})$$

$$+ (u_{i+1,j} - 2u_{i,j} + u_{i-1,j})\},$$

i.e.,

$$-ru_{i+1,j+1} + (2 + 2r)u_{i,j+1} - ru_{i-1,j+1} = ru_{i+1,j} + (2 - 2r)u_{i,j} + ru_{i-1,j}, \tag{3.2.7}$$

where $r = \Delta t/\Delta x^2$, which is known as the Crank–Nicolson (1947) formulae and is denoted by the computational molecule given in Figure 3.2.1. This formula is known to be unconditionally stable for all $r > 0$ and has a principal truncation error of $0((\Delta x)^2 + (\Delta t)^2)$.

(b) If $\theta_i' = 0$, $\theta_i = 1$, $i = 1, 2$, all conditions except (3.2.6) are fulfilled. This leads to the formula

$$\frac{u_{i,j+1} - u_{i,j}}{\Delta t} = \frac{u_{i+1,j+1} - 2u_{i,j+1} + u_{i-1,j+1}}{(\Delta x)^2},$$

i.e.,

$$-ru_{i-1,j+1} + (1 + 2r)u_{i,j+1} - ru_{i+1,j+1} = u_{i,j}, \tag{3.2.8}$$

which is known as the fully implicit formulae. This scheme is also unconditionally stable for all $r > 0$. The principal truncation error is of $0(\Delta t + (\Delta x)^2)$ and the computational molecule is given in Figure 3.2.2.

(c) If $\theta_i = 0$, $\theta_i' = 1$, $i = 1, 2$, again the condition (3.2.6) is not fulfilled. This gives the classical explicit formulae (Fig. 3.2.3)

$$u_{i,j+1} = ru_{i-1,j} + (1 - 2r)u_{i,j} + ru_{i+1,j}, \tag{3.2.9}$$

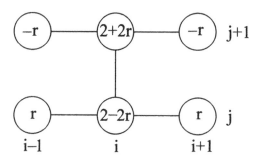

Figure 3.2.1

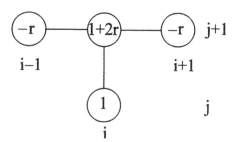

Figure 3.2.2

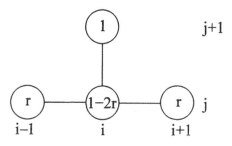

Figure 3.2.3

which is stable for $r \leq 1/2$ and has a principal truncation error of $0(\Delta t + (\Delta x)^2)$ (Smith, 1978).

(d) If $\theta_i = \alpha/2$, $\theta_i' = 1 - \alpha/2$, $i = 1, 2$ where α is a free parameter, then the conditions (3.2.2–3.2.5) are fulfilled and condition (3.2.6) is given by

$$- \theta_1 - \theta_2 + \theta_1' + \theta_2' = -2\alpha + 2.$$

This will also be satisfied if $\alpha = 1$ (as in the case of (a)), and will give the finite difference formula

$$- \alpha r(u_{i-1,j+1} + u_{i+1,j+1}) + 2(1 + \alpha r)u_{i,j+1}$$
$$= (2 - \alpha)r(u_{i-1,j} + u_{i+1,j}) + 2(1 - 2r + \alpha r)u_{i,j}, \quad (3.2.10)$$

which is due to Saul'yev (1964) and the computational molecule is given by Figure 3.2.4. It can be easily seen that equations (3.2.7–3.2.9) are special cases of equation (3.2.10). This formula has the stability condition

$$r \le \frac{1}{2(1-\alpha)}, \tag{3.2.11}$$

and the principal truncation error

$$\text{TE} = \begin{cases} 0(\Delta t + (\Delta x)^2), & \alpha \ne 1, \\ 0(\Delta t)^2 + (\Delta x)^2, & \alpha = 1. \end{cases} \tag{3.2.12}$$

(e) If $\theta_i = \alpha - 1/(6r)$, $\theta_i' = 1 - (\alpha - 1/(6r))$, $i = 1,2$ where α is as given in (d), conditions (3.2.2–3.2.5) are fulfilled and the condition (3.2.6) given by

$$-\theta_1 - \theta_2 + \theta_1' + \theta_2' = -4(\alpha - 1/6r) + 2,$$

will also be satisfied if $\alpha = 1/2 + 1/(6r)$.

This will lead to the cubic spline approximation of Papamichael and Whiteman (1973) which is

$$(1 - 6r\alpha)(u_{i-1,j+1} + u_{i+1,j+1}) + 2(2 + 6r\alpha)u_{i,j+1}$$
$$= \{1 + 6r(1 - \alpha)\}(u_{i-1,j} + u_{i+1,j}) + 2\{2 - 6r(1 - \alpha)\}u_{i,j}. \tag{3.2.13}$$

The scheme (3.2.13) is unconditionally stable for $1/2 \le \alpha \le 1$ and stable when $r \le 1/(6(1 - 2\alpha))$ for $0 \le \alpha \le 1/2$. The principal truncation error is $0(\Delta t + (\Delta x)^2)$ and the computational molecule is given in Figure 3.2.5.

(f) If $\theta_1 = \theta_2' = 1$ and $\theta_2 = \theta_1' = 0$ we see that the condition (3.2.5) is not fulfilled. This gives the formula due to Saul'yev (1964), (Fig. 3.2.6), i.e.,

$$-ru_{i+1,j+1} + (1 + r)u_{i,j+1} = ru_{i-1,j} + (1 - r)u_{i,j}. \tag{3.2.14}$$

(g) Another formulae due to Saul'yev (1964) is found when $\theta_2 = \theta_1' = 1$ and $\theta_1 = \theta_2' = 0$ and again the condition (3.2.5) is not fulfilled. It is

$$(1 + r)u_{i,j+1} - ru_{i-1,j+1} = ru_{i+1,j} + (1 - r)u_{i,j}, \tag{3.2.15}$$

as in Figure 3.2.7.

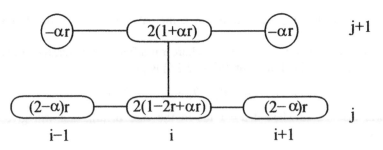

Figure 3.2.4

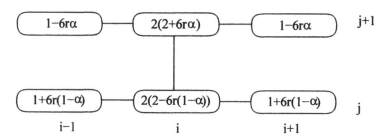

Figure 3.2.5

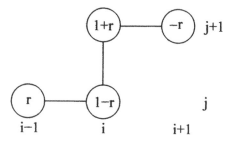

Figure 3.2.6

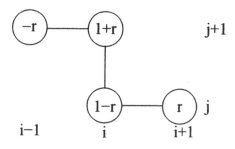

Figure 3.2.7

From the above examples, we can see that from equation (3.2.1) it is possible:

(1) to establish the well-known standard explicit, implicit and Crank–Nicolson formulae (i.e., (3.2.7–3.2.9));
(2) to obtain the weighted six point formula, i.e., (3.2.10);
(3) to obtain asymmetric formulae such as (3.2.14) and (3.2.15);
(4) to establish any six point formula with truncation error of order less or equal to $\{(\Delta t)^2 + (\Delta x)^2\}$, of which one of the examples is equation (3.2.13).

Therefore the approximation (3.2.1) is a general two-level six point finite difference approximation which can be used to develop a new strategy for the numerical solution of equation (3.1.1).

3.3. Truncation Errors for the General Approximation

The general two-time level six point finite difference approximation (3.2.1) can be written as

$$-r\theta_1 u_{i+1,j+1} + [1 + r(\theta_1 + \theta_2)]u_{i,j+1} - r\theta_2 u_{i-1,j+1}$$

$$= r\theta_1' u_{i+1,j} + [1 - r(\theta_1' + \theta_2')]u_{i,j} + r\theta_2' u_{i-1,j}. \quad (3.3.1)$$

To estimate the truncation error for this approximation, we expand each term in a Taylor's series expansion about the point $(i, j + 1/2)$ to obtain the result in the form given by

$$\left(\frac{\partial U}{\partial t} - \frac{\gamma_1}{2}\frac{\partial^2 U}{\partial x^2}\right)_{i,j+1/2} + \gamma_2\left(\frac{1}{\Delta x}\frac{\partial U}{\partial x} + \frac{\Delta x}{6}\frac{\partial^3 U}{\partial x^3} + \frac{1}{8}\frac{\Delta t^2}{\Delta x}\frac{\partial^3 U}{\partial x \partial t^2}\right)_{i,j+1/2}$$

$$- \gamma_1\left(\frac{(\Delta x)^2}{24}\frac{\partial^4 U}{\partial x^4} + \frac{(\Delta t)^2}{16}\frac{\partial^4 U}{\partial x^2 \partial t^2}\right)_{i,j+1/2}$$

$$+ \gamma_3\left(\frac{1}{2}\left(\frac{\Delta t}{\Delta x}\right)\frac{\partial^2 U}{\partial x \partial t} + \frac{1}{12}\Delta x \Delta t\frac{\partial^4 U}{\partial x^3 \partial t} + \frac{1}{48}\frac{\Delta t^3}{\Delta x}\frac{\partial^4 U}{\partial x \partial t^3}\right)_{i,j+1/2}$$

$$+ \frac{\gamma_4}{4}\Delta t\frac{\partial^3 U}{\partial x^2 \partial t}\bigg|_{i,j+1/2} + \frac{\Delta t^2}{24}\frac{\partial^3 U}{\partial t^3}\bigg|_{i,j+1/2} + \frac{1}{\Delta t}0(\Delta x^{\alpha_1}, \Delta t^{\alpha_2}) = 0, \quad (3.3.2)$$

with $\alpha_1 + \alpha_2 = 5$, and $\gamma_1, \gamma_2, \gamma_3$ and γ_4 are given by the left-hand sides of equations (3.2.3), (3.2.4), (3.2.5) and (3.2.6), respectively. Assuming the compulsory conditions are fulfilled (i.e. $\gamma_1 = 2$ and $\gamma_2 = 0$), (3.3.2) can now be written as

$$\left(\frac{\partial U}{\partial t} - \frac{\partial^2 U}{\partial x^2}\right)_{i,j+1/2} = \gamma_3\left(\frac{1}{2}\left(\frac{\Delta t}{\Delta x}\right)\frac{\partial^2 U}{\partial x \partial t} + \frac{1}{12}\Delta x \Delta t\frac{\partial^3 U}{\partial x^3 \partial t} + \frac{1}{48}\frac{\Delta t^3}{\Delta x}\frac{\partial^4 U}{\partial x \partial t^3}\right)_{i,j+1/2}$$

$$- \left(\frac{(\Delta x)^2}{12}\frac{\partial^4 U}{\partial x^4} + \frac{(\partial t)^2}{8}\frac{\partial^4 U}{\partial x^2 \partial t^2}\right)_{i,j+1/2}$$

$$+ \frac{\gamma_4}{4}\Delta t\frac{\partial^3 U}{\partial x^2 \partial t}\bigg|_{i,j+1/2} + \frac{\Delta t^2}{24}\frac{\partial^3 U}{\partial t^3}\bigg|_{i,j+1/2}$$

$$+ \frac{1}{\Delta t}0(\Delta x^{\alpha_1}, \Delta t^{\alpha_2}) = 0. \quad (3.3.3)$$

From this analysis it is clearly evident why it is necessary to define the conditions (3.2.2–3.2.4) as compulsory and the conditions (3.2.5) and (3.2.6) as optional. If the compulsory conditions are not fulfilled, it is clear that the approximation (3.2.1) does not approximate equation (3.1.1) and is inconsistent. However if γ_3 and γ_4 do not vanish, scheme (3.2.1) can still approximate equation (3.1.1) provided some restrictions are put on either Δt, Δx or their ratio.

By using equation (3.3.3) we can easily verify the following truncation errors for equation (3.2.7) as:

$$T_{3.2.7} = -\left(\frac{(\Delta x)^2}{12}\frac{\partial^4 U}{\partial x^4} + \frac{(\Delta t)^2}{8}\frac{\partial^4 U}{\partial x^2 \partial t^2}\right)_{i,j+1/2} + \frac{\Delta t^2}{24}\frac{\partial^3 U}{\partial t^3}\bigg|_{i,j+1/2} + \frac{1}{\Delta t}0(\Delta x^{\alpha_1}, \Delta t^{\alpha_2}),$$

(3.3.4)

for equation (3.2.8):

$$T_{3.2.8} = -\left(\frac{(\Delta x)^2}{12}\frac{\partial^4 U}{\partial x^4} + \frac{(\Delta t)^2}{8}\frac{\partial^4 U}{\partial x^2 \partial t^2}\right)_{i,j+1/2} - \frac{1}{2}\Delta t\frac{\partial^3 U}{\partial x^2 \partial t}\bigg|_{i,j+1/2}$$

$$+ \frac{\Delta t^2}{24}\frac{\partial^3 U}{\partial t^3}\bigg|_{i,j+1/2} + \frac{1}{\Delta t}0(\Delta x^{\alpha_1}, \Delta t^{\alpha_2}),$$

(3.3.5)

for equation (3.2.9):

$$T_{3.2.9} = -\left(\frac{(\Delta x)^2}{12}\frac{\partial^4 U}{\partial x^4} + \frac{(\Delta t)^2}{8}\frac{\partial^4 U}{\partial x^2 \partial t^2}\right)_{i,j+1/2} + \frac{1}{2}\Delta t\frac{\partial^3 U}{\partial x^2 \partial t}\bigg|_{i,j+1/2}$$

$$+ \frac{\Delta t^2}{24}\frac{\partial^3 U}{\partial t^3}\bigg|_{i,j+1/2} + \frac{1}{\Delta t}0(\Delta x^{\alpha_1}, \Delta t^{\alpha_2}),$$

(3.3.6)

for equation (3.2.10):

$$T_{3.2.10} = -\left(\frac{(\Delta x)^2}{12}\frac{\partial^4 U}{\partial x^4} + \frac{(\Delta t)^2}{8}\frac{\partial^4 U}{\partial x^2 \partial t^2}\right)_{i,j+1/2} + \frac{\Delta t}{2}(1-\alpha)\frac{\partial^3 U}{\partial x^2 \partial t}\bigg|_{i,j+1/2}$$

$$+ \frac{\Delta t^2}{24}\frac{\partial^3 U}{\partial t^3}\bigg|_{i,j+1/2} + \frac{1}{\Delta t}0(\Delta x^{\alpha_1}, \Delta t^{\alpha_2}),$$

(3.3.7)

and finally for equation (3.2.13):

$$T_{3.2.13} = -\left(\frac{(\Delta x)^2}{12}\frac{\partial^4 U}{\partial x^4} + \frac{(\Delta t)^2}{8}\frac{\partial^4 U}{\partial x^2 \partial t^2}\right)_{i,j+1/2} + \frac{\Delta t}{2}[1 - 2(\alpha - 1/6r)]\frac{\partial^3 U}{\partial x^2 \partial t}\bigg|_{i,j+1/2}$$

$$+ \frac{\Delta t^2}{24}\frac{\partial^3 U}{\partial t^3}\bigg|_{i,j+1/2} + \frac{1}{\Delta t}0(\Delta x^{\alpha_1}, \Delta t^{\alpha_2}).$$

(3.3.8)

For the asymmetric equations (3.2.14) and (3.2.15) the truncation errors are given by

$$T_{3.2.14} = -\left(\frac{\Delta t}{\Delta x}\frac{\partial^2 U}{\partial x \partial t} + \frac{1}{6}\Delta x \Delta t\frac{\partial^4 U}{\partial x^3 \partial t} + \frac{1}{24}\frac{\Delta t^3}{\Delta x}\frac{\partial^4 U}{\partial x \partial t^3}\right)_{i,j+1/2}$$

$$-\left(\frac{(\Delta x)^2}{12}\frac{\partial^4 U}{\partial x^4} + \frac{(\Delta t)^2}{8}\frac{\partial^4 U}{\partial x^2 \partial t^2}\right)_{i,j+1/2} + \frac{\Delta t^2}{24}\frac{\partial^3 U}{\partial t^3}\bigg|_{i,j+1/2}$$

$$+ \frac{1}{\Delta t}0(\Delta x^{\alpha_1}, \Delta t^{\alpha_2}),$$

(3.3.9)

and

$$T_{3.2.15} = \left(\frac{\Delta t}{\Delta x} \frac{\partial^2 U}{\partial x \partial t} + \frac{1}{6} \Delta x \Delta t \frac{\partial^4 U}{\partial x^3 \partial t} + \frac{1}{24} \frac{\Delta t^3}{\Delta x} \frac{\partial^4 U}{\partial x \partial t^3} \right)_{i,j+1/2}$$

$$- \left(\frac{(\Delta x)^2}{12} \frac{\partial^4 U}{\partial x^4} + \frac{(\Delta t)^2}{8} \frac{\partial^4 U}{\partial x^2 \partial t^2} \right)_{i,j+1/2} + \frac{\Delta t^2}{24} \frac{\partial^3 U}{\partial t^3} \bigg|_{i,j+1/2}$$

$$+ \frac{1}{\Delta t} 0(\Delta x^{\alpha_1}, \Delta t^{\alpha_2}), \tag{3.3.10}$$

respectively. Consequently from equation (3.3.3) the general expression for the truncation error of the approximation is represented by (3.2.1). We can derive the truncation errors for all the approximate formulae under discussion.

3.4. Stability Analysis for the Generalised Equation

To investigate the stability condition of equation (3.3.1), we use the method of Fourier series.

Substitution of the error function at any point (i,j), i.e.,

$$E_{i,j} = e^{\sqrt{-1} \beta x} e^{\alpha t},$$

$$= e^{\sqrt{-1} \beta i \Delta x} e^{\alpha j \Delta t},$$

$$= e^{\sqrt{-1} \beta i \Delta x} \xi^j, \qquad \xi = e^{\alpha \Delta t}, \quad \alpha \text{ complex},$$

into the original approximating equation (3.3.1), will result in

$$\xi = \frac{r\theta_1' e^{\sqrt{-1} \beta \Delta x} + [1 - r(\theta_1' + \theta_2')] + r\theta_2' e^{-\sqrt{-1} \beta \Delta x}}{-r\theta_1 e^{\sqrt{-1} \beta \Delta x} + [1 + r(\theta_1 + \theta_2)] - r\theta_2 e^{-\sqrt{-1} \beta \Delta x}} \tag{3.4.1}$$

$$= \frac{1 - r(1 - \cos \beta \Delta x)(\theta_1' + \theta_2') + \sqrt{-1} r(\theta_1' - \theta_2') \sin \beta \Delta x}{1 - r(\cos \beta \Delta x - 1)(\theta_1 + \theta_2) - \sqrt{-1} r(\theta_1 - \theta_2) \sin \beta \Delta x}, \tag{3.4.2}$$

i.e.,

$$|\xi|^2 = \frac{1 + 16r^2 \theta_1' \theta_2' s^4 - 4r(\theta_1' + \theta_2') s^2 + 4r^2(\theta_1' - \theta_2')^2 s^2}{1 + 16r^2 \theta_1 \theta_2 s^4 + 4r(\theta_1 + \theta_2) s^2 + 4r^2(\theta_1 - \theta_2)^2 s^2}, \tag{3.4.3}$$

where $s = \sin^2(\beta \Delta x)/2$.

For equation (3.3.1) to be stable we need $|\xi| \le 1$. The expression (3.4.3) can be used to verify the stability condition of all the schemes given in Section 3.2. For the equation (3.2.7), we have

$$|\xi| = \frac{1 + 4r^2s^4 - 4rs^2}{1 + 4r^2s^4 + 4rs^2} = \frac{1 + 4rs^2(rs^2 - 1)}{1 + 4rs^2(rs^2 + 1)} \leq 1, \qquad (3.4.4)$$

for all $r > 0$. Similarly, for the equation (3.2.8),

$$|\xi| = \frac{1}{1 + 16\rho^2s^4 + 8rs^2} \leq 1 \quad \text{for all } r > 0. \qquad (3.4.5)$$

Meanwhile the stability of (3.2.9) can be fulfilled if

$$1 + 8rs^2(2r^2 - 1) \leq 1, \qquad (3.4.6)$$

i.e.,

$$r \leq \frac{1}{2s^2} \leq \frac{1}{2}. \qquad (3.4.7)$$

For the equation (3.2.10), (3.4.3) will be less than (or equal) to unity if

$$r \leq \frac{1}{2(1 - \alpha)s^2} \leq \frac{1}{2(1 - \alpha)}, \qquad (3.4.8)$$

which is similar to the original condition derived by Saul'yev.

In the case of equation (3.2.13), the stability condition can be fulfilled if

$$8rs^2 \left[2rs^2 \left\{ 1 - 2\left(\alpha - \frac{1}{6r} \right) \right\} - 1 \right] \leq 0, \qquad (3.4.9)$$

i.e.,

$$r \leq \frac{1}{6(1 - 2\alpha)}. \qquad (3.4.10)$$

For the equation (3.2.14) and (3.2.15), equation (3.4.3) is equal to

$$\frac{1 - 4rs^2 + 4r^2s^2}{1 + 4rs^2 + 4r^2s^2}, \qquad (3.4.11)$$

which is less than (or equal to) unity for all $r > 0$. Therefore both equations are unconditionally stable. Therefore, the following theorems can be established.

Theorem 3.1.
The finite difference approximation (3.2.1) is stable if (3.4.3) is less than (or equal to) unity for any choice of θ_i and θ_i' which satisfy (3.2.2–3.2.4).

Theorem 3.2.
The finite difference approximations (3.2.14) and (3.2.15) are unconditionally stable for all $r > 0$.

Theorem 3.1 gives the stability condition for the general six point two-level finite difference approximation (3.2.1) and Theorem 3.2 is necessary for purposes which will be developed later.

3.5. A New Group Explicit Method

Now consider any group of two points, i.e., $(i\Delta x, (j+1/2)\Delta t)$ and $((i+1)\Delta x, (j+1/2)\Delta t)$, in which equations (3.2.14) and (3.2.15) are used simultaneously to calculate the values of u at these points, respectively. Therefore, at point $(i\Delta x, (j+1/2)\Delta t)$ the solution is approximated by (Evans and Abdullah, 1983)

$$-ru_{i+1,j+1} + (1+r)u_{i,j+1} = ru_{i-1,j} + (1-r)u_{i,j} \quad \text{(i.e., eqn. 3.2.14)}, \qquad (3.5.1)$$

whilst at point $((i+1)\Delta x, (j+1/2)\Delta t))$ the solution is given by

$$-ru_{i,j+1} + (1+r)u_{i+1,j+1} = (1-r)u_{i+1,j} + ru_{i+2,j} \quad \text{(i.e., eqn. 3.2.15)}. \qquad (3.5.2)$$

If we now write equation (3.5.1) and (3.5.2) simultaneously in the matrix form, we have

$$\begin{bmatrix} 1+r & -r \\ -r & 1+r \end{bmatrix} \begin{bmatrix} u_{i,j+1} \\ u_{i+1,j+1} \end{bmatrix} = \begin{bmatrix} 1-r & 0 \\ 0 & 1-r \end{bmatrix} \begin{bmatrix} u_{i,j} \\ u_{i+1,j} \end{bmatrix} + \begin{bmatrix} ru_{i-1,j+1} \\ ru_{i+2,j} \end{bmatrix}, \qquad (3.5.3)$$

whose (2×2) matrix of coefficients can easily be inverted so that the equation can be written in explicit form as

$$\begin{bmatrix} u_{i,j+1} \\ u_{i+1,j+1} \end{bmatrix} = \frac{1}{|A|} \begin{bmatrix} 1+r & r \\ r & 1+r \end{bmatrix} \left\{ \begin{bmatrix} 1-r & 0 \\ 0 & 1-r \end{bmatrix} \begin{bmatrix} u_{i,j} \\ u_{i+1,j} \end{bmatrix} + \begin{bmatrix} ru_{i-1,j+1} \\ ru_{i+2,j} \end{bmatrix} \right\}, \qquad (3.5.4)$$

where $|A| = \det A = 1 + 2r$. This simplifies to the new formulae

$$\begin{bmatrix} u_{i,j+1} \\ u_{i+1,j+1} \end{bmatrix} = \frac{1}{|A|} \begin{bmatrix} r(1+r)u_{i-1,j} + (1-r^2)u_{i,j} + r(1-r)u_{i+1,j} + r^2 u_{i+2,j} \\ r^2 u_{i-1,j} + r(1+r)u_{i,j} + (1-r^2)u_{i+1,j} + r(1+r)u_{i+2,j} \end{bmatrix}, \qquad (3.5.5)$$

where the computational molecule is represented by Figure 3.5.1a.

For the ungrouped (single) points near the right and left boundaries we can use the less accurate equations (3.2.14) and (3.2.15), respectively, i.e., for the right boundary

$$u_{m-1,j+1} = \frac{1}{(1+r)} \{ ru_{m,j+1} + ru_{m-2,j} + (1-r)u_{m-1,j} \}, \qquad (3.5.6)$$

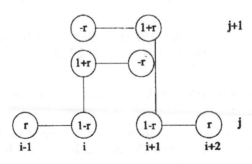

Figure 3.5.1a

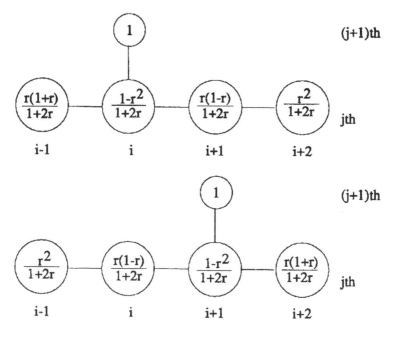

Figure 3.5.1b

and for the left boundary i.e.,

$$u_{1,j+1} = \frac{1}{1+r}\{ru_{0,j+1} + ru_{2,j} + (1-r)u_{1,j}\}.$$ (3.5.7)

Now we will consider a variety of schemes for this class of methods which are possible for solving the equation (3.1.1) for an odd or even number of intervals.

(a) Even Number of Intervals
Assume the line segment $0 \le x \le 1$ is divided into a number m of equal intervals with m even. Therefore, at every time level, the number of internal points is odd, i.e. $(m-1)$. This results in one ungrouped point at either end as shown in Figure 3.5.2.

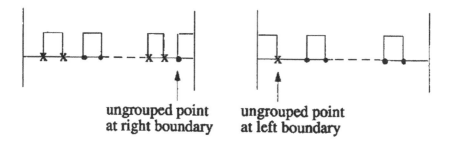

ungrouped point   ungrouped point
at right boundary   at left boundary

Figure 3.5.2

(i) *GER Scheme*   This scheme is obtained by the use of equation (3.5.5) for $(1/2)$ $(m-2)$ times for the first $(m-2)$ points grouped 2 at a time and equation (3.5.6) for the $(m-1)^{th}$ point at every time level. This is called the Group Explicit with Right ungrouped point (GER) scheme.

In implicit matrix form, it is given by

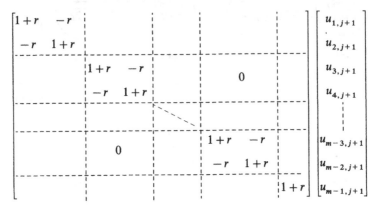

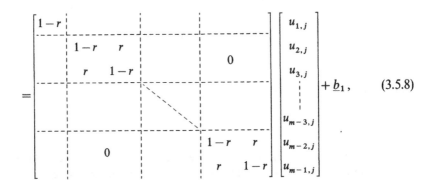

where

$$\underline{b}_1^T = [ru_{o,j}, 0, \ldots, 0, ru_{m,j+1}],$$

and consists of known boundary values, or,

$$(I + rG_1)\underline{u}_{j+1} = (I - rG_2)\underline{u}_j + \underline{b}_1, \qquad (3.5.9)$$

where

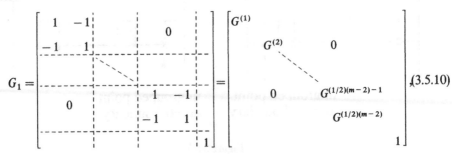

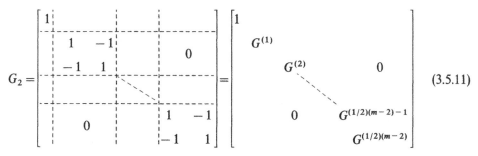

$$(3.5.11)$$

with

$$G^{(i)} = \begin{bmatrix} 1 & -1 \\ -1 & 1 \end{bmatrix}, \quad i = 1, 2, \ldots, (1/2)(m-2), \; j = 0, 1, 2, \ldots \; .$$

The scheme can be described diagrammatically in Figure 3.5.3.

(ii) *GEL Scheme*   The second scheme is obtained by the use of equation (3.4.7) for the first point and $(1/2)(m-2)$ times of equation (3.5.5) for the remainder of the points at every time level. This is called Group Explicit with Left ungrouped point (GEL) scheme. This scheme is given by

$$(I + rG_2)\underline{u}_{j+1} = (I - rG_1)\underline{u}_j + \underline{b}_2, \qquad (3.5.12)$$

$\underline{b}_2^T = [ru_{0,j+1}, 0, \ldots, 0, ru_{m,j}]$ and the 'brick' diagram as in Figure 3.5.4.

(iii) *The SAGE Scheme*   Another variation is the coupled use of the first scheme (3.5.9) and the second scheme (3.5.12) at every alternate time level. This scheme is called the Single Alternating Group Explicit (SAGE) and is given by

$$(I + rG_1)\underline{u}_{j+1} = (I - rG_2)\underline{u}_j + \underline{b}_1,$$

$$(I + rG_2)\underline{u}_{j+2} = (I - rG_1)\underline{u}_{j+1} + \underline{b}_2, \qquad (3.5.13)$$

and the diagram is given by Figure 3.5.5.

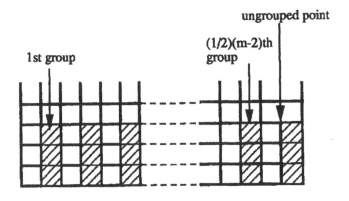

Figure 3.5.3   GER (Group Explicit with Right Ungrouped Point) Method.

ungrouped point

1st group

(1/2)(m-2)th
group

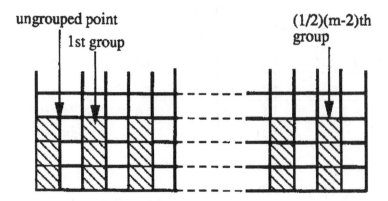

Figure 3.5.4 GEL (Group Explicit with Left Ungrouped Point) Method.

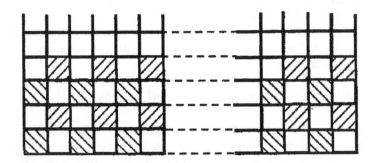

Figure 3.5.5 SAGE (Single Alternating Group Explicit) Method.

(iv) *The DAGE Scheme* The final scheme developed is a periodic rotation of the SAGE two time level as given in the previous scheme resulting in a four time level step process with the second half-cycle in reverse order. The scheme is called the Double Alternating Group Explicit DAGE method and is given by the formulae

$$(I + rG_1)\underline{u}_{j+1} = (I - rG_2)\underline{u}_j + \underline{b}_1,$$

$$(I + rG_2)\underline{u}_{j+2} = (I - rG_1)\underline{u}_{j+1} + \underline{b}_2,$$

$$(I + rG_2)\underline{u}_{j+3} = (I - rG_1)\underline{u}_{j+2} + \underline{b}_2,$$

$$(I + rG_1)\underline{u}_{j+4} = (I - rG_2)\underline{u}_{j+3} + \underline{b}_1. \tag{3.5.14}$$

Diagrammatically it is represented by Figure 3.5.6.

(b) Odd Number of Intervals
In this case, m is assumed to be odd. Therefore at every time level, the number of internal points is even, i.e. $(m-1)$. This gives at every time level either $(m-1/2)$ complete groups of two points or $(m-3/2)$ groups of two points and one ungrouped point, adjacent to the left and right boundaries.

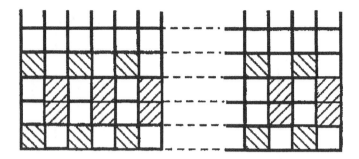

Figure 3.5.6 DAGE (Double Alternating Group Explicit) Method.

The first scheme for this case is obtained by using at every time level, equation (3.5.7) at the left ungrouped point, $(m-3/2)$ times equation (3.5.5) from the second point to $(m-2)^{th}$ point and equation (3.5.6) for the last $(m-1)^{th}$ point which is left ungrouped at the far right of the line adjacent to the boundary. This scheme is written as

$$(I + r\hat{G}_1)\underline{u}_{j+1} = (I - r\hat{G}_2)\underline{u}_j + \underline{b}_3, \tag{3.5.15}$$

where

$$\underline{b}_3^T = [ru_{0,j+1}, 0, \ldots, 0, ru_{m,j+1}],$$

$$\hat{G}_1 = \begin{bmatrix} 1 & & & & & \\ & G^{(1)} & & & 0 & \\ & & G^{(2)} & & & \\ & & & \ddots & & \\ & 0 & & G^{(1/2)(m-3)-1} & \\ & & & & G^{(1/2)(m-3)} & \\ & & & & & 1 \end{bmatrix}, \tag{3.5.16}$$

$$\hat{G}_2 = \begin{bmatrix} G^{(1)} & & & \\ & G^{(2)} & & 0 \\ & & \ddots & \\ & 0 & & G^{(1/2)(m-3)} \\ & & & G^{(1/2)(m-1)} \end{bmatrix}, \tag{3.5.17}$$

and $G^{(i)}$, $i = 1, 2, \ldots, (1/2)(m-1)$ are (2×2) matrices as previously defined. This scheme is denoted as Group Explicit with both Ungrouped ends (GEU) and described by Figure 3.5.7.

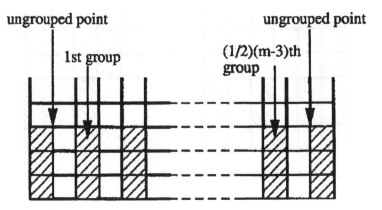

Figure 3.5.7   GEU (Group Explicit with Ungrouped ends) Method.

The second scheme is obtained using $(m - 1/2)$ times equation (3.5.5) for the first to $(m - 1)^{th}$ point (Fig. 3.5.8) with the equation given by

$$(I + r\hat{G}_2)\underline{u}_{j+1} = (I - r\hat{G}_1)\underline{u}_j + \underline{b}_4, \qquad (3.5.18)$$

$\underline{b}_4^T = [ru_{0,j}, 0, \ldots, 0, ru_{m,j}]$, and the scheme is known as the Group Explicit Complete (GEC) method.

The alternating schemes corresponding to the schemes (3.5.13) and (3.5.14) above for the even case are given by

$$(I + r\hat{G}_1)\underline{u}_{j+1} = (I - r\hat{G}_2)\underline{u}_j + \underline{b}_3,$$

$$(I + r\hat{G}_2)\underline{u}_{j+2} = (I - r\hat{G}_1)\underline{u}_{j+1} + \underline{b}_4, \qquad (3.5.19)$$

for SAGE (Fig. 3.5.9) and

$$(I + r\hat{G}_1)\underline{u}_{j+1} = (I - r\hat{G}_2)\underline{u}_j + \underline{b}_3,$$

$$(I + r\hat{G}_2)\underline{u}_{j+2} = (I - r\hat{G}_1)\underline{u}_{j+1} + \underline{b}_4,$$

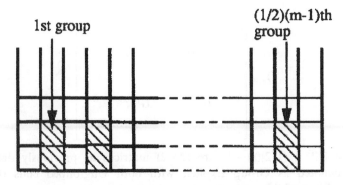

Figure 3.5.8   GEC (Group Explicit Complete) Method.

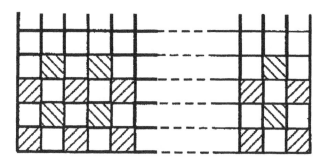

Figure 3.5.9 SAGE (Single Alternating Group Explicit) Method.

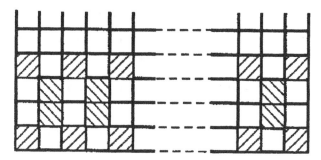

Figure 3.5.10 DAGE (Double Alternating Group Explicit) Method.

$$(I + r\hat{G}_2)\underline{u}_{j+3} = (I - r\hat{G}_1)\underline{u}_{j+2} + \underline{b}_4,$$
$$(I + r\hat{G}_1)\underline{u}_{j+4} = (I - r\hat{G}_2)\underline{u}_{j+3} + \underline{b}_3, \tag{3.5.20}$$

for DAGE (Fig. 3.5.10).

3.6. Group Explicit Methods–Fractional Splitting

Now if the levels j and $(j + 1)$ of the time variable are divided into a small number of fractional or 'artificial' sub-levels, then instead of using the SAGE and DAGE schemes for every two and four sub-time levels, respectively, we can use them as a complete cycle at every time level since we have intermediate 'results' which fall on the artificial levels. Therefore a fractional splitting type of the SAGE scheme equation (3.5.13) is given by

$$\left(I + \frac{r}{2}G_1\right)\underline{u}_{j+1/2} = \left(I - \frac{r}{2}G_2\right)\underline{u}_j + \underline{b}_1,$$

$$\left(I + \frac{r}{2}G_2\right)\underline{u}_{j+1} = \left(I - \frac{r}{2}G_1\right)\underline{u}_{j+1/2} + \underline{b}_2. \tag{3.6.1}$$

Similarly the fractional splitting form of the DAGE scheme equation (3.5.14) is given by

$$\left(I + \frac{r}{2}G_1\right)\underline{u}_{j+1/2} = \left(I - \frac{r}{2}G_2\right)\underline{u}_j + \underline{b}_1,$$

$$\left(I + \frac{r}{2}G_2\right)\underline{u}_{j+1} = \left(I - \frac{r}{2}G_1\right)\underline{u}_{j+1/2} + \underline{b}_2,$$

$$\left(I + \frac{r}{2}G_2\right)\underline{u}_{j+3/2} = \left(I - \frac{r}{2}G_1\right)\underline{u}_j + \underline{b}_2,$$

$$\left(I + \frac{r}{2}G_1\right)\underline{u}_{j+2} = \left(I - \frac{r}{2}G_2\right)\underline{u}_{j+3/2} + \underline{b}_1, \tag{3.6.2}$$

where the complete splitting process takes place every two time levels.

Alternatively if the levels j and $(j+1)$ are divided into four sub-levels, the fractional or quarter-step of the DAGE scheme is then given as:

$$\left(I + \frac{r}{4}G_1\right)\underline{u}_{j+1/4} = \left(I - \frac{r}{4}G_2\right)\underline{u}_j + \underline{b}_1,$$

$$\left(I + \frac{r}{4}G_2\right)\underline{u}_{j+1/2} = \left(I - \frac{r}{4}G_1\right)\underline{u}_{j+1/4} + \underline{b}_2,$$

$$\left(I + \frac{r}{4}G_2\right)\underline{u}_{j+3/4} = \left(I - \frac{r}{4}G_1\right)\underline{u}_{j+1/2} + \underline{b}_2,$$

$$\left(I + \frac{r}{4}G_1\right)\underline{u}_{j+1} = \left(I - \frac{r}{4}G_2\right)\underline{u}_{j+3/4} + \underline{b}_1. \tag{3.6.3}$$

3.7. The Group Explicit Method for Periodic Boundary Conditions

All the schemes discussed in the previous section are applicable only for the Dirichlet boundary conditions. However they can be easily adapted to cope with Neumann and mixed conditions with minor modifications. However in the case where equation (3.1.1) is associated with periodic boundary conditions, the matrix of coefficients involved will be slightly different.

As will now be shown given the problem defined by equation (3.1.1) with the periodic boundary conditions

$$U(0,t) = U(1,t) \quad \text{and} \quad \frac{\partial U}{\partial x}(0,t) = \frac{\partial U}{\partial x}(1,t). \tag{3.7.1}$$

Then, after discretisation on the given grid of n points, the class of group explicit methods can be reformulated as follows.

If it is assumed that the number of intervals is even then the number of unknown points is also even. The Group Explicit Complete (GEC) scheme is given by

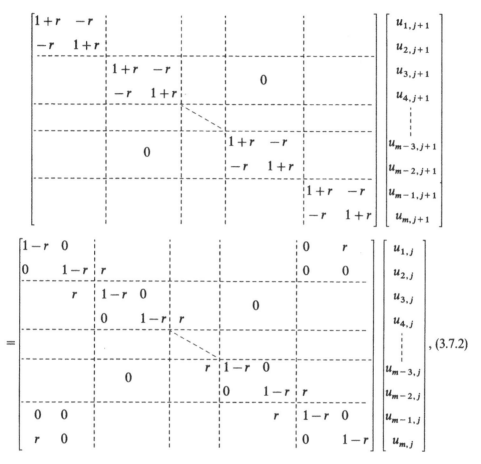

$$, (3.7.2)$$

or

$$(I + r\tilde{G}_2)\underline{u}_{j+1} = (I - r\tilde{\tilde{G}}_1)\underline{u}_j,$$ (3.7.3)

where $\tilde{G}_2$ is given by (3.5.17), and,

$$\tilde{\tilde{G}}_1 = \begin{bmatrix} 1 & & & & & -1 \\ & G^{(1)} & & & & \\ & & G^{(2)} & & 0 & \\ & & & \ddots & & \\ & 0 & & & G^{(m-2/2)} & \\ -1 & & & & & 1 \end{bmatrix},$$ (3.7.4)

and the $G^{(i)}$, $i = 1, 2, \ldots, (m-2)/2$, are as previously defined.

Meanwhile the Group Explicit with both Ungrouped end points (GEU) is given by

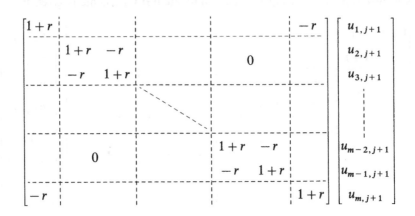

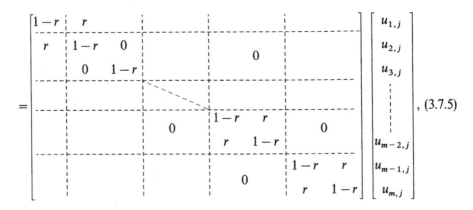

$$, (3.7.5)$$

or

$$(I + r\overset{\approx}{G}_1)\underline{u}_{j+1} = (I - r\tilde{G}_2)\underline{u}_j. \tag{3.7.6}$$

Now from the Group Explicit Complete (GEC) and Group Explicit with both Ungrouped ends (GEU) methods, the following fractional step schemes can be developed:

(i) Single Alternating Group Explicit (SAGE), i.e.,

$$(I + r\overset{\approx}{G}_2)\underline{u}_{j+1} = (I - r\tilde{G}_1)\underline{u}_j,$$
$$(I + r\overset{\approx}{G}_1)\underline{u}_{j+2} = (I - r\tilde{G}_2)\underline{u}_{j+1}. \tag{3.7.7}$$

(ii) Double Alternating Group Explicit (DAGE), i.e.,

$$(I + r\tilde{G}_2)\underline{u}_{j+1} = (I - r\overset{\approx}{G}_1)\underline{u}_j,$$
$$(I + r\overset{\approx}{G}_1)\underline{u}_{j+2} = (I - r\tilde{G}_2)\underline{u}_{j+1},$$
$$(I + r\overset{\approx}{G}_1)\underline{u}_{j+3} = (I - r\tilde{G}_2)\underline{u}_{j+2},$$
$$(I + r\tilde{G}_2)\underline{u}_{j+4} = (I - r\overset{\approx}{G}_1)\underline{u}_{j+3}. \tag{3.7.8}$$

(iii) Half-step single AGE, i.e.,

$$\left(I+\frac{r}{2}\tilde{G}_2\right)\underline{u}_{j+1/2}=\left(I-\frac{r}{2}\tilde{\tilde{G}}_1\right)\underline{u}_j,$$

$$\left(I+\frac{r}{2}\tilde{\tilde{G}}_1\right)\underline{u}_{j+1}=\left(I-\frac{r}{2}\tilde{G}_2\right)\underline{u}_{j+1/2}. \tag{3.7.9}$$

(iv) Half-step double AGE, i.e.,

$$\left(I+\frac{r}{4}\tilde{G}_2\right)\underline{u}_{j+1/2}=\left(I-\frac{r}{4}\tilde{\tilde{G}}_1\right)\underline{u}_j,$$

$$\left(I+\frac{r}{4}\tilde{\tilde{G}}_1\right)\underline{u}_{j+1}=\left(I-\frac{r}{4}\tilde{G}_2\right)\underline{u}_{j+1/2},$$

$$\left(I+\frac{r}{4}\tilde{\tilde{G}}_1\right)\underline{u}_{j+3/2}=\left(I-\frac{r}{4}\tilde{G}_2\right)\underline{u}_{j+1},$$

$$\left(I+\frac{r}{4}\tilde{G}_2\right)\underline{u}_{j+2}=\left(I-\frac{r}{4}\tilde{\tilde{G}}_1\right)\underline{u}_{j+3/2}. \tag{3.7.10}$$

(v) Quarter-step double AGE, i.e.,

$$\left(I+\frac{r}{4}\tilde{G}_2\right)\underline{u}_{j+1/4}=\left(I-\frac{r}{4}\tilde{\tilde{G}}_1\right)\underline{u}_j,$$

$$\left(I+\frac{r}{4}\tilde{\tilde{G}}_1\right)\underline{u}_{j+1/2}=\left(I-\frac{r}{4}\tilde{G}_2\right)\underline{u}_{j+1/4},$$

$$\left(I+\frac{r}{4}\tilde{\tilde{G}}_1\right)\underline{u}_{j+3/4}=\left(I-\frac{r}{4}\tilde{G}_2\right)\underline{u}_{j+1/2},$$

$$\left(I+\frac{r}{4}\tilde{G}_2\right)\underline{u}_{j+1}=\left(I-\frac{r}{4}\tilde{\tilde{G}}_1\right)\underline{u}_{j+3/4}. \tag{3.7.11}$$

Now if it is assumed that the number of intervals is odd, then the number of unknowns is also odd, i.e., $u_{0,j}, u_{1,j}, \ldots, u_{m-1,j}$ for $j = 1, 2, \ldots$. Therefore the Group Explicit with right ungrouped point (GER) scheme is given by

$$(I+r\check{G}_1)\underline{u}_{j+1}=(I-r\check{G}_2)\underline{u}_j, \tag{3.7.12}$$

where

$$\check{G}_1 = \left[\begin{array}{ccccc}
1 & -1 & & & \\
-1 & 1 & & 0 & \\
& & & & \\
& & 1 & -1 & \\
0 & & -1 & 1 & \\
& & & & 1
\end{array}\right], \tag{3.7.13}$$

and

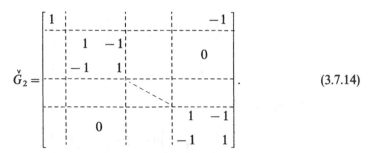

$$\check{G}_2 = \begin{bmatrix} 1 & & & & -1 \\ & 1 & -1 & & \\ & -1 & 1 & & 0 \\ & & & & \\ & & & 1 & -1 \\ & 0 & & -1 & 1 \end{bmatrix}. \qquad (3.7.14)$$

While the scheme corresponding to the Group Explicit method with Left ungrouped point (GEL) is given by

$$(I + r\check{G}_2)\underline{u}_{j+1} = (I - r\check{G}_1)\underline{u}_j. \qquad (3.7.15)$$

Finally from the two schemes (3.7.12) and (3.7.15) we can derive in an exactly similar manner the corresponding fractional scheme methods to equations (3.7.7–3.7.11).

3.8. The Truncation Errors for the Group Explicit Method

To estimate the truncation errors for the schemes in the class of Group Explicit method, firstly, we need to estimate the truncation errors for the approximations (3.5.5–3.5.7).

From (3.3.9) and (3.3.10), the estimate of the truncation errors for (3.5.6) and (3.5.7) are, respectively, given by

$$T_{3.5.6} = -\left(\frac{\Delta t}{\Delta x}\frac{\partial^2 U}{\partial x \partial t} + \frac{1}{6}\Delta x \Delta t \frac{\partial^4 U}{\partial x^3 \partial t} + \frac{1}{24}\frac{\Delta t^3}{\Delta x}\frac{\partial^4 U}{\partial x \partial t^3}\right)_{m-1,j+1/2}$$
$$-\left(\frac{(\Delta x)^2}{12}\frac{\partial^4 U}{\partial x^4} + \frac{(\Delta t)^2}{8}\frac{\partial^4 U}{\partial x^2 \partial t^2}\right)_{m-1,j+1/2} + \frac{\Delta t^2}{24}\frac{\partial^3 U}{\partial t^3}\Big|_{m-1,j+1/2}$$
$$+\frac{1}{\Delta t}0(\Delta x^{\alpha_1}, \Delta t^{\alpha_2}), \qquad (3.8.1)$$

and

$$T_{3.5.7} = \left(\frac{\Delta t}{\Delta x}\frac{\partial^2 U}{\partial x \partial t} + \frac{1}{6}\Delta x \Delta t \frac{\partial^4 U}{\partial x^3 \partial t} + \frac{1}{24}\frac{\Delta t^3}{\Delta x}\frac{\partial^4 U}{\partial x \partial t^3}\right)_{1,j+1/2}$$
$$-\left(\frac{(\Delta x)^2}{12}\frac{\partial^4 U}{\partial x^4} + \frac{(\Delta t)^2}{8}\frac{\partial^4 U}{\partial x^2 \partial t^2}\right)_{1,j+1/2} + \frac{\Delta t^2}{24}\frac{\partial^3 U}{\partial t^3}\Big|_{1,j+1/2}$$
$$+\frac{1}{\Delta t}0(\Delta x^{\alpha_1}, \Delta t^{\alpha_2}). \qquad (3.8.2)$$

Expanding each term of both equations in (3.5.5) about $(i\Delta x, (j+1/2)\Delta t)$ and $((i+1)\Delta x, (j+1/2)\Delta t)$, respectively, using a Taylor's series expansion will give the following:

For the first equation of (3.5.5) results in the expression

$$\left(\frac{\partial U}{\partial t} - \frac{\partial^2 U}{\partial x^2}\right)_{i,j+1/2} - \frac{\Delta t}{\Delta x}\left(\frac{1}{1+2r}\right)\frac{\partial^3 U}{\partial x^2 \partial t}\bigg|_{i,j+1/2} + \Delta t \frac{\partial^3 U}{\partial x^2 \partial t}\bigg|_{i,j+1/2}$$

$$- \frac{\Delta t^2}{24}\frac{\partial^3 U}{\partial t^3}\bigg|_{i,j+1/2} - \frac{(\Delta x)^2}{24}\frac{(r+8)}{(1+2r)}\frac{\partial^4 U}{\partial x^4}\bigg|_{i,j+1/2}$$

$$+ \frac{\Delta t \Delta x}{2}\frac{r}{1+2r}\frac{\partial^4 U}{\partial x^3 \partial t}\bigg|_{i,j+1/2} - \frac{\Delta t^2}{8}\frac{\partial^4 U}{\partial x^2 \partial t^2}\bigg|_{i,j+1/2} + \frac{1}{\Delta t}0(\Delta x^{\alpha_1}, \Delta t^{\alpha_2}) = 0, \quad (3.8.3)$$

while the second equation of (3.5.5) results in the expression

$$\left(\frac{\partial U}{\partial t} - \frac{\partial^2 U}{\partial x^2}\right)_{i+1,j+1/2} + \frac{\Delta t}{\Delta x}\left(\frac{1}{1+2r}\right)\frac{\partial^3 U}{\partial x^2 \partial t}\bigg|_{i,j+1/2} + \Delta t \frac{\partial^3 U}{\partial x^2 \partial t}\bigg|_{i+1,j+1/2}$$

$$- \frac{\Delta t^2}{24}\frac{\partial^3 U}{\partial t^3}\bigg|_{i+1,j+1/2} - \frac{(\Delta x)^2}{24}\frac{(r+8)}{(1+2r)}\frac{\partial^4 U}{\partial x^4}\bigg|_{i+1,j+1/2}$$

$$- \frac{\Delta t \Delta x}{2}\frac{r}{1+2r}\frac{\partial^4 U}{\partial x^3 \partial t}\bigg|_{i+1,j+1/2} - \frac{\Delta t^2}{8}\frac{\partial^4 U}{\partial x^2 \partial t^2}\bigg|_{i,j+1/2} + \frac{1}{\Delta t}0(\Delta x^{\alpha_1}, \Delta t^{\alpha_2}) = 0. \quad (3.8.4)$$

Now having found the truncation error terms for equations (3.5.5–3.5.7) which are the basic equations for all the schemes used, we can now proceed to examine the truncation error for the class of Group Explicit methods.

(i) *GER Scheme* The truncation errors at a group of two points for any time level are, respectively, given by the error terms of equations (3.8.3) and (3.8.4) namely,

$$\left\{ -\frac{\Delta t}{\Delta x}\left(\frac{1}{1+2r}\right)\frac{\partial^3 U}{\partial x^2 \partial t} + \Delta t \frac{\partial^3 U}{\partial x^2 \partial t} - \frac{\Delta t^2}{24}\frac{\partial^3 U}{\partial t^3} - \frac{(\Delta x)^2}{24}\frac{(r+8)}{(1+2r)}\frac{\partial^4 U}{\partial x^4} \right.$$

$$\left. + \frac{\Delta t \Delta x}{2}\left(\frac{r}{1+2r}\right)\frac{\partial^4 U}{\partial x^3 \partial t} - \frac{\Delta t^2}{8}\frac{\partial^4 U}{\partial x^2 \partial t^2} + \frac{1}{\Delta t}0(\Delta x^{\alpha_1}, \Delta t^{\alpha_2}) \right\}_{i,j+1/2}, \quad (3.8.5)$$

and

$$\left\{ \frac{\Delta t}{\Delta x}\left(\frac{1}{1+2r}\right)\frac{\partial^3 U}{\partial x^2 \partial t} + \Delta t \frac{\partial^3 U}{\partial x^2 \partial t} - \frac{\Delta t^2}{24}\frac{\partial^3 U}{\partial t^3} - \frac{(\Delta x)^2}{24}\frac{(r+8)}{(1+2r)}\frac{\partial^4 U}{\partial x^4} \right.$$

$$\left. - \frac{\Delta t \Delta x}{2}\left(\frac{r}{1+2r}\right)\frac{\partial^4 U}{\partial x^3 \partial t} - \frac{\Delta t^2}{8}\frac{\partial^4 U}{\partial x^2 \partial t^2} + \frac{1}{\Delta t}0(\Delta x^{\alpha_1}, \Delta t^{\alpha_2}) \right\}_{i+1,j+1/2}, \quad (3.8.6)$$

$$i = 1, 3, 5, \ldots, m-3,$$

and at the last ungrouped point the error term is given by equation (3.8.1).

(ii) *GEL Scheme* The truncation error at the first point from the left-hand side boundary is given by the equation (3.8.2) for any time level. Similarly the truncation errors at the remaining groups of two points are given by equations (3.8.5) and (3.8.6), respectively, for $i = 2, 4, \ldots, m - 2$.

(iii) *GEU Scheme* This scheme which occurs in the case of an odd number of intervals has the truncation errors at the first point given by the equation (3.8.2), and at every group of two points from the second point up to $(m - 2)^{th}$ point given by the equations (3.8.5) and (3.8.6), respectively, and at the last point, i.e., the $(m - 1)^{th}$ point given by the equation (3.8.1).

(iv) *GEC Scheme* Since the number of points along each line form $(m - 1)/2$ complete groups of two points each, then the truncation error for this scheme is given by the equation (3.8.5) and (3.8.6), respectively, for $i = 1, 3, \ldots, m - 4, m - 2$, where m is odd.

(v) *SAGE Scheme* For an even number of intervals, the SAGE schemes constitute the alternating use of the GER and GEL schemes. If it is assumed that for the jth and $(j + 1)^{th}$ time levels the GER and GEL schemes are used, then the truncation error along the jth and $(j + 1)^{th}$ time level is given by the truncation errors of the GER and GEL schemes, respectively. Due to the difference in signs of the truncation errors, these alternating errors will tend to cancel the effect of the $(\Delta t/\Delta x)$ and $\Delta t \Delta x$ terms at most internal points, thus leaving the truncation error to have the approximate value $0(\Delta t + (\Delta x)^2)$.

(vi) *DAGE Scheme* By a similar reasoning this scheme will tend to have the approximate error of $0(\Delta t + (\Delta x)^2)$ also. However the numerical experiments will show in a later section that this scheme produces more accurate results than the SAGE scheme.

3.9. Stability of the Group Explicit Method

All schemes in the class of the group explicit methods above are formed independently by the combination of the equations (3.5.5–3.5.7). Theorem 3.2 has already concluded that the schemes (3.5.6) and (3.5.7) are unconditionally stable for $r > 0$. Now we try to establish the stability of each of the formulae (3.5.5).

Equation (3.5.5) can be written individually as

$$u_{i,j+1} = \frac{1}{\Delta}\{r(1+r)u_{i-1,j} + (1-r^2)u_{i,j} + r(1-r)u_{i+1,j} + r^2 u_{i+2,j}\}, \quad (3.9.1)$$

and

$$u_{i+1,j+1} = \frac{1}{\Delta}\{r^2 u_{i-1,j} + r(1-r)u_{i,j} + (1-r^2)u_{i+1,j} + r(1+r)u_{i+2,j}\}, \quad (3.9.2)$$

where $\Delta = 1 + 2r$.

If we let E_j denote the modulus of the maximum error along the jth time row then by omitting the truncation error, then we have

$$|e_{i,j+1}| < \frac{r(1+r)}{1+2r}|e_{i-1,j}| + \frac{(1-r^2)}{1+2r}|e_{ij}| + \frac{r(1-r)}{1+2r}|e_{i+1,j}| + \frac{r^2}{1+2r}|e_{i+2,j}|,$$

it can be seen that for $r \leq 1$ all coefficients are zero or positive thus,

$$E_{j+1} \leq \frac{r(1+r)}{1+2r}E_j + \frac{(1-r^2)}{1+2r}E_j + \frac{r(1-r)}{1+2r}E_j + \frac{r^2}{1+2r}E_j < E_j,$$

which is true for all values of i. Thus equation (3.5.5) is convergent for $r \leq 1$.

A similar result holds for equation (3.9.2) for each of equations (3.9.1) and (3.9.2).

Thus by using the maximum norm principle it can be shown that the stability condition is $r \leq 1$.

It would also be expected that the other schemes in the class of Group Explicit methods, which are formed from the combination of (3.5.5–3.5.7) should also be conditionally stable.

The step by step computations involving GER of equation (3.5.9), GEL of equation (3.5.12), GEU of equation (3.5.15) and GEC of equation (3.5.18) can all be written explicitly in difference form as

$$\underline{u}_{j+1} = T\underline{u}_j + \underline{b}, \tag{3.9.3}$$

where the amplification matrices T for each case are defined as follows:

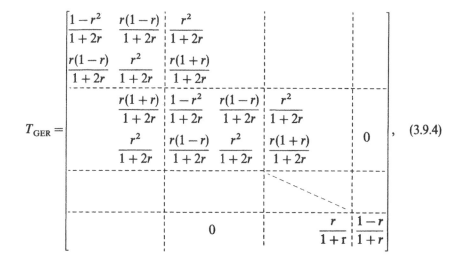

$$T_{GER} = \begin{bmatrix} \frac{1-r^2}{1+2r} & \frac{r(1-r)}{1+2r} & \frac{r^2}{1+2r} & & & \\ \frac{r(1-r)}{1+2r} & \frac{r^2}{1+2r} & \frac{r(1+r)}{1+2r} & & & \\ & \frac{r(1+r)}{1+2r} & \frac{1-r^2}{1+2r} & \frac{r(1-r)}{1+2r} & \frac{r^2}{1+2r} & \\ & \frac{r^2}{1+2r} & \frac{r(1-r)}{1+2r} & \frac{r^2}{1+2r} & \frac{r(1+r)}{1+2r} & 0 \\ & & & & & \\ & & 0 & & & \frac{r}{1+r} \quad \frac{1-r}{1+r} \end{bmatrix}, \tag{3.9.4}$$

$$T_{\text{GEL}} = \begin{bmatrix}
\dfrac{1-r}{1+r} & \dfrac{r}{1+r} & & & & & \\[2mm]
\dfrac{r(1+r)}{1+2r} & \dfrac{1-r^2}{1+2r} & \dfrac{r(1-r)}{1+2r} & \dfrac{r^2}{1+2r} & & 0 & \\[2mm]
\dfrac{r^2}{1+2r} & \dfrac{r(1-r)}{1+2r} & \dfrac{1-r^2}{1+2r} & \dfrac{r(1+r)}{1+2r} & & & \\[2mm]
& & & & & & \\[2mm]
& 0 & & & \dfrac{r(1+r)}{1+2r} & \dfrac{1-r^2}{1+2r} & \dfrac{r(1-r)}{1+2r} \\[2mm]
& & & & \dfrac{r^2}{1+2r} & \dfrac{r(1-r)}{1+2r} & \dfrac{1-r^2}{1+2r}
\end{bmatrix}, \qquad (3.9.5)$$

$$T_{\text{GEU}} = \begin{bmatrix}
\dfrac{1-r}{1+r} & \dfrac{r}{1+r} & & & & & \\[2mm]
\dfrac{r(1+r)}{1+2r} & \dfrac{1-r^2}{1+2r} & \dfrac{r(1-r)}{1+2r} & \dfrac{r^2}{1+2r} & & 0 & \\[2mm]
\dfrac{r^2}{1+2r} & \dfrac{r(1-r)}{1+2r} & \dfrac{1-r^2}{1+2r} & \dfrac{r(1+r)}{1+2r} & & & \\[2mm]
& & & \dfrac{r(1+r)}{1+2r} & \dfrac{1-r^2}{1+2r} & \dfrac{r(1-r)}{1+2r} & \dfrac{r^2}{1+2r} \\[2mm]
& 0 & & \dfrac{r^2}{1+2r} & \dfrac{r(1-r)}{1+2r} & \dfrac{1-r^2}{1+2r} & \dfrac{r(1+r)}{1+2r} \\[2mm]
& & & & & \dfrac{r}{1+r} & \dfrac{1-r}{1+r}
\end{bmatrix}, \qquad (3.9.6)$$

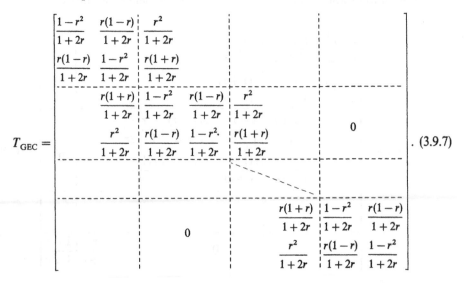

$$T_{\text{GEC}} = \begin{bmatrix}
\dfrac{1-r^2}{1+2r} & \dfrac{r(1-r)}{1+2r} & \dfrac{r^2}{1+2r} & & & & \\[2mm]
\dfrac{r(1-r)}{1+2r} & \dfrac{1-r^2}{1+2r} & \dfrac{r(1+r)}{1+2r} & & & & \\[2mm]
& \dfrac{r(1+r)}{1+2r} & \dfrac{1-r^2}{1+2r} & \dfrac{r(1-r)}{1+2r} & \dfrac{r^2}{1+2r} & & \\[2mm]
& \dfrac{r^2}{1+2r} & \dfrac{r(1-r)}{1+2r} & \dfrac{1-r^2}{1+2r} & \dfrac{r(1+r)}{1+2r} & 0 & \\[2mm]
& & & & & & \\[2mm]
& & 0 & & & \dfrac{r(1+r)}{1+2r} & \dfrac{1-r^2}{1+2r} & \dfrac{r(1-r)}{1+2r} \\[2mm]
& & & & & \dfrac{r^2}{1+2r} & \dfrac{r(1-r)}{1+2r} & \dfrac{1-r^2}{1+2r}
\end{bmatrix}. \qquad (3.9.7)$$

Now the system (3.9.3) will be stable against growth of rounding errors if

$$\| T \|_\infty \leq 1, \qquad\qquad (3.9.8)$$

where we define

$$\| T \|_\infty = \max_{1 \leq i \leq m-1} \left\{ \sum_{j=1}^{m-1} |t_{i,j}| \right\}$$

$$= \text{maximum absolute row sum of the}$$

$$(m-1) \times (m-1) \text{ matrix } T \text{ with elements } t_{i,j}.$$

Thus, from (3.9.4–3.9.7) we have

$$\| T_{\text{GER}} \|_\infty = \max \left\{ \frac{1}{1+2r}(|1-r^2|+|r(1-r)|+r^2), \frac{1}{1+2r}(|r(1-r)|+|1-r^2| \right.$$

$$+ |r(1+r)|), \frac{1}{1+2r}(|r(1+r)|+|1-r^2|+|r(1-r)|+r^2),$$

$$\left. \frac{1}{1+r}(|r|+|1-r|) \right\} \leq 1 \quad \text{for all } r \leq 1.$$

Similarly, we find that $\| T_{\text{GEL}} \|_\infty, \| T_{\text{GEU}} \|_\infty$ and $\| T_{\text{GEC}} \|_\infty$ fulfill the same condition when $r \leq 1$. Therefore all the single time step group explicit methods GER, GEL, GEU and GEC are conditionally stable for $r \leq 1$. This result can be easily confirmed from the numerical experiments given later.

Alternatively, if we preserve the implicit form of (3.5.9), (3.5.12) and (3.5.15), i.e.,

$$(I + rG_A)\underline{u}_{j+1} = (I - rG_B)\underline{u}_j + \underline{b},$$

we can still reach a similar stability condition using the following lemma of Kellogg (1964).

Lemma 3.3.
If $r > 0$ and $(B + B^*)$ is non-negative (or positive) definite, then $(rI + B)^{-1}$ is bounded and $\|(rI + B)^{-1}\|_2 \leq r^{-1}$.

From equations (3.5.10), (3.5.11), (3.5.16) and (3.5.17), it is easily verified that $(rG_A + rG_A^*)$ is non-negative (or positive) definite, then $(I + rG_A)^{-1}$ is bounded and $\|(I + rG_A)^{-1}\|_2 \leq 1$.

For stability, from (3.9.8) we need,

$$\| T \| = \|(I + rG_A)^{-1}(I - rG_B)\| \leq 1.$$

Thus

$$\|(I + rG_A)^{-1}(I - rG_B)\|_2 \leq \|(I + rG_A)^{-1}\|_2 \, \|(I - rG_B)\|_2$$

$$\leq \|(I - rG_B)\|_2$$

$$= \max \{|1-r|, |1-2r|, 1\}.$$

It can be easily shown that the eigenvalues of the matrices G_A and G_B, irrespective of the schemes described, are $1, 1-2r$ and $1-r$ (with 1 and $1-2r$ corresponding to the multiplicity of group). Therefore,

$$\|(I + rG_A)^{-1}(I - rG_B)\|_2 \leq 1,$$

for all $r \leq 1$, i.e., similar to that of the earlier condition. This leads to the following theorem.

Theorem 3.4.
The GER scheme of (3.5.9), GEL scheme of (3.5.12), GEU scheme of (3.5.15) and the GEC scheme of (3.5.18) are conditionally stable for all $r \leq 1$.

We now consider the stability of the two-step process Single Alternating Group Explicit (SAGE) method as in the equations (3.5.13), (3.5.19) and (3.6.1). Generally, these equations are written as

$$(I + r\theta G_A)\underline{u}_{j+\theta} = (I - r\theta G_B)\underline{u}_j + \underline{b},$$
$$(I + r\theta G_B)\underline{u}_{j+2\theta} = (I - r\theta G_A)\underline{u}_{j+\theta} + \underline{b}, \tag{3.9.9}$$

where

$$\theta = \begin{cases} 1, & \text{for equations (3.5.13) and (3.5.19),} \\ 1/2, & \text{for equation (3.6.1).} \end{cases}$$

Eliminating $\underline{u}_{j+\theta}$ from (3.9.9) gives

$$\underline{u}_{j+2\theta} = T\underline{u}_j + \underline{b}', \tag{3.9.10}$$

where $\underline{b}'$ is independent of the $\underline{u}$'s and the amplification matrix T is

$$T = (I + r\theta G_B)^{-1}(I - r\theta G_A)(I + r\theta G_A)^{-1}(I - r\theta G_B). \tag{3.9.11}$$

In proving the stability condition we use another lemma of Kellogg (1964).

Lemma 3.5.
If $r > 0$ and $(B + B^*)$ is non-negative (or positive) definite, then $(rI - B)(rI + B)^{-1}$ is bounded and $\|(rI - B)(rI + B)^{-1}\|_2 \leq 1$.

Now we define the matrix

$$\tilde{T} = (I + r\theta G_B)T (I + r\theta G_B)^{-1}, \tag{3.9.12}$$

which is similar to T and thus has the same eigenvalues as T. With (3.9.11) it follows that we can express $\tilde{T}$ as

$$\tilde{T} = \{(I - r\theta G_A) (I + r\theta G_A)^{-1}\} \{(I - r\theta G_B)(I + r\theta G_B)^{-1}\}. \tag{3.9.13}$$

From our knowledge of matrix norms and spectral radii, it is evident that,

$$\rho(T) = \rho(\tilde{T}) = \|\tilde{T}\|_2 \leq \|(I - r\theta G_A)(I + r\theta G_A)^{-1}\|_2 \; \|(I + r\theta G_B)(I - r\theta G_B)^{-1}\|_2. \tag{3.9.14}$$

Since $r\theta(G_A + G_A^*)$ is non-negative (or positive definite for all cases of (3.5.13), (3.5.19) and (3.6.1), and $r\theta(G_B + G_B^*)$ can also be easily shown then from (3.9.14) and Lemma 3.2 we get the unconditional stability condition for the schemes governed by the equation (3.9.9). Hence we can state the following theorem.

Theorem 3.6.
The single Alternating Group Explicit (SAGE) method (3.5.13), (3.5.19) and (3.6.1) for the solution of the heat conduction equation (3.1.1) on the region R of the open rectangle is unconditionally stable for all $r > 0$.

The analysis of stability for the four-step process Double Alternating Group Explicit method as in equations (3.5.14), (3.5.20), (3.6.2) and (3.6.4) is evident in a similar manner. In general, these equations are written as:

$$(I + r\theta G_A)\underline{u}_{j+\theta} = (I - r\theta G_B)\underline{u}_j + \underline{b},$$
$$(I + r\theta G_B)\underline{u}_{j+2\theta} = (I - r\theta G_A)\underline{u}_{j+\theta} + \underline{b},$$
$$(I + r\theta G_B)\underline{u}_{j+3\theta} = (I - r\theta G_A)\underline{u}_{j+2\theta} + \underline{b}, \tag{3.9.15}$$
$$(I + r\theta G_A)\underline{u}_{j+4\theta} = (I - r\theta G_B)\underline{u}_{j+3\theta} + \underline{b},$$

where

$$\theta = \begin{cases} 1, & \text{for equations (3.5.14) and (3.5.20),} \\ 1/2, & \text{for equation (3.6.2),} \\ 1/4, & \text{for equation (3.6.4).} \end{cases}$$

Eliminating $\underline{u}_{j+\theta}, \underline{u}_{j+2\theta}$ and $\underline{u}_{j+3\theta}$ from (3.9.15) gives

$$\underline{u}_{j+4\theta} = T'\underline{u}_j + \underline{b}'', \tag{3.9.16}$$

where $\underline{b}''$ is independent of the $\underline{u}$'s and where

$$T' = (I + r\theta G_A)^{-1}(I - r\theta G_B)(I + r\theta G_B)^{-1}(I - r\theta G_A)(I + r\theta G_B)^{-1}$$
$$(I - r\theta G_A)(I + r\theta G_A)^{-1}(I - r\theta G_B). \tag{3.9.17}$$

Let us define

$$T_1' = (I + r\theta G_A)^{-1}(I - r\theta G_B)(I + r\theta G_B)^{-1}(I - r\theta G_A)$$

and

$$T_2' = (I + r\theta G_B)^{-1}(I - r\theta G_A)(I + r\theta G_A)^{-1}(I - r\theta G_B), \tag{3.9.18}$$

then

$$T' = T_1' T_2'. \tag{3.9.19}$$

If $\tilde{T}_1'$ and $\tilde{T}_2'$ are defined such that

$$\tilde{T}_1' = (I + r\theta G_A)T_1'(I + r\theta G_A)^{-1}$$

and

$$\tilde{T}_2' = (I + r\theta G_B)T_2'(I + r\theta G_B)^{-1}, \tag{3.9.20}$$

then $\tilde{T}_1'$ and $\tilde{T}_2'$ are similar to T_1' and T_2' respectively. Therefore T_1' and T_2' have the same eigenvalues (and spectral radius) as $\tilde{T}_1'$ and $\tilde{T}_2'$, respectively. For stability we need $\|T'\| \leq 1$, i.e.,

$$\|T'\|_2 = \|T_1' T_2'\|_2$$
$$\leq \|T_1'\|_2 \|T_2'\|_2$$
$$= \rho(T_1')\rho(T_2')$$
$$= \rho(T_1')\rho(T_2')$$
$$= \|(I - r\theta G_B)(I + r\theta G_B)^{-1}(I - r\theta G_A)(I + r\theta G_A)^{-1}\|$$
$$\times \|(I - r\theta G_A)(I + r\theta G_A)^{-1}(I - r\theta G_B)(I + r\theta G_B)^{-1}\|$$
$$\leq \|(I - \rho\theta G_A)(I + r\theta G_A)^{-1}\|^2 \|(I - r\theta G_B)(I + r\theta G_B)^{-1}\|^2. \tag{3.9.21}$$

Since in all cases $r\theta(G_A + G_A^*)$ and $r\theta(G_B + G_B^*)$ are positive (or non-negative) definite matrices, therefore from Lemma 3.2 and (3.9.21) we get

$$\|T'\|_2 \le 1, \tag{3.9.22}$$

which means the schemes governed by (3.9.15) are unconditionally stable for all $r > 0$. Thus the theorem below follows immediately.

Theorem 3.7.
The Double Alternating Group Explicit (DAGE) method for the solution of the heat conduction equation (3.1.1) on the region R of the open rectangle is unconditionally stable for all $r > 0$.

It is observed that the two-step process single AGE and four-step process double AGE, either in fractional form or not are unconditionally stable. While the one step processes, i.e., the GER, GEL, GEU and GEC schemes, still retain a larger stability condition compared to the ordinary explicit method. However, this conditional stability, is only to be expected since these schemes are formed from the combination of the unconditionally stable formulae (3.5.6) and (3.5.7) and the conditionally stable formula (3.5.5).

At this stage it is important to mention that the unconditional stability condition does not mean that the time increment can be taken indefinitely large, particularly at small time levels if the solution is to satisfactorily approximate that of the exact solution of equation (3.1.1). This behaviour is common to all unconditionally stable explicit or semi-explicit methods.

Periodic Case
We now establish the stability for the periodic case. Equations (3.7.3) and (3.7.6) can be explicitly written as

$$\underline{u}_{j+1} = T\underline{u}_j, \tag{3.9.23}$$

where for (3.7.3),

$$T = (I + rG_2)^{-1}(I - rG_1), \tag{3.9.24}$$

and for (3.7.6) and the $(j-1)^{th}$ step we have, $\underline{u}_j = \tilde{T}\underline{u}_{j-1}$ with,

$$\tilde{T} = (I + rG_1)^{-1}(I - rG_2). \tag{3.9.25}$$

From equations (3.7.2) and (3.7.5) we can easily determine the amplification matrices for equation (3.7.3) where T is given by

$$T = \frac{1}{(1+2r)}
\begin{bmatrix}
1-r^2 & r(1-r) & -r^2 & & & & r(1+r) \\
r(1-r) & 1-r^2 & r(1+r) & & & & -r^2 \\
0 & r(1+r) & 1-r^2 & r(1-r) & -r^2 & & 0 \\
 & -r^2 & r(1-r) & 1-r^2 & r(1+r) & & \\
 & & & & & & \\
-r^2 & & & & r(1+r) & 1-r^2 & r(1-r) \\
r(1+r) & & & & -r^2 & r(1-r) & 1-r^2
\end{bmatrix}, \tag{3.9.26}$$

and for equation (3.7.6) where $\tilde{T}$ is given by

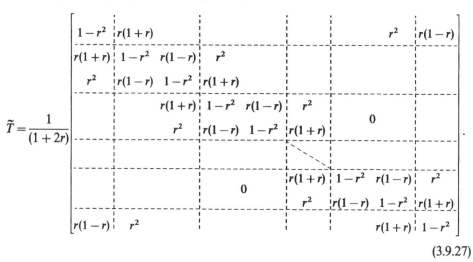

$$(3.9.27)$$

In both the cases of (3.9.26) and (3.9.27) we can easily verify that

$$\|T\|_\infty \le 1.$$

To show the unconditional stability of the two step single AGE method (eqn. 3.7.7) and four step double (AGE) (eqn. 3.7.8) processes, we first prove $r(G_2 + G_2^*)$ and $r(G_1 + G_1^*)$ are non-negative definite matrices (where $*$ denotes the transpose matrix), and, integer m even.

Proof. Since the eigenvalues of the matrices G_2 and G_2^* are 0 and 2 (non-negative) with multiplicity $m/2$ therefore the eigenvalues of $r(G_2 + G_2^*)$ are also non-negative. Hence, by using the theorem below, it is proven that the matrix $r(G_2 + G_2^*)$ is non-negative definite.

Theorem 3.8.
A real matrix is positive (non-negative) definite if and only if it is symmetric and all its eigenvalues are positive (non-negative), with at least one eigenvalue equal to zero.

For the matrix G_1, it can be easily shown that

$$\det G_1 = (\lambda^2 - 2\lambda) \begin{bmatrix} 1-\lambda & -1 & & & 0 \\ -1 & 1-\lambda & & & \\ & & & & \\ & 0 & & 1-\lambda & -1 \\ & & & -1 & 1-\lambda \end{bmatrix}_{(m-2)(m-2)}$$

$$= (\lambda^2 - 2\lambda)(\lambda^2 - 2\lambda)^{(m-2)/2}$$

$$= (\lambda^2 - 2\lambda)^{m/2}, \qquad (3.9.28)$$

where the λ are eigenvalues of the matrix G_1 where $\det G_1 = 0$. Therefore, the eigenvalues of G_1 of order m are 0 and 2 with multiplicity $m/2$. Since the matrix G_1

is a symmetric matrix and its eigenvalues are non-negative, therefore the matrix $r(G_1 + G_1^*)$ which is also a symmetric matrix having non-negative eigenvalues is clearly a non-negative definite matrix. Hence, by following the lemma of Kellog (1964) it can be shown that the schemes governed by equations (3.7.7) and (3.7.8) are unconditionally stable for all $r > 0$.

3.10. The Single Term Alternating Group Explicit (STAGE) Method

In this section we show that the Single Alternating Group Explicit method (SAGE) given by equations (3.7.7) comprising of 2 alternate sweeps of the group explicit method can be combined into the single term method (STAGE) comprising of 1 sweep of a more complicated formula, i.e.,

$$\underline{u}_{j+1} = T\,\tilde{T}\,\underline{u}_{j-1}, \tag{3.10.1}$$

where T and $\tilde{T}$ are given by equations (3.9.26) and (3.9.27), respectively.

The STAGE method can be shown to be given by the single equation

$$\underline{u}_{j+1} = (I + rG_2)^{-1}(I - rG_1)(I + rG_1)^{-1}(I - rG_2)\underline{u}_{j-1}, \tag{3.10.2}$$

with the amplification matrix $T\,\tilde{T}$ determined by matrix multiplication of the simple (2×2) component matrices of G_1 and G_2. We can easily show that $T\,\tilde{T}$ has the form,

$$\Delta^{-2}\begin{bmatrix}
1 & 2r & 2r^2(1-r) & 2r^3 & & & 2r^2(1+r) & 2r(1-r^2) \\
2r & 1 & 2r(1-r^2) & 2r^2(1+r) & & & 2r^3 & 2r^2(1-r) \\
2r^2(1+r) & 2r(1-r^2) & 1 & 2r & 2r^2(1-r) & 2r^3 & & \\
2r^3 & 2r^2(1-r) & 2r & 1 & 2r(1-r^2) & 2r^2(1+r) & & \\
& & & & & 0 & & \\
2r^2(1-r) & 2r^3 & & & 2r^2(1+r) & 2r(1-r^2) & 1 & 2r \\
2r(1-r^2) & 2r^2(1+r) & \multicolumn{2}{c}{0} & 2r^3 & 2r^2(1-r) & 2r & 1
\end{bmatrix}$$

with $\Delta^2 = (1 + 2r)^2$.

The new method (STAGE) can be shown to be numerically stable, i.e., $\|T\,\tilde{T}\|_\infty < 1$ and when represented by the following computational molecule the method can be implemented in 6 mults + 5 adds for each double step. This represents a saving of 25% in the computational work of the SAGE method represented by equation (3.7.7) and Figure 3.10.1.

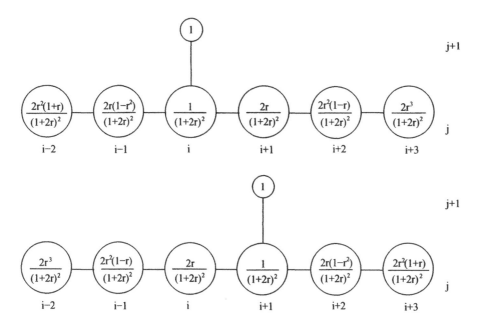

Figure 3.10.1 Computational Molecule of the STAGE Method at Time $j\Delta t$.

3.11. The Group Explicit Method for the Two-Space Dimensional Problem

In this section, the concept of the Group Explicit method is extended to the case of a two-space dimensional problem of the form

$$\frac{\partial U}{\partial t} = \frac{\partial^2 U}{\partial x^2} + \frac{\partial^2 U}{\partial y^2} + g(x, y, t).$$ (3.11.1)

The approximate solution u will be required in the cylinder $R \times [0 \le t \le T]$ where $R = \{(x, y) | 0 \le x, y \le 1\}$. Appropriate initial condition and boundary data are given on $t = 0$ and $B \times [0 \le t \le T], B = \{(x, y) | x = 0$ or $1, y = 0$ or $1\}$, respectively. For example, in this simple case we assume

$$U(x, y, 0) = e(x, y), \quad x, y \in R,$$ (3.11.2)

and

$$U(x, y, t) = f(x, y, t), \quad x, y \in B, \quad 0 \le t \le T,$$ (3.11.3)

where $e(x, y)$ and $f(x, y, t)$ are given known values for the prescribed values of x, y, t. We let $u_{i,j,k}$ denote the approximate solution of (3.11.1) at the point $(i, j, k) \equiv (i\Delta x, j\Delta y, k\Delta t)$, $i, j = 0, 1, 2, \ldots, m$, $m\Delta x = m\Delta y = 1$, $k = 0, 1, \ldots$. For simplicity, we assume that $\Delta x = \Delta y = \Delta s$ and hence $r = \Delta t (\Delta s)^{-2}$.

Consider at any time level, the group of four points A, B, C, D, i.e. (i,j,k), $(i+1,j,k)$, $(i,j+1,k)$ and $(i+1,j+1,k)$. Then at each of these points and at $t = (k+1/2)\Delta t$, we approximate the equation (3.11.1) by:

at point $(i,j,k+1/2)$, A:

$$\frac{u_{i,j,k+1} - u_{i,j,k}}{\Delta t} = \frac{u_{i+1,j,k+1} - u_{i,j,k+1} - u_{i,j,k} + u_{i-1,j,k}}{(\Delta x)^2}$$

$$+ \frac{u_{i,j,k+1} - u_{i,j,k+1} - u_{i,j,k} + u_{i,j-1,k}}{(\Delta y)^2} + g(i,j,k+1/2); \quad (3.11.4)$$

at point $(i+1,j,k+1/2)$, B:

$$\frac{u_{i+1,j,k+1} - u_{i+1,j,k}}{\Delta t} = \frac{u_{i,j,k+1} - u_{i+1,j,k+1} - u_{i+1,j,k} + u_{i+2,j,k}}{(\Delta x)^2}$$

$$+ \frac{u_{i+1,j+1,k+1} - u_{i+1,j,k+1} - u_{i+1,j,k} + u_{i+1,j-1,k}}{(\Delta y)^2}$$

$$+ g(i+1,j,k+1/2); \quad (3.11.5)$$

at point $(i,j+1,k+1/2)$, C:

$$\frac{u_{i,j+1,k+1} - u_{i,j+1,k}}{\Delta t} = \frac{u_{i+1,j+1,k+1} - u_{i,j+1,k+1} - u_{i,j+1,k} + u_{i-1,j+1,k}}{(\Delta x)^2}$$

$$+ \frac{u_{i,j,k+1} - u_{i,j+1,k+1} - u_{i,j+1,k} + u_{i,j+2,k}}{(\Delta y)^2}$$

$$+ g(i,j+1,k+1/2); \quad (3.11.6)$$

and at point $(i+1,j+1,k+1/2)$, D:

$$\frac{u_{i+1,j+1,k+1} - u_{i+1,j+1,k}}{\Delta t} = \frac{u_{i,j+1,k+1} - u_{i+1,j+1,k+1} - u_{i+1,j+1,k} + u_{i+2,j+1,k}}{(\Delta x)^2}$$

$$+ \frac{u_{i+1,j,k+1} - u_{i+1,j+1,k+1} - u_{i+1,j+1,k} + u_{i+1,j+2,k}}{(\Delta y)^2}$$

$$+ g(i+1,j+1,k+1/2); \quad (3.11.7)$$

respectively (Fig. 3.11.1). Upon simplification the following equations can be obtained, at point A:

$$-ru_{i+1,j,k+1} + (1+2r)u_{i,j,k+1} - ru_{i,j+1,k+1}$$

$$= ru_{i-1,j,k} + (1-2r)u_{i,j,k} + ru_{i,j-1,k} + \Delta tg(i,j,k+1/2), \quad (3.11.8)$$

at point B:

$$-ru_{i,j,k+1} + (1+2r)u_{i+1,j,k+1} - ru_{i+1,j+1,k+1}$$

$$= ru_{i+2,j,k} + (1-2r)u_{i+1,j,k} + ru_{i+1,j-1,k} + \Delta tg(i+1,j,k+1/2), \quad (3.11.9)$$

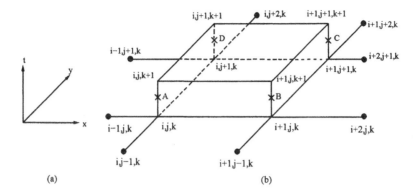

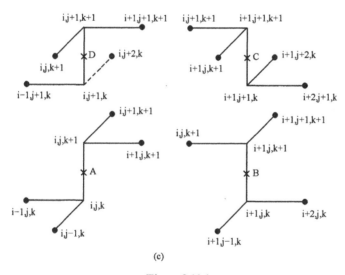

(c)

Figure 3.11.1

at point C:

$$-ru_{i,j,k+1} + (1+2r)u_{i,j+1,k+1} - ru_{i+1,j+1,k+1}$$

$$= ru_{i-1,j+1,k} + (1-2r)u_{i,j+1,k} + ru_{i,j+2,k} + \Delta t g(i,j+1,k+1/2); \quad (3.11.10)$$

and at point D:

$$-ru_{i,j+1,k+1} + (1+2r)u_{i+1,j+1,k+1} - ru_{i+1,j,k+1}$$

$$= ru_{i+2,j+1,k} + (1-2r)u_{i+1,j+1,k} + ru_{i+1,j+2,k} + \Delta t g(i+1,j+1,k+1/2).$$

$$(3.11.11)$$

Therefore the equations (3.11.8–3.11.11) when ordered cyclically can be grouped together to give the (4 × 4) implicit system whose matrix form is given by

$$\begin{bmatrix} (1+2r) & -r & 0 & -r \\ -r & (1+2r) & -r & 0 \\ 0 & -r & (1+2r) & -r \\ -r & 0 & -r & (1+2r) \end{bmatrix} \begin{bmatrix} u_{i,j,k+1} \\ u_{i+1,j,k+1} \\ u_{i+1,j+1,k+1} \\ u_{i,j+1,k+1} \end{bmatrix} = \begin{bmatrix} f_{i,j,k} \\ f_{i+1,j,k} \\ f_{i+1,j+1,k} \\ f_{i,j+1,k} \end{bmatrix}, \qquad (3.11.12)$$

where the values are defined by the right-hand sides of (3.11.8–3.11.11) and since Δt is normally small enough g can be taken as

$$g_{i,j,k+1/2} = \begin{cases} g_{i,j,k}, \text{ or,} \\ g_{i,j,k+1}, \text{ or,} \\ \tfrac{1}{2}(g_{i,j,k} + g_{i,j,k+1}). \end{cases} \qquad (3.11.13)$$

The system (3.11.12) can be written in explicit form as

$$\underline{u}_{k+1} = A^{-1}\underline{f}_k, \qquad (3.11.14)$$

where A^{-1} is the inverse of the coefficient matrix in (3.11.12) and when evaluated explicitly is given by

$$A^{-1} = \frac{1}{(1+2r)(1+4r)} \begin{bmatrix} 1+4r+2r^2 & r(1+2r) & 2r^2 & r(1+2r) \\ r(1+2r) & 1+4r+2r^2 & r(1+2r) & 2r^2 \\ 2r^2 & r(1+2r) & 1+4r+2r^2 & r(1+2r) \\ r(1+2r) & 2r^2 & r(1+2r) & 1+4r+2r^2 \end{bmatrix}. \qquad (3.11.15)$$

Special treatment will have to be made for points near the boundary (see Fig. 3.11.2), depending on the position. For the position 1, the solutions at any two points are

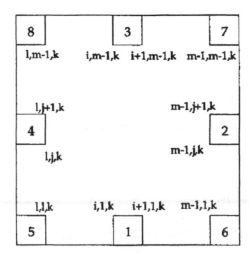

Figure 3.11.2   The position of the points near the boundary for any time level k.

approximated by (Fig. 3.11.3a)

$$\begin{bmatrix} 1+2r & -r \\ -r & 1+2r \end{bmatrix} \begin{bmatrix} u_{i,1,k+1} \\ u_{i+1,1,k+1} \end{bmatrix}$$

$$= \begin{bmatrix} ru_{i,0,k+1} + ru_{i-1,1,k} + (1-2r)u_{i,1,k} + ru_{i,2,k} + \Delta t g_{i,1,k+1/2} \\ ru_{i+1,0,k+1} + ru_{i+2,1,k} + (1-2r)u_{i+1,1,k} + ru_{i+1,2,k} + \Delta t g_{i+1,1,k+1} \end{bmatrix},$$

$$\text{(3.11.16a)}$$

which can be reduced to

$$\begin{bmatrix} u_{i,1,k+1} \\ u_{i+1,1,k+1} \end{bmatrix} = \frac{1}{(1+4r+3r^2)} \begin{bmatrix} 1+2r & -r \\ -r & 1+2r \end{bmatrix} \begin{bmatrix} f_{i,1,k} \\ f_{i+1,1,k} \end{bmatrix}, \quad \text{(3.11.16b)}$$

with f corresponding to the right-hand sides of equation (3.11.16a).

For position 2, the system reduces to

$$\begin{bmatrix} u_{m-1,j,k+1} \\ u_{m-1,j+1,k+1} \end{bmatrix} = \frac{1}{(1+4r+3r^2)} \begin{bmatrix} 1+2r & r \\ r & 1+2r \end{bmatrix} \begin{bmatrix} f_{m-1,j,k} \\ f_{m-1,j+1,k} \end{bmatrix}, \quad \text{(3.11.17)}$$

with corresponding f, where

$$f_{m-1,j,k} = ru_{m-2,j,k} + (1-2r)u_{m-1,j,k} + ru_{m-1,j-1,k} + ru_{m,j,k+1} + \Delta t g_{m-1,j,k+1/2}$$

and

$$f_{m-1,j+1,k} = ru_{m-2,j+1,k} + (1-2r)u_{m-1,j+1,k} + ru_{m-1,j+2,k}$$

$$+ ru_{m,j+1,k+1} + \Delta t g_{m-1,j+1,k+1/2},$$

(see Fig. 3.11.3b).

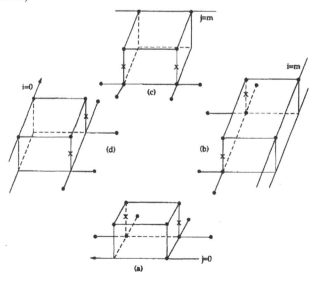

Figure 3.11.3

For position 3, the system reduces to

$$\begin{bmatrix} u_{i,m-1,k+1} \\ u_{i+1,m-1,k+1} \end{bmatrix} = \frac{1}{(1+4r+3r^2)} \begin{bmatrix} (1+2r) & r \\ r & (1+2r) \end{bmatrix} \begin{bmatrix} f_{i,m-1,k} \\ f_{i+1,m-1,k} \end{bmatrix}, \quad (3.11.18)$$

where

$$f_{i,m-1,k} = ru_{i,m-2,k} + (1-2r)u_{i,m-1,k} + ru_{i-1,m-1,k}$$
$$+ ru_{i,m,k+1} + \Delta t g_{i,m-1,k+1/2},$$

and

$$f_{i+1,m-1,k} = ru_{i+1,m-2,k} + (1-2r)u_{i+1,m-1,k} + ru_{i+2,m-1,k}$$
$$+ ru_{i+1,m,k+1} + \Delta t g_{i+1,m-1,k+1/2},$$

(see Fig. 3.11.3c).

Finally for position 4 the system results in

$$\begin{bmatrix} u_{1,j,k+1} \\ u_{1,j+1,k+1} \end{bmatrix} = \frac{1}{(1+4r+3r^2)} \begin{bmatrix} (1+2r) & r \\ r & (1+2r) \end{bmatrix} \begin{bmatrix} f_{1,j,k} \\ f_{1,j+1,k} \end{bmatrix}, \quad (3.11.19)$$

where

$$f_{1,j,k} = ru_{0,j,k+1} + ru_{1,j-1,k} + (1-2r)u_{1,j,k,} + ru_{2,j,k} + \Delta t g_{1,j,k+1/2},$$

and

$$f_{1,j+1,k} = ru_{0,j+1,k+1} + ru_{1,j+2,k} + (1-2r)u_{1,j+1,k} + ru_{2,j+1,k} + \Delta t g_{1,j+1,k+1/2},$$

(see Fig. 3.11.3d).

The solutions at the corner points of the x-y plane, i.e., positions 5, 6, 7 and 8 in Figure 3.11.2 are given by:

$$u_{1,1,k+1} = \frac{1}{(1+2r)} \{ ru_{0,1,k+1} + ru_{1,0,k+1} + ru_{1,2,k} + (1-2r)u_{1,1,k}$$
$$+ ru_{2,1,k} + \Delta t g_{1,1,k+1/2} \}, \quad (3.11.20)$$

$$u_{m-1,1,k+1} = \frac{1}{(1+2r)} \{ ru_{m-1,0,k+1} + ru_{m,1,k+1} + ru_{m-1,2,k} + (1-2r)u_{m-1,1,k}$$
$$+ ru_{m-2,1,k} + \Delta t g_{m-1,1,k+1/2} \}, \quad (3.11.21)$$

$$u_{m-1,m-1,k+1} = \frac{1}{(1+2r)} \{ ru_{m-2,m-1,k} + (1-2r)u_{m-1,m-1,k} + ru_{m-1,m-2,k}$$
$$+ ru_{m,m-1,k+1} + ru_{m-1,m,k+1}$$
$$+ \Delta t g_{m-1,m-1,k+1/2} \}, \quad (3.11.22)$$

and

$$u_{1,m-1,k} = \frac{1}{(1+2r)} \{ ru_{2,m-1,k} + (1-2r)u_{1,m-1,k} + ru_{1,m-2,k}$$
$$+ ru_{0,m-1,k+1} + ru_{1,m,k+1} + \Delta t g_{1,m-1,k+1/2} \}, \quad (3.11.23)$$

respectively. These are shown in Figure 3.11.4.

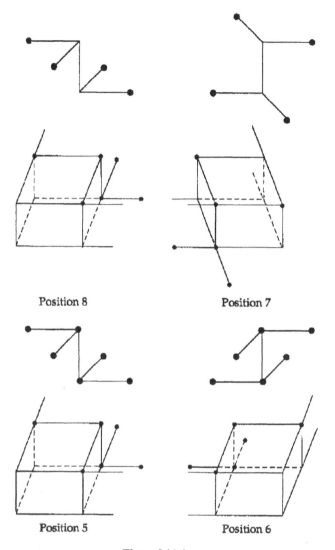

Figure 3.11.4

These equations can be derived from the computational molecule as illustrated in Figure 3.11.5. In addition, similar explicit equations can be derived for all the corner and edge point positions given by equations (3.11.16–3.11.23).

Group Explicit Method
Similar to the one-dimensional problem, now we can develop various types of GE methods using equations (3.11.14), (3.11.16–3.11.23). To simplify the discussion, we assume that the x-y square is divided into an even number of squares in both directions, hence the number of unknown points in each direction are odd. Therefore corresponding to the one-dimensional case, there are many possible types of GE methods available based on the GER, GEL, SAGE, DAGE schemes and their respective fractional splitting versions.

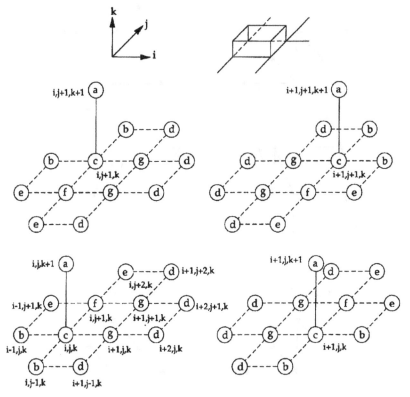

$a = (1+2r)(1+4r), b = (1+4r+2r^2), c = (1-2r)(1+4r+2r^2) , d = r^2(1+2r), e = 2r^3 , f = 2r^2(1-2r), g = r(1-4r^2)$

Figure 3.11.5

GER Scheme (*x*-direction)
This scheme at any time level is obtained by using equation (3.11.14) for the first $((m-2)/2)^2$ group of four points, starting from $(1,1,k)$, equation (3.11.17) at points $(m-1,j,k)$, $j = 1,2,3,\ldots,m-2$, equation (3.11.22) at point $(m-1,m-1,k)$ and equation (3.11.18) at points $(i,m-1,k)$, $i = 1,2,3,\ldots,m-2$ (see Fig. 3.11.6a).

GEL Scheme (*x*-direction)
This scheme at any time level is obtained by using equation (3.11.20) at point $(1,1,k)$, equation (3.10.16) at points $(i,1,k), i = 2,3,\ldots,m-1$, equation (3.11.19) at points $(1,j,k)$, $j = 2,3,\ldots,m-1$ and equation (3.11.14) for the remaining $((m-2)/2)^2$ group of four points starting from $(2,2,k)$. This method is described in Figure 3.11.6b.

SAGE and DAGE Schemes
The concept of the SAGE (Fig. 3.11.6) and DAGE schemes for the two-dimensional problem are similar to the one-dimensional case. For the SAGE scheme it is the alternate use of the GER and GEL schemes at alternate time levels. Meanwhile the DAGE scheme is the alternate use of the GER and GEL schemes with the direction of the alternation changed at every third time step level.

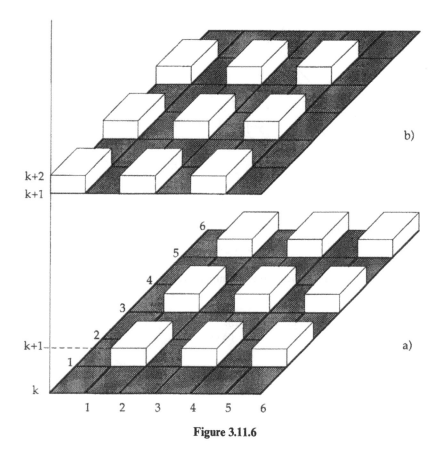

Figure 3.11.6

(b) The Extension to Multi-Space Dimensional Problems
In this section we will extend the GE formulation to a three space dimensional problem for the heat equation of the form

$$\frac{\partial U}{\partial t} = \frac{\partial^2 U}{\partial x^2} + \frac{\partial^2 U}{\partial y^2} + \frac{\partial^2 U}{\partial z^2} + g(x, y, z, t), \qquad (3.11.24)$$

where the appropriate initial and boundary conditions for a valid numerical solution are given. For this case the group explicit simulation will consist of 8 grid points taken to form a cube instead of the 4 points in a plane as for the two-dimensional problem (Fig. 3.11.7) and 2 points for the one-dimensional problem (Fig. 3.5.3–3.5.6).

At each of points $a, b, c, \ldots, g, h$ and $t = k + \Delta t/2$, we approximate equation (3.11.24) by the following equations:

$$\frac{u_{i,j,\ell}^{k+1} - u_{i,j,\ell}^{k}}{\Delta t} = \frac{u_{i+1,j,\ell}^{k+1} - u_{i,j,\ell}^{k+1} - u_{i,j,\ell}^{k} + u_{i-1,j,\ell}^{k}}{(\Delta x)^2} + \frac{u_{i,j+1,\ell}^{k+1} - u_{i,j,\ell}^{k+1} - u_{i,j,\ell}^{k} + u_{i,j-1,\ell}^{k}}{(\Delta y)^2}$$

$$+ \frac{u_{i,j,\ell+1}^{k+1} - u_{i,j,\ell}^{k+1} - u_{i,j,\ell}^{k} + u_{i,j,\ell-1}^{k}}{(\Delta z)^2} + g_{i,j,\ell}^{k+1/2}, \qquad (3.11.25)$$

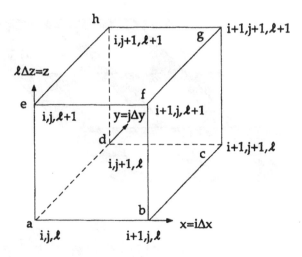

Figure 3.11.7   A cube which forms an implicit group at any time level k.

$$\frac{u_{i+1,j,\ell}^{k+1} - u_{i+1,j,\ell}^{k}}{\Delta t} = \frac{u_{i+2,j,\ell}^{k} - u_{i+1,j,\ell}^{k} - u_{i+1,j,\ell}^{k+1} + u_{i,j,\ell}^{k+1}}{(\Delta x)^2}$$

$$+ \frac{u_{i+1,j+1,\ell}^{k+1} - u_{i+1,j,\ell}^{k+1} - u_{i+1,j,\ell}^{k} + u_{i+1,j-1,\ell}^{k}}{(\Delta y)^2}$$

$$+ \frac{u_{i+1,j,\ell+1}^{k+1} - u_{i+1,j,\ell}^{k+1} - u_{i+1,j,\ell}^{k} + u_{i+1,j,\ell-1}^{k}}{(\Delta z)^2} + g_{i+1,j,\ell}^{k+1/2},$$

$$(3.11.26)$$

$$\frac{u_{i+1,j+1,\ell}^{k+1} - u_{i+1,j+1,\ell}^{k}}{\Delta t} = \frac{u_{i,j+1,\ell}^{k+1} - u_{i+1,j+1,\ell}^{k+1} - u_{i+1,j+1,\ell}^{k} + u_{i+2,j+1,\ell}^{k}}{(\Delta x)^2}$$

$$+ \frac{u_{i+2,j+1,\ell}^{k} - u_{i+1,j+1,\ell}^{k} - u_{i+1,j+1,\ell}^{k+1} - u_{i+1,j+1,\ell}^{k+1} - u_{i+1,j,\ell}^{k+1}}{(\Delta y)^2}$$

$$+ \frac{u_{i+1,j+1,\ell+1}^{k+1} - u_{i+1,j+1,\ell}^{k+1} - u_{i+1,j+1,\ell}^{k} + u_{i+1,j+1,\ell-1}^{k}}{(\Delta z)^2}$$

$$+ g_{i+1,j+1,\ell}^{k+1/2},$$

$$(3.11.27)$$

$$\frac{u_{i,j+1,\ell}^{k+1} - u_{i,j+1,\ell}^{k}}{\Delta t} = \frac{u_{i+1,j+1,\ell}^{k+1} - u_{i,j+1,\ell}^{k+1} - u_{i,j+1,\ell}^{k} + u_{i-1,j+1,\ell}^{k}}{(\Delta x)^2}$$

$$+ \frac{u_{i,j+2,\ell}^{k} - u_{i,j+1,\ell}^{k} - u_{i,j+1,\ell}^{k+1} + u_{i,j,\ell}^{k+1}}{(\Delta y)^2}$$

$$+ \frac{u_{i,j+1,\ell+1}^{k+1} - u_{i,j+1,\ell}^{k+1} - u_{i,j+1,\ell}^{k} + u_{i,j+1,\ell-1}^{k}}{(\Delta z)^2} + g_{i,j+1,\ell}^{k+1/2},$$

$$(3.11.28)$$

$$\frac{u_{i,j,\ell+1}^{k+1} - u_{i,j,\ell+1}^{k}}{\Delta t} = \frac{u_{i+1,j,\ell+1}^{k+1} - u_{i,j,\ell+1}^{k+1} - u_{i,j,\ell+1}^{k} + u_{i-1,j,\ell+1}^{k}}{(\Delta x)^2}$$

$$+ \frac{u_{i,j+1,\ell+1}^{k+1} - u_{i,j,\ell+1}^{k+1} - u_{i,j,\ell+1}^{k} + u_{i,j-1,\ell+1}^{k}}{(\Delta y)^2}$$

$$+ \frac{u_{i,j,\ell}^{k+1} - u_{i,j,\ell+1}^{k+1} - u_{i,j,\ell+1}^{k} + u_{i,j,\ell+2}^{k}}{(\Delta z)^2} + g_{i,j,\ell+1}^{k+1/2}, \qquad (3.11.29)$$

$$\frac{u_{i+1,j,\ell+1}^{k+1} - u_{i+1,j,\ell+1}^{k}}{\Delta t} = \frac{u_{i,j,\ell+1}^{k+1} - u_{i+1,j,\ell+1}^{k+1} - u_{i+1,j,\ell+1}^{k} + u_{i+2,j,\ell+1}^{k}}{(\Delta x)^2}$$

$$+ \frac{u_{i+1,j+1,\ell+1}^{k+1} - u_{i+1,j,\ell+1}^{k+1} - u_{i+1,j,\ell+1}^{k} + u_{i+1,j-1,\ell+1}^{k}}{(\Delta y)^2}$$

$$+ \frac{u_{i+1,j,\ell}^{k+1} - u_{i+1,j,\ell+1}^{k+1} - u_{i+1,j,\ell+1}^{k} + u_{i+1,j,\ell+2}^{k}}{(\Delta z)^2} + g_{i+1,j,\ell+1}^{k+1/2},$$

$$(3.11.30)$$

$$\frac{u_{i+1,j+1,\ell+1}^{k+1} - u_{i+1,j+1,\ell+1}^{k}}{\Delta t} = \frac{u_{i,j+1,\ell+1}^{k+1} - u_{i+1,j+1,\ell+1}^{k+1} - u_{i+1,j+1,\ell+1}^{k} + u_{i+2,j+1,\ell+1}^{k}}{(\Delta x)^2}$$

$$+ \frac{u_{i+1,j+1,\ell+1}^{k+1} - u_{i+1,j+1,\ell+1}^{k+1} - u_{i+1,j+1,\ell+1}^{k} + u_{i+1,j+2,\ell+1}^{k}}{(\Delta y)^2}$$

$$+ \frac{u_{i+1,j+1,\ell}^{k+1} - u_{i+1,j+1,\ell+1}^{k+1} - u_{i+1,j+1,\ell+1}^{k} + u_{i+1,j+1,\ell+2}^{k}}{(\Delta z)^2}$$

$$+ g_{i+1,j+1,\ell+1}^{k+1/2}, \qquad (3.11.31)$$

and

$$\frac{u_{i,j+1,\ell+1}^{k+1} - u_{i,j+1,\ell+1}^{k}}{\Delta t} = \frac{u_{i+1,j+1,\ell+1}^{k+1} - u_{i,j+1,\ell+1}^{k+1} - u_{i,j+1,\ell+1}^{k} + u_{i-1,j+1,\ell+1}^{k}}{(\Delta x)^2}$$

$$+ \frac{u_{i,j+1,\ell+1}^{k+1} - u_{i,j+1,\ell+1}^{k+1} - u_{i,j+1,\ell+1}^{k} + u_{i,j+2,\ell+1}^{k}}{(\Delta y)^2}$$

$$+ \frac{u_{i,j+1,\ell}^{k+1} - u_{i,j+1,\ell+1}^{k+1} - u_{i,j+1,\ell+1}^{k} + u_{i,j+1,\ell+2}^{k}}{(\Delta z)^2}$$

$$+ g_{i,j+1,\ell+1}^{k+1/2}, \qquad (3.11.32)$$

respectively.

Assuming $\Delta x = \Delta y = \Delta z = \Delta s$ and $r = \Delta t/(\Delta s)^2$, as before equations (3.11.25–3.11.32) can be reduced to the set of implicit difference equations

$$-r[u_{i+1,j,\ell}^{k+1} + u_{i,j+1,\ell}^{k+1} + u_{i,j,\ell+1}^{k+1}] + (1 + 3r)u_{i,j,\ell}^{k+1}$$

$$= r(u_{i-1,j,\ell}^{k} + u_{i,j-1,\ell}^{k} + u_{i,j,\ell-1}^{k}) + (1 - 3r)u_{i,j,\ell}^{k} + g_{i,j,\ell}^{1/2}, \quad (3.11.33)$$

$$-r[u_{i,j,\ell}^{k+1} + u_{i+1,j+1,\ell}^{k+1} + u_{i+1,j,\ell+1}^{k+1}] + (1 + 3r)u_{i+1,j,\ell}^{k+1}$$

$$= r(u_{i+2,j,\ell}^{k} + u_{i+1,j-1,\ell}^{k} + u_{i+1,j,\ell-1}^{k}) + (1 - 3r)u_{i+1,j,\ell}^{k} + g_{i+1,j,\ell}^{1/2}, \quad (3.11.34)$$

$$-r[u_{i,j+1,\ell}^{k+1} + u_{i+1,j,\ell}^{k+1} + u_{i+1,j+1,\ell+1}^{k+1}] + (1+3r)u_{i+1,j+1,\ell}^{k+1}$$

$$= r(u_{i+2,j+1,\ell}^{k} + u_{i+1,j+2,\ell}^{k} + u_{i+1,j+1,\ell-1}^{k}) + (1-3r)u_{i+1,j+1,\ell}^{k}$$

$$+ g_{i+1,j+1,\ell}^{1/2}, \tag{3.11.35}$$

$$-r[u_{i+1,j+1,\ell}^{k+1} + u_{i,j,\ell}^{k+1} + u_{i,j+1,\ell+1}^{k+1}] + (1+3r)u_{i,j+1,\ell}^{k+1}$$

$$= r(u_{i-1,j+1,\ell}^{k} + u_{i,j+2,\ell}^{k} + u_{i,j+1,\ell-1}^{k}) + (1-3r)u_{i,j+1,\ell}^{k} + g_{i,j+1,\ell}^{1/2}, \tag{3.11.36}$$

$$-r[u_{i+1,j,\ell+1}^{k+1} + u_{i,j+1,\ell+1}^{k+1} + u_{i,j,\ell}^{k+1}] + (1+3r)u_{i,j,\ell+1}^{k+1}$$

$$= r(u_{i-1,j,\ell+1}^{k} + u_{i,j-1,\ell+1}^{k} + u_{i,j,\ell+2}^{k}) + (1-3r)u_{i,j,\ell+1}^{k} + g_{i,j,\ell+1}^{1/2}, \tag{3.11.37}$$

$$-r[u_{i,j,\ell+1}^{k+1} + u_{i+1,j+1,\ell+1}^{k+1} + u_{i+1,j,\ell}^{k+1}] + (1+3r)u_{i+1,j,\ell+1}^{k+1}$$

$$= r(u_{i+2,j,\ell+1}^{k} + u_{i+1,j-1,\ell+1}^{k} + u_{i+1,j,\ell+2}^{k}) + (1-3r)u_{i+1,j,\ell+1}^{k}$$

$$+ g_{i+1,j,\ell+1}^{1/2}, \tag{3.11.38}$$

$$-r[u_{i,j+1,\ell+1}^{k+1} + u_{i+1,j,\ell+1}^{k+1} + u_{i+1,j+1,\ell}^{k+1}] + (1+3r)u_{i+1,j+1,\ell+1}^{k+1}$$

$$= r(u_{i+2,j+1,\ell+1}^{k} + u_{i+1,j+2,\ell+1}^{k} + u_{i+1,j+1,\ell+2}^{k}) + (1-3r)u_{i+1,j+1,\ell+1}^{k}$$

$$+ g_{i+1,j+1,\ell+1}^{1/2}, \tag{3.11.39}$$

and

$$-r[u_{i+1,j+1,\ell+1}^{k+1} + u_{i,j,\ell+1}^{k+1} + u_{i,j+1,\ell}^{k+1}] + (1+3r)u_{i,j+1,\ell+1}^{k+1}$$

$$= r(u_{i-1,j+1,\ell+1}^{k} + u_{i,j+2,\ell+1}^{k} + u_{i,j+1,\ell+2}^{k}) + (1-3r)u_{i,j+1,\ell+1}^{k}$$

$$+ g_{i,j+1,\ell+1}^{1/2}, \tag{3.11.40}$$

respectively.

The implicit system formed by equations (3.11.33–3.11.40) can then be written in matrix form as

$$A\underline{u}^{k+1} = \underline{f}^{k}, \tag{3.11.41}$$

where $\underline{f}^{k}$ is a vector whose elements are defined by the right-hand side of equations (3.11.33–3.11.40), with A given by

$$A = \begin{bmatrix}
1+3r & -r & 0 & -r & -r & 0 & 0 & 0 \\
-r & 1+3r & -r & 0 & 0 & -r & 0 & 0 \\
0 & -r & 1+3r & -r & 0 & 0 & -r & 0 \\
-r & 0 & -r & 1+3r & 0 & 0 & 0 & -r \\
-r & 0 & 0 & 0 & 1+3r & -r & 0 & -r \\
0 & -r & 0 & 0 & -r & 1+3r & -r & 0 \\
0 & 0 & -r & 0 & 0 & -r & 1+3r & -r \\
0 & 0 & 0 & -r & -r & 0 & -r & 1+3r
\end{bmatrix}, \tag{3.11.42}$$

and

$$\underline{u}^{k+1} = [u_{i,j,\ell}, u_{i+1,j,\ell}, u_{i+1,j+1,\ell}, u_{i,j+1,\ell}, u_{i,j,\ell+1}, u_{i+1,j,\ell+1}, u_{i+1,j+1,\ell+1}, u_{i,j+1,\ell+1}]^T.$$

Similar to the one- and two-dimensional cases to form the GE method we have to invert the matrix A to make the equations explicit and form the GE method. The inversion of this matrix can be obtained using its special form. The procedure is as follows:

Let

$$A = \begin{bmatrix} A_1 & A_2 \\ A_2 & A_1 \end{bmatrix},$$

we rewrite $A = (1 + 3r)A'$ to obtain

$$A_1 = \begin{bmatrix} 1 & -\dfrac{r}{1+3r} & 0 & -\dfrac{r}{1+3r} \\ -\dfrac{r}{1+3r} & 1 & -\dfrac{r}{1+3r} & 0 \\ 0 & -\dfrac{r}{1+3r} & 1 & -\dfrac{r}{1+3r} \\ -\dfrac{r}{1+3r} & 0 & -\dfrac{r}{1+3r} & 1 \end{bmatrix}, \quad A_2 = \begin{bmatrix} -\dfrac{r}{1+3r} & & & 0 \\ & -\dfrac{r}{1+3r} & & \\ & & -\dfrac{r}{1+3r} & \\ 0 & & & -\dfrac{r}{1+3r} \end{bmatrix},$$

and let

$$(A')^{-1} = \begin{bmatrix} D & B \\ B & D \end{bmatrix}.$$

From the identity $A'(A')^{-1} = I$ we can obtain the following relationships:

$$D = -BA_1 A_2^{-1} \text{ and } B = [A_2 - A_1 A_2^{-1} A_1]^{-1}.$$

We evaluate B first from

$$A_1 A_2^{-1} A_1 = \begin{bmatrix} \dfrac{1+2\alpha^2}{\alpha} & 2 & 2\alpha & 2 \\ 2 & \dfrac{1+2\alpha^2}{\alpha} & 2 & 2\alpha \\ 2\alpha & 2 & \dfrac{1+2\alpha^2}{\alpha} & 2 \\ 2 & 2\alpha & 2 & \dfrac{1+2\alpha^2}{\alpha} \end{bmatrix},$$

with $\alpha = -r(1+3r)$. Therefore

$$A_2 - A_1 A_2^{-1} A_1 = \begin{bmatrix} -\dfrac{1+\alpha^2}{\alpha} & -2 & -2\alpha & -2 \\ -2 & \dfrac{1+\alpha^2}{\alpha} & -2 & -2\alpha \\ -2\alpha & -2 & \dfrac{1+\alpha^2}{\alpha} & -2 \\ -2 & -2\alpha & -2 & -\dfrac{1+\alpha^2}{\alpha} \end{bmatrix},$$

and after some algebraic manipulation we obtain the result

$$B = (\alpha/k) \begin{bmatrix} 3\alpha^2 - 1 & 2\alpha & -6\alpha^2 & 2\alpha \\ 2\alpha & 3\alpha^2 - 1 & 2\alpha & -6\alpha^2 \\ -6\alpha^2 & 2\alpha & 3\alpha^2 - 1 & 2\alpha \\ 2\alpha & -6\alpha^2 & 2\alpha & 3\alpha^2 - 1 \end{bmatrix},$$

and from which D can easily be obtained in the form

$$\begin{bmatrix} 7\alpha^2 - 1 & \alpha(1 - 3\alpha^2) & \dfrac{-2\alpha^2(1 + 3\alpha^2)}{1 - \alpha^2} & \alpha(1 - 3\alpha^2) \\ \alpha(1 - 3\alpha^2) & 7\alpha^2 - 1 & \alpha(1 - 3\alpha^2) & \dfrac{-2\alpha^2(1 + 3\alpha^2)}{1 - \alpha^2} \\ \dfrac{-2\alpha^2(1 + 3\alpha^2)}{1 - \alpha^2} & \alpha(1 - 3\alpha^2) & 7\alpha^2 - 1 & \alpha(1 - 3\alpha^2) \\ \alpha(1 - 3\alpha^2) & \dfrac{-2\alpha^2(1 + 3\alpha^2)}{1 - \alpha^2} & \alpha(1 - 3\alpha^2) & 7\alpha^2 - 1 \end{bmatrix},$$

where $k = (1 - \alpha^2)(1 - 9\alpha^2)$. Hence the matrix A^{-1} can now be assembled in the form

$$A^{-1} = \frac{-1}{k} \left[\begin{array}{cccc|cccc} \dfrac{(1-7\alpha^2)r}{\alpha} & (3\alpha^2-1)r & \dfrac{2\alpha(1+3\alpha^2)r}{1-\alpha^2} & (3\alpha^2-1)r & (3\alpha^2-1)r & 2\alpha r & -6\alpha^2 r & 2\alpha r \\ (3\alpha^2-1)r & \dfrac{(1-7\alpha^2)r}{\alpha} & (3\alpha^2-1)r & \dfrac{2\alpha(1+3\alpha^2)r}{1-\alpha^2} & 2\alpha r & (3\alpha^2-1)r & 2\alpha r & -6\alpha^2 r \\ \dfrac{2\alpha(1+3\alpha^2)r}{1-\alpha^2} & (3\alpha^2-1)r & \dfrac{(1-7\alpha^2)r}{\alpha} & (3\alpha^2-1)r & -6\alpha^2 r & 2\alpha r & (3\alpha^2-1)r & 2\alpha r \\ (3\alpha^2-1)r & \dfrac{2\alpha(1+3\alpha^2)r}{1-\alpha^2} & (3\alpha^2-1)r & \dfrac{(1-7\alpha^2)r}{\alpha} & 2\alpha r & -6\alpha^2 r & 2\alpha r & (3\alpha^2-1)r \\ \hline (3\alpha^2-1)r & 2\alpha r & -6\alpha^2 r & 2\alpha r & \dfrac{(1-7\alpha^2)r}{\alpha} & (3\alpha^2-1)r & \dfrac{2\alpha(1+3\alpha^2)r}{1-\alpha^2} & (3\alpha^2-1)r \\ 2\alpha r & (3\alpha^2-1)r & 2\alpha r & -6\alpha^2 r & (3\alpha^2-1)r & \dfrac{(1-7\alpha^2)r}{\alpha} & (3\alpha^2-1)r & \dfrac{2\alpha(1+3\alpha^2)r}{1-\alpha^2} \\ -6\alpha^2 r & 2\alpha r & (3\alpha^2-1)r & 2\alpha r & \dfrac{2\alpha(1+3\alpha^2)r}{1-\alpha^2} & (3\alpha^2-1)r & \dfrac{(1-7\alpha^2)r}{\alpha} & (3\alpha^2-1)r \\ 2\alpha r & -6\alpha^2 r & 2\alpha r & (3\alpha^2-1)r & (3\alpha^2-1)r & \dfrac{2\alpha(1+3\alpha^2)r}{1-\alpha^2} & (3\alpha^2-1)r & \dfrac{(1-7\alpha^2)r}{\alpha} \end{array} \right]$$

Near the boundary, the points are treated either as a group of 4 points or a group of 2 points or an ungrouped point depending on the location of the points in the space. These special cases can be treated in a similar manner as the ordinary points of the two-space dimensional case or the ordinary point of the one-space dimension case. To avoid mathematical repetition, these will not be given here.

We can express the following conclusions for the new Group Explicit Method:

(a) The generalised finite difference approximation (3.2.1) is an equation from which it is possible to derive explicit, implicit, symmetrical and asymmetrical types of finite difference approximation.

(b) All valid numerical schemes of order upto $((\Delta x)^2 + (\Delta t)^2)$ have to fulfill the compulsory conditions (3.2.2–3.2.4) whilst the fulfilment of optional conditions (3.2.5–3.2.6) varies according to the chosen scheme.

(c) The GE methods are a class of methods which are derived from the combination of two implicit low-order asymmetrical formulae, i.e., Figures 3.2.6 and 3.2.7 which results in two explicit formulae of low-order accuracy, i.e., Figure 3.5.1. These two formulae are implemented in such a way that sometimes their truncation error signs will tend to cancel each other, therefore upgrading the order of accuracy of the formulae.

(d) In general, all the GE schemes have better stability conditions as compared to the classical explicit scheme.

(e) The GER, GEL, GEC and GEU schemes are best used for values of $r \approx 0.5$.

(f) For $0.5 \le r \le 1.0$, the alternating formulae are recommended for use, i.e., (S)AGE and (D)AGE.

(g) For $1.0 \le r \le 2.0$, the splitting type of half-step is recommended, i.e., (S)AGE or (D)AGE half-splitting.

(h) For $r \ge 2.0$, the fractional splitting type of quarter step (D)AGE is recommended.

(i) For $r \le 0.1$, the classical explicit scheme gives the better accuracy.

(j) Most of the numerical results using the recommended schemes are as accurate as the results obtained from the CN method.

(k) One important fact from this class of method is its avoidance of solving tridiagonal systems of equations.

(l) The extension of the method to two- and higher-space dimensions is fairly straightforward and obviously the unconditional stability of the explicit scheme is a great advantage.

(m) To some extent we are able to indicate the best weighting factor to choose for certain values of r in the Weighted Group Explicit method.

(n) Although only examples in rectangular space-domains were considered the schemes are usable for non-rectangular domains.

(o) Since the schemes are totally explicit, they are also suitable for use on the parallel processing systems of the future.

3.12. Fast Group Explicit Methods

(1) The Special Case of $r = 1$

Let us now consider a very special and practical case, that is the case when $r = 1$. For this case equation (3.5.5) becomes

$$\begin{bmatrix} u_{i,j+1} \\ u_{i+1,j+1} \end{bmatrix} = \frac{1}{3} \begin{bmatrix} 2u_{i-1,j} + u_{i+2,j} \\ u_{i-1,j} + 2u_{i+2,j} \end{bmatrix}, \tag{3.12.1}$$

with the corresponding computational molecule given by Figures 3.12.1a and 3.12.1b, while equations (3.5.9), (3.5.12–3.5.14) become:

$$(I + G_1)\underline{u}_{j+1} = (I - G_2)\underline{u}_j + \underline{b}_1, \tag{3.12.2}$$

$$(I + G_2)\underline{u}_{j+1} = (I - G_1)\underline{u}_j + \underline{b}_2, \tag{3.12.3}$$

$$(I + G_3)\underline{u}_{j+1} = (I - G_4)\underline{u}_j + \underline{b}_3, \tag{3.12.4}$$

$$(I + G_4)\underline{u}_{j+1} = (I - G_3)\underline{u}_j + \underline{b}_4, \tag{3.12.5}$$

respectively. Diagrammatically the schemes (3.12.2–3.12.5) are described by Figures 3.12.2–3.12.5 respectively.

From each of these figures, a special feature that is common to all is that we can always choose either nodes

and advance the time level by the equations (3.12.2–3.12.5) using only the same nodes and independent from the nodes

□ or ○

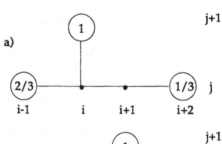

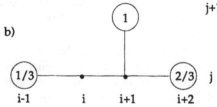

Figure 3.12.1

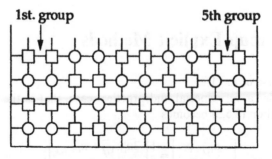

Figure 3.12.2   GEC (Group Explicit Complete) Method for $r = 1$ and $n = 11$.

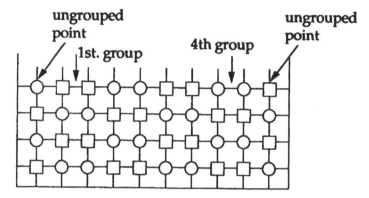

Figure 3.12.3 GEC (Group Explicit with Ungrouped Ends) Method for $r = 1$ and $n = 11$.

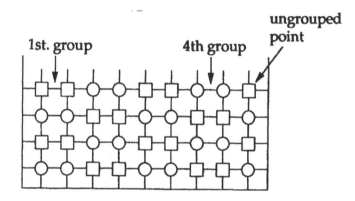

Figure 3.12.4 GER (Group Explicit with Right Ungrouped Points) Method for $r = 1$ and $n = 10$.

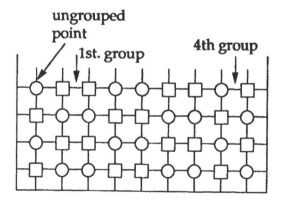

Figure 3.12.5 GEL (Group Explicit with Left Ungrouped Point) Method for $r = 1$ and $n = 10$.

This feature (i.e., Hopscotch) along the 2 vertical lines if exploited and coupled with suitable existing algorithms may lead to an algorithm which in effect can reduce the cost of the computation by half. The following section will discuss such an algorithm.

(2) A Fast Group Explicit (FGE) Algorithm

In the numerical solution of propagated problems like (3.1.1), usually it is not possible to omit calculating some of the points on a certain time level before proceeding to the next time level. This is due to the fact that the information at all points of that time level is required for computing the solution at the next time level. However the following Fast Group Explicit (FGE) algorithms are exceptional in this sense. The formula can advance the time level leaving half of the solution at grid points unknown. The solution at the remaining grid point can then be easily obtained by interpolation.

For example, in the case of equations (3.12.2–3.12.5), at each time level the distribution of the calculated solutions and uncalculated solutions is as in Figure 3.12.6.

We now consider the situation when the solution is required at every point on the jth time level only, then the following finite difference schemes could be used:

(i) *Fully implicit* With reference to Figure 3.12.6 the application of the fully implicit formulae (or Crank–Nicolson formulae) at points (i,j) and $(i+1,j)$ gives

$$\begin{bmatrix} 1+2r & -r \\ -r & 1+2r \end{bmatrix} \begin{bmatrix} u_{i,j} \\ u_{i+1,j} \end{bmatrix} = \begin{bmatrix} u_{i,j-1}+ru_{i-1,j} \\ u_{i+1,j-1}+ru_{i+2,j} \end{bmatrix}, \qquad (3.12.6)$$

which for $r=1$ results in the m decoupled or explicit equations

$$u_{i,j}=\tfrac{1}{8}[3u_{i,j-1}+u_{i+1,j-1}+3u_{i-1,j}+u_{i+2,j}],$$

and

$$u_{i+1,j}=\tfrac{1}{8}[u_{i,j-1}+3u_{i+1,j-1}+u_{i-1,j}+3u_{i+2,j}]. \qquad (3.12.7)$$

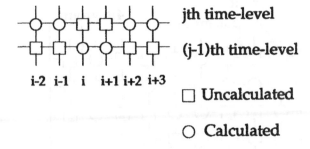

jth time-level

(j-1)th time-level

i-2 i-1 i   i+1 i+2 i+3

☐ Uncalculated

○ Calculated

Figure 3.12.6

Since the information on the right-hand side of equations (3.12.7) are all known therefore the application of equations (3.12.7) enables the solution at the unknown points of the jth time level to be calculated. The associated computational molecules for equations (3.12.7) are then given in Figure 3.12.7.

(ii) *Semi-implicit equations* The application of the Saul'yev equations (3.2.14) and (3.2.15), respectively at the points (i,j) and $(i+1,j)$ of Figure 3.12.6 gives

$$\begin{bmatrix} 1+r & 0 \\ 0 & 1+r \end{bmatrix}\begin{bmatrix} u_{i,j} \\ u_{i+1,j} \end{bmatrix} = \begin{bmatrix} 1-r & r \\ r & 1-r \end{bmatrix}\begin{bmatrix} u_{i,j-1} \\ u_{i+1,j-1} \end{bmatrix} + \begin{bmatrix} ru_{i-1,j} \\ ru_{i+2,j} \end{bmatrix}, \quad (3.12.8)$$

which for $r=1$ results in the less accurate formulae

$$u_{i,j} = \tfrac{1}{2}[u_{i-1,j} + u_{i+1,j-1}],$$

and

$$u_{i+1,j} = \tfrac{1}{2}[u_{i+2,j} + u_{i,j-1}]. \quad (3.12.9)$$

Similarly, as all the information required in the right-hand side of equation (3.12.9) are known, therefore the solution at all points on the jth time level can be calculated. The associated computational molecules are given in Figure 3.12.8.

Equations (3.12.8) and (3.12.9) coupled with the various Group Explicit formula of (3.12.2–3.12.5) can now form the class of **Fast Group Explicit (FGE)** algorithms as follows:

(i) *FGEC Scheme* This scheme uses the Group Explicit Complete (GEC) formula (3.12.2) and at the jth time level it combines with formulae (3.12.7) or (3.12.9) to calculate the solution at all points.

(ii) *FGEU Scheme* This scheme uses the Group Explicit with Ungrouped (GEU) ends formula (3.12.3) for advancing the time level and at the jth time level uses the formulae (3.12.7) or (3.12.9).

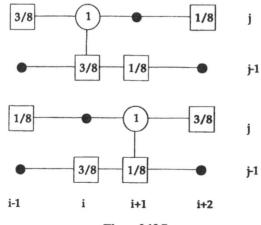

Figure 3.12.7

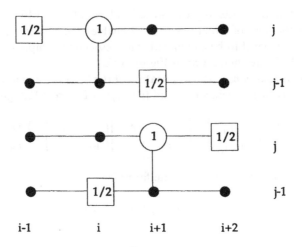

Figure 3.12.8

(iii) *FGER Scheme* The scheme initially advances the calculation using the Group Explicit with Right Ungrouped Point (GER) formula (3.12.4) and finally at the jth time level is incorporated with the formulae (3.12.7) or (3.12.9).

(iv) *FGEL Scheme* The last of the four GE schemes initially used the Group Explicit with Left Ungrouped point (GEL) formula (3.12.5) and lastly combines with formulae (3.12.7) or (3.12.9).

(v) *FSAGE and FDAGE Schemes* These schemes involve the use of the above methods in a single alternating strategy (FGER–FGEL) or (FGEL–FGEU) for FSAGE and a double FSAGE alternating strategy of the form (FGER–FGEL–FGEL–FGER) or (FGEC–FGEU–FGEU–FGEL) for the FDAGE scheme.

Table 3.12.1

Method	No. of Operations at Each Time Level		
	Addition	Multiplication	Division
GEC	$3(n-1)$	$4(n-1)$	$n-1$
GEU	$3n-5$	$4n-2$	$n+1$
GER/GEL	$3n-4$	$4n-6$	n
Explicit	$2(n-1)$	$3(n-1)$	0
Implicit	$5n-6$	$5n-6$	$2n-3$
Crank–Nicolson	$5n-6$	$5n-6$	$2n-3$
Saul'yev	$2(n-1)$	$3(n-1)$	$n-1$
FGE (while advancing)	$1/2(n-1)$	$(n-1)$	0
FGE (coupled with fully implicit) equations (3.12.7)	$2(n-1)$	$3(n-1)$	0
FGE (Coupled with Saul'yev) equations (3.12.9)	$(n-1)$	$2(n-1)$	0

Finally, a similar analogy for using the fast computational schemes in a fractional splitting mode of operation is discussed in Abdullah (1983) is applicable.

As these four schemes enable us to avoid the calculation of the solutions at half of the points on each time level and furthermore the calculation of the solution at each of the other points only involves one of the equations (3.2.14) or (3.2.15) or (3.12.1) or (3.12.7) or (3.12.9), therefore the cost of computation is very economical. Table 3.12.1 gives the comparison of the computational complexity of the given schemes compared to existing algorithms. From the table, it is clear that the FGE schemes when used while advancing the time level is always extremely efficient. Further, the FGE scheme coupled with the fully implicit method is as simple as the classical explicit formulae while coupled with the Saul'yev method is still less computationally expensive than any existing method.

Finally, the new schemes presented are stable formulae for the case $r = 1$, while the truncation error term is inherited from the truncation error of the GE schemes and is of $0(\Delta t + \Delta x^2 + (\Delta t/\Delta x))$.

4. The Alternating Group Explicit (AGE) Methods for Parabolic Equations

4.1. Introduction

From Chapter 2 it was outlined that the ADI method was developed to deal with two-dimensional parabolic problems where the solutions at each (1/2) time step were obtained *implicitly* along the horizontal and vertical directions of the two-dimensional grid. The method can also be extended for applications to higher-dimensional problems. However the method has no analogue for the one-dimensional case.

Here we show that it is possible to derive such an analogue, i.e., the AGE method, the analysis of which is analogous to the ADI scheme. Initially the method is presented for one-dimensional problems but its implementation to higher-dimensional equations can be easily obtained. The *iterative* method employs the fractional splitting strategy which is applied alternately at each half (intermediate) time step on tridiagonal systems of difference schemes and has proved to be stable. Its rate of convergence is governed by the acceleration parameter ρ. The accuracy of this method is, in general, comparable if not better than that of the GE class of problems described in Chapter 3 as well as other existing schemes.

Initially Evans and Abdullah (1983) considered the group explicit method for the numerical solution of parabolic equations. Later Evans (1985) introduced a more accurate but iterative version of the method was introduced for elliptic problems. This new approach is very similar to the ADI method, but since the iterative formulae for the new approach can be easily expressed in explicit form the method is called the Alternating Group Explicit (AGE) method. Since then, the AGE method has been extensively studied and applied to a variety of problems involving parabolic, hyperbolic and elliptic partial differential equations. It has been shown that the method is extremely powerful and flexible and very easy to parallelise.

Further, because of the simplicity of the matrix operators involved, i.e. (2×2) submatrices, a coupled form of the iteration can be derived which involves only 1 sweep of the mesh points at each time step which yields considerable efficiency.

4.2. The Alternating Group Explicit Method to Solve Second Order Parabolic Equations with Dirichlet Boundary Conditions

Consider the following second order linear parabolic equation

$$\frac{\partial U}{\partial t} = \frac{\partial^2 U}{\partial x^2}, \quad 0 \le x \le 1,\ 0 < t \le T, \tag{4.2.1}$$

subject to the initial-boundary value conditions,

$$U(x,0) = f(x), \quad 0 < x < 1,$$

$$U(0,t) = g(t), \quad 0 < t \le T, \tag{4.2.1a}$$

and

$$U(1,t) = h(t).$$

A uniformly-spaced network whose mesh points are $x_i = i\Delta x$, $t_j = j\Delta t$ for $i = 0, 1, \ldots, m, m+1$ and $j = 0, 1, \ldots, n, n+1$ is used with $\Delta x = 1/(m+1), \Delta t = T/(n+1)$ and $r = \Delta t/(\Delta x)^2$, the mesh ratio. The real line $0 \le x \le 1$ is thus divided as illustrated:

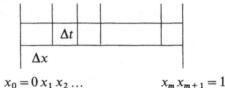

$$x_0 = 0\ x_1\ x_2 \ldots \qquad\qquad x_m\ x_{m+1} = 1$$

From equation (3.3.1), a weighted approximation to the differential equation (4.2.1) at the point $(x_i, t_{j+1/2})$ is given by

$$-r\theta u_{i-1,j+1} + (1 + 2r\theta)u_{i,j+1} - r\theta u_{i+1,j+1}$$

$$= r(1-\theta)u_{i-1,j} + (1 - 2r(1-\theta))u_{i,j} + r(1-\theta)u_{i+1,j}, \quad i = 1, 2, \ldots, m.$$

$$\tag{4.2.2}$$

This approximation can be displayed in a more compact matrix form as

$$\begin{bmatrix} a & b & & & & \\ c & a & b & & & \\ & c & a & b & & \\ & & & & 0 & \\ & 0 & & & & \\ & & & c & a & b \\ & & & & c & a \end{bmatrix}_{(m \times m)} \begin{bmatrix} u_1 \\ u_2 \\ \vdots \\ \vdots \\ u_{m-1} \\ u_m \end{bmatrix}_{j+1} = \begin{bmatrix} f_1 \\ f_2 \\ \vdots \\ \vdots \\ f_{m-1} \\ f_m \end{bmatrix}, \tag{4.2.3}$$

i.e.,

$$A\underline{u} = \underline{f}, \tag{4.2.4}$$

where

$$c = -r\theta, \quad a = (1 + 2r\theta), \quad b = -r\theta;$$

$$f_1 = r(1 - \theta)(u_{0j} + u_{2j}) + r\theta u_{0,j+1} + (1 - 2(1 - \theta))u_{1,j};$$

$$f_i = r(1 - \theta)(u_{i-1,j} + u_{i+1,j}) + (1 - 2r(1 - \theta))u_{ij}, \quad i = 2, 3, \ldots, m-1, m-2;$$

$$f_m = r(1 - \theta)(u_{m-1,j} + u_{m+1,j}) + (1 - 2r(1 - \theta))u_{mj} + r\theta u_{m+1,j+1}, \quad (4.2.4a)$$

and

$$\underline{u} = (u_{1,j+1}, u_{2,j+1}, \ldots, u_{m,j+1})^T \quad \text{and} \quad \underline{f} = (f_1, f_2, \ldots, f_m)^T.$$

We note that $\underline{f}$ is a column vector of order m consisting of the boundary values as well as known u_j values at time level j while u_{j+1} are the values at time level $(j+1)$ which we seek. We also recall that (4.2.4) corresponds to the fully implicit, the Crank–Nicolson, the Douglas and the classical explicit methods when θ takes the values 1, $(1/2, 1/2 - 1/12r)$ and 0 with accuracies of the order $0([\Delta x]^2 + \Delta t)$, $0([\Delta x]^2 + [\Delta t]^2)$, $0([\Delta x]^4 + [\Delta t]^2)$, and $0([\Delta x]^2 + \Delta t)$, respectively.

Let us first assume that we have an *even number of intervals* (corresponding to an odd number of internal points, i.e. m odd) on the real line $0 \leq x \leq 1$. We can then perform the following splitting of the coefficient matrix A:

$$A = G_1 + G_2, \quad (4.2.5)$$

where

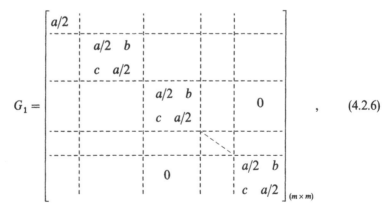

$$\quad (4.2.6)$$

and

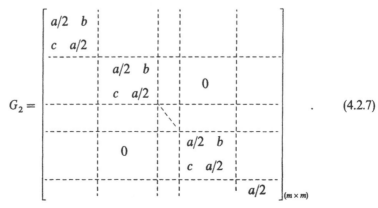

$$\quad (4.2.7)$$

It is assumed that the following conditions are satisfied:

(i) $G_1 + \rho I$ and $G_2 + \rho I$ are non-singular for any $\rho > 0$
(ii) for any vectors $\underline{f}_1$ and $\underline{f}_2$ and for any $\rho > 0$, the systems

$$(G_1 + \rho I)\underline{u}_1 = \underline{f}_1$$

and

$$(G_2 + \rho I)\underline{u}_2 = \underline{f}_2 \qquad (4.2.8)$$

are more easily solved in explicit form since they consist of only (2×2) subsystems.

Thus, with these conditions, system (4.2.4) becomes

$$(G_1 + G_2)\underline{u} = \underline{f}. \qquad (4.2.9)$$

The AGE method consists of writing (4.2.9) as a pair of consistent equations

$$(G_1 + \rho I)\underline{u} = (\rho I - G_2)\underline{u} + \underline{f}$$

and

$$(G_2 + \rho I)\underline{u} = (\rho I - G_1)\underline{u} + \underline{f}. \qquad (4.2.10)$$

The AGE iterative method using *the Peaceman and Rachford variant* for the *stationary case* ($\rho = $ constant) is given by

$$(G_1 + \rho I)\underline{u}^{(p+1/2)} = (\rho I - G_2)\underline{u}^{(p)} + \underline{f}$$

and

$$(G_2 + \rho I)\underline{u}^{(p+1)} = (\rho I - G_1)\underline{u}^{(p+1/2)} + \underline{f}, \quad p \geq 0, \qquad (4.2.11)$$

where $\underline{u}^{(0)}$ is a starting approximation and ρ are positive constants called acceleration parameters whose values are chosen to maximise the rate of convergence. We now seek to analyse the convergence properties of the AGE method. From (4.2.11) we can write

$$\underline{u}^{(p+1)} = M(\rho)\underline{u}^{(p)} + \underline{q}(\rho), \quad p \geq 0, \qquad (4.2.12)$$

where *the AGE iteration matrix* is given by

$$M(\rho) = (G_2 + \rho I)^{-1}(\rho I - G_1)(G_1 + \rho I)^{-1}(\rho I - G_2). \qquad (4.2.13)$$

If $\underline{e}$ denotes the error vector and $\underline{U}$ the exact solution of (4.2.1) then, $\underline{e}^{(p)} = \underline{u}^{(p)} - \underline{U}$ and $\underline{e}^{(p+1)} = M(\rho)\underline{e}^{(p)}$. Hence, we have

$$\underline{e}^{(p)} = M^p(\rho)\underline{e}^{(0)}, \quad p \geq 1. \qquad (4.2.14)$$

We now prove the following theorem.

Theorem 4.1.
If G_1 and G_2 are real positive definite matrices and if $\rho > 0$ then,

$$\lambda(M(\rho)) \leq 1. \qquad (4.2.15)$$

Proof. If we let $M(\rho) = (G_2 + \rho I)M(\rho)(G_2 + \rho I)^{-1}$, then by a similarity transformation, $M(\rho)$ and $\tilde{M}(\rho)$ have the same eigenvalues. Hence, from (4.2.13) we find that

$$\lambda(M(\rho)) = \lambda(\tilde{M}(\rho))$$

$$\leq \|\tilde{M}(\rho)\|$$

$$\leq \|(\rho I - G_1)(G_1 + \rho I)^{-1}\| \; \|(\rho I - G_2)(G_2 + \rho I)^{-1}\|, \tag{4.2.16}$$

where $\lambda(M(\rho))$ is the spectral radius of $M(\rho)$. But since G_1 and G_2 are symmetric and $(\rho I - G_1)$ commutes with $(G_1 + \rho I)^{-1}$, then in the L_2 norm we have

$$\|(\rho I - G_1)(G_1 + \rho I)^{-1}\|_2 = \lambda((\rho I - G_1)(G_1 + \rho I)^{-1})$$

$$= \max_{1 \leq i \leq m} \left| \frac{\rho - \mu_i}{\rho + \mu_i} \right|, \tag{4.2.17}$$

where μ_i are the eigenvalues of G_1. But since G_1 is positive definite, its eigenvalues are positive. Therefore,

$$\|(\rho I - G_1)(G_1 + \rho I)^{-1}\|_2 < 1. \tag{4.2.18}$$

Similarly,

$$\|(\rho I - G_2)(G_2 + \rho I)^{-1}\|_2 < 1, \tag{4.2.18}$$

and we have

$$\lambda(M(\rho)) = \lambda(\tilde{M}(\rho)) \leq \|M(\rho)\|_2 < 1, \tag{4.2.19}$$

and convergence is assured. We note that to establish the condition (4.2.15) for unsymmetric matrices G_1 and G_2 may require us to evaluate directly the eigenvalues of $M(\rho)$ which can be difficult.

It is possible to determine the optimum parameter $\hat{\rho}$ such that the bound for $\lambda(M(\rho))$ is minimised. To investigate this we assume that G_1 and G_2 are real positive definite matrices and that bounds for their eigenvalues μ_i and η_i are available, i.e.,

$$0 < \alpha \leq \mu_i, \qquad \eta_i \leq \beta. \tag{4.2.20}$$

In the L_2 norm, (4.2.16) implies

$$\lambda(M(\rho)) \leq \lambda((\rho I - G_1)(G_1 + \rho I)^{-1})\lambda((\rho I - G_2)(G_2 + \rho I)^{-1})$$

$$= \left\{ \max_{1 \leq i \leq m} \left| \frac{\rho - \mu_i}{\rho + \mu_i} \right| \right\} \left\{ \max_{1 \leq i \leq m} \left| \frac{\rho - \eta_i}{\rho + \eta_i} \right| \right\}$$

$$\leq \left\{ \max_{\alpha \leq z \leq \beta} \left| \frac{\rho - z}{\rho + z} \right| \right\}^2 = \phi(\alpha, \beta; \rho). \tag{4.2.21}$$

But $(\rho - z)/(\rho + z)$ is a decreasing function of z. Therefore, we find that

$$\max_{\alpha \leq z \leq \beta} \left| \frac{\rho - z}{\rho + z} \right| = \max\left(\left| \frac{\rho - \alpha}{\rho + \alpha} \right|, \left| \frac{\rho - \beta}{\rho + \beta} \right| \right). \tag{4.2.22}$$

When $\rho = \sqrt{\alpha\beta}$ we have

$$\left|\frac{\rho-\alpha}{\rho+\alpha}\right| = \left|\frac{\rho-\beta}{\rho+\beta}\right| = \frac{\sqrt{\beta}-\sqrt{\alpha}}{\sqrt{\alpha}+\sqrt{\beta}}. \qquad (4.2.23)$$

For $0 < \rho < \sqrt{\alpha\beta}$ we obtain

$$\left|\frac{\rho-\beta}{\rho+\beta}\right| - \frac{(\sqrt{\beta}-\sqrt{\alpha})}{(\sqrt{\alpha}+\sqrt{\beta})} = \frac{2\sqrt{\beta}(\sqrt{\alpha\beta}-\rho)}{(\rho+\beta)(\sqrt{\alpha}+\sqrt{\beta})} > 0,$$

i.e.,

$$\left|\frac{\rho-\beta}{\rho+\beta}\right| > \frac{(\sqrt{\beta}-\sqrt{\alpha})}{(\sqrt{\alpha}+\sqrt{\beta})}. \qquad (4.2.24)$$

Similarly, for $\sqrt{\alpha\beta}$, we get

$$\left|\frac{\rho-\alpha}{\rho+\alpha}\right| - \frac{(\sqrt{\beta}-\sqrt{\alpha})}{(\sqrt{\alpha}+\sqrt{\beta})} = \frac{2\sqrt{\alpha}(\rho-\sqrt{\alpha\beta})}{(\rho+\alpha)(\sqrt{\alpha}+\sqrt{\beta})} > 0,$$

i.e.,

$$\left|\frac{\rho-\alpha}{\rho+\alpha}\right| > \frac{(\sqrt{\beta}-\sqrt{\alpha})}{(\sqrt{\alpha}+\sqrt{\beta})}. \qquad (4.2.25)$$

Hence, using $(4.2.21 - 4.2.25)$ we deduce that the bound $\phi(\alpha,\beta;\rho)$ for $\lambda(M(\rho))$ is minimised when $\rho = \hat{\rho} = \sqrt{\alpha\beta}$ and $\lambda(M(\hat{\rho})) \le \phi(\alpha,\beta;\hat{\rho}) = ((\sqrt{\beta}-\sqrt{\alpha})/(\sqrt{\alpha}+\sqrt{\beta}))^2$.

For an efficient implementation of the AGE algorithm, it is essential to vary the acceleration parameters ρ_p from iteration to iteration – *the non-stationary case*. This may result in an improvement in the rate of convergence of the AGE method as when the Peaceman–Rachford ADI variant is employed. The Peaceman–Rachford AGE formula (4.2.11) will now become

$$(G_1 + \rho_{p+1}I)\underline{u}^{(p+1/2)} = (\rho_{p+1}I - G_2)\underline{u}^{(p)} + \underline{f}$$

and

$$(G_2 + \rho_{p+1}I)\underline{u}^{(p+1)} = (\rho_{p+1}I - G_1)\underline{u}^{(p+1/2)} + \underline{f}, \quad p \ge 0. \qquad (4.2.26)$$

The best values of ρ_p can be ascertained provided G_1 and G_2 are *commutative* – a property which is not possessed by our model problem. However, the matrices do commute if the boundary conditions are *periodic* and of order 4, that is, when the conditions (4.2.1a) are replaced by

$$U(0,t) = U(1,t), \qquad \frac{\partial U}{\partial x}(0,t) = \frac{\partial U}{\partial x}(1,t). \qquad (4.2.27)$$

For the general application of (4.2.26), $(\rho_p > 0)$ we assume first of all that the positive definite matrices G_1 and G_2 commute. Thus, G_1 and G_2 have a common set of (orthonormal) eigenvectors. Let $(\mu_j, v_j)_{j=1}^m$ and $(\eta_j, v_j)_{j=1}^m$ be the eigensystem of G_1 and G_2, respectively.

For p iterations of (4.2.26), relation (4.2.13) yields, for $1 \le j \le m$,

$$\left(\prod_{i=1}^{p} M(\rho_i)\right)v_j = \left\{\prod_{i=1}^{p}\left(\frac{\rho_i-\mu_j}{\rho_i+\mu_j}\right)\left(\frac{\rho_i-\eta_j}{\rho_i+\eta_j}\right)\right\}v_j. \qquad (4.2.28)$$

It follows that, $\Pi_{i=1}^{p} M(\rho_i)$ is symmetric and therefore,

$$
\left\| \prod_{i=1}^{p} M(\rho_i) \right\|_2 = \lambda \left(\prod_{i=1}^{p} M(\rho_i) \right)
$$

$$
= \max_{1 \leq j \leq m} \prod_{i=1}^{p} \left| \frac{\rho - \mu_j}{\rho + \mu_j} \right| \left| \frac{\rho - \eta_j}{\rho + \eta_j} \right| < 1, \tag{4.2.29}
$$

and convergence of the iterative process is achieved. Now, it is clear that,

$$
\max_{1 \leq j \leq m} \prod_{i=1}^{p} \left| \frac{\rho_i - \mu_j}{\rho_i + \mu_j} \right| \left| \frac{\rho_i - \eta_j}{\rho_i + \eta_j} \right| \leq \max_{1 \leq j \leq m} \prod_{i=1}^{p} \left| \frac{\rho_i - \mu_j}{\rho_i + \mu_j} \right| \max_{1 \leq j \leq m} \prod_{i=1}^{p} \left| \frac{\rho_i - \eta_j}{\rho_i + v_j} \right|
$$

$$
\leq \left\{ \max_{1 \leq j \leq m} \prod_{i=1}^{p} \left| \frac{\rho_i - z}{\rho_i + z} \right| \right\}^2, \tag{4.2.30}
$$

where we have used the bounds for the eigenvalues given by (4.2.20). Hence,

$$
\lambda \left(\prod_{i=1}^{p} M(\rho_i) \right) \leq \max_{\alpha \leq z \leq \beta} |R_p(z)|^2 = \phi(\alpha, \beta; \rho_1, \ldots, \rho_p), \tag{4.2.31}
$$

where $R_p(z) = \Pi_{i=1}^{p} M(\rho_i) (\rho_i - z)/(\rho_i + z)$. The difficulty of determining the optimum parameters by minimising $\phi(\alpha, \beta; \rho_1, \ldots, \rho_p)$ has led to a number of alternative sequences being devised. For example, the parameters used by Peaceman and Rachford (1955) are

$$
\hat{\rho}_j = \beta \left(\frac{\alpha}{\beta} \right)^{(2j-1)/(2p)}, \quad j = 1, 2, \ldots, p, \tag{4.2.32}
$$

from which we obtain the result

$$
\lambda \left(\prod_{i=1}^{p} M(\rho_i) \right) \leq \left\{ \frac{1 - (\alpha/\beta)^{1/(2p)}}{1 + (\alpha/\beta)^{1/(2p)}} \right\}^2. \tag{4.2.33}
$$

Wachpress (1966) on the other hand, solved the minimisation problem analytically in terms of elliptic functions and arrived at the result

$$
\hat{\rho}_j = \beta \left(\frac{\alpha}{\beta} \right)^{(j-1)/(p-1)}, \quad p \geq 2, j = 1, 2, \ldots, p. \tag{4.2.34}
$$

It must be noted that the requirement that G_1 and G_2 be commutative can be very restrictive indeed. However, in far more general situations we can expect convergence of the AGE iteration for a *fixed acceleration parameter* without the condition $G_1 G_2 = G_2 G_1$.

We conclude this section by considering the case when we have an *odd number of intervals* (corresponding to an even number of internal points, i.e. m even) on $0 \le x \le 1$. The coefficient matrix A will be split as in (4.2.5) but G_1 and G_2 now take the form

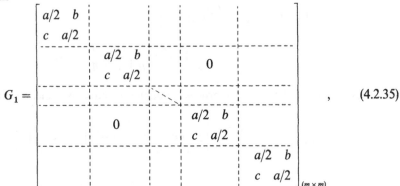

$$G_1 = \qquad , \qquad (4.2.35)$$

and

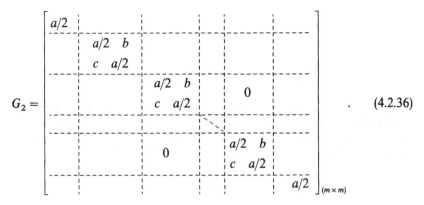

$$G_2 = \qquad . \qquad (4.2.36)$$

All the preceding conditions and arguments regarding convergence and the choice of the optimal acceleration parameter for both the stationary and non-stationary cases remain valid.

4.3. Variants of the AGE Scheme and its Computation

Many variants of the basic AGE scheme can be proposed. For example, we have on modifying the second stage of (4.2.26) (the non-stationary case)

$$(G_1 + \rho_{p+1}I)\underline{u}^{(p+1/2)} = (\rho_{p+1}I - G_2)\underline{u}^{(p)} + \underline{f}$$

and

$$(G_2 + \rho_{p+1}I)\underline{u}^{(p+1)} = (G_2 - (1-\omega)\rho_{p+1}I)\underline{u}^{(p)} + (2-\omega)\rho_{p+1}\underline{u}^{(p+1/2)} , \qquad (4.3.1)$$

where ω is a parameter. For $\omega = 0$ we have the scheme similar to the Peaceman–Rachford ADI method (1955) (4.2.26) and for $\omega = 1$, we obtain the scheme similar to the Douglas and Rachford ADI method (1956). For G_1 and G_2 symmetric and positive definite and with a *fixed acceleration parameter* $\rho > 0$, the resulting *generalised AGE scheme* is convergent for any $0 \le \omega \le 2$. As we shall see later, an extension of the AGE algorithm to implement it on higher-dimensional boundary value problems using the schemes similar to the Douglas–Rachford, Douglas and Guittet ADI variants can be easily obtained.

For the purpose of computation, we now attempt to derive explicit equations that are satisfied at each intermediate (half-time) level. For the *Peaceman–Rachford form*, in particular and with fixed parameter ρ, the *AGE method* can be applied to determine $\underline{u}^{(p+1/2)}$ and $\underline{u}^{(p+1)}$ implicitly by

$$(G_1 + \rho I)\underline{u}^{(p+1/2)} = (\rho I - G_2)\underline{u}^{(p)} + \underline{f},$$

and

$$(G_2 + \rho I)\underline{u}^{(p+1)} = (\rho I - G_1)\underline{u}^{(p+1/2)} + \underline{f}, \tag{4.3.2}$$

or explicitly by

$$\underline{u}^{(p+1/2)} = (G_1 + \rho I)^{-1}\{(\rho I - G_2)\underline{u}^{(p)} + \underline{f}\}$$

and

$$\underline{u}^{(p+1)} = (G_2 + \rho I)^{-1}\{(\rho I - G_1)\underline{u}^{(p+1/2)} + \underline{f}\} \tag{4.3.3}$$

If we assume m to be odd (*even number of intervals*) and if we write

$$\hat{G} = \begin{bmatrix} \rho_2 & b \\ c & \rho_2 \end{bmatrix} \tag{4.3.4}$$

where

$$\rho_2 = \rho + \frac{a}{2}, \tag{4.3.5}$$

then from (4.2.6) and (4.2.7) we have

$$(G_1 + \rho I) = \begin{bmatrix} \rho_2 & & & & \\ & \hat{G} & & & \\ & & \hat{G} & & 0 \\ & & & \ddots & \\ & & 0 & & \hat{G} \end{bmatrix}_{(m \times m)} \tag{4.3.6}$$

and

$$(G_2 + \rho I) = \begin{bmatrix} \hat{G} & & & & \\ & \hat{G} & & 0 & \\ & & \ddots & & \\ & 0 & & \hat{G} & \\ & & & & \rho_2 \end{bmatrix}_{(m \times m)} \tag{4.3.7}$$

It is clear that $(G_1 + \rho I)$ and $(G_2 + \rho I)$ are block diagonal matrices. All the diagonal elements except the first (or the last for $(G_2 + \rho I)$) are (2×2) submatrices. Therefore, $(G_1 + \rho I)$ and $(G_2 + \rho I)$ can be easily inverted by merely inverting their block diagonal entries. Hence,

$$(G_1 + \rho I)^{-1} = \begin{bmatrix} \dfrac{1}{\rho_2} & & & \\ & \hat{G}^{-1} & & \\ & & \hat{G}^{-1} & 0 \\ & & & \ddots \\ & 0 & & \hat{G}^{-1} \end{bmatrix}_{(m \times m)} \tag{4.3.8}$$

$$= \frac{1}{\Delta} \begin{bmatrix} \dfrac{\Delta}{\rho_2} & & & \\ & \begin{matrix} \rho_2 & -b \\ -c & \rho_2 \end{matrix} & & 0 \\ & & \ddots & \\ & 0 & & \begin{matrix} \rho_2 & -b \\ -c & \rho_2 \end{matrix} \end{bmatrix}_{(m \times m)} , \tag{4.3.9}$$

where

$$\Delta = \rho_2^2 - bc. \tag{4.3.10}$$

Similarly, we obtain

$$(G_2 + \rho I)^{-1} = \frac{1}{\Delta} \begin{bmatrix} \begin{matrix} \rho_2 & -b \\ -c & \rho_2 \end{matrix} & & & \\ & \begin{matrix} \rho_2 & -b \\ -c & \rho_2 \end{matrix} & & 0 \\ & & \ddots & \\ & 0 & & \begin{matrix} \rho_2 & -b \\ -c & \rho_2 \end{matrix} & \\ & & & & \dfrac{\Delta}{\rho_2} \end{bmatrix}_{(m \times m)} . \tag{4.3.11}$$

From (4.3.3), $\underline{u}^{(p+1/2)}$ and $\underline{u}^{(p+1)}$ are given by

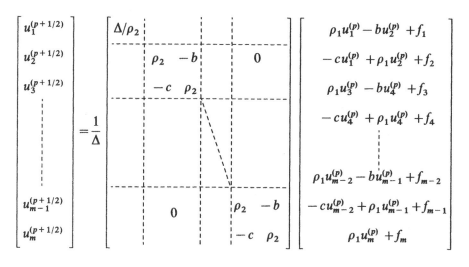

$$(4.3.12)$$

and

$$(4.3.13)$$

where

$$\rho_1 = \rho - a/2, \quad \rho_2 = \rho + a/2 \quad \text{and} \quad \Delta = \rho_2^2 - bc. \qquad (4.3.14)$$

The *alternating implicit* nature of the (2×2) groups in the equations (4.3.2) is shown in Figure 4.3.1 where the implicit/explicit values are given on the forward/backward levels for sweeps on the $(p + 1/2)^{th}$ and $(p + 1)^{th}$ levels.

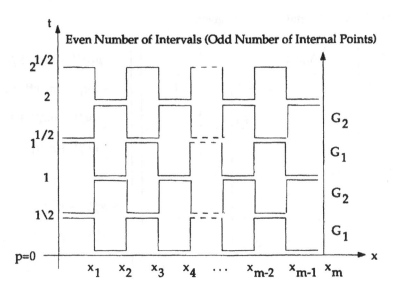

Figure 4.3.1   The AGE Method (Implicit).

The corresponding explicit expressions for the AGE equations are obtained by carrying out the multiplications in (4.3.12) and (4.3.13). Thus we have:

(i) *at level* $(p + 1/2)$

$$u_1^{(p+1/2)} = (\rho_1 u_1^{(p)} - b u_2^{(p)} + f_1)/\rho_2,$$

$$u_i^{(p+1/2)} = (A u_{i-1}^{(p)} + B u_i^{(p)} + C u_{i+1}^{(p)} + D u_{i+2}^{(p)} + E_i)/\Delta$$

$$u_{i+1}^{(p+1+1/2)} = (\tilde{A} u_{i-1}^{(p)} + \tilde{B} u_i^{(p)} + \tilde{C} u_{i+1}^{(p)} + \tilde{D} u_{i+2}^{(p)} + \tilde{E}_i)/\Delta$$
$$\left.\right\} \; i = 2, 4, \ldots, m-1,$$
$$(4.3.15)$$

where

$$A = -c\rho_2, \quad B = \rho_1\rho_2, \quad C = -b\rho_1, \quad E_i = \rho_2 f_i - b f_{i+1}, \quad D = \begin{cases} 0, & \text{for } i = m-1, \\ b^2, & \text{otherwise} \end{cases}$$
$$(4.3.15a)$$

and

$$\tilde{A} = c^2, \quad \tilde{B} = -c\rho_1, \quad \tilde{C} = \rho_1\rho_2, \quad \tilde{E}_i = \rho_2 f_{i+1} - c f_i, \quad \tilde{D} = \begin{cases} 0, & \text{for } i = m-1, \\ -b\rho_2, & \text{otherwise} \end{cases}$$

with the following computational molecules (Fig. 4.3.2)

(ii) *at level* $(p + 1)$

$$u_i^{(p+1)} = (P u_{i-1}^{(p+1/2)} + Q u_i^{(p+1/2)} + R u_{i+1}^{(p+1/2)} + S u_{i+2}^{(p+1/2)} + T_i)/\Delta,$$

$$u_{i+1}^{(p+1)} = (\tilde{P} u_{i-1}^{(p+1/2)} + \tilde{Q} u_i^{(p+1/2)} + \tilde{R} u_{i+1}^{(p+1/2)} + \tilde{S} u_{i+2}^{(p+1/2)} + \tilde{T}_i)/\Delta,$$

$$i = 1, 3, \ldots, m-2$$

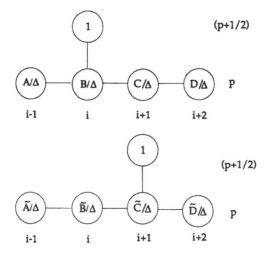

Figure 4.3.2 The AGE Method (Explicit) at Level $(p + 1/2)$.

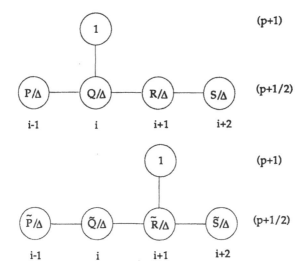

Figure 4.3.3 The AGE Method (Explicit) at Level $(p + 1)$.

and

$$u_m^{(p+1)} = (-cu_{m-1}^{(p+1/2)} + \rho_1 u_m^{(p+1/2)} + f_m)/\rho_2 , \qquad (4.3.16)$$

where

$$P = \begin{cases} 0 & \text{for } i = 1, \\ -c\rho_2 & \text{for } i \neq 1, \end{cases} \quad Q = \rho_1\rho_2, \ R = -b\rho_1, \ S = b^2, \ T_i = \rho_2 f_i - bf_{i+1}$$

and $\qquad (4.3.16a)$

$$\tilde{P} = \begin{cases} 0 & \text{for } i = 1, \\ c^2 & \text{for } i \neq 1, \end{cases} \quad \tilde{Q} = -c\rho_1, \ \tilde{R} = Q = \rho_1\rho_2, \ \tilde{S} = -b\rho_2, \ \tilde{T}_i = -cf_i + \rho_2 f_{i+1},$$

with its computational molecules given by Figure (4.3.3).

For the generalised AGE scheme (4.3.1) with fixed acceleration parameter ρ, the relevant equations at level $(p + 1/2)$ remain the same as in (4.3.15/4.3.15a). The equations at level $(p + 1)$ are, however, now replaced by

$$
\begin{bmatrix} u_1^{(p+1)} \\ u_2^{(p+1)} \\ u_3^{(p+1)} \\ \vdots \\ u_{m-2}^{(p+1)} \\ u_{m-1}^{(p+1)} \\ u_m^{(p+1)} \end{bmatrix}
= \frac{1}{\Delta}
\begin{bmatrix}
\rho_2 & -b & & & & & \\
-c & \rho_2 & & & & 0 & \\
& & \rho_2 & -b & & & \\
& & -c & \rho_2 & & & \\
& & & & \ddots & & \\
& 0 & & & \rho_2 & -b_2 & \\
& & & & -c & \rho_2 & \\
& & & & & & \Delta/\rho_2
\end{bmatrix}
$$

$$
\times
\begin{bmatrix}
\rho_3 u_1^{(p)} + b u_2^{(p)} + \rho_4 u_1^{(p+1/2)} \\
c u_1^{(p)} + \rho_3 u_2^{(p)} + \rho_4 u_2^{(p+1/2)} \\
\rho_3 u_3^{(p)} + b u_4^{(p)} + \rho_4 u_3^{(p+1/2)} \\
c u_3^{(p)} + \rho_3 u_4^{(p)} + \rho_4 u_4^{(p+1/2)} \\
\vdots \\
\rho_3 u_{m-2}^{(p)} + b u_{m-1}^{(p)} + \rho_4 u_{m-2}^{(p+1/2)} \\
c u_{m-2}^{(p)} + \rho_3 u_{m-1}^{(p)} + \rho_4 u_{m-1}^{(p+1/2)} \\
\rho_3 u_m^{(p)} + \rho_4 u_m^{(p+1/2)}
\end{bmatrix},
\qquad (4.3.17)
$$

where ρ_1, ρ_2 and Δ are given by (4.3.14) and $\rho_3 = a/2(1 - \omega)\rho$ and $\rho_4 = (2 - \omega)\rho$. This leads to

$$
u_i^{(p+1)} = (P u_i^{(p+1/2)} + Q u_{i+1}^{(p+1/2)} + R u_i^{(p)} + S u_{i+1}^{(p)})/\Delta,
$$
$$
u_{i+1}^{(p+1)} = (\tilde{P} u_i^{(p+1/2)} + \tilde{Q} u_{i+1}^{(p+1/2)} + \tilde{R} u_i^{(p)} + \tilde{S} u_{i+1}^{(p)})/\Delta, \quad i = 1, 3, \ldots, m - 2
$$

and

$$
u_m^{(p+1)} = (\rho_3 u_m^{(p)} + \rho_4 u_m^{(p+1/2)} + f_m)/\rho_2, \qquad (4.3.18)
$$

where

$$
P = \rho_2 \rho_4, \quad Q = -b \rho_4, \quad R = \rho_2 \rho_3 - bc, \quad S = b(\rho_2 - \rho_3),
$$

and

$$
\tilde{P} = -c \rho_4, \quad \tilde{Q} = P = \rho_2 \rho_4, \quad \tilde{R} = c(\rho_2 - \rho_3)
$$

and

$$\tilde{S} = \rho_2\rho_3 - bc, \tag{4.3.19}$$

and the computational molecules are given by Figure 4.3.4.

Finally, let us now consider the case when *m is even (corresponding to an odd number of intervals)*. We shall then have from (4.2.35) and (4.2.36)

$$(G_1 + \rho I)^{-1} = \begin{bmatrix} \hat{G}^{-1} & & & \\ & \hat{\hat{G}}^{-1} & & \\ & & & 0 \\ & & & \\ & 0 & & \hat{G}^{-1} \end{bmatrix}_{(m \times m)} \tag{4.3.20}$$

$$= \frac{1}{\Delta} \begin{bmatrix} \rho_2 & -b & & & \\ -c & \rho_2 & & & \\ & & \rho_2 & -b & 0 \\ & & -c & \rho_2 & \\ & & & & \rho_2 & -b \\ & 0 & & & -c & \rho_2 \end{bmatrix}_{(m \times m)} \tag{4.3.21}$$

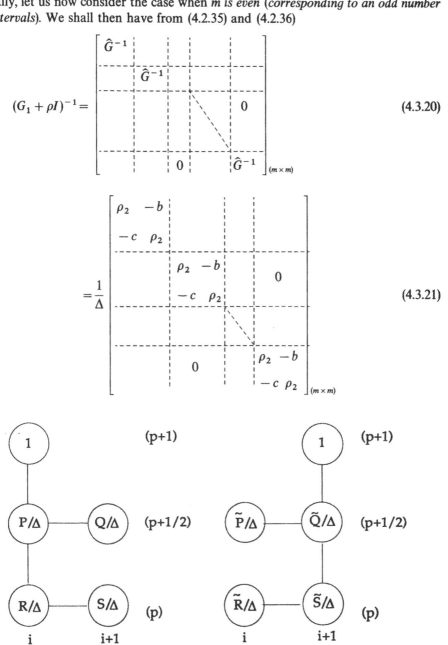

Figure 4.3.4 The Generalised AGE Method (Explicit) at Level $(p+1)$.

and

$$(G_2 + \rho I)^{-1} = \begin{bmatrix} 1/\rho_2 & & & & \\ & \hat{G}^{-1} & & 0 & \\ & & & & \\ & 0 & & \hat{G}^{-1} & \\ & & & & 1/\rho_2 \end{bmatrix}_{(m \times m)} \tag{4.3.22}$$

The Peaceman–Rachford variant in its implicit form (4.3.2) can be pictorially represented by Figure 4.3.5.

By means of equations (4.2.35–4.2.36), (4.3.3) and (4.3.20–4.3.22) we obtain the following explicit expressions for the AGE scheme:

(i) *at level* $(p + 1/2)$

$$u_i^{(p + 1/2)} = (A u_{i-1}^{(p)} + B u_i^{(p)} + C u_{i+1}^{(p)} + D u_{i+2}^{(p)} + E_i)/\Delta$$

and

$$u_{i+1}^{(p + 1/2)} = (\tilde{A} u_{i-1}^{(p)} + \tilde{B} u_i^{(p)} + \tilde{C} u_{i+1}^{(p)} + \tilde{D} u_{i+2}^{(p)} + \tilde{E}_i)/\Delta, \quad i = 1, 3, \dots, m - 1, \tag{4.3.23}$$

where

$$A = \begin{cases} 0 & \text{for } i = 1, \\ -c\rho_2 & \text{otherwise,} \end{cases} \qquad B = \rho_1 \rho_2, \quad C = -b\rho_1,$$

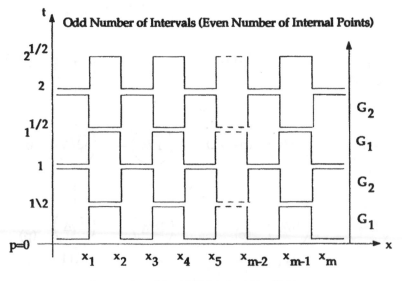

Figure 4.3.5 The AGE Method (Implicit).

$$D = \begin{cases} 0 & \text{for } i = m-1, \\ b^2 & \text{otherwise,} \end{cases} \quad E_i = \rho_2 f_i - b f_{i+1}$$

and

$$\tilde{A} = \begin{cases} 0 & \text{for } i = 1, \\ -c^2 & \text{otherwise,} \end{cases} \quad \tilde{B} = -c\rho_1, \quad \tilde{C} = \rho_1 \rho_2,$$

$$\tilde{D} = \begin{cases} 0 & \text{for } i = m-1, \\ -b\rho_2 & \text{otherwise,} \end{cases} \quad \tilde{E} = -cf_i + \rho_2 f_{i+1}; \qquad (4.3.23a)$$

(ii) *at level* $(p+1)$

$$u_1^{(p+1)} = (\rho_1 u_1^{(p+1/2)} - b u_2^{(p+1/2)} + f_1)/\rho_2,$$

$$u_i^{(p+1)} = (P u_{i-1}^{(p+1/2)} + Q u_i^{(p+1/2)} + R u_{i+1}^{(p+1/2)} + S u_{i+2}^{(p+1/2)} + T_i)/\Delta,$$

$$u_{i+1}^{(p+1)} = (\tilde{P} u_{i-1}^{(p+1/2)} + \tilde{Q} u_i^{(p+1/2)} + \tilde{R} u_{i+1}^{(p+1/2)} + \tilde{S} u_{i+2}^{(p+1/2)} + \tilde{T}_i)/\Delta,$$

$$i = 2, 4, \ldots, m-2,$$

and

$$u_m^{(p+1)} = (-c u_{m-1}^{(p+1/2)} + \rho_1 u_m^{(p+1/2)} + f_m)/\rho_2, \qquad (4.3.24)$$

where

$$P = -c\rho_2, \quad Q = \rho_1 \rho_2, \quad R = -b\rho_1, \quad S = b^2, \quad T_i = \rho_2 f_i - b f_{i+1}$$

and

$$\tilde{P} = c^2, \quad \tilde{Q} = -c\rho_1, \quad \tilde{R} = Q = \rho_1 \rho_2, \quad \tilde{S} = -b\rho_2, \quad \tilde{T}_i = -cf_i + \rho_2 f_{i+1}.$$

$$(4.3.24a)$$

For *the generalised AGE scheme* (4.3.1) with fixed acceleration parameter ρ, the same equations in (4.3.23) still apply for level $(p+1/2)$. At level $(p+1)$, we have the following equations:

$$u_1^{(p+1)} = (\rho_3 u_1^{(p)} + \rho_4 u_1^{(p+1/2)})/\rho_2,$$

$$u_i^{(p+1)} = (P u_i^{(p)} + Q u_{i+1}^{(p)} + R u_i^{(p+1/2)} + S u_{i+1}^{(p+1/2)})/\Delta,$$

$$u_{i+1}^{(p+1)} = (\tilde{P} u_i^{(p)} + \tilde{Q} u_{i+1}^{(p)} + \tilde{R} u_i^{(p+1/2)} + \tilde{S} u_i^{(p+1/2)})/\Delta, \quad i = 2, 4, \ldots, m-2,$$

and

$$u_m^{(p+1)} = (\rho_3 u_m^{(p)} + \rho_4 u_m^{(p+1/2)})/\rho_2, \qquad (4.3.25)$$

where

$$P = \rho_2 \rho_3 - bc, \quad Q = b(\rho_2 - \rho_3), \quad R = \rho_2 \rho_4, \quad S = -b\rho_4,$$

and

$$\tilde{P} = c(\rho_2 - \rho_3), \quad \tilde{Q} = \rho_2 \rho_3 - bc, \quad \tilde{R} = -c\rho_4, \quad \tilde{S} = r_2 \rho_4. \qquad (4.3.25a)$$

The AGE algorithm is completed *explicitly* by using the required equations at levels $(p+1/2)$ and $(p+1)$ in *alternate sweeps* along all the points in the interval $(0,1)$ until a specified convergence criterion is satisfied.

4.4. The Coupled Alternating Group Explicit (CAGE) Method

Consider the following second order parabolic equation

$$\frac{\partial U}{\partial t} = \frac{\partial^2 U}{\partial x^2} + k(x,t), \quad 0 \le x \le 1, \ 0 \le t \le T, \tag{4.4.1}$$

subject to the initial condition

$$U(x,0) = f(x), \quad 0 < x < 1,$$

and the periodic boundary condition

$$U(0,t) = U(1,t), \quad \frac{\partial U}{\partial x}(0,t) = \frac{\partial U}{\partial x}(1,t). \tag{4.4.1a}$$

As developed in Section 4.2, the weighted approximation to the differential equation (4.4.1) at the point $(x_i, t_{j+1/2})$ is

$$-r\theta u_{i-1,j+1} + (1 + 2r\theta) u_{i,j+1} - r\theta u_{i+1,j+1}$$
$$= r(1-\theta)u_{i-1,j} + (1 - 2r(1-\theta))u_{i,j} + r(1-\theta)u_{i+1,j} + \Delta t k_{i,j+1/2},$$

for $i = 1, 2, \ldots, m$. $\tag{4.4.2}$

When written in matrix form and taking account of (4.4.1a) we obtain the system (4.4.3) where

$$c = -r\theta, \quad a = (1 + 2r\theta), \quad b = -r\theta;$$

$$f_1 = (1 - 2r)(1 - \theta)u_{1j} + r(1-\theta)u_{2j} + \lambda(1-\theta)u_{mj} + \Delta t k_{1,j+1/2};$$

$$f_i = r(1-\theta)u_{i-1,j} + (1 - 2r(1-\theta))u_{ij} + r(1-\theta)u_{i+1,j} + \Delta t k_{i,j+1/2},$$

$$i = 2, \ldots, m-1,$$

$$f_m = r(1-\theta)u_{1j} + r(1-\theta)u_{m-1,j} + (1 - 2r)(1 - \theta)u_{mj} + \Delta t k_{m,j+1/2}.$$

To implement the AGE algorithm, we split the coefficient matrix A where A is given by

$$A = \begin{bmatrix} a & b & & & & & c \\ c & a & b & & & & \\ & c & a & b & & 0 & \\ & & & \ddots & & & \\ & & & & & & \\ & & 0 & & c & a & b \\ b & & & & & c & a \end{bmatrix}, \tag{4.4.3}$$

and to ascertain the forms of G_1 and G_2, we treat two different cases of m as before.

(a) *m even* (even number of points since $u_0 = u_m$ at every level with $x_0 = 0$ and $x_m = 1$). We have

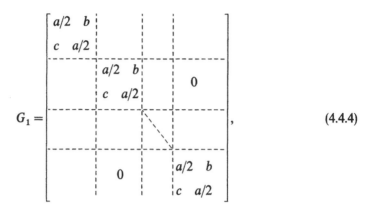

$$(4.4.4)$$

$$(4.4.5)$$

and

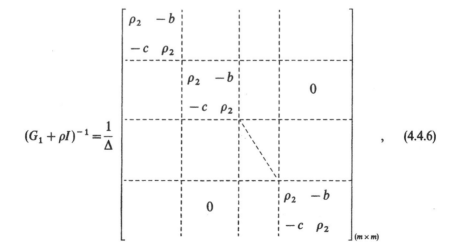

$$(4.4.6)$$

and

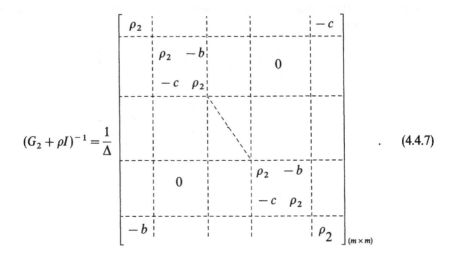

$$(G_2 + \rho I)^{-1} = \frac{1}{\Delta} \qquad\qquad (4.4.7)$$

Hence, using (4.3.1), (4.4.4–4.4.7) we obtain the following equations for the computation of *the generalised AGE scheme*:

(i) *at level* $(p + 1/2)$

$$
\begin{bmatrix} u_1^{(p+1/2)} \\ u_2^{(p+1/2)} \\ u_3^{(p+1/2)} \\ \vdots \\ u_{m-1}^{(p+1/2)} \\ u_m^{(p+1/2)} \end{bmatrix}
= \frac{1}{\Delta}
\begin{bmatrix} \rho_2 & -b & & & \\ -c & \rho_2 & & & \\ & & \rho_2 & -b & & 0 \\ & & -c & \rho_2 & \\ & & & & \ddots \\ & 0 & & & \rho_2 & -b \\ & & & & -c & \rho_2 \end{bmatrix}
\begin{bmatrix} \rho_1 u_1^{(p)} - c u_m^{(p)} + f_1 \\ \rho_1 u_2^{(p)} - b u_3^{(p)} + f_2 \\ -c u_2^{(p)} + \rho_1 u_3^{(p)} + f_3 \\ \rho_1 u_4^{(p)} - b u_5^{(p)} + f_4 \\ -c u_4^{(p)} + \rho_1 u_5^{(p)} + f_5 \\ \vdots \\ \rho_1 u_{m-2}^{(p)} - b u_{m-1}^{(p)} + f_{m-2} \\ -c u_{m-2}^{(p)} + \rho_1 u_{m-1}^{(p)} + f_{m-1} \\ -b u_1^{(p)} + \rho u_m^{(p)} + f_m \end{bmatrix},
$$

$$(4.4.8)$$

which leads to

$$u_i^{(p+1/2)} = (A u_{i-1}^{(p)} - B u_i^{(p)} + C u_{i+1}^{(p)} + D u_{i+2}^{(p)} + E_i)/\Delta$$

and

$$u_{i+1}^{(p+1/2)} = (\tilde{A} u_{i-1}^{(p)} + \tilde{B} u_i^{(p)} + \tilde{C} u_{i+1}^{(p)} + \tilde{D} u_{i+2}^{(p)} + \tilde{E}_i)/\Delta,$$

$$\text{for } i = 1, 3, \dots, m-1, \qquad (4.4.9)$$

with

$$u_0^{(p)} = u_m^{(p)} \text{ and } u_{m+1}^{(p)} = u_1^{(p)},$$

where

$$A = -c\rho_2, \quad B = \rho_1\rho_2, \quad C = -b\rho_1, \quad D = b^2 \quad E_i = \rho_2 f_i - b f_{i+1}$$

and

$$\tilde{A} = c^2, \quad \tilde{B} = -c\rho_1, \quad \tilde{C} = \rho_1\rho_2, \quad \tilde{D} = -b\rho_2, \quad \tilde{E}_i = -cf_i + \rho_2 f_{i+1}.$$

(4.4.9a)

(ii) *at level* $(p+1)$

$$
\begin{bmatrix} u_1^{(p+1)} \\ u_2^{(p+1)} \\ \vdots \\ \\ \\ \\ \\ \vdots \\ u_{m-1}^{(p+1)} \\ u_m^{(p+1)} \end{bmatrix}
= \frac{1}{\Delta}
\begin{bmatrix}
\rho_2 & & & & & -c \\
& \rho_2 & -b & & 0 & \\
& -c & \rho_2 & & & \\
& & & \diagdown & & \\
& & & & & \\
& 0 & & \rho_2 & -b & \\
& & & -c & \rho_2 & \\
-b & & & & & \rho_2
\end{bmatrix}
\begin{bmatrix}
\rho_3 u_1^{(p)} + c u_m^{(p)} + \rho_4 u_1^{(p+1/2)} \\
\rho_3 u_2^{(p)} + b u_3^{(p)} + \rho_4 u_2^{(p+1/2)} \\
c u_2^{(p)} + \rho_3 u_3^{(p)} + \rho_4 u_2^{(p+1/2)} \\
\rho_3 u_4^{(p)} + b u_5^{(p)} + \rho_4 u_4^{(p+1/2)} \\
c u_4^{(p)} + \rho_3 u_5^{(p)} + \rho_4 u_5^{(p+1/2)} \\
\vdots \\
\rho_3 u_{m-2}^{(p)} + b u_{m-1}^{(p)} + \rho_4 u_{m-2}^{(p+1/2)} \\
c u_4^{(p)} + \rho_3 u_{m-1}^{(p)} + \rho_4 u_{m-1}^{(p+1/2)} \\
b u_1^{(p)} + \rho_3 u_m^{(p)} + \rho_4 u_m^{(p+1/2)}
\end{bmatrix}.
$$

(4.4.10)

This gives

$$u_i^{(p+1)} = (P u_{i-1}^{(p)} + Q u_i^{(p)} + R u_{i-1}^{(p+1/2)} + S u_i^{(p+1/2)})/\Delta$$

and

$$u_{i+1}^{(p+1)} = (\tilde{P} u_{i+1}^{(p)} + \tilde{Q} u_{i+2}^{(p)} + \tilde{R} u_{i+1}^{(p+1/2)} + \tilde{S} u_{i+2}^{(p+1/2)})/\Delta,$$
$$i = 1, 3, \ldots, m-1,$$

(4.4.11)

with, $u_0 = u_m$ and $u_1 = u_{m+1}$ at both alternating levels, where

$$P = c(\rho_2 - \rho_3), \quad Q = \rho_2\rho_3 - bc, \quad R = -c\rho_4, \quad S = \rho_2\rho_4$$

and

$$\tilde{P} = \rho_2\rho_3 - bc, \quad \tilde{Q} = b(\rho_2 - \rho_3), \quad \tilde{R} = \rho_2\rho_4, \quad \tilde{S} = -b\rho_4.$$

(4.4.11a)

(b) *m odd* (odd number of points).

We have

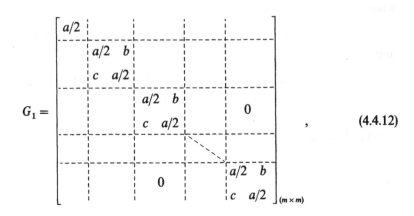

$$, \qquad (4.4.12) $$

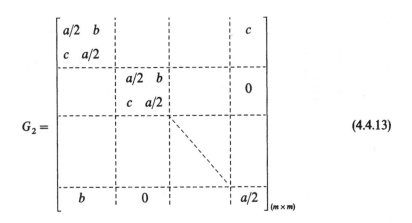

$$ (4.4.13) $$

and

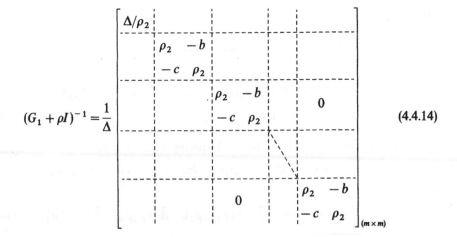

$$ (4.4.14) $$

and

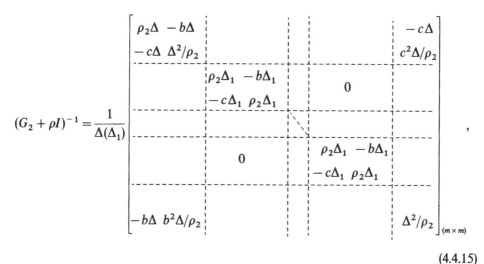

$$(4.4.15)$$

where

$$\Delta_1 = \rho_2^2 - 2bc . \tag{4.4.16}$$

We shall now derive the AGE equations using (4.3.1) and (4.4.12–4.4.15):

(i) *level* $(p + 1/2)$

$$
\begin{bmatrix}
u_1^{(p+1/2)} \\
u_2^{(p+1/2)} \\
\vdots \\
\vdots \\
u_{m-1}^{(p+1/2)} \\
u_m^{(p+1/2)}
\end{bmatrix}
= \frac{1}{\Delta}
\begin{bmatrix}
\Delta/\rho_2 & & & \\
& \rho_2 & -b & \\
& -c & \rho_2 & \quad 0 \\
& & & \\
& & & \\
0 & & \rho_2 & -b \\
& & -c & \rho_2
\end{bmatrix}
\begin{bmatrix}
\rho_1 u_1^{(p)} - b u_2^{(p)} - c u_m^{(p)} + f_1 \\
-c u_1^{(p)} + \rho_1 u_2^{(p)} + f_2 \\
\rho_1 u_3^{(p)} - b u_4^{(p)} + f_3 \\
-c u_4^{(p)} + \rho_1 u_4^{(p)} + f_4 \\
\vdots \\
\rho_1 u_{m-2}^{(p)} - b u_{m-1}^{(p)} + f_{m-2} \\
-c u_{m-2}^{(p)} + \rho_1 u_{m-1}^{(p)} + f_{m-1} \\
\rho_1 u_m^{(p)} + f_m - b u_1^{(p)}
\end{bmatrix}
$$

$$(4.4.17)$$

Hence, we obtain

$$u_1^{(p+1/2)} = (\rho_1 u_1^{(p)} - b u_2^{(p)} - c m_m^{(p)} + f_1)/\rho_2,$$

$$u_i^{(p+1/2)} = (A u_{i-1}^{(p)} + B u_i^{(p)} + C u_{i+1}^{(p)} + D u_{i+2}^{(p)} + E_i)/\Delta,$$

$$i = 2, 4, \ldots, m-3, m-1,$$

and

$$u_{i+1}^{(p+1/2)} = (\tilde{A} u_{i-1}^{(p)} + \tilde{B} u_i^{(p)} + \tilde{C} u_{i+1}^{(p)} + \tilde{D} u_{i+2}^{(p)} + \tilde{E}_i)/\Delta), \tag{4.4.17a}$$

with $u_1^{(p)} = u_{m+1}^{(p)}$, where

$$A = -c\rho_2, \quad B = \rho_1\rho_2, \quad C = -b\rho_1, \quad D = b^2, \quad E_i = \rho_2 f_i - b f_{i+1}$$

and

$$\tilde{A} = c^2, \quad \tilde{B} = -c\rho_1 \quad \tilde{C} = \rho_1\rho_2, \quad \tilde{D} = -b\rho_2, \text{ and } \tilde{E}_i = -cf_i + \rho_2 f_{i+1}.$$

$$(4.4.17b)$$

(ii) *level* $(p+1)$

$$
\begin{bmatrix}
u_1^{(p+1)} \\[4pt]
u_2^{(p+1)} \\
\vdots \\
\\
\\
u_{m-1}^{(p+1)} \\[4pt]
u_m^{(p+1)}
\end{bmatrix}
= \frac{1}{\Delta(\Delta_1)}
\begin{bmatrix}
\rho_2\Delta & -b\Delta & & & & & -c\Delta \\
-c\Delta & \Delta^2/\rho_2 & & & & & c^2\Delta/\rho_2 \\
\hline
& & \rho_2\Delta_1 & -b\Delta_1 & & & \\
& & -c\Delta_1 & \rho_2\Delta_1 & \mathbf{0} & & \\
\hline
& & & & \rho_2\Delta_1 & -b\Delta_1 & \\
& \mathbf{0} & & & -c\Delta_1 & \rho_2\Delta_1 & \\
\hline
-b\Delta & b^2\Delta/\rho_2 & & & & & \Delta^2/\rho_2
\end{bmatrix}_{(m \times m)} ,
$$

$$
\begin{bmatrix}
\rho_3 u_1^{(p)} + b u_2^{(p)} + c u_m^{(p)} + \rho_4 u_1^{(p+1/2)} \\[4pt]
c u_1^{(p)} + \rho_3 u_2^{(p)} + \rho_4 u_2^{(p+1/2)} \\[4pt]
\rho_3 u_3^{(p)} + b u_4^{(p)} + \rho_4 u_3^{(p+1/2)} \\[4pt]
c u_3^{(p)} + \rho_3 u_4^{(p)} + \rho_4 u_4^{(p+1/2)} \\
\vdots \\
\rho_3 u_{m-2}^{(p)} + b u_{m-1}^{(p)} + \rho_4 u_{m-2}^{(p+1/2)} \\[4pt]
c u_{m-2}^{(p)} + \rho_3 u_{m-1}^{(p)} + \rho_4 u_{m-1}^{(p+1/2)} \\[4pt]
b u_1^{(p)} + \rho_3 u_m^{(p)} + \rho_4 u_m^{(p+1/2)}
\end{bmatrix} ,
$$

$$(4.4.18)$$

which leads to

$$u_1^{(p+1)} = (P_1 u_1^{(p)} + P_2 u_2^{(p)} + P_3 u_m^{(p)} + P_4 u_1^{(p+1/2)} + P_5 u_2^{(p+1/2)} + P_6 u_m^{(p+1/2)})/\Delta_1,$$

$$u_2^{(p+1)} = (Q_1 u_1^{(p)} + Q_2 u_2^{(p)} + Q_3 u_m^{(p)} + Q_4 u_1^{(p+1/2)} + Q_5 u_2^{(p+1/2)} + Q_6 u_m^{(p+1/2)})/\Delta_1,$$

$$u_i^{(p+1)} = (P u_i^{(p)} + Q u_{i+1}^{(p)} + R u_i^{(p+1/2)} + S u_{i+1}^{(p+1/2)})/\Delta,$$

and

$$u_{i+1}^{(p+1)} = (\tilde{P}u_i^{(p)} + \tilde{Q}u_{i+1}^{(p)} + \tilde{R}u_i^{(p+1/2)} + \tilde{S}u_{i+1}^{(p+1/2)})/\Delta, \quad i = 3, 5, \ldots, m-2,$$

with

$$u_m^{(p+1)} = (R_1 u_1^{(p)} + R_2 u_2^{(p)} + R_3 u_m^{(p)} + R_4 u_1^{(p+1/2)}$$
$$+ R_5 u_2^{(p+1/2)} + R_6 u_m^{(p+1/2)})/\Delta_1, \tag{4.4.19}$$

where

$$P_1 = \rho_2\rho_3 - 2bc, \quad P_2 = b(\rho_2 - \rho_3), \quad P_3 = c(\rho_2 - \rho_3),$$
$$P_4 = \rho_2\rho_4, \quad P_5 = -b\rho_4, \quad P_6 = -c\rho_4,$$
$$Q_1 = c(\rho_2 - \rho_3), \quad Q_2 = -bc + \rho_3\Delta/r_2, \quad Q_3 = -c^2(\rho_2 - \rho_3)/r_2,$$
$$Q_4 = -c\rho_4, \quad Q_5 = \rho_4\Delta/\rho_2, \quad Q_6 = \rho_4 c^2/r_2,$$
$$P = \rho_2\rho_3 - bc, \quad Q = b(\rho_2 - \rho_3), \quad R = \rho_2\rho_4, \quad S = -b\rho_4,$$
$$\tilde{P} = c(\rho_2 - \rho_3), \quad \tilde{Q} = \rho_2\rho_3 - bc, \quad \tilde{R} = -c\rho_4, \quad \tilde{S} = \rho_2\rho_4,$$
$$R_1 = b(\rho_2 - \rho_3), \quad R_2 = b^2(\rho_3 - \rho_2)/r_2, \quad R_3 = -bc + \rho_3\Delta/\rho_2,$$
$$R_4 = -b\rho_4, \quad R_5 = b^2\rho_4/\rho_2, \quad R_6 = \Delta\rho_4/\rho_2. \tag{4.4.19a}$$

The iterative process is continued for each alternate sweep until convergence is reached.

The computational molecules for the equations (4.4.9) or (4.4.17) are given for $i = 1, 3, 5, \ldots, m-1 \pmod{m}$ by

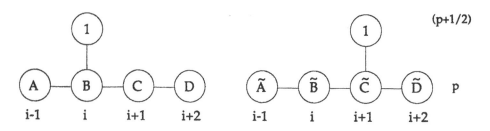

Similarly the computational molecules for the equations. (4.4.11) or (4.4.19) are given for $i = 2, 4, 6, \ldots, m \pmod{m}$ by

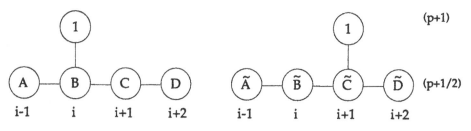

Figure 4.4.1 Computational Molecules for the 2 Stage AGE Method.

Now we show that the 2 stages of the Alternating Group Explicit method given by equation (4.4.11) can be combined into a coupled single stage method. We write equation (4.4.9) in explicit form

$$\underline{u}^{(p+1/2)} = (\rho I + G_1)^{-1} (\rho I - G_2) \underline{u}^{(p)} + (\rho I + G_1)^{-1} \underline{f}, \tag{4.4.20}$$

and substitute, into equation (4.4.11) written in explicit form

$$\underline{u}^{(p+1)} = (\rho I + G_2)^{-1} (\rho I - G_1) \underline{u}^{(p+1/2)} + (\rho I - G_2)^{-1} \underline{f}. \tag{4.4.21}$$

Thus, the coupled (CAGE) method can be written as

$$\underline{u}^{(p+1)} = (\rho I + G_2)^{-1} (\rho I - G_1) (\rho I + G_1)^{-1} (\rho I - G_2) \underline{u}^{(p)}$$
$$+ (\rho I + G_2)^{-1} (\rho I - G_1) (\rho I + G_2)^{-1} \underline{f} + (\rho I + G_2)^{-1} \underline{f}. \tag{4.4.22}$$

Since the iteration matrix $(\rho I + G_1)^{-1} (\rho I - G_2)$ of equation (4.4.9) can be written in matrix form as

$$\begin{bmatrix}
B & C & D & & & & & & A \\
\tilde{B} & \tilde{C} & \tilde{D} & & & & & & \tilde{A} \\
 & & A & B & C & D & & & \\
 & & \tilde{A} & \tilde{B} & \tilde{C} & \tilde{D} & & 0 & \\
 & & & & & & & & \\
 & & & & & & & & \\
D & & & & & A & B & C & \\
\tilde{D} & 0 & & & & \tilde{A} & \tilde{B} & \tilde{C} &
\end{bmatrix} \tag{4.4.23}$$

and the iteration matrix $(\rho I + G_2)^{-1} (\rho I - G_1)$ of equation (4.4.11) as

$$\begin{bmatrix}
\tilde{C} & \tilde{D} & & & & \tilde{A} & \tilde{B} \\
A & B & C & D & & & \\
\tilde{A} & \tilde{B} & \tilde{C} & \tilde{D} & & 0 & \\
 & & & & & & \\
 & & & A & B & C & D \\
 & 0 & & \tilde{A} & \tilde{B} & \tilde{C} & \tilde{D} \\
C & D & & & & A & B
\end{bmatrix}, \tag{4.4.24}$$

where A, $\tilde{A}$, etc. are defined by equation (4.4.9a) or (4.4.11a).

The iteration matrix of the coupled method can thus be determined by matrix multiplication of (4.4.23) and (4.4.24) to give a single iteration matrix given by equation (4.4.25).

$$\frac{1}{\Delta^2}\begin{bmatrix} \tilde{\delta} & \tilde{\varepsilon} & \tilde{\theta} & & & & \tilde{\alpha} & \tilde{\beta} & \tilde{\gamma} \\ \beta & \gamma & \delta & \varepsilon & \theta & & & & \alpha \\ \tilde{\beta} & \tilde{\gamma} & \tilde{\delta} & \tilde{\varepsilon} & \tilde{\theta} & & & & \tilde{\alpha} \\ & \alpha & \beta & \gamma & \delta & \varepsilon & \theta & & \\ & \tilde{\alpha} & \tilde{\beta} & \tilde{\gamma} & \tilde{\delta} & \tilde{\varepsilon} & \tilde{\theta} & & 0 \\ & 0 & & & & & & & \\ \theta & & & & \alpha & \beta & \gamma & \delta & \varepsilon \\ \tilde{\theta} & & & & \tilde{\alpha} & \tilde{\beta} & \tilde{\gamma} & \tilde{\delta} & \tilde{\varepsilon} \\ \delta & \varepsilon & \theta & & & \alpha & \beta & & \gamma \end{bmatrix}, \qquad (4.4.25)$$

where,

$$\alpha = \rho_2 c^2(\rho_1 + \rho_2), \qquad\qquad \tilde{\alpha} = -c^3(\rho_1 + \rho_2),$$

$$\beta = -c\rho_1\rho_2(\rho_1 + \rho_2), \qquad\quad \tilde{\beta} = c^2\rho_1(\rho_1 + \rho_2),$$

$$\gamma = (\rho_1\rho_2 + bc)^2, \qquad\qquad\quad \tilde{\gamma} = -(\rho_1 + \rho_2)c(bc + \rho_1\rho_2),$$

$$\delta = -c(\rho_1 + \rho_2)(bc + \rho_1\rho_2), \quad \tilde{\delta} = (\rho_1\rho_2 + bc)^2, \qquad\qquad (4.4.26)$$

$$\varepsilon = \rho_1 b^2(\rho_1 + \rho_2), \qquad\qquad \tilde{\varepsilon} = -\beta\rho_1\rho_2(\rho_1 + \rho_2),$$

$$\theta = -b^3(\rho_1 + \rho_2), \qquad\qquad \tilde{\theta} = b^2\rho_2(\rho_1 + \rho_2),$$

and $\Delta = (1 + 2r)$, whilst the computational molecule for the CAGE method can be shown to be given by the 6 term formula, i.e.,

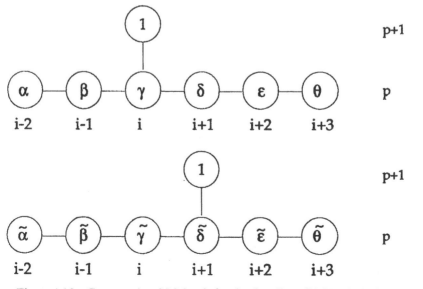

Figure 4.4.2 Computational Molecule for the One Step CAGE Method.

Finally the matrix equation (4.4.22) for the CAGE method written in component form at the points i and $(i+1)$ can be expressed as

$$u_i = (\alpha u_{i-2} + \beta u_{i-1} + \gamma u_i + \delta u_{i+1} + \gamma u_{i+2} + \varepsilon u_{i+3})\Delta^2$$

$$+ (-\rho_2 c E_{i-1} + r_1 r_2 E_i - br E_{i+1} + b^2 E_{i+2})/\Delta^2 + T_i/\Delta, \qquad (4.4.27)$$

and

$$u_{i+1} = (\tilde{\alpha} u_{i-2} + \tilde{\beta} u_{i-1} + \tilde{\gamma} u_i + \tilde{\delta} u_{i+1} + \tilde{\gamma} u_{i+1} + \tilde{\varepsilon} u_{i+3})/\Delta^2$$

$$+ (c^2 \tilde{E}_{i-1} - cr_1 \tilde{E}_i + r_1 r_2 \tilde{E}_{i+1} - br_2 \tilde{E}_{i+2})/\Delta^2 + T_i/\Delta,$$

for $i = 2, 4, 6, \ldots, m \pmod{m}$ with,

$$T_i = \rho_2 f_i - b f_{i+1}, \qquad \tilde{T}_{i+1} = c f_i + \rho_2 f_{i+1},$$

$$E_i = -c f_i + \rho_2 f_{i+1} \quad \text{and} \quad \tilde{E}_{i+1} = \rho_2 f_{i+2} - b f_{i+3}$$

and all indices i interpreted as modulo m.

The new method is numerically stable and from equations (4.4.26) and Figure 4.4.2 can be implemented in 6 multiplications and 6 additions for each double step. This represents a 25% saving in computational work over the AGE algorithm given by equations (4.4.9) and (4.4.11).

Numerical experiments on a model problem confirm the viability of the CAGE method while the accuracy of the results are comparable to existing methods (i.e., AGE and ADI).

4.5. The AGE Method to Solve Non-Linear Parabolic Equations

The concept of AGE method is now shown to be easily extended to a variety of *non-linear* problems.

(i) The equation:

$$\frac{\partial U}{\partial t} = \frac{\partial^2 U^n}{\partial x^2}, \quad n \geq 2.$$

We shall now consider implementing the AGE algorithm on the following non-linear parabolic problem of the form

$$\frac{\partial U}{\partial t} = \frac{\partial^2 U^n}{\partial x^2}, \quad n \geq 2 \qquad (4.5.1)$$

given the initial conditions

$$U(x, 0) = f(x), \quad 0 < x < 1,$$

and the boundary conditions

$$U(0, t) = g(t),$$

and

$$U(1,t) = h(t), \quad 0 < t \leq T. \tag{4.5.1a}$$

The equation (4.5.1) is approximated at the grid points by implicit finite-difference schemes and the approach of Lees (1962) to linearise them is adopted which results in tridiagonal systems of equations which can be solved by the AGE algorithm as before.

(a) *Richtmyer's linearisation* The non-linear equation (4.5.1) is approximated by the implicit weighted average difference formula

$$\frac{1}{\Delta t}(u_{i,j+1} - u_{i,j}) = \frac{1}{(\Delta x)^2}[\theta \delta_x^2 \, u_{i,j+1}^n + (1-\theta) \, \delta_x^2 \, u_{i,j}^n],$$

$$\text{with } i = 1, 2, \ldots, m. \tag{4.5.2}$$

As in the linear case $n = 1$, the above formula corresponds to the fully implicit, the Crank–Nicolson and the Douglas schemes with $\theta = 1$, $1/2$ and $(6r - 1)/12\lambda$, respectively, with $r = \Delta t/(\Delta x)^2$. By using the Taylor series expansion of $u_{i,j+1}^n$ about the point (x_i, t_j) we have

$$u_{i,j+1}^n = u_{i,j}^n + (\Delta t) \frac{\partial u_{i,j}^n}{\partial t} + \cdots$$

$$= u_{i,j}^n + (\Delta t) \frac{\partial u_{i,j}^n}{\partial u_{i,j}^n} \frac{\partial u_{i,j}^n}{\partial t} + \cdots.$$

Hence, to first order terms of order n, the approximation

$$u_{i,j+1}^n = u_{i,j}^n + n u_{i,j}^{n-1}(u_{i,j+1} - u_{i,j}) \tag{4.5.3}$$

replaces the non-linear unknown $u_{i,j+1}^n$ by an appproximation which is linear in $u_{i,j+1}$. Now, if we let

$$w_i = u_{i,j+1} - u_{i,j}, \tag{4.5.4}$$

then using (4.5.2) and (4.5.4) we obtain

$$\frac{1}{\Delta t} w_i = \frac{1}{(\Delta x)^2}[\theta \delta_x^2 (u_{i,j}^n + n u_{i,j}^{n-1} w_i) + (1-\theta) \, \delta_x^2 \, u_{i,j}^n]$$

$$= \frac{1}{(\Delta x)^2}[n\theta \delta_x^2 \, u_{i,j}^{n-1} \, w_i + \delta_x^2 \, u_{i,j}^n]$$

$$= \frac{1}{(\Delta x)^2}[n\theta \, (u_{i-1,j}^{n-1} \, w_{i-1} - 2 u_{i,j}^{n-1} \, w_i + u_{i+1,j}^{n-1} \, w_{i+1})$$

$$+ (u_{i-1,j}^n - 2 u_{i,j}^n + u_{i+1,j}^n)], \tag{4.5.5}$$

which gives a set of linear equations for the w_i. Now the system of equations (4.5.5) can be written in the more compact matrix form

$$A\underline{w} = \underline{f}, \tag{4.5.6}$$

which is solved for $\underline{w}$ and by means of (4.5.4), the solution at time level $(j+1)$ is given by

$$\underline{u}_{j+1} = \underline{w} + \underline{u}_j. \tag{4.5.7}$$

The coefficient matrix A takes the following tridiagonal form for $n = 2$:

$$A = \begin{bmatrix} a_{1j} & b_{1j} & & & & \\ c_{2j} & a_{2j} & b_{2j} & & \mathbf{0} & \\ & & & & & \\ \mathbf{0} & & & c_{m-1,j} & a_{m-1,j} & b_{m-1,j} \\ & & & & c_{mj} & a_{mj} \end{bmatrix}_{(m \times m)}, \tag{4.5.8}$$

where

$$a_{i,j} = 1 + 4r\theta u_{i,j}, \quad i = 1, 2, \dots, m;$$

$$b_{i,j} = -2r\theta u_{i+1,j}, \quad i = 1, 2, \dots, m-1;$$

and

$$c_{i,j} = -2r\theta u_{i-1,j}, \quad i = 2, 3, \dots, m.$$

The entries of the right-hand side vector of (4.5.6) are given by

$$f_1 = (-2ru_{1,j})u_{1,j} + (ru_{2,j})u_{2,j} + ru_{0,j}\left[u_{0,j} + 2\theta(u_{0,j+1} - u_{0,j})\right];$$

$$f_i = (ru_{i-1,j})u_{i-1,j} + (-2ru_{i,j})u_{i,j} + (ru_{i+1,j})u_{i+1,j},$$

$$i = 2, 3, \dots, m-1;$$

and

$$f_m = (ru_{m-1,j})u_{m-1,j} + (-2ru_{m,j})u_{m,j}$$

$$+ ru_{m+1,j}\left[u_{m+1,j} + 2\theta(u_{m+1,j+1} - u_{m+1,j})\right].$$

If we assume m odd then when A is split into the sum of the matrices G_1 and G_2, these constituent matrices take the forms

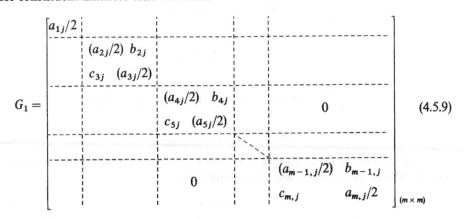

$$G_1 = \begin{bmatrix} a_{1j}/2 & & & & & \\ & (a_{2j}/2) & b_{2j} & & & \\ & c_{3j} & (a_{3j}/2) & & & \\ & & & (a_{4j}/2) & b_{4j} & & \mathbf{0} \\ & & & c_{5j} & (a_{5j}/2) & & \\ & & & & & & \\ & & \mathbf{0} & & & (a_{m-1,j}/2) & b_{m-1,j} \\ & & & & & c_{m,j} & a_{m,j}/2 \end{bmatrix}_{(m \times m)} \tag{4.5.9}$$

and

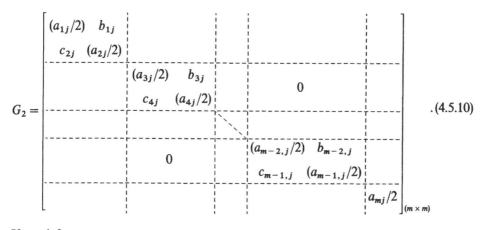

$$. (4.5.10)$$

If we define

$$
G^{(i)} = \begin{bmatrix} (\tfrac{1}{2})a_{2i,j} + \rho & b_{2i,j} \\ c_{2i+1,j} & (\tfrac{1}{2})a_{2i+1,j} + \rho \end{bmatrix},
$$

$$(4.5.11)$$

then

$$
\alpha_i = |G^{(i)}|
$$

$$
= ((\tfrac{1}{2})a_{2i,j} + \rho)\,((\tfrac{1}{2})a_{2i+1,j} + \rho) - b_{2i,j}\,c_{2i+1,j},
$$

$$(4.5.12)$$

and

$$
(G_1 + \rho I)^{-1} = \begin{bmatrix} \dfrac{1}{(\rho + (\tfrac{1}{2})a_{1j})} & & & \\ & \hat{G}^{(1)} & & \\ & & \hat{G}^{(2)} & 0 \\ & & & \\ & & 0 & \hat{G}^{(m-1)/2} \end{bmatrix}_{(m \times m)},
$$

$$, \qquad (4.5.13)$$

where

$$
\hat{G}^{(i)} = (G^{(i)})^{-1}
$$

$$
= \frac{1}{\alpha_i} \begin{bmatrix} (\tfrac{1}{2})a_{2i+1,j} + \rho & -b_{2i,j} \\ -c_{2i+1,j} & (\tfrac{1}{2})a_{2i,j} + \rho \end{bmatrix}, \quad i = 1, 2, \ldots, \tfrac{1}{2}(m-1).
$$

$$(4.5.14)$$

Similarly, we have

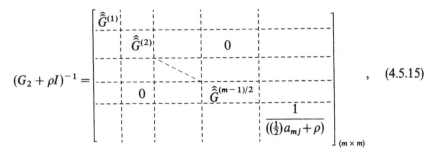

$$, \quad (4.5.15)$$

where

$$\hat{\hat{G}}^{(i)} = \frac{1}{\hat{\alpha}_i} \begin{bmatrix} (\tfrac{1}{2})a_{2i,j} + \rho & -b_{2i-1,j} \\ -c_{2i,j} & (\tfrac{1}{2})a_{2i-1,j} + \rho \end{bmatrix}, \qquad (4.5.16)$$

with

$$\hat{\alpha}_i = ((\tfrac{1}{2})a_{2i,j} + \rho)((\tfrac{1}{2})a_{2i-1,j} + \rho) - b_{2i-1,j} c_{2i,j}, \quad i = 1,2,\ldots,\tfrac{1}{2}(m-1). \quad (4.5.17)$$

We therefore, arrive at the following equations for the computation of the solution of the non-linear problem using the generalised AGE scheme:

(1) *at level (iterate)* $(p + \tfrac{1}{2})$

$$w_1^{(p+1/2)} = (\bar{s}_1 \, w_1^{(p)} - b_1 \, w_2^{(p)} + f_1)/\bar{\rho}_1,$$

$$w_i^{(p+1/2)} = (A_i \, w_{i-1}^{(p)} + B_i \, w_i^{(p)} + C_i \, w_{i+1}^{(p)} + D_i \, w_{i+2}^{(p)} + E_i)/\alpha_{i/2},$$

$$i = 2,4,\ldots,m-1,$$

$$w_{i+1}^{(p+1/2)} = (\tilde{A}_i \, w_{i-1}^{(p)} + \tilde{B}_i \, w_i^{(p)} + \tilde{C}_i \, w_{i+1}^{(p)} + \tilde{D}_i \, w_{i+2}^{(p)} + \tilde{E}_i)/\alpha_{i/2}, \qquad (4.5.18)$$

where

$$A_i = -c_i \bar{\rho}_{i+1}, \quad B_i = \bar{\rho}_{i+1} \bar{s}_i, \quad C_i = -b_i \bar{s}_{i+1}, \quad D_i = \begin{cases} 0 & \text{for } i = m-1, \\ b_i b_{i+i} & \text{otherwise}, \end{cases}$$

$$E_i = -\bar{\rho}_{i+1} f_i - b_i f_{i+1},$$

and

$$\tilde{A}_i = c_i c_{i+1}, \quad \tilde{B}_i = -c_{i+1} \bar{s}_i, \quad \tilde{C}_i = \bar{\rho}_i \bar{s}_{i+1}, \quad \tilde{D}_i = \begin{cases} 0 & \text{for } i = m-1, \\ -b_i \bar{\rho}_i, & \text{otherwise}, \end{cases}$$

$$\tilde{E}_i = -c_{i+1} f_i + \bar{\rho}_i f_{i+1},$$

with

$$\bar{r}_i = \rho + (\tfrac{1}{2})a_i$$

and

$$\bar{s}_i = \rho - (\tfrac{1}{2})a_i, \quad i = 1,2,\ldots,m. \qquad (4.5.18a)$$

(2) *at level (iterate)* $(p+1)$

$$w_i^{(p+1)} = (P_i\, w_i^{(p)} + Q_i\, w_{i+1}^{(p)} + R_i\, w_i^{(p+1/2)} + S_i\, w_{i+1}^{(p+1/2)})/\hat{\alpha}_{(i+1)/2},$$

$$w_{i+1}^{(p+1)} = (\tilde{P}_i\, w_i^{(p)} + \tilde{Q}_i\, w_{i+1}^{(p)} + \tilde{R}_i\, w_i^{(p+1/2)} + \tilde{S}_i\, w_{i+1}^{(p+1/2)})/\hat{\alpha}_{(i+1)/2},$$

$$i = 1, 3, \ldots, m-2,$$

$$w_m^{(p+1)} = (\bar{q}_m\, w_m^{(p)} + dw_m^{(p+1/2)})/\bar{r}_m, \qquad (4.5.19)$$

where

$$P_i = \bar{p}_{i+1}\,\bar{q}_i - b_i c_{i+1}, \quad Q_i = b_i(\bar{p}_{i+1} - \bar{q}_{i+1}), \quad R_i = \bar{p}_{i+1}d, \quad S_i = -b_i d$$

and

$$\tilde{P}_i = c_{i+1}(\bar{p}_i - \bar{q}_i), \quad \tilde{Q}_i = -b_i c_{i+1} + \bar{p}_i\,\bar{q}_{i+1}, \quad \tilde{R}_i = -c_{i+1}d, \quad \tilde{S}_i = \bar{p}_i d,$$

with

$$\bar{r}_i = \rho + (1/2)a_i, \quad \bar{q}_i = (1/2)a_i - (1-\omega)\rho, \quad d = (2-\omega)\rho, \quad i = 1, 2, \ldots, m. \quad (4.5.19a)$$

If on the other hand, we assume that m *even* then we obtain the following forms of the constituent matrices

$$G_1 = \begin{bmatrix} (a_{1j}/2) & b_{1j} & & & & & \\ c_{2j} & (a_{2j}/2) & & & & & \\ & & (a_{3j}/2) & b_{3j} & & & \\ & & c_{4j} & (a_{4j}/2) & & 0 & \\ & & & & \ddots & & \\ & & & & & (a_{m-1,j}/2) & b_{m-1,j} \\ & 0 & & & & c_{m,j} & (a_{m,j}/2) \end{bmatrix}_{(m \times m)}, \qquad (4.5.20)$$

$$G_2 = \begin{bmatrix} (a_{1j}/2) & & & & & & \\ & (a_{2j}/2) & b_{2j} & & & & \\ & c_{3j} & (a_{3j}/2) & & & & \\ & & & \ddots & & 0 & \\ & & & & (a_{m-1,j}/2) & b_{m-1,j} & \\ & 0 & & & c_{mj} & (a_{mj}/2) & \\ & & & & & & (a_{m,j}/2) \end{bmatrix}_{(m \times m)}, \qquad (4.5.21)$$

$$(G_1 + \rho I)^{-1} = \begin{bmatrix} \hat{\hat{G}}^{(1)} & & & \\ & \hat{\hat{G}}^{(2)} & & \mathbf{0} \\ & & \ddots & \\ & \mathbf{0} & & \hat{\hat{G}}^{(m-2)} \end{bmatrix}_{(m \times m)} , \qquad (4.5.22)$$

$$(G_2 + \rho I)^{-1} = \begin{bmatrix} \dfrac{1}{(r + (\frac{1}{2})a_{1j})} & & & \\ & \hat{G}^{(1)} & & \mathbf{0} \\ & & \ddots & \\ & \mathbf{0} & & \hat{G}^{(m-1)/2} & \\ & & & & \dfrac{1}{(r + (\frac{1}{2})a_{mj})} \end{bmatrix}_{(m \times m)} , \qquad (4.5.22)$$

where $\hat{G}^{(i)}$ and $\hat{\hat{G}}^{(i)}$ are given by (4.5.14) and (4.5.16), respectively. The AGE equations can be derived along similar lines as before and are given by

(1) *at level* $(p + \frac{1}{2})$

$$w_i^{(p+1/2)} = (A_i\, w_{i-1}^{(p)} + B_i\, w_i^{(p)} + C_i\, w_{i+1}^{(p)} + D_i\, w_{i+2}^{(p)} + E_i)/\hat{\alpha}_{(i+1)/2},$$

$$w_{i+1}^{(p+1/2)} = (\tilde{A}_i\, w_{i-1}^{(p)} + \tilde{B}_i\, w_i^{(p)} + \tilde{C}_i\, w_{i+1}^{(p)} + \tilde{D}_i\, w_{i+2}^{(p)} + \tilde{E}_i)/\hat{\alpha}_{(i+1)/2},$$

$$i = 1, 3, \ldots, m-1, \qquad\qquad (4.5.24)$$

where

$$A_i = \begin{cases} 0 & \text{for } i = 1, \\ -c_i \bar{\rho}_{i+1} & \text{otherwise,} \end{cases} \qquad B_i = \bar{\rho}_{i+1}\bar{s}_i, \quad C_i = -b_i \bar{s}_{i+1},$$

$$D_i = \begin{cases} 0 & \text{for } i = m-1, \\ b_i b_{i+1} & \text{otherwise,} \end{cases} \qquad E_i = \bar{\rho}_{i+1} f_i - b_i f_{i+1},$$

$$\tilde{A}_i = \begin{cases} 0 & \text{for } i = 1, \\ c_i c_{i+1} & \text{otherwise,} \end{cases} \qquad \tilde{B}_i = -c_{i+1}\bar{s}_i, \quad \tilde{C}_i = \bar{\rho}_i \bar{s}_{i+1},$$

$$\tilde{D}_i = \begin{cases} 0 & \text{for } i = m-1, \\ -\bar{\rho}_i b_{i+1} & \text{otherwise,} \end{cases} \qquad \tilde{E}_i = -c_{i+1} f_i + \bar{\rho}_i f_{i+1}; \qquad (4.5.24a)$$

(2) *at level* $(p + 1)$

$$w_1^{(p+1)} = (\bar{q}_1\, w_1^{(p)} + dw_1^{(p+1/2)})/\bar{\rho}_1,$$

$$w_i^{(p+1)} = (P_i\, w_i^{(p)} + Q_i\, w_{i+1}^{(p)} + R_i\, w_i^{(p+1/2)} + S_i\, w_{i+1}^{(p+1/2)})/\alpha_{i/2},$$

$$w_{i+1}^{(p+1)} = (\bar{P}_i w_i^{(p)} + \bar{Q}_i\, w_{i+1}^{(p)} + \tilde{R}_i\, w_i^{(p+1/2)} + \tilde{S}_i\, w_{i+1}^{(p+1/2)})/\alpha_{i/2},$$

$$i = 2, 4, \ldots, m-2,$$

$$w_m^{(p+1)} = (\bar{q}_m\, w_m^{(p)} + dw_m^{(p+1/2)})/\bar{p}_m, \tag{4.5.25}$$

where

$$P_i = (\bar{p}_{i+1}\, \bar{q}_i - b_i c_{i+1}), \quad Q_i = b_i(\bar{p}_{i+1} - \bar{q}_{i+1}), \quad R_i = d\bar{p}_{i+1}, \quad S_i = -db_i.$$

and

$$\bar{P}_i = c_{i+1}(\bar{p}_i - \bar{q}_i), \quad \bar{Q}_i = \bar{p}_i \bar{q}_{i+1} - c_{i+1}b_i, \quad \tilde{R}_i = -dc_{i+1}d, \quad \tilde{S}_i = d\bar{p}_i.$$

$$\tag{4.5.25a}$$

The iterative process is continued until the convergence requirement is met.

(b) *Lees three-level linearisation* Lees (1966) considered the non-linear equation

$$b(U)\frac{\partial U}{\partial t} = \frac{\partial}{\partial x}\left\{ a(U)\frac{\partial U}{\partial x}\right\}, \quad a(U) > 0, \ b(U) > 0, \tag{4.5.26}$$

and investigated a difference scheme that:

(i) achieved *linearity* in the unknowns $u_{i,j+1}$ by evaluating all coefficients of $u_{i,j+1}$ at a time level of known solution values,

(ii) preserved *stability* by averaging $u_{i,j}$ over three time levels, and

(iii) maintained *accuracy* by using central-difference approximation.

Since by central differences approximation we have

$$\left(\frac{\partial U}{\partial x}\right)_{i,j} \approx \frac{1}{\Delta x}(u_{i+1/2,j} - u_{i-1/2,j})$$

$$= \frac{1}{\Delta x}\delta_x u_{i,j},$$

then a central-difference approximation to (4.5.26) is given by

$$b(u_{i,j})\frac{1}{2\Delta t}(u_{i,j+1} - u_{i,j-1}) = \frac{1}{\Delta x}\delta_x\left\{ a(u_{i,j})\frac{1}{\Delta x}\delta_x u_{i,j}\right\}$$

$$= \frac{1}{(\Delta x)^2}\delta_x\{a(u_{i,j})\delta_x\}u_{i,j}, \tag{4.5.27}$$

which reduces to the well-known Richardson formula when $a(u) = b(u) = 1$ and is therefore unconditionally unstable. However, in *the linear constant coefficient case* (see Mitchell and Griffiths, 1980), unconditional stability is obtained by replacing $\delta_x^2 u_{i,j}$ by $\frac{1}{3}\delta_x^2(u_{i,j+1} + u_{i,j} + u_{i,j-1})$. Following this procedure, equation (4.5.27) is rewritten as

$$b(u_{i,j})(u_{i,j+1} - u_{i,j-1}) = 2r[a(u_{i+1/2,j})(u_{i+1,j} - u_{i,j}) - a(u_{i-1/2,j})(u_{i,j} - u_{i-1,j})]$$

and then $u_{i+1,j}$, $u_{i,j}$ and $u_{i-1,j}$ are replaced by

$$\tfrac{1}{3}(u_{i+1,j+1} + u_{i+1,j} + u_{i+1,j-1}), \quad \tfrac{1}{3}(u_{i,j+1} + u_{i,j} + u_{i,j-1})$$

and

$$\tfrac{1}{3}(u_{i-1,j+1} + u_{i-1,j} + u_{i-1,j-1}),$$

respectively. Furthermore, since $u_{i\pm 1/2,j}$ do not fall on the grid points, we replace $a(u_{i+1/2,j})$ and $a(u_{i-1/2,j})$ by $a(u_{i+1,j} + u_{i,j}/2)$ and $a(u_{i,j} + u_{i-1,j}/2)$, respectively. This leads to *the linearised three-level formula*

$$b(u_{i,j})(u_{i,j+1} - u_{i,j-1}) = \tfrac{2}{3}r[\beta^+\{(u_{i+1,j+1} - u_{i,j+1}) + (u_{i+1,j} - u_{i,j})$$
$$+ (u_{i+1,j-1} - u_{i,j-1})\} - \beta^-\{(u_{i,j+1} - u_{i-1,j+1})$$
$$+ (u_{i,j} - u_{i-1,j}) + (u_{i,j-1} - u_{i-1,j-1})\}], \qquad (4.5.28)$$

where

$$\beta^+ = a\left(\frac{u_{i+1,j} + u_{i,j}}{2}\right) \quad \text{and} \quad \beta^- = a\left(\frac{u_{i,j} + u_{i-1,j}}{2}\right). \qquad (4.5.29)$$

Lees (1966) proved the convergence result for (4.5.28) by showing that for sufficiently small values of Δx and Δt

$$\max_{i,j} |U_{i,j} - u_{i,j}| \le K((\Delta x)^2 + (\Delta t)^2),$$

where K is a constant. For this method to be applied to (4.5.1) it is necessary to write the equation in *self-adjoint form* as

$$\frac{\partial U}{\partial t} = \frac{\partial^2 U^n}{\partial x^2}, \quad n \ge 2,$$
$$= \frac{\partial}{\partial x}\left(nU^{n-1}\frac{\partial U}{\partial x}\right).$$

On comparing this equation with (4.5.26), we find that for the particular value of $n = 2$, $a(U) = 2U$ and $b(U) = 1$ and from (4.5.29)

$$\beta^+ = u_{i+1,j} + u_{i,j}, \quad \beta^- = u_{i,j} + u_{i-1,j}. \qquad (4.4.30)$$

Hence the formula (4.5.28) becomes

$$\tfrac{2}{3}r\beta^- u_{i-1,j+1} + (1 + \tfrac{2}{3}r(\beta^+ + \beta^-))u_{i,j+1} - \tfrac{2}{3}r\beta^+ u_{i+1,j+1}$$
$$= \tfrac{2}{3}r\beta^- u_{i-1,j} - \tfrac{2}{3}r(\beta^+ + \beta^-)u_{i,j} + \tfrac{2}{3}r\beta^+ u_{i+1,j}$$
$$+ \tfrac{2}{3}r\beta^- u_{i-1,j-1} + (1 - \tfrac{2}{3}r(\beta^+ + \beta^-))u_{i,j-1} + \tfrac{2}{3}r\beta^+ u_{i+1,j-1},$$
$$\text{for } i = 1, 2, \ldots, m, \qquad (4.5.31)$$

which is a tridiagonal system of equations that can be written in the matrix form (4.5.6) (with $\underline{u}$ replacing $\underline{w}$) when A takes the form (4.5.8) and

$$a_{i,j} = 1 + \tfrac{2}{3}r(\beta^+ + \beta^-)$$
$$= 1 + \tfrac{2}{3}r(u_{i-1,j} + 2u_{i,j} + u_{i+1,j}), \quad i = 1, 2, \ldots, m;$$
$$b_{i,j} = -\tfrac{2}{3}r\beta^+$$
$$= -\tfrac{2}{3}r(u_{i+1,j} + u_{i,j}), \quad i = 1, 2, \ldots, m-1;$$

and

$$c_{i,j} = -\tfrac{2}{3}r\beta^{-}u_{i-1,j+1}$$
$$= -\tfrac{2}{3}r(u_{i,j} + u_{i-1,j}), \quad i = 2,3,\ldots,m.$$

The components of the right-hand side vector f are given by

$$f_1 = \tfrac{2}{3}r[(u_{1,j} + u_{0,j})(u_{0,j-1} + u_{0,j} + u_{0,j+1}) + (u_{2,j} + u_{1,j})(u_{2,j-1} + u_{2,j})$$
$$- (u_{0,j} + 2u_{1,j} + u_{2,j})u_{1,j}] + [1 - \tfrac{2}{3}r(u_{0,j} + 2u_{1,j} + u_{2,j})]u_{1,j-1}$$

$$f_i = \tfrac{2}{3}r[\beta^{+}(u_{i+1,j} + u_{i+1,j-1}) + \beta^{-}(u_{i-1,j} + u_{i-1,j-1}) - (\beta^{+} + \beta^{-})u_{i,j}]$$
$$+ [1 - \tfrac{2}{3}r(\beta^{+} + \beta^{-})]u_{i,j-1}, \quad \text{for } i = 2,3,\ldots,m-1$$

and

$$f_m = \tfrac{2}{3}r[(u_{m+1,j} + u_{m,j})(u_{m+1,j-1} + u_{m+1,j} + u_{m+1,j+1})$$
$$+ (u_{m,j} + u_{m-1,j})(u_{m-1,j-1} + u_{m-1,j}) - (u_{m+1,j} + 2u_{m,j} + u_{m-1,j})u_{m,j}]$$
$$+ [1 - \tfrac{2}{3}r(u_{m+1,j} + 2u_{m,j} + u_{m-1,j})]u_{m,j-1}.$$

When the AGE procedure is implemented on the above tridiagonal system of equations, we arrive at the same computational formulae (with $\underline{w}$ replaced by $\underline{u}$) at the $(p + 1/2)$ and $(p + 1)$ iterates as that for Richtmyer's linearisation. This implies that equations (4.5.18–4.5.19) (for the case m odd) and the equations (4.5.24–4.5.25) (when m is even) will be used for our iterative process.

(ii) *Solving Burgers' Equation:*

$$\varepsilon\frac{\partial^2 U}{\partial x^2} = \frac{\partial U}{\partial t} + U\frac{\partial U}{\partial x}, \quad \varepsilon > 0.$$

The *general non-linear parabolic equation* for the initial boundary value problems is given by

$$\frac{\partial U}{\partial x} = \phi\left(x, t, U, \frac{\partial U}{\partial x}, \frac{\partial^2 U}{\partial x^2}\right), \quad 0 < x < 1, \ 0 < t \leq T, \tag{4.5.32}$$

subject to smooth initial and boundary conditions. This problem is well posed in the region (see, for example, Friedman, 1964) if

$$\frac{\partial\phi}{\partial U_{xx}} \geq a > 0. \tag{4.5.33}$$

If this holds, then the implicit relation (4.5.32) may be solved for $\partial^2 U/\partial x^2$. Thus, we assume the partial differential equation to have the form

$$\frac{\partial\psi}{\partial x^2} = \psi\left(x, t, U, \frac{\partial U}{\partial x}, \frac{\partial U}{\partial t}\right), \tag{4.5.34}$$

where the properly posed requirement is

$$\frac{\partial\psi}{\partial U_t} \geq a > 0. \tag{4.5.35}$$

In some instances, (4.5.34) may be written in the quasi-linear form

$$\frac{\partial^2 U}{\partial x^2} + f(x, t, U)\frac{\partial U}{\partial x} + g(x, t, U) = p(x, t, U)\frac{\partial U}{\partial t}. \qquad (4.5.36)$$

It can be seen that Burgers' equation $\partial^2 U/\partial x^2 = \partial U/\partial t + (U)\,\partial U/\partial x$ is of this form.

(a) *The fully implicit form*   At the point $(i, j + 1)$, (4.5.36) can be approximated by the formula

$$\frac{1}{(\Delta x)^2}\delta_x^2 u_{i,j+1} + \frac{1}{2\Delta x}f[i\Delta x, (j+1)\Delta t, u_{ij}]\mu\delta_x u_{i,j+1} + g[i\Delta x, (j+1)\Delta t, u_{i,j}]$$

$$= p[i\Delta x, (j+1)\Delta t, u_{i,j}]\frac{(u_{i,j+1} - u_{i,j})}{\Delta t}, \qquad (4.5.37)$$

which contains $u_{i,j,+1}$ only linearly and where the difference operators δ and μ are defined as usual by

$$\delta y_n = y_{n+1/2} - y_{n-1/2} \quad \text{(central)},$$

and

$$\mu y_n = \tfrac{1}{2}[y_{n+1/2} + y_{n-1/2}] \quad \text{(averaging)}.$$

Thus, the algebraic problem is *linear and tridiagonal* at each time step. For Burgers' equation, we have the analogue

$$\frac{\varepsilon}{(\Delta x)^2}\delta_x^2 u_{i,j+1} + \frac{1}{2\Delta x}[-u_{i,j}]\mu\delta_x u_{i,j+1} = \frac{u_{i,j+1} - u_{i,j}}{\Delta t},$$

which leads to the equations

$$-r\left(\varepsilon + \frac{\Delta x}{4}u_{i,j}\right)u_{i-1,j+1} + (1 + 2\varepsilon r)u_{i,j+1} - r\left(\varepsilon - \frac{\Delta x}{4}u_{i,j}\right)u_{i+1,j+1} = u_{i,j},$$

$$i = 1, \ldots, m. \qquad (4.5.38)$$

(b) *The Crank–Nicolson form*   The application of the Crank–Nicolson concept to the equation (4.5.36) now gives

$$\frac{1}{2(\Delta x)^2}\delta_x^2(u_{i,j+1} + u_{i,j}) + \frac{1}{2\Delta x}f[i\Delta x, (j + \tfrac{1}{2})\Delta t, \tfrac{1}{2}(u_{i,j+1} + u_{i,j})$$

$$\mu\delta_x(u_{i,j+1} + u_{i,j})] + g[i\Delta x, (j + \tfrac{1}{2})\delta t, \tfrac{1}{2}(u_{i,j+1} + u_{i,j})]$$

$$= p[i\Delta x, (j + \tfrac{1}{2})\Delta t, \tfrac{1}{2}(u_{i,j+1} + u_{i,j})]\frac{(u_{i,j+1} - u_{i,j})}{\Delta t}, \qquad (4.5.39)$$

which are *non-linear*. For Burgers' equation, these simplify to

$$u_{i,j+1} - u_{i,j} = \frac{\varepsilon r}{2}[(u_{i-1,j} - 2u_{i,j} + u_{i+1,j}) + (u_{i-1,j+1} - 2u_{i,j+1} + u_{i+1,j+1})]$$

$$-\frac{r\Delta x}{4}[(u_{i+1,j} - u_{i-1,j})\alpha_{i,j} + (u_{i+1,j+1} - u_{i-1,j+1})\beta_{i,j}], \qquad (4.5.40)$$

where $\alpha_{i,j} = \beta_{i,j} = (u_{i,j+1} + u_{i,j})/2$. This equation, however, can be *linearised* if we replace $\alpha_{i,j}$ by $u_{i,j+1}$ and $\beta_{i,j}$ by $u_{i,j}$. Thus, (4.5.40) becomes

$$-\left(\frac{\varepsilon r}{2} + \frac{\Delta x}{4} r u_{i,j}\right) u_{i-1,j+1} + \left(1 + \varepsilon r + \frac{\Delta x}{4} \lambda(u_{i+1,j} - u_{i-1,j})\right) u_{i,j+1}$$

$$-\left(\frac{\varepsilon r}{2} - \frac{\Delta x}{4} r u_{i,j}\right) u_{i+1,j+1} = \frac{\varepsilon r}{2} u_{i-1,j} + (1 - \varepsilon r) u_{i,j} + \frac{\varepsilon r}{2} u_{i+1,j},$$

$$i = 1, 2, \ldots, m. \tag{4.5.41}$$

(c) *The Predictor–Corrector form* Non-linear algebraic equations can be avoi-ded if two-step methods called Predictor–Corrector (PC) methods are used. The 'predictor' gives a first approximation to the solution and the 'corrector' is used repeatedly, if necessary, to provide the final result. If ψ of (4.5.34) assumes the form

$$\psi = f_1(x, t, U)\frac{\partial U}{\partial t} + f_2(x, t, U)\frac{\partial U}{\partial x} + f_3(x, t, U) \tag{4.5.42a}$$

and

$$\psi = g_1\left(x, t, U\frac{\partial U}{\partial x}\right)\frac{\partial U}{\partial t} + g_2\left(x, t, U\frac{\partial U}{\partial x}\right), \tag{4.5.42b}$$

then a Predictor–Corrector modification of the Crank–Nicolson procedure is possible so that the resulting algebraic problem is linear. The class of equation (4.5.42a) includes Burgers' equation and if ψ is of the form (4.5.34), then one possible form for the predictor is,

$$\frac{1}{(\Delta x)^2}\delta_x^2 u_{i,j+1/2}$$

$$= \psi\left[i\Delta x, (j + \tfrac{1}{2})\Delta t, u_{i,j}, \frac{1}{\Delta x}\mu\delta_x u_{i,j}, \frac{2}{\Delta t}(u_{i,j+1/2} - u_{i,j})\right], \tag{4.5.43}$$

for $i = 1, 2, \ldots, m$ followed by the corrector

$$\frac{1}{2(\Delta x)^2}\delta_x^2[u_{i,j+1} + u_{i,j}] = \psi\left[i\Delta x, (j + \tfrac{1}{2})\Delta t, u_{i,j+1/2}, \frac{1}{2\Delta x}\mu\delta_x(u_{i,j+1} + u_{i,j}),\right.$$

$$\left.\frac{1}{\Delta t}(u_{i,j+1} - u_{i,j})\right]. \tag{4.5.44}$$

For Burgers' equation, the corresponding Predictor–Corrector (PC) pair are given by

$$P: \varepsilon r u_{i-1,j+1/2} - 2(1 + \varepsilon r)u_{i,j+1/2} + \varepsilon r u_{i+1,j+1/2}$$

$$= [\tfrac{1}{2}r(\Delta x)(u_{i+1,j} - u_{i-1,j}) - 2u_{i,j}] \tag{4.5.45}$$

and

$$C: -\frac{r}{2}\left(\varepsilon + \frac{\Delta x}{2} u_{i,j+1/2}\right)u_{i-1,j+1} + (1 + \varepsilon r)u_{i,j+1} - \frac{r}{2}\left(\varepsilon - \frac{\Delta x}{2} u_{i,j+1/2}\right)u_{i+1,j+1}$$

$$= \frac{\varepsilon r}{2}(u_{i-1,j} - 2u_{i,j} + u_{i+1,j}) + u_{i,j} - \frac{r\Delta x}{4}u_{i,j+1/2}(u_{i+1,j} - u_{i-1,j}). \tag{4.5.46}$$

The above Predictor–Corrector formulae are known to have second-order accuracy in both space and time (Douglas and Jones 1963). Notice that (4.5.43) is a backward difference equation. One may also use the following modified Crank–Nicolson predictor

$$\frac{1}{2(\Delta x)^2}\delta_x^2(u_{i,j+1/2}+u_{i,j})$$

$$=\psi\left[i\Delta x,(j+\tfrac{1}{2})\Delta t,u_{i,j},\frac{1}{\Delta x}\mu\delta_x u_{i,j},\frac{2}{\Delta t}(u_{i,j+1/2}-u_{i,j})\right]. \quad (4.5.46a)$$

While the procedure (4.5.43) and (4.5.44) leads to a set linear algebraic equations to solve for ψ when it is of the form (4.5.42a), it does not for ψ of the form (4.5.42b). If (4.5.44) is replaced by

$$\frac{1}{2(\Delta x)^2}\delta_x^2(u_{i,j+1}+u_{i,j})$$

$$=\psi\left[i\Delta x,(j+\tfrac{1}{2})\Delta t,u_{i,j+1/2},\frac{1}{\Delta x}\mu\delta_x u_{i,j+1/2},\frac{1}{\Delta t}(u_{i,j+1}-u_{i,j})\right]. \quad (4.5.47)$$

the system (4.5.43) and (4.5.47) does produce the desired linear algebraic equations with local truncation error $0((\Delta x)^2+0((\Delta t)^{3/2})$.

All of the equations (4.5.38), (4.5.41) and (4.5.46) generate, as before, tridiagonal systems of the form

$$A\underline{u}=\underline{f}, \quad (4.5.47a)$$

where A takes the same form as (4.5.8), $\underline{u}=(u_{1,j+1},u_{2,j+1},\ldots,u_{m,j+1})^T$ and $\underline{f}=(f_1,f_2,\ldots,f_m)^T$. Hence we have

  (1) *for the implicit form* (4.5.38)

$$a_{i,j}=1+2\varepsilon r,\quad i=1,2,\ldots,m,$$

$$b_{i,j}=-r\left(\varepsilon-\frac{\Delta x}{4}u_{i,j}\right),\quad i=1,2,\ldots,m-1,$$

$$c_{i,j}=-r\left(\varepsilon+\frac{\Delta x}{4}u_{i,j}\right),\quad i=2,3,\ldots,m,$$

$$f_1=r\left(\varepsilon+\frac{\Delta x}{4}u_{1,j}\right)u_{0,j+1}+u_{1,j},$$

$$f_i=u_{i,j},\quad i=2,3,\ldots,m-1,$$

and

$$f_m=r\left(\varepsilon-\frac{\Delta x}{4}u_{m,j}\right)u_{m+1,j+1}+u_{m,j};$$

(2) *for the Crank–Nicolson form* (4.5.41):

$$a_{i,j} = 1 + r\left[\varepsilon + \frac{\Delta x}{4}(u_{i+1,j} - u_{i-1,j})\right], \quad i = 1, 2, \ldots, m,$$

$$b_{i,j} = -\frac{r}{2}\left(\varepsilon - \frac{\Delta x}{2}u_{i,j}\right), \quad i = 1, 2, \ldots, m-1, m,$$

$$c_{i,j} = -\frac{r}{2}\left(e + \frac{\Delta x}{2}u_{i,j}\right), \quad i = 2, 3, \ldots, m,$$

$$f_1 = (1 - \varepsilon r)\, u_{1,j} + \frac{r}{2}\left[\varepsilon(u_{0,j} + u_{2,j}) + \left(\varepsilon + \frac{\Delta x}{2}u_{1,j}\right)u_{0,j+1}\right],$$

$$f_i = \frac{\varepsilon r}{2}(u_{i-1,j} + u_{i+1,j}) + (1 - \varepsilon\lambda)u_{i,j}, \quad i = 2, 3, \ldots, m-1,$$

and

$$f_m = (1 - \varepsilon r)u_{m,j} + \frac{r}{2}\left[\varepsilon(u_{m-1,j} + u_{m+1,j}) + \left(\varepsilon - \frac{\Delta x}{2}u_{m,j}\right)u_{m+1,j+1}\right];$$

(3) *for the Predictor–Corrector form* (4.5.46):

$$c_{i,j} = -\frac{r}{2}\left(\varepsilon + \frac{\Delta x}{2}u_{i,j+1/2}\right), \quad i = 1, 2, \ldots, m,$$

$$a_{i,j} = 1 + \varepsilon r, \quad i = 1, 2, \ldots, m-1,$$

$$b_{i,j} = -\frac{r}{2}\left(\varepsilon - \frac{\Delta x}{2}u_{i,j+1/2}\right), \quad i = 2, 3, \ldots, m,$$

$$f_1 = \frac{r}{2}\left[\varepsilon(u_{0,j} - 2u_{1,j} + u_{2,j}) + \left(\varepsilon + \frac{\Delta x}{2}u_{1,j+1/2}\right)u_{0,j+1}\right.$$
$$\left. - \frac{\Delta x}{2}u_{1,j+1/2}\,(u_{2,j} - u_{0,j})\right] + u_{1,j},$$

$$f_i = \frac{r}{2}\left[\varepsilon(u_{i-1,j} - 2u_{i,j} + u_{i+1,j}) - \frac{\Delta x}{2}u_{i,j+1/2}\,(u_{i+1,j} - u_{i-1,j})\right] + u_{i,j},$$
$$i = 2, 3, \ldots, m-1,$$

and

$$f_m = \frac{r}{2}\left[\varepsilon(u_{m-1,j} - 2u_{m,j} + u_{m+1,j}) + \left(\varepsilon - \frac{\Delta x}{2}u_{m,j+1/2}\right)u_{m+1,j+1}\right.$$
$$\left. - \frac{\Delta x}{2}u_{m,j+1/2}\,(u_{m+1,j} - u_{m-1,j})\right] + u_{m,j}.$$

When the AGE algorithm is implemented, the iterative process will require the same equations for computation as in (4.5.18) and (4.5.19) (for the case, *m* odd) and (4.5.24) and (4.5.25) (for the case, *m* even) with w̲ replaced by u̲. We note, however, that for the Predictor–Corrector form, the solutions at the *predictor* stage are obtained using a direct method, i.e., the Thomas elimination algorithm.

(iii) *A non-linear example for the reaction–diffusion equation* We shall now consider the following one-dimensional *reaction–diffusion equation* taken from Ramos (1985)

$$\frac{\partial U}{\partial t} = \frac{\partial^2 U}{\partial x^2} + S, \tag{4.5.48}$$

where $S = U^2(1 - U)$ and $-\infty < x < \infty$ and $t \geq 0$. This equation has an exact travelling wave solution given by

$$U(x, t) = 1/(1 + \exp(V(x - Vt))), \tag{4.5.49}$$

where

$$U(-\infty, t) = 1, \quad U(\infty, t) = 0, \tag{4.5.50}$$

and V is the *steady-state wave speed* which is equal to $1/\sqrt{2}$. In our numerical experiments, however equation (4.4.48) was solved in a truncated domain $-50 \leq x \leq 400$, where the locations of the boundaries were selected so that they did not influence the wave propagation. In the truncated domain, the following initial condition was used

$$U(x, 0) = 1/(1 + \exp(Vx)), \tag{4.5.51}$$

i.e., the initial condition corresponds to the exact solution.

Varioius schemes in the GE and AGE class of methods are now developed to solve (4.5.48)

(a) *GE schemes involving an explicit evaluation of the source term: GE-EXP*
A generalised approximation to (4.5) at the point $(x_j, t_{j+1/2})$ is

$$\frac{(u_{i,j+1} - u_{i,j})}{\Delta t} = \frac{1}{(\Delta x)^2} \{\theta_1 \delta_x u_{i+1/2, j+1} - \theta_2 \delta_x u_{i-1/2, j+1}$$

$$+ \theta_3 \delta_x u_{i+1/2, j} - \theta_4 \delta_x u_{i-1/2, j}\} + s_{i,j}. \tag{4.5.52}$$

By letting $\theta_1 = \theta_4 = 1$ and $\theta_2 = \theta_3 = 0$ in (4.5.52), we obtain the following asymmetric LR approximation:

$$-ru_{i+1,j+1} + (1 + r)u_{i,j+1} = ru_{i-1,j} + (1 - r)u_{i,j} + \delta t s_{i,j}, \tag{4.5.53}$$

where $r = \Delta t/(\Delta x)^2$. If we choose $\theta_2 = \theta_3 = 1$, and $\theta_1 = \theta_4 = 0$ we arrive at the following RL formula

$$(1 + r)u_{i,j+1} - ru_{i-1,j+1} = ru_{i+1,j} + (1 - r)u_{i,j} + \Delta t s_{i,j},$$

or equivalently, at the point $(i + 1, j + 1/2)$

$$-ru_{i,j+1} + (1 + r)u_{i+1,j+1} = (1 - r)u_{i+1,j} + ru_{i+2,j} + \Delta t s_{i+1,j}. \tag{4.5.54}$$

When we couple equations (4.5.53) and (4.5.54) we obtain

$$\begin{bmatrix} (1+r) & -r \\ -r & (1+r) \end{bmatrix} \begin{bmatrix} u_{i,j+1} \\ u_{i+1,j+1} \end{bmatrix} = \begin{bmatrix} 1-r & 0 \\ 0 & 1-r \end{bmatrix} \begin{bmatrix} u_{i,j} \\ u_{i+1,j} \end{bmatrix} + \begin{bmatrix} ru_{i-1,j} + \Delta t \ s_{i,j} \\ ru_{i+2,j} + \Delta t \ s_{i+1,j} \end{bmatrix},$$

i.e.,

$$A\underline{u}_{j+1} = B\underline{u}_j + \hat{\underline{u}}_j, \qquad \underline{u}_{j+1} = A^{-1}(B\underline{u}_j + \hat{\underline{u}}_j),$$

giving

$$\begin{bmatrix} u_{i,j+1} \\ u_{i+1,j+1} \end{bmatrix} = \frac{1}{(1+2r)} \times$$

$$\begin{bmatrix} r(1+r)u_{i-1,j} + (1-r^2)u_{i,j} + r(1-r)u_{i+1,j} + r^2u_{i+2,j} + \Delta t[(1+r)s_{i,j} + rs_{i+1,j}] \\ r^2u_{i-1,j} + r(1-r)u_{i,j} + (1-r^2)u_{i+1,j} + r(1+r)u_{i+2,j} + \Delta t[rs_{i,j} + (1+r)s_{i+1,j}] \end{bmatrix}.$$

Let

$$a_1 = \frac{r(1+r)}{(1+2r)}, \quad a_2 = \frac{(1-r^2)}{(1+2r)}, \quad a_3 = \frac{r(1-r)}{(1+2r)},$$

$$a_4 = \frac{r^2}{(1+2r)}, \quad a_5 = \frac{\Delta t(1+r)}{(1+2r)} \quad \text{and} \quad a_6 = \frac{\Delta tr}{(1+2r)},$$

we have the following set of explicit equations defining the GE schemes at the general points (i,j):

$$u_{i,j+1} = a_1u_{i-1,j} + a_2u_{i,j} + a_3u_{i+1,j} + a_4u_{i+2,j} + a_5s_{i,j} + a_6s_{i+1,j}, \quad (4.5.55)$$

and

$$u_{i+1,j+1} = a_4u_{i-1,j} + a_3u_{i,j} + a_2u_{i+1,j} + a_1u_{i+2,j} + a_6s_{i,j} + a_5s_{i+1,j}. \quad (4.5.56)$$

For the ungrouped point at the right end boundary, we have used (4.5.53),

$$u_{m-1,j+1} = \frac{1}{(1+r)}(ru_{m-2,j} + u_{m,j+1}) + (1-r)(u_{m-1,j} + \Delta ts_{m-1,j}),$$

i.e.,

$$u_{m-1,j+1} = a_7(u_{m-2,j} + u_{m,j+1}) + a_8u_{m-1,j} + a_9s_{m-1,j}, \quad (4.5.57)$$

where

$$a_7 = r(1+r), \quad a_8 = (1-r)/(1+r), \quad a_9 = \Delta t/(1+r).$$

Similarly, using (4.5.54) the ungrouped point at the left boundary is given by

$$u_{1,j+1} = a_8u_{1,j} + a_7(u_{2,j} + u_{0,j+1}) + a_9s_{1,j}. \quad (4.5.58)$$

With equations (4.5.55–4.5.58) we can then construct the alternating schemes of SAGE/DAGE-EXP schemes based on their constituent GER and GEL formulae as shown in Chapter 3.

(b) *GE methods employing the Predictor–Corrector technique: GE-PC* The formulation is the same as in iii(a) except that the source term S is now approximated by $(s_{i,j} + \bar{s}_i)/2$ where $\bar{s}_i = S(\bar{u}_i)$ and the value $\bar{u}_i$ in the predictor step is determined by the solution of the system of differential equations (the explicit method of lines),

$$\frac{du_i}{dt} = \frac{1}{(\Delta x)^2}[u_{i+1,j} - 2u_{i,j} + u_{i-1,j}] + s_i, \quad i = 1,2,\ldots,m-1. \quad (4.5.59)$$

(The system (4.5.59) is derived by discretising the spatial derivatives in (4.5.48) and keeping the time as a continuous variable. The diffusion terms are evaluated at the previous time step.) These equations can be solved by means of *an explicit, fourth order accurate Runge-Kutta method.*

The solution is then employed in the corrector step by the appropriate GE schemes whose set of explicit equations could be similarly derived as in iii(a) and given by

$$u_{i,j+1} = a_1 u_{i-1,j} + a_2 u_{i,j} + a_3 u_{i+1,j} + a_4 u_{i+2,j}$$

$$+ a_5(s_{i,j} + \bar{s}_i)/2 + a_6(s_{i+1,j} + \bar{s}_{i+1})/2 \qquad (4.5.60)$$

and

$$u_{i+1,j+1} = a_4 u_{i-1,j} + a_3 u_{i,j} + a_2 u_{i+1,j} + a_1 u_{i+2,j}$$

$$+ a_6(s_{i,j} + \bar{s}_i)/2 + a_5(s_{i+1,j} + \bar{s}_{i+1})/2. \qquad (4.5.61)$$

In the same way, the ungrouped point at the right end boundary is given as

$$u_{m-1,j+1} = a_7(u_{m-2,j} + u_{m,j+1}) + a_8 u_{m-1,j} + a_9(s_{m-1,j} + \bar{s}_{m-1})/2 \qquad (4.5.62)$$

and at the left end boundary by

$$u_{1,j+1} = a_8 u_{1,j} + a_7(u_{2,j} + u_{0,j+1}) + a_9(s_{1,j} + \bar{s}_1)/2. \qquad (4.5.63)$$

From the basic equation of the GE schemes given by (4.5.60–4.5.63) the SAGE/DAGE-PC methods can be constructed.

(c) *The AGE methods employing the Predictor–Corrector technique on the implicit approximation*: AGE-PC   As in iii(b), before applying the AGE algorithm on the corrector, the explicit method of lines is again employed to determine the values of u_i from the system of ordinary differential equations

$$\frac{du_i}{dt} = \frac{1}{(\Delta x)^2}[u_{i+1,j} - 2u_{i,j} + u_{i-1,j}] + s_i, \quad i = 1, 2, \ldots, m,$$

which is solved by means of an explicit, fourth-order accurate, Runge-Kutta method.

The solution $\bar{u}_i$ (say), is then employed in the following implicit approximation to (4.5.48) in the corrector step

$$\frac{(u_{i,j+1} - u_{i,j})}{\Delta t} = \frac{1}{2(\Delta x)^2}[(u_{i+1,j} - 2u_{i,j} + u_{i-1,j})$$

$$+ (u_{i+1,j+1} - 2u_{i,j+1} + u_{i-1,j+1})] + (s_{i,j} + \bar{s}_i)/2,$$

where $\bar{s}_i = S(\bar{u}_i)$. This C-N like scheme can be written in tridiagonal matrix form as

$$-ru_{i-1,j+1} + (2 + 2r)u_{i,j+1} - ru_{i+1,j+1}$$

$$= 2u_{i,j} + r(u_{i+1,j} - 2u_{i,j} + u_{i-1,j}) + \Delta t(s_{i,j} + \bar{s}_i), \qquad (4.5.64)$$

or

$$A\underline{u} = \underline{f},$$

where A takes the same form as in (4.2.3) $\underline{u} = (u_{1,j+1}, u_{2,j+1}, \ldots, u_{m,j+1})^T$ and

$\underline{f} = (f_1, f_2, \ldots, f_m)^T$, with

$$c = b = -r, \quad a = 2(1 + r),$$

$$f_1 = 2u_{1,j} + r(u_{0,j} + u_{0,j+1} + u_{2,j} - 2u_{1,j}) + \Delta t(s_{1,j} + \bar{s}_i),$$

$$f_i = 2u_{i,j} + r(u_{i+1,j} - 2u_{i,j} + u_{i-1,j}) + \Delta t(s_{i,j} + \bar{s}_i), \quad i = 2, 3, \ldots, m - 1,$$

and

$$f_m = 2u_{m,j} + r(u_{m+1,j} - u_{m+1,j+1} - 2u_{m,j} + u_{m-1,j}) + \Delta t(s_{m,j} + \bar{s}_m).$$

The AGE equations required to solve the system (4.5.64) are given by (4.3.15) and (4.3.18) for the case m odd and (4.3.23) and (4.5.25) for the case m even.

(d) *The AGE methods employing time linearisation techniques on the implicit approximation*: *AGE-TL* Four different time linearisation schemes can be derived as approximations to the differential equation (4.5.48).

(1) *First time linearisation scheme* (1TL) The familar fully implicit approximation to (4.5.48) is given by

$$u_{i,j+1} = u_{ij} + r(u_{i-1,j+1} - 2u_{i,j+1} + u_{i+1,j+1})$$
$$+ \Delta t s_{i,j+1}, \quad i = 1, 2, \ldots, m. \tag{4.5.65}$$

The source term $s_{i,j+1}$ can be linearised around the previous time step by means of Taylor's series as

$$s_{i,j+1} = s_{i,j} + \left(\frac{\partial S}{\partial U}\right)_{ij} (u_{i,j+1} - u_{i,j}). \tag{4.5.66}$$

The substitution of the expression in (4.5.66) into (4.5.65) yields the following 3 term equation at (i, j)

$$-ru_{i-1,j+1} + \left(1 + 2r - \Delta t \left(\frac{\partial S}{\partial U}\right)_{ij}\right) u_{i,j+1} - ru_{i+1,j+1}$$

$$= u_{i,j} + \Delta t \left(s_{i,j} - \left(\frac{\partial S}{\partial U}\right)_{ij} u_{i,j}\right). \tag{4.5.67}$$

(2) *Second time linearisation scheme* (2TL) By applying the Crank–Nicolson concept to the differential equation (4.5.48) we obtain the following approximation

$$u_{i,j+1} = u_{i,j} + r(u_{i+1,j+1} - 2u_{i,j+1} + u_{i-1,j+1})/2$$
$$+ r(u_{i,j} - 2u_{i-1,j})/2 + \Delta t(s_{ij} + s_{i,j+1})/2, \quad \text{for } i = 1, 2, \ldots, m. \tag{4.5.68}$$

The substitution of (4.5.66) into (4.5.68) leads to

$$-ru_{i-1,j+1} + \left(2 + 2r - \Delta t \left(\frac{\partial S}{\partial U}\right)_{ij}\right) u_{i,j+1} - ru_{i+1,j+1}$$

$$= 2u_{ij} + r(u_{i+1,j} - 2u_{ij} + u_{i-1,j}) + \Delta t \left(2s_{ij} - \left(\frac{\partial S}{\partial U}\right)_{ij} u_{ij}\right), \tag{4.5.69}$$

again a 3 term equation at (i, j).

(3) *Third time linearisation scheme (3TL)* This technique employs a first-order accurate time discretisation and a three-point compact formula for the diffusion terms and can be written as

$$u_{i,j+1} = u_{i,j} + r\frac{\delta_x^2}{(1+\delta_x^2/12)}u_{i,j+1} + \Delta t s_{i,j+1}, \quad i=1,2,\ldots,m,$$

or

$$\left(1+\frac{\delta_x^2}{12}\right)u_{i,j+1} = \left(1+\frac{\delta_x^2}{12}\right)u_{ij} + r\delta_x^2 u_{i,j+1} + \Delta t\left(1+\frac{\delta_x^2}{12}\right)s_{i,j+1}, \tag{4.5.70}$$

where δ denotes the usual central difference operator. By substituting (4.5.66) into equation (4.5.70) we obtain the 3 term equation at (i,j)

$$\left[-r+\frac{1}{12}\left\{1-\Delta t\left(\frac{\partial S}{\partial U}\right)_{i-1,j}\right\}\right]u_{i-1,j+1} + \left[2\lambda+\frac{5}{6}\left\{1-\Delta t\left(\frac{\partial S}{\partial U}\right)_{i,j}\right\}\right]u_{i,j+1}$$

$$+\left[-r+\frac{1}{12}\left\{1-\Delta t\left(\frac{\partial S}{\partial U}\right)_{i+1,j+1}\right\}\right]$$

$$=\frac{1}{12}\left[1-\Delta t\left(\frac{\partial S}{\partial U}\right)_{i-1,j}\right]u_{i-1,j} + \frac{5}{6}\left[1-\Delta t\left(\frac{\partial S}{\partial U}\right)_{i,j}\right]u_{i,j} + \frac{1}{12}\left[1-\Delta t\left(\frac{\partial S}{\partial U}\right)_{i+1,j}\right]$$

$$+\frac{1}{6}\Delta t\left[5s_{i,j}+\frac{1}{2}(s_{i-1,j}+s_{i+1,j})\right], \quad i=1,2,\ldots,m. \tag{4.5.71}$$

(4) *Fourth time linearisation scheme (4TL)* This scheme employs a second-order accurate time discretisation and a three-point compact formula for the diffusion terms which is fourth-order accurate in space. The finite-difference form of the 4TL method can be written as

$$u_{i,j+1} = u_{ij} + r\frac{\delta_x^2}{(1+\delta_x^2/12)}[u_{i,j+1}+u_{i,j+1}]/2$$

$$+ \Delta t[s_{i,j}+s_{i,j+1}]/2, \quad i=1,2,\ldots,m,$$

or

$$\left(2+\frac{\delta_x^2}{6}\right)u_{i,j+1} = \left(2+\frac{\delta_x^2}{6}\right)u_{i,j} + r\delta_x^2(u_{i,j}+u_{i,j+1}) + \Delta t\left(1+\frac{\delta_x^2}{12}\right)(s_{i,j}+s_{i,j+1}). \tag{4.5.72}$$

The substitution of (4.5.66) into (4.5.72) yields

$$\left\{\frac{1}{6}\left[1-\frac{\Delta t}{2}\left(\frac{\partial S}{\partial U}\right)_{i-1,j}\right]-r\right\}u_{i-1,j+1} + \left\{\frac{5}{6}\left[2-\Delta t\left(\frac{\partial S}{\partial U}\right)_{i,j}\right]+2r\right\}u_{i,j+1}$$

$$+\left\{\frac{1}{6}\left[1-\frac{\Delta t}{2}\left(\frac{\partial S}{\partial U}\right)_{i+1,j}\right]-r\right\}u_{i+1,j+1}$$

$$= \frac{1}{6}\left\{1 - \frac{\Delta t}{2}\left(\frac{\partial S}{\partial U}\right)_{i-1,j}\right] + r\right\}u_{i-1,j} + \left\{\frac{5}{6}\left[2 - \Delta t\left(\frac{\partial S}{\partial U}\right)_{i,j}\right] - 2r\right\}u_{i,j}$$

$$+ \left\{\frac{1}{6}\left[1 - \frac{\Delta t}{2}\left(\frac{\partial S}{\partial U}\right)_{i+1,j}\right] + r\right\}u_{i+1,j} + \frac{\Delta t}{6}[S_{i-1,j} + 10S_{i,j} + S_{i+1,j}]. \quad (4.5.73)$$

We note that equations (4.5.67), (4.5.69), (4.5.71) and (4.5.73) generate tridiagonal systems of the form (4.5.47a) which can be solved by the AGE algorithm. Hence, we have:

(1) *for 1TL equation (4.5.67)*:

$$c_{i,j} = -r, \quad i = 1,2,\ldots,m,$$

$$a_{i,j} = 1 - 2r - \Delta t\left(\frac{\partial S}{\partial U}\right)_{i,j}, \quad i = 1,2,\ldots,m,$$

$$b_{i,j} = -r, \quad i = 1,2,\ldots,m-1,$$

$$f_1 = u_{1,j} + \Delta t\left(s_{1,j} - \left(\frac{\partial S}{\partial U}\right)_{1,j} u_{1,j}\right) + ru_{0,j+1},$$

$$f_i = u_{i,j} + \Delta t\left(s_{ij} - \left(\frac{\partial S}{\partial U}\right)_{i,j} u_{i,j}\right), \quad i = 2,3,\ldots,m-1,$$

and

$$f_m = u_{m,j} + \Delta t\left(s_{m,j} - \left(\frac{\partial S}{\partial U}\right)_{m,j} u_{m,j}\right) + ru_{m+1,j+1};$$

(2) *for 2TL equation (4.5.69)*:

$$c_{i,j} = -r, \quad i = 2,3,\ldots,m,$$

$$a_{i,j} = 2 + 2r - \Delta t\left(\frac{\partial S}{\partial U}\right)_{i,j}, \quad i = 1,2,\ldots,m,$$

$$b_{i,j} = -r, \quad i = 1,2,\ldots,m,$$

$$f_1 = 2u_{1,j} + r(u_{2,j} - 2u_{1,j}) + \Delta t\left(2s_{1,j} - \left(\frac{\partial S}{\partial U}\right)_{1,j} u_{1,j}\right)$$

$$+ r(u_{0,j} + u_{0,j+1}),$$

$$f_i = 2u_{i,j} + r(u_{i-1,j} - 2u_{i,j} + u_{i+1,j}) + \Delta t\left(2s_{i,j} - \left(\frac{\partial S}{\partial U}\right)_{i,j} u_{i,j}\right),$$

$$i = 1,2,\ldots,m-1,$$

and

$$f_m = 2u_{m,j} + r(u_{m-1,j} - 2u_{m,j}) + \Delta t\left(2s_{m,j} - \left(\frac{\partial S}{\partial U}\right)_{m,j} u_{m,j}\right) + r(u_{m+1,j} + u_{m+1,j+1});$$

(3) *for 3TL equation* (4.5.71):

$$c_{i,j} = -r + \frac{1}{12}\left[1 - \Delta t\left(\frac{\partial S}{\partial U}\right)_{i-1,j}\right], \quad i = 2,3,\dots,m,$$

$$a_{i,j} = 2r + \frac{5}{6}\left[1 - \Delta t\left(\frac{\partial S}{\partial U}\right)_{i,j}\right], \quad i = 1,2,\dots,m,$$

$$b_{i,j} = -r + \frac{1}{12}\left[1 - \Delta t\left(\frac{\partial S}{\partial U}\right)_{i+1,j}\right], \quad i = 1,2,\dots,m-1,$$

$$f_1 = \frac{1}{12}\left[1 - \Delta t\left(\frac{\partial S}{\partial U}\right)_{0,j}\right][u_{0,j} - u_{0,j+1}] + ru_{0,j+1} + \frac{5}{6}\left[1 - \Delta t\left(\frac{\partial S}{\partial U}\right)_{1,j}\right]u_{1,j}$$

$$+ \frac{1}{6}\Delta t\left[5s_{1,j} + \frac{1}{2}(s_{0,j} + s_{2,j})\right],$$

$$f_i = \frac{1}{12}\left[1 - \Delta t\left(\frac{\partial S}{\partial U}\right)_{i-1,j}\right]u_{i-1,j} + \frac{5}{6}\left[1 - \Delta t\left(\frac{\partial S}{\partial U}\right)_{i,j}\right]$$

$$+ \frac{1}{12}\left[1 - \Delta t\left(\frac{\partial S}{\partial U}\right)_{i+1,j}\right]u_{i+1,j} + \frac{1}{6}\Delta t\left[5s_{i,j} + \frac{1}{2}(s_{i-1,j} + s_{i+1,j})\right],$$

$$i = 2,\dots,m-1,$$

and

$$f_m = \frac{1}{12}\left[1 - \Delta t\left(\frac{\partial S}{\partial U}\right)_{m+1,j}\right][u_{m+1,j} - u_{m+1,j+1}] + ru_{m+1,j+1}$$

$$+ \frac{1}{12}\left[1 - \Delta t\left(\frac{\partial S}{\partial U}\right)_{m-1,j}\right]u_{m-1,j} + \frac{1}{6}5\left[1 - \Delta t\left(\frac{\partial S}{\partial U}\right)_{m,j}\right]u_{m,j}$$

$$+ \Delta 5\left[5s_{m,j} + \frac{1}{2}(s_{m-1,j} + s_{m+1,j})\right];$$

(4) *for 4TL equation* (4.5.73):

$$c_{i,j} = \frac{1}{6}\left[1 - \frac{\Delta t}{2}\left(\frac{\partial S}{\partial U}\right)_{i+1,j}\right] - r, \quad i = 2,3,\dots,m,$$

$$a_{i,j} = \frac{5}{6}\left[2 - \Delta t\left(\frac{\partial S}{\partial U}\right)_{i,j}\right] + 2r, \quad i = 1,2,\dots,m,$$

$$b_{i,j} = \frac{1}{6}\left[1 - \frac{\Delta t}{2}\left(\frac{\partial S}{\partial U}\right)_{i+1,j}\right] - r, \quad i = 1,2,\dots,m,$$

$$f_1 = \frac{1}{6}\left[1 - \frac{\Delta t}{2}\left(\frac{\partial S}{\partial U}\right)_{0,j}\right](u_{0,j} - u_{0,j+1}) + r(u_{0,j} + u_{0,j+1})$$

$$+ \left[\frac{5}{6}\left(2 - \Delta t\frac{\partial S}{\partial U}\right)_{1,j} - 2r\right]u_{1,j} + \left[\frac{1}{6}\left(1 - \frac{\Delta t}{2}\frac{\partial S}{\partial U}\right)_{2,j} + r\right]u_{2,j}$$

$$+ \frac{\Delta t}{6}[s_{0,j} + 10s_{1,j} + s_{2,j}],$$

$$f_i = \left[\frac{1}{6}\left[1 - \frac{\Delta t}{2}\left(\frac{\partial S}{\partial U}\right)_{i-1,j}\right] + r\right]u_{i-1,j} + \frac{5}{6}\left(2 - \Delta t\frac{\partial S}{\partial U}\right)_{i,j} - 2r\right]u_{i,j}$$

$$+ \left[\frac{1}{6}\left(1 - \frac{\Delta t}{2}\frac{\partial S}{\partial U}\right)_{i+1,j} + r\right]u_{i+1,j} + \frac{\Delta t}{6}[s_{i-1,j} + 10s_{i,j} + s_{i+1,j}],$$

$$i = 2, \dots, m - 1,$$

and

$$f_m = \frac{1}{6}\left[1 - \frac{\Delta t}{2}\left(\frac{\partial S}{\partial U}\right)_{m+1,j}\right](u_{m+1,j} - u_{m+1,j+1})$$

$$+ r(u_{m+1,j} + u_{m+1,j+1}) + \left[\frac{1}{6}\left(1 - \frac{\Delta t}{2}\frac{\partial S}{\partial U}\right)_{m-1,j} + r\right]u_{m-1,j}$$

$$+ \left[\frac{5}{6}\left(2 - \Delta t\frac{\partial S}{\partial U}\right)_{m,j} - 2r\right]u_{m,j} + \frac{\Delta t}{6}[s_{m-1,j} + 10s_{m,j} + s_{m+1,j}].$$

Again we find that the equations governing the convergence of the AGE iterative process that utilises each of the above time linearisation schemes are given by (4.5.18) and (4.5.19) (for the case m odd) and (4.5.24) and (4.5.25) (for the case m even) with $\underline{w}$ replaced by $\underline{u}$.

4.6. The AGE Method for Two-Dimensional Problems

The AGE method can be readily extended to higher-space dimensions. To ensure unconditional stability, the Douglas–Rachford (DR) variant is used instead of the Peaceman–Rachford (PR) formula.

In two space dimensions, for example, the specific problem under consideration is the diffusion equation

$$\frac{\partial U}{\partial t} = \frac{\partial^2 U}{\partial x^2} + \frac{\partial^2 U}{\partial y^2} + h(x, y, t), \quad (x, y, t) \in R \times (0, T], \qquad (4.6.1)$$

with the initial condition,

$$U(x, y, 0) = F(x, y) \quad (x, y, t) \in R \times \{0\}, \qquad (4.6.1a)$$

and $U(x, y, t)$ is specified on the boundary of R, ∂R by

$$U(x, y, t) = G(x, y, t), \quad (x, y, t) \in \partial R \times (0, T], \qquad (4.6.1b)$$

where for simplicity we assume that the region R of the xy-plane is a rectangle. Similarly, the three-dimensional heat equation is given by

$$\frac{\partial U}{\partial t} = \frac{\partial^2 U}{\partial x^2} + \frac{\partial^2 U}{\partial y^2} + \frac{\partial^2 U}{\partial z^2} + h(x, y, z, t), \qquad (4.6.2)$$

with the initial and boundary conditions specified on R which is now a cube. Based on the AGE concept for the one-dimensional case, the formulation for higher-dimensional problems can be done in very much the same way by employing the *fractional splitting* strategy introduced by Yanenko (1971).

(a) *Two-Dimensional Parabolic Problems* Consider the two-dimensional heat equation (4.6.1) with the auxiliary conditions (4.6.1a) and (4.6.1b). The region R is a rectangle defined by

$$R = \{(x, y): \; 0 \leq x \leq L, \; 0 \leq y \leq M\}.$$

At the point $P(x_i, y_j, t_k)$ in the solution domain, the value of $U(x, y, t)$ is denoted by $u_{i,j,k}$ where $x_i = i\Delta x$, $y_j = j\Delta y$, for $0 \leq i \leq (m + 1)$, $0 \leq i \leq (n + 1)$ and $\Delta x = L/(m + 1)$, $\Delta y = M/(n + 1)$. The increment in the time t, Δt is chosen such that $t_k = k\Delta t$, for $k = 0, 1, 2, \ldots$. For simplicity of presentation, we assume that m and n are chosen so that $\Delta x = \Delta y$ and consequently the mesh ratio is defined by $r = \Delta t/(\Delta x)^2 = \Delta t/(\Delta y^2)$. Analogous to the heat equation in one space dimension, a weighted finite-difference approximation to (4.6.1) at the point $(i, j, k + 1/2)$ is given by (with $0 \leq \theta \leq 1$)

$$\frac{\Delta_t u_{i,j,k}}{\Delta t} = \frac{1}{(\Delta x)^2} \{\theta(\delta_x^2 + \delta_y^2)u_{i,j,k+1} + (1 - \theta)(\delta_x^2 + \delta_y^2)u_{i,j,k}\} = h_{i,j,k+1/2}, \quad (4.6.3)$$

which leads to the *five-point formula*

$$-r\theta u_{i-1,j,k+1} + (1 + 4r\theta) \, u_{i,j,k+1} - r\theta u_{i+1,j,k+1} - r\theta u_{i,j-1,k+1} - r\theta u_{i,j+1,k+1}$$

$$= r(1 - \theta)u_{i-1,j,k} + (1 - 4r)(1 - \theta)u_{i,j,k} + r(1 - \theta)u_{i+1,j,k}$$

$$+ r(1 - \theta)u_{i,j-1,k} + r(1 - \theta)u_{i,j+1,jk} + \Delta t h_{i,j,k+1/2},$$

$$\text{for } i = 1, 2, \ldots, m; \; j = 1, 2, \ldots, n. \qquad (4.6.4)$$

We note that when θ takes the values 0, 1/2 and 1, we obtain the classical explicit, the Crank–Nicolson and the fully implicit schemes whose truncation errors are $0([\Delta x]^2 + \Delta t)$, $0([\Delta x]^2 + [\Delta t]^2)$ and $0([\Delta x]^2 + \Delta t)$, respectively. The explicit scheme is stable only for $r \leq 1/4$ (if $\Delta x \neq \Delta y$, we need $\Delta t/[(\Delta x)^2 + (\Delta y)^2] \leq 1/8$). The fully implicit and the Crank–Nicolson schemes are, however, unconditionally stable. The computational molecules of these schemes are presented in Figures 4.6.1–4.6.3.

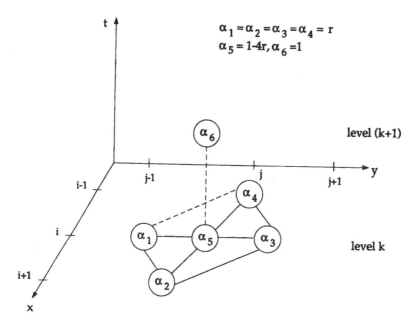

Figure 4.6.1 The classical explicit scheme.

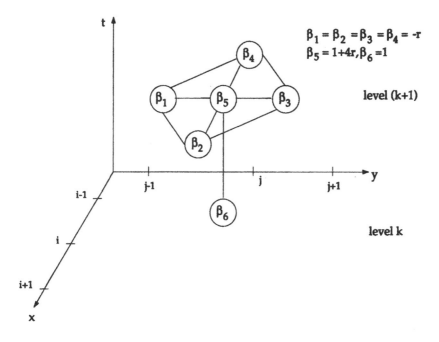

Figure 4.6.2 The fully implicit scheme.

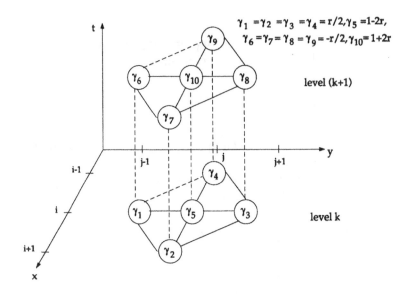

Figure 4.6.3 The Crank–Nicolson scheme.

The weighted finite-difference equations (4.6.4) can be expressed in compact matrix form as

$$A\underline{u}_{(r)}^{[k+1]} = B\underline{u}_{(r)}^{[k]} + \underline{b} + \underline{g}, \tag{4.6.5}$$

$$= \underline{f}, \tag{4.6.6}$$

where, $\underline{u}_{(r)}^{[k]}$ are the known u-values at time level k and,

$$\underline{u}_{(r)} = (\underline{u}_1, m\underline{u}_2, \ldots, \underline{u}_n)^T$$

with

$$\underline{u}_j = (u_{1,j}, u_{2,j}, \ldots, u_{m,j})^T, \quad j = 1, 2, \ldots, n.$$

Thus the mn internal mesh points on the rectangular grid system R are ordered *row-wise* (hence the suffix r) as shown in Figure 4.6.4.

The vector $\underline{b}$ consists of the boundary values where

$$\underline{b} = (\underline{b}_1, \underline{b}_2, \ldots, \underline{b}_n)^T$$

with

$$\underline{b}_1 = (r(1 - \theta)[u_{0,1,k} + u_{1,0,k}] + r\theta[u_{0,1,k+1} + u_{1,0,k+1}],$$

$$r(1 - \theta)u_{2,0,k} + r\theta u_{2,0,k+1}, \ldots, r(1 - \theta)u_{m-1,0,k}, + r\theta u_{m-1,0,k+1},$$

$$r(1 - \theta)[u_{m,0,k} + u_{m+1,1,k}] + r\theta[u_{m,0,k+1} + u_{m+1,1,k+1}])^T;$$

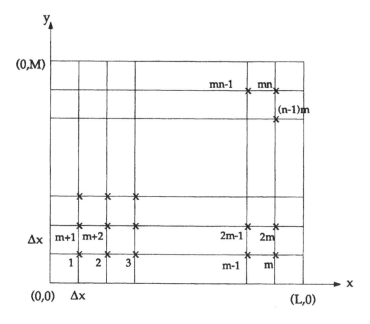

Figure 4.6.4

$$\underline{b}_j = (r(1-\theta)u_{0,j,k} + r\theta u_{0,j,k+1}, 0, \ldots, 0, r(1-\theta)u_{m+1,j,k} + r\theta u_{m+1,j,k+1})^T,$$

for $j = 2, 3, \ldots, n-1$;

and

$$\underline{b}_n = (r(1-\theta)[u_{0,n,k} + u_{1,n+1,k}] + r\theta[u_{0,n,k+1} + u_{1,n+1,k+1}], r(1-\theta)u_{2,n+1,k}$$

$$+ r\theta u_{2,n+1,k+1}, \ldots, r(1-\theta)u_{m-1,n+1,k} + r\theta u_{m-1,n+1,k+1}, r(1-\theta)$$

$$[u_{m,n+1,k} + u_{m+1,n,k}] + r\theta[u_{m,n+1,k+1} + u_{m+1,n,k+1}])^T,$$

and the vector $\underline{g}$ contains the source term of (4.6.4) given by

$$\underline{g} = (\underline{g}_1, \underline{g}_2, \ldots, \underline{g}_n)^T$$

with

$$\underline{g}_j = (\underline{g}_{1,j}, \underline{g}_{2,j}, \ldots, \underline{g}_{m,j})^T$$

$$= \Delta t(h_{1,j,k+1/2}, h_{2,j,k+1/2}, \ldots, h_{m,j,k+1/2})^T, \quad \text{for } j = 1, 2, \ldots, n.$$

The coefficient matrix A in (4.6.5) takes the block tridiagonal form as

$$A = \begin{bmatrix} A_1 & A_2 & & & & & \\ A_2 & A_1 & A_2 & & & & \\ & A_2 & A_1 & A_2 & & 0 & \\ & & & \ddots & & & \\ & & 0 & & A_2 & A_1 & A_2 \\ & & & & & A_2 & A_1 \end{bmatrix}_{(mn \times mn)}, \tag{4.6.7}$$

with

$$A_1 = \begin{bmatrix} c & a_1 & & & \\ a_1 & c & a_1 & & 0 \\ & & \ddots & & \\ 0 & & a_1 & c & a_1 \\ & & & a_1 & c \end{bmatrix}_{(m \times m)}, \tag{4.6.7a}$$

where

$$c = 1 + 4r\theta \quad \text{and} \quad a_1 = -r\theta, \tag{4.6.7b}$$

and

$$A_2 = \text{diag}(a_1) \text{ of order } (m \times m). \tag{4.6.7c}$$

Similarly, the matrix B is of the form

$$B = \begin{bmatrix} B_1 & B_2 & & & & & \\ B_2 & B_1 & B_2 & & & & \\ & B_2 & B_1 & B_2 & & 0 & \\ & & & \ddots & & & \\ & & 0 & & B_2 & B_1 & B_2 \\ & & & & & B_2 & B_1 \end{bmatrix}_{(mn \times mn)}, \tag{4.6.8}$$

with

$$B_1 = \begin{bmatrix} d & e_1 & & & \\ e_1 & d & e_1 & & 0 \\ & & \ddots & & \\ 0 & & e_1 & d & e_1 \\ & & & e_1 & d \end{bmatrix}_{(m \times m)}, \tag{4.6.8a}$$

where

$$d = 1 - 4r(1 - \theta) \quad \text{and} \quad e_1 = r(1 - \theta), \tag{4.6.8b}$$

and

$$B_2 = \text{diag}(e_1) \text{ of order } (m \times m).$$ (4.6.8c)

We observe from (4.6.5) that the matrix A is of the form

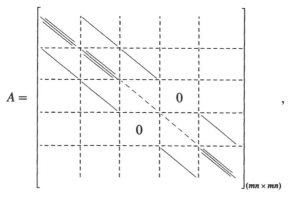

$A =$, $(mn \times mn)$

If we split A into the sum of its constituent symmetric and positive definite matrices G_1, G_2, G_3 and G_4, we have

$$A = G_1 + G_2 + G_3 + G_4,$$ (4.6.9)

where

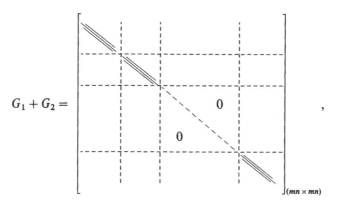

$G_1 + G_2 =$, $(mn \times mn)$

with $\text{diag}(G_1 + G_2) = (1/2)$ with $\text{diag}(A)$, and,

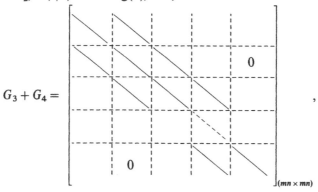

$G_3 + G_4 =$, $(mn \times mn)$

with $\text{diag}(G_3 + G_4) = (1/2) \, \text{diag}(A)$.

In particular, if we let

$$
\hat{A}_1 =
\begin{bmatrix}
(c/2) & a_1 & & & & \\
a_1 & (c/2) & a_1 & & 0 & \\
& a_1 & & & & \\
& & & & & \\
& 0 & & a_1 & (c/2) & a_1 \\
& & & & a_1 & (c/2)
\end{bmatrix}_{(m \times m)} ,
\tag{4.6.10}
$$

and

$$
A_4 = \mathrm{diag}(c/2) \text{ of order } (m \times m),
$$

we have

$$
G_1 + G_2 =
\begin{bmatrix}
\hat{A}_1 & & & \\
& \hat{A}_1 & & 0 \\
& & & \\
& 0 & & \hat{A}_1
\end{bmatrix}_{(mn \times mn)} ,
\tag{4.6.11}
$$

and

$$
G_3 + G_4 =
\begin{bmatrix}
A_4 & A_2 & & & & & \\
A_2 & A_4 & A_2 & & 0 & & \\
& & & & & & \\
& & & & & & \\
& & 0 & & A_2 & A_4 & A_2 \\
& & & & & A_2 & A_4
\end{bmatrix}_{(mn \times mn)}
\tag{4.6.12}
$$

The Douglas–Rachford formula for the AGE fractional scheme for equation (4.6.6) then takes the form:

$$
(G_1 + \rho I)\underline{u}_{(r)}^{(p+1/4)} = (\rho I - G_1 - 2G_2 - 2G_3 - 2G_4)\underline{u}_{(r)}^{(p)} + 2\underline{f},
$$

$$
(G_2 + \rho I)\underline{u}_{(r)}^{(p+1/2)} = G_2\underline{u}_{(r)}^{(p)} + \rho\underline{u}_{(r)}^{(p+1/4)},
$$

$$
(G_3 + \rho I)\underline{u}_{(r)}^{(p+3/4)} = G_3\underline{u}_{(r)}^{(p)} + \rho\underline{u}_{(r)}^{(p+1/2)},
\tag{4.6.13}
$$

$$
(G_4 + \rho I)\underline{u}_{(r)}^{(p+1)} = G_4\underline{u}_{(r)}^{(p)} + \rho\underline{u}_{(r)}^{(p+3/4)},
$$

for the $(p+1)^{th}$ iterate. The iteration proceeds until convergence is achieved.

We now consider the above iterative formulae at each of the four intermediate levels:

(i) *At the first intermediate level (the $(p + 1/4)^{th}$ iterate) in the $(k + 1)^{th}$ time step*:

Since

$$A = G_1 + G_2 + G_3 + G_4,$$

then using the first expression of (4.6.13) and (4.6.6), we obtain

$$(G_1 + \rho I)\underline{u}_{(r)}^{(p + 1/4)} = ((\rho I + G_1) - 2A)\underline{u}_{(r)}^{(p)} + 2B\underline{u}_{(r)}^{[k]} + 2(\underline{b} + \underline{g}),$$

or

$$\underline{u}_{(r)}^{(p + 1/4)} = (G_1 + \rho I)^{-1}[((\rho I + G_1) - 2A)\underline{u}_{(r)}^{(p)} + 2B\underline{u}_{(r)}^{[k]}]. \tag{4.6.14}$$

Without loss of generality we assume that the *order of the matrix A is odd*. We find that

$$(\rho I + G_1) = \begin{bmatrix} C_1 & & & & & \\ & C_2 & & & & \\ & & C_1 & & 0 & \\ & & & \ddots & & \\ & 0 & & & C_2 & \\ & & & & & C_1 \end{bmatrix}_{(mn \times mn)} \tag{4.6.15}$$

and

$$(\rho I + G_1)^{-1} = \begin{bmatrix} C_1^{-1} & & & & & \\ & C_2^{-1} & & & & \\ & & C_1^{-1} & & 0 & \\ & & & \ddots & & \\ & 0 & & & C_2^{-1} & \\ & & & & & C_1^{-1} \end{bmatrix}_{(mn \times mn)} \tag{4.6.16}$$

where

$$C_1 = \begin{bmatrix} \rho_1 & & & & & & \\ & \rho_1 & a_1 & & & & \\ & a_1 & \rho_1 & & & & \\ & & & \rho_1 & a_1 & & \\ & & & a_1 & \rho_1 & 0 & \\ & & & & & \ddots & \\ & & & & & \rho_1 & a_1 \\ & & 0 & & & a_1 & \rho_1 \end{bmatrix}_{(m \times m)}, \tag{4.6.17}$$

$$C_2 = \begin{bmatrix} \rho_1 & a_1 & & & & & \\ a_1 & \rho_1 & & & & & \\ & & \rho_1 & a_1 & & 0 & \\ & & a_1 & \rho_1 & & & \\ & & & & \rho_1 & a_1 & \\ & 0 & & & a_1 & \rho_1 & \\ & & & & & & \rho_1 \end{bmatrix}_{(m \times m)}$$ (4.6.18)

and

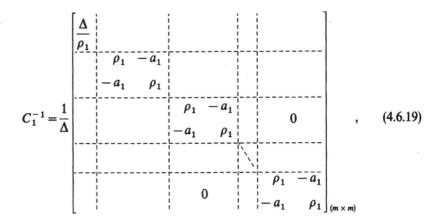

$$C_1^{-1} = \frac{1}{\Delta} \begin{bmatrix} \dfrac{\Delta}{\rho_1} & & & & \\ & \rho_1 & -a_1 & & \\ & -a_1 & \rho_1 & & \\ & & & \rho_1 & -a_1 & & 0 \\ & & & -a_1 & \rho_1 & & \\ & & & & & \rho_1 & -a_1 \\ & 0 & & & & -a_1 & \rho_1 \end{bmatrix}_{(m \times m)}$$,   (4.6.19)

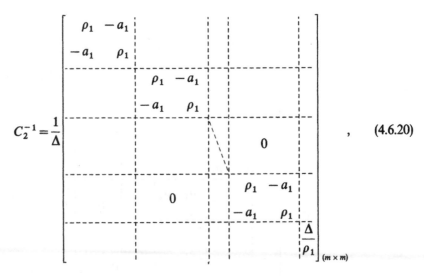

$$C_2^{-1} = \frac{1}{\Delta} \begin{bmatrix} \rho_1 & -a_1 & & & & \\ -a_1 & \rho_1 & & & & \\ & & \rho_1 & -a_1 & & \\ & & -a_1 & \rho_1 & & 0 \\ & & & & \rho_1 & -a_1 \\ & 0 & & & -a_1 & \rho_1 \\ & & & & & & \dfrac{\Delta}{\rho_1} \end{bmatrix}_{(m \times m)}$$,   (4.6.20)

with

$$\rho_1 = \rho + \frac{c}{4} \quad \text{and} \quad \Delta = (\rho_1 + a_1)(\rho_1 - a_1).$$ (4.6.21)

By writing $D_1 = C_1 - 2A_1$, $D_2 = C_2 - 2A_1$, $E_1 = -2A_2$, $F_1 = 2B_1$ and $F_2 = 2B_2$ we get

$$((\rho I + G_1) - 2A)\underline{u}_{(r)}^{(p)} + 2B\underline{u}_{(r)}^{[k]} + 2(\underline{b} + \underline{g})$$

$$= \begin{bmatrix} D_1 & E_1 & & & & & \\ E_1 & D_2 & E_1 & & & & \\ & E_1 & D_1 & E_1 & & 0 & \\ & & & \ddots & & & \\ & & 0 & & E_1 & D_1 & E_1 \\ & & & & & E_1 & D_1 \end{bmatrix} \begin{bmatrix} \underline{u}_{1(r)}^{(p)} \\ \underline{u}_{2(r)}^{(p)} \\ \underline{u}_{3(r)}^{(p)} \\ \vdots \\ \underline{u}_{n-1(r)}^{(p)} \\ \underline{u}_{n(r)}^{(p)} \end{bmatrix}$$

$$+ \begin{bmatrix} F_1 & F_2 & & & & & \\ F_2 & F_1 & F_2 & & & & \\ & F_2 & F_1 & F_2 & & 0 & \\ & & & \ddots & & & \\ & & 0 & & F_2 & F_1 & F_2 \\ & & & & & F_2 & F_1 \end{bmatrix} \begin{bmatrix} \underline{u}_{1(r)}^{[k]} \\ \underline{u}_{2(r)}^{[k]} \\ \underline{u}_{3(r)}^{[k]} \\ \vdots \\ \underline{u}_{n-1(r)}^{[k]} \\ \underline{u}_{n(r)}^{[k]} \end{bmatrix} + 2 \begin{bmatrix} \underline{b}_1 + \underline{g}_1 \\ \underline{b}_2 + \underline{g}_2 \\ \underline{b}_3 + \underline{g}_3 \\ \vdots \\ \underline{b}_{n-1} + \underline{g}_{n-1} \\ \underline{b}_n + \underline{g}_n \end{bmatrix}.$$

$$(4.6.22)$$

Hence, using (4.6.14) we obtain the following set of equations for the computation of the AGE algorithm at the $(p + 1/4)^{th}$ iterate of the $(k + 1)$ time step:

$$\underline{u}_{1(r)}^{(p+1/4)} = C_1^{-1} (D_1 \underline{u}_{1(r)}^{(p)} + E_1 \underline{u}_{2(r)}^{(p)} + F_1 \underline{u}_{1(r)}^{[k]} + F_2 \underline{u}_{2(r)}^{[k]} + 2(\underline{b} + \underline{g}_1)); \quad (4.6.23a)$$

$$\underline{u}_{j(r)}^{(p+1/4)} = C_2^{-1} (E_1 \underline{u}_{j-1(r)}^{(p)} + \underline{u}_{j+1(r)}^{(p)} + D_2 \underline{u}_{j(r)}^{(p)} + F_2 (\underline{u}_{j-1(r)}^{[k]} + \underline{u}_{j+1(r)}^{[k]})$$
$$+ F_1 \underline{u}_{j(r)}^{[k]} + 2(\underline{b}_j + \underline{g}_j)), \quad \text{for } j = 2, 4, \dots, n-1; \quad (4.6.23b)$$

$$\underline{u}_{j(r)}^{(p+1/4)} = C_2^{-1} (E_1 \underline{u}_{j-1(r)}^{(p)} + \underline{u}_{j+1(r)}^{(p)} + D_1 \underline{u}_{j(r)}^{(p)} + F_2 (\underline{u}_{j-1(r)}^{[k]} + \underline{u}_{j+1(r)}^{[k]})$$
$$+ F_1 \underline{u}_{j(r)}^{[k]} + 2(\underline{b}_j + \underline{g}_j)), \quad \text{for } j = 3, 5, \dots, n-2; \quad (4.6.23c)$$

$$\underline{u}_{n(r)}^{(p+1/4)} = C_1^{-1} (E_1 \underline{u}_{n-1(r)}^{(p)} + D_1 \underline{u}_{n(r)}^{(p)} + F_2 \underline{u}_{n-1(r)}^{[k]}$$
$$+ F_1 \underline{u}_{n(r)}^{[k]} + 2(\underline{b}_n + \underline{g}_n)). \quad (4.6.23d)$$

Let $\alpha_1 = \rho_1 - 2c$, $\alpha_2 = -2a_1$, $\alpha_3 = 2d$ and $\alpha_4 = 2e_1$. When the above equations are written component-wise, we have:

(a) For equation (4.6.23a):

$$u_{11}^{(p+1/4)} = [\alpha_1 u_{11}^{(p)} + \alpha_2 (u_{21}^{(p)} + u_{12}^{(p)}) + \alpha_3 u_{11}^{[k]} + \alpha_4 (u_{21}^{[k]} + u_{12}^{[k]}) + 2(b_{11} + g_{11})]/\rho_1,$$

$$u_{i1}^{(p+1/4)} = [\rho_1 v_i - a_i \bar{v}_i]/\Delta, \quad (4.6.24a)$$

$$u_{i+1,1}^{(p+1/4)} = [-a_1 v_i + \rho_1 \bar{v}_i]/\Delta, \quad i = 2, 4, \dots, m-1,$$

where

$$v_i = -a_1 u^{(p)}_{i+1,1} + \alpha_1 u^{(p)}_{i1} + \alpha_2(u^{(p)}_{i-1,1} + u^{(p)}_{i2}) + \alpha_3 u^{[k]}_{i1}$$
$$+ \alpha_4(u^{[k]}_{i-1,1} + u^{[k]}_{i+1,1} + u^{[k]}_{i2}) + 2(g_{i1} + b_{i1})$$

and

$$\bar{v}_i = -a_1 u^{(p)}_{i1} + \alpha_1 u^{(p)}_{i+1,1} + \alpha_2(u^{(p)}_{i+2,1} + u^{(p)}_{i+1,2}) + \alpha_3 u^{[k]}_{i+1,1}$$
$$+ \alpha_4(u^{[k]}_{i1} + u^{[k]}_{i+2,1} + u^{[k]}_{i+1,2}) + 2(b_{i+1,1} + g_{i+1,1}),$$

with $u_{i1} = 0$ for $i > m$;

(b) for equation (4.6.23b):

$$u^{(p+1/4)}_{i,j} = [\rho_1 v_{ij} - a_i \bar{v}_{ij}]/\Delta,$$
$$u^{(p+1/4)}_{i+1,j} = [-a_1 v_{ij} + \rho_1 \bar{v}_{ij}]/\Delta, \quad i = 2,4,\ldots,n-1, \; i = 1,3,\ldots,m-2,$$
$$u^{(p+1/4)}_{m,j} = [\alpha_1 u^{(p)}_{mj} + \alpha_2(u^{(p)}_{m-1,j} + u^{(p)}_{m,j-1} + u^{(p)}_{m,j+1}) + \alpha_3 u^{[k]}_{mj}$$
$$+ \alpha_4(u^{[k]}_{m-1,j} + u^{[k]}_{m,j-1} + u^{[k]}_{m,j+1})) + 2(b_{mj} + g_{mj})]/\rho_1,$$

$$j = 2,4,\ldots,n-1, \tag{4.6.24b}$$

where

$$v_{ij} = \alpha_1 u^{(p)}_{ij} + \alpha_2(u^{(p)}_{i,j-1} + u^{(p)}_{i,j+1} + u^{(p)}_{i-1,j}) + \alpha_3 u^{[k]}_{ij} + \alpha_4(u^{[k]}_{i,j-1} + u^{[k]}_{i,j+1}$$
$$+ u^{[k]}_{i+1,j} + u^{[k]}_{i-1,j}) - a_1 u^{(p)}_{i+1,j} + 2(b_{i+1,j} + g_{i+1,j}),$$

and

$$\bar{v}_{ij} = \alpha_1 u^{(p)}_{i+1,j} + \alpha_2(u^{(p)}_{i+1,j-1} + u^{(p)}_{i+1,j+1} + u^{(p)}_{i-2,j}) + \alpha_3 u^{[k]}_{i+1,j}$$
$$+ \alpha_4(u^{[k]}_{ij} + u^{[k]}_{i+1,j-1} + u^{[k]}_{i+1,j+1} + u^{[k]}_{i+2,j}) - a_1 u^{(p)}_{ij} + 2(b_{i+1,j} + g_{i+1,j}),$$

with $u_{0j} = 0$;

(c) for equation (4.6.23c):

$$u^{(p+1/4)}_{1,j} = [\alpha_1 u^{(p)}_{1j} + \alpha_2(u^{(p)}_{1,j-1} + u^{(p)}_{1,j+1} + u^{(p)}_{2j}) + \alpha_3 u^{[k]}_{1j}$$
$$+ \alpha_4(u^{[k]}_{1,j-1} + u^{[k]}_{1,j+1} + u^{[k]}_{2j}) + 2(b_{1j} + g_{1j})]/\rho_1, \quad j = 3,5,\ldots,n-2.$$
$$u^{(p+1/4)}_{i,j} = [\rho_1 w_{ij} - a_i \bar{w}_{ij}]/\Delta,$$
$$u^{(p+1/4)}_{i+1,j} = [-a_1 w_{ij} + \rho_1 \bar{w}_{ij}]/\Delta, \quad \text{for } j = 3,5,\ldots,n-2, \; i = 2,4,\ldots,m-1,$$

$$\tag{4.6.24c}$$

where

$$w_{ij} = \alpha_1 u^{(p)}_{ij} + \alpha_2(u^{(p)}_{i-1,j} + u^{(p)}_{i,j-1} + u^{(p)}_{i,j+1}) + \alpha_3 u^{[k]}_{ij}$$
$$+ \alpha_4(u^{[k]}_{i-1,j} + u^{[k]}_{i,j-1} + u^{[k]}_{i,j+1} + u^{[k]}_{i+1,j}) - a_1 u^{(p)}_{i+1,j} + 2(b_{i,j} + g_{i,j})$$

and

$$\bar{w}_{ij} = \alpha_1 u^{(p)}_{i+1,j} + \alpha_2(u^{(p)}_{i+1,j-1} + u^{(p)}_{i+1,j+1} + u^{(p)}_{i+2,j}) + \alpha_3 u^{[k]}_{i+1,j}$$
$$+ \alpha_4(u^{[k]}_{ij} + u^{[k]}_{i+1,j-1} + u^{[k]}_{i+1,j+1} + u^{[k]}_{i+2,j}) - a_1 u^{(p)}_{ij} + 2(b_{i+1,j} + g_{i+1,j}),$$

with $u_{ij} = 0$ for $i > m$;

(d) for equation (4.6.23d):

$$u_{1n}^{(p+1/4)} = [\alpha_1 u_{1n}^{(p)} + \alpha_2(u_{1,n-1}^{(p)} + u_{2n}^{(p)}) + \alpha_3 u_{1n}^{[k]}$$

$$+ \alpha_4(u_{1,n-1}^{[k]} + u_{2n}^{[k]}) + 2(b_{1n} + g_{1n})]/\rho_1,$$

$$u_{i,n}^{(p+1/4)} = [\rho_1 z_i - a_1 \bar{z}_i]/\Delta,$$

$$u_{i+1,n}^{(p+1/4)} = [-a_1 z_i + \rho_1 \bar{z}_i]/\Delta, \quad i = 2, 4, \ldots, m-1, \tag{4.6.24d}$$

where

$$z_{ij} = \alpha_1 u_{in}^{(p)} + \alpha_2(u_{i-1,n}^{(p)} + u_{i,n-1}^{(p)}) + \alpha_3 u_{in}^{[k]}$$

$$+ \alpha_4(u_{i-1,n}^{[k]} + u_{i,n-1}^{[k]} + u_{i,n+1}^{[k]}) - a_1 u_{i+1,n}^{(p)} + 2(b_{i,n} + g_{i,n})$$

and

$$\bar{z}_{ij} = \alpha_1 u_{i+1,n}^{(p)} + \alpha_2(u_{i+1,n-1}^{(p)} + u_{i+2,n}^{(p)}) + \alpha_3 u_{i+1,n}^{[k]}$$

$$+ \alpha_4(u_{in}^{[k]} + u_{i+1,n-1}^{[k]} + u_{i+2,n}^{[k]}) - a_1 u_{in}^{(p)} + 2(b_{i+1,n} + g_{i+1,n}),$$

with $u_{in} = 0$ for $i > m$.

(ii) *At the second intermediate level (the $(p+1/2)^{th}$ iterate)*: From the second equation of (4.6.13) we have

$$\underline{u}_{(r)}^{(p+1/2)} = (g_2 + \rho I)^{-1} [G_2 \underline{u}_{(r)}^{(p)} + r \underline{u}_{(r)}^{(p+1/4)}]. \tag{4.6.25}$$

Let $\hat{C}_1 \equiv C_1$ as in (4.6.17) with the diagonal elements r_1 replaced by $c/4$ and $\hat{C}_2 \equiv C_2$ as in (4.6.18) with the diagonal elements r_1 replaced by $c/4$.

We also have

$$G_2 + \rho I = \begin{bmatrix} C_2 & & & & & \\ & C_1 & & & & \\ & & C_2 & & 0 & \\ & & & \ddots & & \\ & 0 & & & C_1 & \\ & & & & & C_2 \end{bmatrix}_{(mn \times mn)} \tag{4.6.26}$$

and

$$(G_2 + \rho I)^{-1} = \begin{bmatrix} C_2^{-1} & & & & & \\ & C_1^{-1} & & & & \\ & & C_2^{-1} & & 0 & \\ & & & \ddots & & \\ & 0 & & & C_1^{-1} & \\ & & & & & C_2^{-1} \end{bmatrix}_{(mn \times mn)}. \tag{4.6.27}$$

Hence, equation (4.6.25) yields

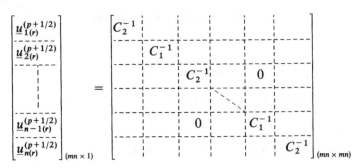

$$(4.6.28)$$

For computational purposes, we will then have

$$\underline{u}_{j(r)}^{(p+1/2)} = C_2^{-1}[\hat{C}_2\underline{u}_{j(r)}^{(p)} + \rho\underline{u}_{j(r)}^{(p+1/4)}], \tag{4.6.29a}$$

$$\underline{u}_{j+1(r)}^{(p+1/2)} = C_1^{-1}[\hat{C}_1\underline{u}_{j+1(r)}^{(p)} + \rho\underline{u}_{j+1(r)}^{(p+1/4)}], \quad j=1,3,\ldots,n-2 \tag{4.6.29b}$$

and

$$\underline{u}_{n(r)}^{(p+1/2)} = C_2^{-1}[\hat{C}_2\underline{u}_{n(r)}^{(p)} + \rho\underline{u}_{n(r)}^{(p+1/4)}]. \tag{4.6.29c}$$

By denoting $\rho_2 = c/4\rho_1$ and $\rho_3 = \rho/\rho_1$ in the above equations can be written component-wise as follows:

(a) for equations (4.6.29a) and (4.6.29c)

$$u_{ij}^{(p+1/2)} = [\rho_1 v_{ij} - a_1\bar{v}_{ij}]/\Delta,$$

$$u_{i+1,j}^{(p+1/2)} = [-a_1 v_{ij} + \rho_1\bar{v}_{ij}]/\Delta, \quad \text{for } 1,3,\ldots,n; \quad i=1,3,\ldots,m-2, \tag{4.6.30a}$$

$$u_{mj}^{(p+1/2)} = \rho_2 u_{mj}^{(p)} + \rho_3 u_{mj}^{(p+1/4)}, \quad j=1,3,\ldots,n,$$

where

$$v_{ij} = \frac{c}{4}u_{ij}^{(p)} + a_1 u_{i+1,j}^{(p)} + \rho u_{i+1,j}^{(p+1/4)}$$

and

$$\bar{v}_{ij} = a_1 u_{ij}^{(p)} + \frac{c}{4}u_{i+1,j}^{(p)} + \rho u_{i+1,j}^{(p+1/4)};$$

(b) for equation (4.6.29b):

$$u_{1j}^{(p+1/2)} = p_2 u_{1j}^{(p)} + p_3 u_{1j}^{(p+1/4)}, \quad j = 2, 4, \ldots, n-1,$$

$$u_{ij}^{(p+1/2)} = [\rho_1 v_{ij} - a_1 \bar{v}_{ij}]/\Delta,$$

$$u_{i+1,j}^{(p+1/2)} = [-a_1 v_{ij} + \rho_1 \bar{v}_{ij}]/\Delta, \quad \text{for } j = 2, 4, \ldots, n-1; \ i = 2, 4, \ldots, m-1,$$

where v_{ij} and $\bar{v}_{ij}$ are given as in (4.6.30a).

(iii) *At the third intermediate level (the $(p+3/4)^{th}$ iterate)*: If we *reorder* the mesh points *column-wise* parallel to the y-axis, we have,

$$\underline{u}_{(c)} = (\underline{u}_1, \underline{u}_2, \ldots, \underline{u}_m)^T \quad \text{with } \underline{u}_i = (u_{i1}, u_{i2}, \ldots, u_{in})^T, \quad \text{for } i = 1, 2, \ldots, m.$$

we also find that

$$(G_3 + G_4)\underline{u}_r = (\hat{G}_1 + \hat{G}_2)\underline{u}_{(c)},$$

and

$$G_3 \underline{u}_{(r)} = \hat{G}_1 \underline{u}_{(c)}, \qquad G_4 \underline{u}_{(r)} = \hat{G}_2 \underline{u}_{(c)}. \tag{4.6.31}$$

Hence the third equation of (4.6.13) is transformed to

$$(\hat{G}_1 + \rho I)\underline{u}_{(c)}^{(p+3/4)} = \hat{G}_1 \underline{u}_{(c)}^{(p)} + \rho \underline{u}_{(c)}^{(p+1/2)},$$

or

$$\underline{u}_{(c)}^{(p+3/4)} = (\hat{G}_1 + \rho I)^{-1}[\hat{G}_1 \underline{u}_{(c)}^{(p)} + \rho \underline{u}_{(c)}^{(p+1/2)}]. \tag{4.6.32}$$

Let the matrices P_1 and P_2 be exactly of the same form as C_1 and C_2 of (4.6.17) and (4.6.18) but of order $(n \times n)$. We will then have

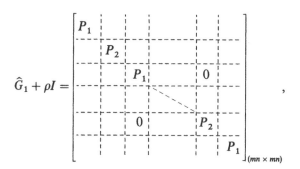

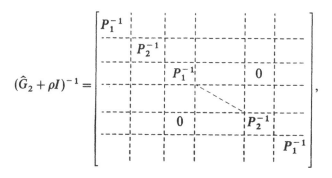

and

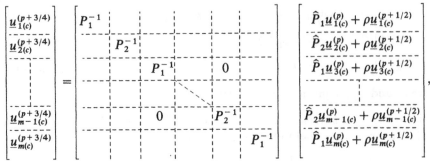

,

where

$$\hat{P}_1 \equiv P_1 \text{ with } \rho_1 \text{ replaced by } \frac{c}{4}$$

and

$$\hat{P}_2 \equiv P_2 \text{ with } \rho_1 \text{ replaced by } \frac{c}{4}.$$

The following equations are therefore obtained for computation at the $(p+3/4)^{th}$ level:

$$\underline{u}_{i(c)}^{(p+3/4)} = P_1^{-1}[\hat{P}_1 \underline{u}_{i(c)}^{(p)} + \rho \underline{u}_{i(c)}^{(p+1/2)}], \quad \text{for } i = 1, 3, \ldots, m \qquad (4.6.33a)$$

and

$$\underline{u}_{i(c)}^{(p+3/4)} = P_2^{-1}[\hat{P}_2 \underline{u}_{i(c)}^{(p)} + \rho \underline{u}_{i(c)}^{(p+1/2)}], \quad \text{for } i = 2, 4, \ldots, m-1, \qquad (4.6.33b)$$

which component-wise yields:

(a) for equation (4.6.33a):

$$u_{i1}^{(p+3/4)} = \rho_2 u_{i1}^{(p)} + \rho_3 u_{i1}^{(p+1/2)}, \quad i = 1, 3, \ldots, m,$$

$$u_{ij}^{(p+3/4)} = [\rho_1 w_{ij} - a_1 \bar{w}_{ij}]/\Delta,$$

$$u_{i+1,j}^{(p+3/4)} = [-a_1 w_{ij} + \rho_1 \bar{w}_{ij}]/\Delta, \quad i = 1, 3, \ldots, m; \ j = 2, 4, \ldots, n-1,$$

$$\qquad (4.6.34a)$$

where

$$w_{ij} = \frac{c}{4} u_{ij}^{(p)} + a_1 u_{i,j+1}^{(p)} + \rho u_{i,j}^{(p+1/2)}$$

and

$$\bar{w}_{ij} = a_1 u_{ij}^{(p)} + \frac{c}{4} u_{i,j+1}^{(p)} + \rho u_{i,j+1}^{(p+1/2)};$$

(b) for equation (4.6.33b):

$$u_{ij}^{(p+3/4)} = [\rho_1 w_{ij} - a_1 \bar{w}_{ij}]/\Delta,$$

$$u_{i,j+1}^{(p+3/4)} = [-a_1 w_{ij} + \rho_1 \bar{w}_{ij}]/\Delta, \quad \text{for } i = 2, 4, \ldots, m-1; \ j = 1, 3, \ldots, n-2,$$

$$u_{n1}^{(p+3/4)} = \rho_2 u_{in}^{(p)} + \rho_3 u_{in}^{(p+1/2)}, \qquad (4.6.34b)$$

where w_{ij} and $\bar{w}_{ij}$ are given as in (4.6.34a).

(iv) *At the fourth intermediate level (the $(p+1)^{th}$ iterate):* By virtue of (4.6.31) the last equation of (4.6.13) is transformed to

$$(\hat{G}_2 + \rho I)\underline{u}_{(c)}^{(p+1)} = \hat{G}_2\underline{u}_{(c)}^{(p)} + \rho\underline{u}_{(c)}^{(p+3/4)},$$

or

$$\underline{u}_{(c)}^{(p+1)} = (\hat{G}_2 + \rho I)^{-1}[\hat{G}_2\underline{u}_{(c)}^{(p)} + \rho\underline{u}_{(c)}^{(p+3/4)}],$$

which leads to the following formulae:

$$\underline{u}_{i(c)}^{(p+1)} = P_2^{-1}[\hat{P}_2\underline{u}_{i(c)}^{(p)} + \rho\underline{u}_{i(c)}^{(p+3/4)}], \quad \text{for } i = 1, 3, \ldots, m \tag{4.6.35a}$$

and

$$\underline{u}_{i(c)}^{(p+1)} = P_1^{-1}[\hat{P}_1\underline{u}_{i(c)}^{(p)} + \rho\underline{u}_{i(c)}^{(p+3/4)}], \quad \text{for } i = 2, 4, \ldots, m-1. \tag{4.6.35b}$$

For computational purposes, we have:

(a) for equation (4.6.35a):

$$u_{ij}^{(p+1)} = [\rho_1 z_{ij} - a_1\bar{z}_{ij}]/\Delta,$$

$$u_{i,j+1}^{(p+1)} = [-a_1 z_{ij} + \rho_1\bar{z}_{ij}]/\Delta, \quad i = 1, 3, \ldots, m; \; j = 1, 3, \ldots, n-2,$$

$$u_{n1}^{(p+1)} = \rho_2 u_{in}^{(p)} + \rho_3 u_{in}^{(p+1)}, \quad i = 1, 3, \ldots, m; \tag{4.6.36a}$$

and

(b) for equation (4.6.35b):

$$u_{i1}^{(p+1)} = \rho_2 u_{i1}^{(p)} + \rho_3 u_{i1}^{(p+3/4)}, \quad i = 2, 4, \ldots, m-1,$$

$$u_{ij}^{(p+1)} = [\rho_1 z_{ij} - a_1\bar{z}_{ij}]/\Delta,$$

$$u_{i,j+1}^{(p+1)} = [-a_1 z_{ij} + \rho_1\bar{z}_{ij}]/\Delta, \quad i = 2, 4, \ldots, m-1; \; j = 2, 4, \ldots, n-1, \tag{4.6.36b}$$

where

$$z_{ij} = \frac{c}{4}u_{ij}^{(p)} + a_1 u_{i,j+1}^{(p)} + \rho u_{i,j}^{(p+3/4)}$$

and

$$\bar{z}_{ij} = a_1 u_{ij}^{(p)} + \frac{c}{4}u_{i,j+1}^{(p)} + \rho u_{i,j+1}^{(p+3/4)}.$$

Hence the AGE scheme corresponds to sweeping through the mesh parallel to the coordinate x and y axes involving at each stage the solution of 2×2 block systems in explicit form. The iterative procedure is continued until convergence is reached, that is when the requirement $|u_{ij}^{(p+1)} - u_{ij}^{(p)}| \le \varepsilon$ is obtained where ε is the convergence criterion.

4.7. The AGE Method to Solve Three-Dimensional Parabolic Problems

The AGE method for three-dimensional problems can be developed in exactly the same manner as for two-dimensional equations. We now consider the following heat equation in three dimensions

$$\frac{\partial U}{\partial t} = \frac{\partial^2 U}{\partial x^2} + \frac{\partial^2 U}{\partial y^2} + \frac{\partial^2 U}{\partial z^2} + h(x,y,z,t), \quad (x,y,z,t) \in R \times (0,T], \tag{4.7.1}$$

with the initial condition

$$U(x,y,z,0) = F(x,y,z), \quad (x,y,z,t) \in R \times \{0\} \tag{4.7.1a}$$

and the boundary conditions

$$U(x,y,z,t) = G(x,y,z,t), \quad (x,y,z,t) \in \partial R \times (0,T], \tag{4.7.1b}$$

where R is the cube $0 < x,y,z < 1$ and ∂R its boundary. Let i,j,k and N be the indices in the x,y,z and t-direction, respectively, with increments $\Delta x, \Delta y, \Delta z$ and Δt (for a cube, $0 \le i,j,k \le (m+1)$, $N = 0,1,\ldots$, and $\Delta x = \Delta y = \Delta z = 1/(m+1)$). At the point $P(x_i, y_j, z_k, t_N)$ in the solution domain, the value of $U(x_i, y_j, z_k, t_N)$ is denoted by $U_{i,j,k}^{[N]}$. A weighted finite-difference approximation to (4.7.1) at the point $(x_i, y_j, z_k, t_{N+1/2})$ is given by (with $0 \le \theta \le 1$)

$$\frac{(u_{i,j,k}^{[N+1]} - u_{i,j,k}^{[N]})}{\Delta t} = \frac{1}{(\Delta x)^2}\{\theta(\delta_x^2 + \delta_y^2 + \delta_z^2)u_{i,j,k}^{[N+1]} + (1-\theta)(\delta_x^2 + \delta_y^2 + \delta_z^2)u_{i,j,k}^{[N]}\}$$

$$+ hu_{i,j,k}^{[N+1/2]}, \quad i,j,k, = 1,2,\ldots,m, \tag{4.7.2}$$

which leads to the *seven-point formula*

$$-r\theta u_{i-1,j,k}^{[N+1]} + (1+6r\theta)u_{i,j,k}^{[N+1]} - r\theta u_{i+1,j,k}^{[N+1]} - r\theta u_{i,j-1,k}^{[N+1]} - r\theta u_{i,j,k-1}^{[N+1]}$$

$$-r\theta u_{i,j+1,k}^{[N+1]} - r\theta u_{i,j,k+1}^{[N+1]} = r(1-\theta)u_{i-1,j,k}^{[N]} + (1-6r(1-\theta))u_{i,j,k}^{[N]}$$

$$+ r(1-\theta)u_{i+1,j,k}^{[N]} + r(1-\theta)u_{i,j-1,k}^{[N]} + r(1-\theta)u_{i,j,k-1}^{[N]}$$

$$+ r(1-\theta)u_{i,j+1,k}^{[N]} + r(1-\theta)u_{i,j,k+1}^{[N]} + \Delta t h_{i,j,k}^{[N+1/2]}. \tag{4.7.3}$$

As before the formula (4.7.2) corresponds to the explicit, the Crank–Nicolson and the fully implicit schemes when θ is equal to 0, 1/2 and 1, respectively. By considering our approximations as sweeps parallel to the xy-plane of the cube $R = \{(x,y,z:0 < x,y,z < 1\}$, equation (4.7.3) can be written in matrix form, as

$$A\underline{u}^{[N+1]} = B\underline{u}^{[N]} + \underline{b} + \underline{g},$$

$$= \underline{f}, \tag{4.7.4}$$

where b consists of the boundary values

$$\underline{g} = (g_1, g_2, \ldots, g_m)^T$$

and

$$\underline{g}_k = (g_{1,1,k}, g_{2,1,k}, \ldots, g_{m,1,k}, g_{1,2,k}, g_{2,2,k}, \ldots, g_{m,2,k}, g_{1,m,k}, g_{2,m,k}, \ldots, g_{m,m,k})^T,$$

$$\text{for } k = 1,2,\ldots,m,$$

and $g_{i,j,k} = \Delta t h_{i,j,k}^{[N+1/2]}$. The vector $\underline{u}^{[N]}$ denotes the u-values at time level N and the coefficient matrix A is given by

$$A = \begin{bmatrix} A_1 & A_2 & & & & & & \\ A_2 & A_1 & A_2 & & & 0 & & \\ & & & & & & & \\ & & & & & & & \\ & & & 0 & A_2 & A_1 & A_2 \\ & & & & & A_2 & A_1 \end{bmatrix}_{(m^3 \times m^3)} , \qquad (4.7.5)$$

where the block matrices A_1 of size $m^2 \times m^2$ takes the form

$$A_1 = \begin{bmatrix} H_1 & H_2 & & & & & & \\ H_2 & H_1 & H_2 & & & 0 & & \\ & & & & & & & \\ & & & & & & & \\ & & & 0 & H_2 & H_1 & H_2 \\ & & & & & H_2 & H_1 \end{bmatrix}_{(m^2 \times m^2)} , $$

$$A_2 = \mathrm{diag}(a_1) \text{ of order } (m^2 \times m^2),$$

with

$$H_1 = \begin{bmatrix} c & a_1 & & & \\ a_1 & c & a_1 & & 0 \\ & & & & \\ 0 & & a_1 & c & a_1 \\ & & & a_1 & c \end{bmatrix} , $$

$$H_2 = \mathrm{diag}(a_1) \text{ of order } (m \times m)$$

and

$$c = 1 + 6r\theta, \quad a_1 = -r\theta, \quad r = \frac{\Delta t}{(\Delta x)^2}.$$

Similarly, the block matrix B is defined by

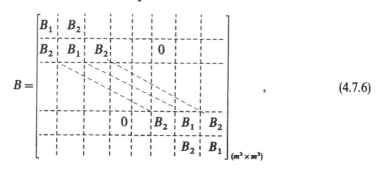

$$B = \begin{bmatrix} B_1 & B_2 & & & & & & \\ B_2 & B_1 & B_2 & & & 0 & & \\ & & & & & & & \\ & & & 0 & B_2 & B_1 & B_2 \\ & & & & & B_2 & B_1 \end{bmatrix}_{(m^3 \times m^3)} , \qquad (4.7.6)$$

where

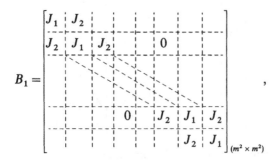

$$B_1 = \qquad ,$$

$$B_2 = \text{diag}(e_1) \text{ of order } (m^2 \times m^2),$$

with

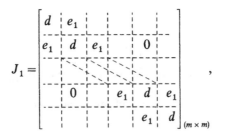

$$J_1 = \qquad ,$$

and

$$J_2 = \text{diag}(e_1) \text{ of order } (m \times m),$$

with

$$d = 1 - 6r(1 - \theta) \quad \text{and} \quad e_1 = r(1 - \theta).$$

The approximations are taken as sweeps parallel to the xy-plane, the u values are evaluated at points lying on planes which are parallel to the xy-plane and perpendicular to the z-axis and on each of these planes the points are ordered row-wise (parallel to the x-axis) as depicted in Figure 4.7.1.

Equation (4.7.4) is therefore written as

$$A\underline{u}_{[xy]}^{[N+1]} = B\underline{u}_{[xy]}^{[N]} + \underline{b} + \underline{g},$$

$$= \underline{f}. \qquad (4.7.7)$$

By splitting A in to the sum of its constituent symmetric and positive definite matrices G_1, G_2, G_3, G_4, G_5 and G_6, we have

$$A = G_1 + G_2 + G_3 + G_4 + G_5 + G_6. \qquad (4.7.8)$$

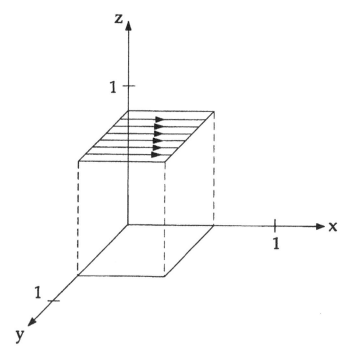

Figure 4.7.1

As a natural extension to Evans and Sahimi (1988) the Douglas–Rachford formula for the AGE fractional scheme for the $(N+1)^{th}$ time step takes the form:

$$(G_1 + rI)\underline{u}_{[xy]}^{(p+1/6)} = (rI - G_1 - 2G_2 - 2G_3 - 2G_4 - 2G_5 - 2G_6)\underline{u}_{[xy]}^{(p)} + 2\underline{f},$$

$$(G_2 + rI)\underline{u}_{[xy]}^{(p+1/3)} = G_2\underline{u}_{[xy]}^{(p)} + r\underline{u}_{[xy]}^{(p+1/6)},$$

$$(G_3 + rI)\underline{u}_{[xy]}^{(p+1/2)} = G_3\underline{u}_{[xy]}^{(p)} + r\underline{u}_{[xy]}^{(p+1/3)},$$

$$(G_4 + rI)\underline{u}_{[xy]}^{(p+2/3)} = G_4\underline{u}_{[xy]}^{(p)} + r\underline{u}_{[xy]}^{(p+1/2)},$$

$$(G_5 + rI)\underline{u}_{[xy]}^{(p+5/6)} = G_5\underline{u}_{[xy]}^{(p)} + r\underline{u}_{[xy]}^{(p+2/3)},$$

$$(G_6 + rI)\underline{u}_{[xy]}^{(p+1)} = G_6\underline{u}_{[xy]}^{(p)} + r\underline{u}_{[xy]}^{(p+5/6)}. \tag{4.7.9}$$

The scheme is applied as 6 sweeps per iteration which is continued until convergence is achieved to obtain the solution of equation (4.7.7) at the $(N+1)^{th}$ time step.

We now consider the above iterative formula at each of the six intermediate levels:

(i) *At the first intermediate level $(p+1/6)^{th}$ iterate of the $(N+1)^{th}$ time step):*

By virtue of (4.7.8) the first equation of (4.7.9) may be written as,

$$(G_1 + rI)\underline{u}_{[xy]}^{(p+1/6)} = ((rI - G_1) - 2A)\underline{u}_{[xy]}^{(p)} + 2\underline{f}.$$

(4.7.10)

By assuming that m *is odd*, we find that

$$(rI + G_1) = \begin{bmatrix} \bar{C}_1^- & & & & & \\ & \bar{C}_2 & & & & \\ & & \bar{C}_1 & & 0 & \\ & & & \ddots & & \\ & & 0 & & \bar{C}_2 & \\ & & & & & \bar{C}_1 \end{bmatrix}_{(m^3 \times m^3)},$$

(4.7.10a)

where

$$\bar{C}_1 = \begin{bmatrix} C_1 & & & & & \\ & C_2 & & & & \\ & & C_1 & & 0 & \\ & & & C_2 & & \\ & & 0 & & C_2 & \\ & & & & & C_1 \end{bmatrix}_{(m^2 \times m^2)},$$

(4.7.11)

$$\bar{C}_2^- = \begin{bmatrix} C_2 & & & & & \\ & C_1 & & & & \\ & & C_2 & & 0 & \\ & & & C_1 & & \\ & & 0 & & C_1 & \\ & & & & & C_2 \end{bmatrix}_{(m^2 \times m^2)},$$

(4.7.12)

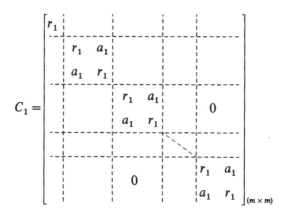

$$
C_1 = \begin{bmatrix}
r_1 & & & & & & \\
& r_1 & a_1 & & & & \\
& a_1 & r_1 & & & & \\
& & & r_1 & a_1 & & \\
& & & a_1 & r_1 & & \\
& & & & & r_1 & a_1 \\
& & & & & a_1 & r_1
\end{bmatrix}_{(m \times m)}
$$

and

$$
C_2 = \begin{bmatrix}
r_1 & a_1 & & & & \\
a_1 & r_1 & & & & \\
& & r_1 & a_1 & & \\
& & a_1 & r_1 & & \\
& & & & r_1 & a_1 \\
& & & & a_1 & r_1 \\
& & & & & & r_1
\end{bmatrix}_{(m \times m)}
. \qquad (4.7.13)
$$

Therefore,

$$
(rI + G_1)^{-1} = \begin{bmatrix}
\bar{C}_1^{-1} & & & & \\
& \bar{C}_2^{-1} & & & \\
& & \bar{C}_1^{-1} & & \\
& & & \bar{C}_2^{-1} & \\
& & & & \bar{C}_1^{-1}
\end{bmatrix}_{(m^3 \times m^3)} , \qquad (4.7.14)
$$

where

$$
\bar{C}_1^{-1} = \begin{bmatrix}
C_1^{-1} & & & & \\
& C_2^{-1} & & & \\
& & C_1^{-1} & & \\
& & & C_2^{-1} & \\
& & & & C_1^{-1}
\end{bmatrix}_{(m^2 \times m^2)} , \qquad (4.7.15)
$$

and similarly for C_2^{-1} with

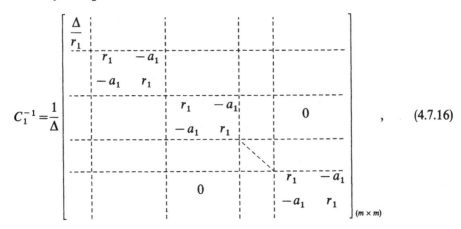

$$C_1^{-1} = \frac{1}{\Delta} \qquad , \qquad (4.7.16)$$

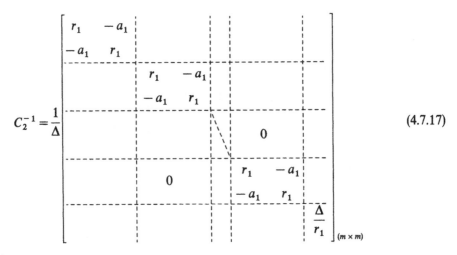

$$C_2^{-1} = \frac{1}{\Delta} \qquad (4.7.17)$$

and

$$r_1 = r + \frac{c}{6}, \qquad \Delta = (r_1 + a_1)(r_1 - a_1).$$

For simplicity we assume that on the boundary the value $b = 0$.
Consequently, we have

$$\underline{f} = B\underline{u}_{[xy]}^{[N]} + \underline{g}$$

and equation (4.7.10) becomes

$$(G_1 + rI)\underline{u}_{[xy]}^{(p+1/6)} = ((rI + G_2) - 2A)\underline{u}_{[xy]}^{(p)} + 2B\underline{u}_{[xy]}^{[N]} + 2\underline{g},$$

or

$$\underline{u}_{[xy]}^{(p+1/6)} = (G_1 + rI)^{-1}[((rI + HG_2) - 2A)\underline{u}_{[xy]}^{(p)}$$
$$+ 2B\underline{u}_{[xy]}^{[N]} + 2\underline{g}]. \qquad (4.7.18)$$

If we write

$$D_1 = \bar{C}_1 - 2A_1, \quad D_2 = \bar{C}_2 - 2A_1, \quad E_1 = -2A_2, \quad F_1 = 2B_1 \text{ and } F_2 = 2B_2,$$

then $((rI + G_1) - 2A)\underline{u}^{(p)}_{[xy]} + 2B\underline{u}^{[N]}_{[xy]} + 2\underline{g}$ becomes

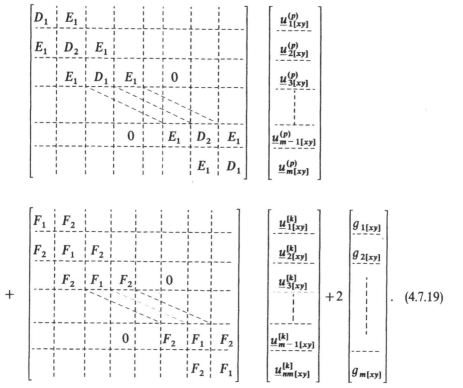

$$(4.7.19)$$

Hence, using (4.6.18) we obtain the following set of equations for the computation of the AGE scheme at the $(p + 1/6)^{th}$ iterate:

$$u^{(p+1/6)}_{1[xy]} = \bar{C}_1^{-1}(D_1 u^{(p)}_{1[xy]} + E_1 u^{[N]}_{1[xy]} + F_1 u^{[N]}_{1[xy]} + F_2 u^{[N]}_{2[xy]} + 2g_{1[xy]}), \quad (4.7.20a)$$

$$u^{(p+1/6)}_{k[xy]} = \bar{C}_2^{-1}(E_1(u^{(p)}_{k-1[xy]} + u^{(p)}_{k+1[xy]}) + D_2 u^{(p)}_{k[xy]}$$

$$+ F_2(u^{[N]}_{k-1[xy]} + u^{[N]}_{k+1[xy]} + F_1 u^{[N]}_{k[xy]} + 2g_{k[xy]}), \quad k = 2, 4, \ldots, m-1,$$

$$(4.7.20b)$$

$$u^{(p+1/6)}_{k[xy]} = \bar{C}_1^{-1}(E_1(u^{(p)}_{k-1[xy]} + u^{(p)}_{k+1[xy]}) + D_1 u^{(p)}_{k[xy]}$$

$$+ F_2(u^{[N]}_{k-1[xy]} + u^{[N]}_{k+1[xy]} + F_1 u^{[N]}_{k[xy]} + 2g_{k[xy]}), \quad k = 3, 5, \ldots, m-2,$$

$$(4.7.20c)$$

$$u_{m[xy]}^{(p+1/6)} = \bar{C}_1^{-1}(E_1 u_{m-1[xy]}^{(p)} + D_1 u_{m[xy]}^{[N]} + F_2 u_{m-1[xy]}^{[N]} + F_1 u_{m[xy]}^{[N]} + 2g_{m[xy]}).$$

$$(4.7.20d)$$

Let $r_2 = r_1 - 2c$. When the above equations are written component-wise, we have:

(a) for equation (4.7.20a):

(1) $u_{1,j,1}^{(p+1/6)} = [r_2 u_{1,j,1}^{(p)} - 2a_1(u_{2,j,1}^{(p)} + u_{1,j,2}^{(p)} + u_{1,j+1,1}^{(p)}) + 2du_{1,j,1}^{[N]}$

$\qquad\qquad + 2e_1(u_{2,j,1}^{[N]} + u_{1,j,2}^{[N]} + u_{1,j+1,1}^{[N]}) + 2g_{1,j,1}]/r_1, \quad$ for $j = 1, 3, \ldots, m$,

$u_{i,j,1}^{(p+1/6)} = (r_1 a_{i,j,1} - a_1 b_{i,j,1})/\Delta,$

$u_{i+1,j,1}^{(p+1/6)} = (-a_1 a_{i,j,1} + r_1 b_{i,j,1})/\Delta, \quad j = 1, 3, \ldots, m; \; i = 2, 4, \ldots, m-1,$

$$(4.7.21a)$$

with,

$$u_{1,m+1,1}^{(p)} = u_{1,m-1,1}^{(p)}, \qquad u_{1,m+1,1}^{[N]} = u_{1,m-1,1}^{[N]};$$

$$u_{i,m+1,1}^{(p)} = u_{i,m-1,1}^{(p)}, \qquad u_{i,m+1,1}^{[N]} = u_{i,m-1,1}^{[N]},$$

and

$$a_{i,j,1} = r_2 u_{i,j,1}^{(p)} - 2a_1(u_{i,j,2}^{(p)} + u_{i-1,j,1}^{(p)} + u_{i,j+1,1}^{(p)}) - a_1 u_{i+1,j,1}^{(p)}$$

$$+ 2e_1(u_{i,j,2}^{[N]} + u_{i-1,j,1}^{[N]} + u_{i+1,j,1}^{[N]} + u_{i,j+1,1}^{[N]}) + 2du_{i,j,1}^{[N]} + 2g_{i,j,1},$$

$$b_{i,j,1} = r_2 u_{i+1,j,1}^{(p)} - 2a_1(u_{i+1,j+1,1}^{(p)} + u_{i+1,j,2}^{(p)}) - a_1 u_{i,j,1}^{(p)}$$

$$+ 2e_1(u_{i,j,1}^{[N]} + u_{i+1,j+1,1}^{[N]} + u_{i+1,j,2}^{[N]}) + 2du_{i+1,j,1}^{[N]} + 2g_{i+1,j,1};$$

(2) $u_{i,j,1}^{(p+1/6)} = (r_1 \bar{a}_{i,j,1} - a_1 \bar{b}_{i,j,1})/\Delta,$

$u_{i+1,j,1}^{(p+1/6)} = (-a_1 \bar{a}_{i,j,1} + r_1 \bar{b}_{i,j,1})/\Delta, \quad j = 2, 4, \ldots, m-1; \; i = 1, 3, \ldots, m-2,$

$u_{m,j,1}^{(p+1/6)} = [r_2 u_{m,j,1}^{(p)} - 2a_1(u_{m,j-1,1}^{(p)} + u_{1,j,2}^{(p)} + u_{m-1,j,1}^{(p)} + u_{m,j+1,1}^{(p)} + u_{m,j,2}^{(p)})$

$\qquad\qquad + 2e_1(u_{m,j-1,1}^{[N]} + u_{m-1,j,2}^{[N]} + u_{1,j,1}^{[N]} + u_{m,j+1,1}^{[N]} + u_{m,j,2}^{[N]})$

$\qquad\qquad + 2du_{m,j,1}^{[N]} + 2g_{m,j,1}]/r_1, \quad$ for $j = 2, 4, \ldots, m-1,$ $\qquad$ (4.7.21b)

where

$$\bar{a}_{i,j,1} = r_2 u_{i,j,1}^{(p)} - 2a_1(u_{i,j-1,1}^{(p)} + u_{i,j+1,1}^{(p)} + u_{i,j,2}^{(p)}) - a_1 u_{i+1,j,1}$$

$$+ 2e_1(u_{1,j-1,1}^{[N]} + u_{i+1,j,1}^{[N]} + u_{i,j+1,1}^{[N]} + u_{i,j,2}^{[N]}) + 2du_{i,j,1}^{[N]} + 2g_{i,j,1},$$

$$\bar{b}_{i,j,1} = r_2 u_{i+1,j,1}^{(p)} - 2a_1(u_{i+1,j-1,1}^{(p)} + u_{i+2,j,1}^{(p)} + u_{i+1,j+1,1}^{(p)} + u_{i+1,j,2}^{(p)})$$

$$- a_1 u_{i,j,1} + 2e_1(u_{i+1,j-1,1}^{[N]} + u_{i,j,1}^{[N]} + u_{i+2,j,1}^{[N]} + u_{i+1,j+1,1}^{[N]} + u_{i+1,j,2}^{[N]})$$

$$+ 2du_{i+1,j,1}^{[N]} + 2g_{i+1,j,1}.$$

(b) for (4.7.20b):

(1) $u_{i,j,k}^{(p+1/6)} = (r_1 c_{i,j,k} - a_1 d_{i,j,k})/\Delta, \quad k = 2, 4, \ldots, m-1, \; j = 1, 3, \ldots, m,$

$$u_{i+1,j,k}^{(p+1/6)} = (-a_1 c_{i,j,k} + r_1 d_{i,j,k})/\Delta, \quad i = 1,3,\ldots,m-2,$$

$$u_{m,j,k}^{(p+1/6)} = [r_2 u_{m,j,k}^{(p)} - 2a_1(u_{m,j,k-1}^{(p)} + u_{1,j,k+1}^{(p)} + u_{m,j,k+1}^{(p)} + u_{m-1,j,k}^{(p)}$$

$$+ u_{m,j+1,k}^{(p)}) + 2e_1(u_{m,j,k-1}^{[N]} + u_{m,j,k+1}^{[N]} + u_{m-1,j,k}^{[N]} + u_{m,j+1,k}^{[N]})$$

$$+ 2du_{m,j,k}^{[N]} + 2g_{m,j,k}]/r_1, \quad \text{for } k = 2,4,\ldots,m-1; \; j = 1,3,\ldots,m,$$

$$(4.7.22a)$$

with,

$$u_{m,m+1,k}^{(p)} = u_{m,m-1,k}^{(p)}, \qquad u_{m,m+1,k}^{[N]} = u_{m,m-1,k}^{[N]};$$

$$u_{i,m+1,k}^{(p)} = u_{i,m-1,k}^{(p)}, \qquad u_{i,m+1,k}^{[N]} = u_{i,m-1,k}^{[N]},$$

$$c_{i,j,1} = r_2 u_{i,j,k}^{(p)} - 2a_1(u_{i,j,k-1}^{(p)} + u_{i,j,k+1}^{(p)} + u_{i,j+1,k}^{(p)}) - a_1 u_{i+1,j,k}^{(p)}$$

$$+ 2e_1(u_{i,j,k}^{[N]} + u_{i,j,k+1}^{[N]} + u_{i+1,j,k}^{[N]} + u_{i,j+1,k}^{[N]}) + 2du_{i,j,k}^{[N]} + 2g_{i,j,k},$$

and

$$d_{i,j,1} = r_2 u_{i+1,j,k}^{(p)} - 2a_1(u_{i+1,j,k-1}^{(p)} + u_{i+1,j,k+1}^{(p)} + u_{i+2,j,k}^{(p)} + u_{i+1,j+1,k}^{(p)})$$

$$- a_1 u_{i,j,k}^{(p)} + 2e_1(u_{i+1,j,k-1}^{[N]} + u_{i+1,j+1,k+1}^{[N]} + u_{i,j,k}^{[N]} + u_{i+2,j,k}^{[N]} + u_{i+1,j+1,k}^{[N]})$$

$$+ 2du_{i+1,j,k}^{[N]} + 2g_{i+1,j,k};$$

(2) $\quad u_{1,j,k}^{(p+1/6)} = [r_2 u_{1,j,k}^{(p)} - 2a_1(u_{1,j,k-1}^{(p)} + u_{1,j,k+1}^{(p)} + u_{m,j-1,k}^{(p)} + u_{2,j,k}^{(p)} + u_{1,j+1,k}^{(p)})$

$$+ 2e_1(u_{1,j,k-1}^{[N]} + u_{1,j,k+1}^{[N]} + u_{1,j-1,k}^{[N]} + u_{2,j,k}^{[N]} + u_{1,j+1,k}^{[N]})$$

$$+ 2du_{1,j,k}^{[N]} + 2g_{1,j,k}]/r_1,$$

$$u_{i,j,k}^{(p+1/6)} = (r_1 \bar{c}_{i,j,k} - a_1 \bar{d}_{i,j,k})/\Delta, \quad k = 2,4,\ldots,m-1, \; j = 2,4,\ldots,m-1,$$

$$u_{i+1,j,k}^{(p+1/6)} = (-a_1 \bar{c}_{i,j,k} - r_1 \bar{d}_{i,j,k})/\Delta, \quad i = 2,4,\ldots,m-1, \tag{4.7.22b}$$

where

$$\bar{c}_{i,j,k} = r_2 u_{i,j,k}^{(p)} - 2a_1(u_{i,j,k-1}^{(p)} + u_{i,j+1,2}^{(p)} + u_{i,j-1,k}^{(p)} + u_{i-1,j,k}^{(p)} + u_{i,j+1,k}^{(p)})$$

$$- a_1 u_{i+1,j,k}^{(p)} + 2e_1(u_{i,j,k-1}^{[N]} + u_{i,j,k+1}^{[N]} + u_{i,j-1,k}^{[N]} + u_{i-1,j,k}^{[N]}$$

$$+ u_{i+1,j,k}^{[N]} + u_{i,j+1,k}^{[N]}) + 2du_{i,j,k}^{[N]} + 2g_{i,j,k},$$

$$\bar{d}_{i,j,k} = r_2 u_{i+1,j,k}^{(p)} - 2a_1(u_{i+1,j,k-1}^{(p)} + u_{i+1,j,k+1}^{(p)} + u_{i+1,j-1,k}^{(p)} + u_{i+1,j+1,k}^{(p)})$$

$$- a_1 u_{i,j,k}^{(p)} + 2e_1(u_{i+1,j,k-1}^{[N]} + u_{i+1,j,k+1}^{[N]} + u_{i+1,j-1,k}^{[N]}$$

$$+ u_{i,j,k}^{[N]} + u_{i+1,j+1,k}^{[N]}) + 2du_{i+1,j,k}^{[N]} + 2g_{i+1,j,k}.$$

(c) for equation (4.7.20c):

(1)　$u_{1,j,k}^{(p+1/6)} = [r_2 u_{1,j,k}^{(p)} - 2a_1(u_{1,j,k-1}^{(p)} + u_{1,j,k+1}^{(p)} + u_{2,j,k}^{(p)} + u_{1,j+1,k}^{(p)})$

$$+ 2e_1(u_{1,j,k-1}^{[N]} + u_{1,j,k+1}^{[N]} + u_{2,j,k}^{[N]} + u_{1,j+1,k}^{[N]})$$

$$+ 2du_{1,j,k}^{[N]} + 2g_{1,j,k}]/r_1,$$

$u_{i,j,k}^{(p+1/6)} = (r_1 e_{i,j,1} - a_1 f_{i,j,1})/\Delta, \quad k = 3,5,\ldots,m-1, \; j = 1,3,\ldots,m,$

$u_{i+1,j,k}^{(p+1/6)} = (-a_1 e_{i,j,1} + r_1 f_{i,j,1})/\Delta, \quad i = 2,4,\ldots,m-1,$　　　　　(4.7.23a)

with

$$u_{1,m+1,k}^{(p)} = u_{1,m-1,k}^{(p)}, \qquad u_{1,m+1,k}^{[N]} = u_{1,m-1,k}^{[N]};$$

$$u_{i,m+1,k}^{(p)} = u_{i,m-1,k}^{(p)}, \qquad u_{i,m+1,k}^{[N]} = u_{i,m-1,k}^{[N]},$$

$$e_{i,j,k} = r_2 u_{i,j,k}^{(p)} - 2a_1(u_{i,j,k-1}^{(p)} + u_{i,j,k+1}^{(p)} + u_{i-1,j,k}^{(p)} + u_{i,j+1,k}^{(p)}) - a_1 u_{i+1,j,k}^{(p)}$$

$$+ 2e_1(u_{i,j,k-1}^{[N]} + u_{i,j,k+1}^{[N]} + u_{i-1,j,k}^{[N]} + u_{i+1,j,k}^{[N]} + u_{i,j+1,k}^{[N]})$$

$$+ 2du_{i,j,k}^{[N]} + 2g_{i,j,k}$$

and

$$f_{i,j,k} = r_2 u_{i+1,j,k}^{(p)} - 2a_1(u_{i+1,j,k-1}^{(p)} + u_{i+1,j,k+1}^{(p)} + u_{i+1,j+1,k}^{(p)}) - a_1 u_{i,j,k}^{(p)}$$

$$+ 2e_1(u_{i+1,j,k-1}^{[N]} + u_{i+1,j,k+1}^{[N]} + u_{i,j,k}^{[N]} + u_{i+1,j+1,k}^{[N]})$$

$$+ 2du_{i+1,j,k}^{[N]} + 2g_{i+1,j,k};$$

(2)　$u_{m,j,k}^{(p+1/6)} = [r_2 u_{m,j,k}^{(p)} - 2a_1(u_{m,j,k-1}^{(p)} + u_{m,j,k+1}^{(p)} + u_{m,j-1,k}^{(p)} + u_{m-1,j,k}^{(p)}$

$$+ u_{m,j+1,k}^{(p)}) + 2e_1(u_{m,j,k-1}^{[N]} + u_{m,j,k+1}^{[N]} + u_{m-1,j-1,k}^{[N]}$$

$$+ u_{m,j+1,k}^{[N]} + u_{1,j+1,k}^{[N]}) + 2du_{m,j,k}^{[N]} + 2g_{m,j,k}]/r_1,$$

$u_{i,j,k}^{(p+1/6)} = (r_1 \bar{e}_{i,j,k} - a_1 \bar{f}_{i,j,k})/\Delta, \quad k = 3,5,\ldots,m-2, \; j = 2,4,\ldots,m-1,$

$u_{i+1,j,k}^{(p+1/6)} = (-a_1 \bar{e}_{i,j,k} + r_1 \bar{f}_{i,j,k})/\Delta, \quad i = 1,3,\ldots,m-2,$　　　　　(4.7.23b)

where

$$\bar{e}_{i,j,k} = r_2 u_{i,j,k}^{(p)} - 2a_1(u_{i,j,k-1}^{(p)} + u_{i,j,k+1}^{(p)} + u_{i,j-1,k}^{(p)} + u_{i,j+1,k}^{(p)}) - a_1 u_{i+1,j,k}^{(p)}$$

$$+ 2e_1(u_{i,j,k-1}^{[N]} + u_{i,j,k+1}^{[N]} + u_{i,j-1,k}^{[N]} + u_{i+1,j,k}^{[N]} + u_{i,j+1,k}^{[N]})$$

$$+ 2du_{i,j,k}^{[N]} + 2g_{i,j,k},$$

and

$$\bar{f}_{i,j,k} = r_2 u^{(p)}_{i+1,j,k}$$

$$- 2a_1(u^{(p)}_{i+1,j,k-1} + u^{(p)}_{i+1,j,k+1} + u^{(p)}_{i+1,j-1,k} + u^{(p)}_{i+2,j,k} + u^{(p)}_{i+1,j-1,k}) - a_1 u^{(p)}_{i,j,k}$$

$$+ 2e_1(u^{[N]}_{i+1,j,k-1} + u^{[N]}_{i+1,j,k+1} + u^{[N]}_{i+1,j-1,k} + u^{[N]}_{i,j,k} + u^{[N]}_{i+2,j,k} + u^{[N]}_{i+1,j+1,k})$$

$$+ 2du^{[N]}_{i+1,j,k} + 2g_{i+1,j,k}.$$

(d) for equation (4.7.20d):

(1) $\quad u^{(p+1/6)}_{1,j,k} = [r_2 u^{(p)}_{1,j,m} - 2a_1(u^{(p)}_{1,j,m-1} + u^{(p)}_{2,j,m} + u^{(p)}_{1,j+1,m})$

$$+ 2e_1(u^{[N]}_{1,j,m-1} + u^{[N]}_{2,j,m} + u^{[N]}_{1,j+1,m})$$

$$+ 2du^{[N]}_{1,j,m} + 2g_{1,j,m}]/r_1, \quad j = 1,3,\dots,m,$$

$$u^{(p+1/6)}_{i,j,m} = (r_1 p_{i,j,m} - a_1 q_{i,j,m})/\Delta,$$

$$u^{(p+1/6)}_{i+1,j,m} = (-a_1 p_{i,j,m} + r_1 q_{i,j,m})/\Delta, \quad i = 1,3,\dots,m, \ i = 2,4,\dots,m-1,$$

$$(4.7.24a)$$

with

$$u^{(p)}_{1,m+1,m} = u^{(p)}_{1,m-1,m}, \qquad u^{[N]}_{1,m+1,m} = u^{[N]}_{1,m-1,m};$$

$$u^{(p)}_{i,m+1,m} = u^{(p)}_{i,m-1,m}, \qquad u^{[N]}_{i,m+1,m} = u^{[N]}_{i,m-1,m},$$

$$p_{i,j,m} = r_2 u^{(p)}_{i,j,m} - 2a_1(u^{(p)}_{i,j,m-1} + u^{(p)}_{i-1,j,m} + u^{(p)}_{i,j+1,m}) - a_1 u^{(p)}_{i+1,j,m}$$

$$+ 2e_1(u^{[N]}_{i,j,m-1} + u^{[N]}_{i+1,j,m} + u^{[N]}_{i-1,j,m} + u^{[N]}_{i,j+1,m})$$

$$+ 2du^{[N]}_{i,j,m} + 2g_{i,j,m}$$

and

$$q_{i,j,m} = r_2 u^{(p)}_{i+1,j,m} - 2a_1(u^{(p)}_{i+1,j,m-1} + u^{(p)}_{i+1,j+1,m})$$

$$+ 2e_1(u^{[N]}_{i+1,j,n-1} + u^{[N]}_{i,j,m} + u^{[N]}_{i+1,j+1,m})$$

$$+ 2du^{[N]}_{i+1,j,m} + 2g_{i+1,j,m};$$

(2) $\quad u^{(p+1/6)}_{m,j,m} = [r_2 u^{(p)}_{m,j,m} - 2a_1(u^{(p)}_{m,j,m-1} + u^{(p)}_{m,j-1,m} + u^{(p)}_{m-1,j,m} + u^{(p)}_{m,j+1,m})$

$$+ 2e_1(u^{[N]}_{m,j,m-1} + u^{[N]}_{m,j-1,m} + u^{[N]}_{m-1,j,m} + u^{[N]}_{m,j+1,m})$$

$$+ 2du^{[N]}_{m,j,m} + 2g_{m,j,m}]/r_1,$$

$$u^{(p+1/6)}_{i,j,m} = (r_1 \bar{p}_{i,j,m} - a_1 \bar{q}_{i,j,m})/\Delta, \quad j = 2,4,\dots,m-1,$$

$$u^{(p+1/6)}_{i+1,j,m} = (-a_1 \bar{p}_{i,j,m} + r_1 \bar{q}_{i,j,m})/\Delta, \quad i = 1,3,\dots,m-2, \qquad (4.7.24b)$$

where

$$\bar{p}_{i,j,m} = r_2 u_{i,j,m}^{(p)} - 2a_1(u_{i,j,m-1}^{(p)} + u_{i,j-1,m}^{(p)} + u_{i,j+1,m}^{(p)}) - a_1 u_{i+1,j,m}^{(p)}$$

$$+ 2e_1(u_{i,j,m-1}^{[N]} + u_{i,j-1,m}^{[N]} + u_{i+1,j,m}^{[N]} + u_{i,j+1,m}^{[N]})$$

$$+ 2du_{i,j,m}^{[N]} + 2g_{i,j,m}$$

and

$$\bar{q}_{i,j,m} = r_2 u_{i+1,j,m}^{(p)} - 2a_1(u_{i+1,j,m-1}^{(p)} + u_{i+1,j-1,m}^{(p)} + u_{i+2,j,m}^{(p)}$$

$$+ u_{i+1,j+1,m}^{(p)}) - a_1 u_{i,j,m}^{(p)} + 2e_1(u_{i+1,j,m-1}^{[N]} + u_{i+1,j-1,m}^{[N]}$$

$$+ u_{i+1,j,m}^{[N]} + u_{i+1,j,m}^{[N]} + u_{i+1,j+1,m}^{[N]})$$

$$+ 2du_{i+1,j,m}^{[N]} + 2g_{i+1,j,m}.$$

(ii) *At the second intermediate level (the* $(p+1/3)^{th}$ *iterate*): From the second equation of (4.7.9) we have

$$(G_2 + rI)\underline{u}_{[xy]}^{(p+1/3)} = G_2 \underline{u}_{[xy]}^{(p)} + r\underline{u}_{[xy]}^{(p+1/6)},$$

which gives

$$\underline{u}_{[xy]}^{(p+1/3)} = (G_2 + rI)^{-1}\{G_2 \underline{u}_{[xy]}^{(p)} + r\underline{u}_{[xy]}^{(p+1/6)}\}. \qquad (4.7.25)$$

Now

$$(rI + G_2) = \begin{bmatrix} \bar{C}_2 & & & & & \\ & \bar{C}_1 & & & & \\ & & \bar{C}_2 & & 0 & \\ & & & & & \\ & & 0 & & \bar{C}_2 & \\ & & & & & \bar{C}_1 & \\ & & & & & & \bar{C}_2 \end{bmatrix}_{(m^3 \times m^3)}$$

By letting $\hat{\bar{C}}_2 \equiv \bar{C}_2$ with diagonal elements r_1 be replaced by $c/6$ and $\hat{\bar{C}}_1 \equiv \bar{C}_1$ with diagonal elements r_1 be replaced by $c/6$ we find from (4.7.25) that

$$u_{k[xy]}^{(p+1/3)} = \begin{cases} \bar{C}_2^{-1}\{\hat{\bar{C}}_2 u_{k[xy]}^{(p)} + r u_{k[xy]}^{(p+1/6)}\} & \text{for } k = 1, 3, \ldots, m, \qquad (4.7.27a) \\[2mm] \bar{C}_1^{-1}\{\hat{\bar{C}}_1 u_{k[xy]}^{(p)} + r u_{k[xy]}^{(p+1/6)}\} & \text{for } k = 2, 4, \ldots, m-1. \qquad (4.7.27b) \end{cases}$$

The computation of the AGE algorithm is carried out as follows:

(1) for equation (4.7.27a):

$$u_{m,j,k}^{(p+1/3)} = \left(\frac{c}{6}u_{m,j,k}^{(p)} + ru_{m,j,k}^{(p+1/6)}\right)\Big/r_1, \quad \text{for } k = 1, 3, \ldots, m; \ j = 1, 3, \ldots, m,$$

$$u_{i,j,k}^{(p+1/3)} = (r_1 a_{i,j,k} - a_1 \bar{a}_{i,j,k})/\Delta,$$

$$u_{i+1,j,k}^{(p+1/3)} = (-a_1 a_{i,j,k} + r_1 \bar{a}_{i,j,k})/\Delta, \quad \text{for } k = 1, 3, \ldots, m; \ j = 1, 3, \ldots, m, \quad (4.7.28a)$$

$$u_{1,j,k}^{(p+1/3)} = \left(\frac{c}{6}u_{1,j,k}^{(p)} + ru_{1,j,k}^{(p+1/6)}\right)\Big/r_1, \quad \text{for } k = 1, 3, \ldots, m; \ j = 2, 4, \ldots, m-1,$$

$$u_{i,j,k}^{(p+1/3)} = (r_1 a_{i,j,k} - a_1 \bar{a}_{i,j,k})/\Delta, \quad k = 1, 3, \ldots, m; \ j = 2, 4, \ldots, m-1,$$

$$u_{i+1,j,k}^{(p+1/3)} = (-a_1 a_{i,j,k} + r_1 \bar{a}_{i,j,k})/\Delta, \quad i = 2, 4, \ldots, m-1.$$

and

(2) for equation (4.7.27b):

$$u_{i,j,k}^{(p+1/3)} = \left(\frac{c}{6}u_{1,j,k}^{(p)} + ru_{1,j,k}^{(p+1/6)}\right)\Big/r_1, \quad \text{for } k = 2, 4, \ldots, m-1; \ j = 1, 3, \ldots, m,$$

$$u_{i,j,k}^{(p+1/3)} = (r_1 a_{i,j,k} - a_1 \bar{a}_{i,j,k})/\Delta, \quad k = 2, 4, \ldots, m-1; \ j = 2, 4, \ldots, m-1,$$

$$u_{i+1,j,k}^{(p+1/3)} = (-a_1 a_{i,j,k} + r_1 \bar{a}_{i,j,k})/\Delta, \quad \text{for } k = 1, 3, \ldots, m-2, \quad (4.7.28b)$$

$$u_{m,j,k}^{(p+1/3)} = \left(\frac{c}{6}u_{m,j,k}^{(p)} + ru_{m,j,k}^{(p+1/6)}\right)\Big/r_1, \quad \text{for } k = 2, 4, \ldots, m-1; \ j = 2, 4, \ldots, m-1,$$

$$u_{i,j,k}^{(p+1/3)} = (r_1 a_{i,j,k} - a_1 \bar{a}_{i,j,k})/\Delta, \quad k = 2, 4, \ldots, m-1; \ j = 2, 4, \ldots, m-1,$$

$$u_{i+1,j,k}^{(p+1/3)} = (-a_1 a_{i,j,k} + r_1 \bar{a}_{i,j,k})/\Delta, \quad i = 1, 3, \ldots, m-2,$$

where

$$a_{i,j,k} = \frac{c}{6}u_{i,j,k}^{(p)} + a_1 u_{i+1,j,k}^{(p)} + ru_{i,j,k}^{(p+1/6)}$$

and

$$\bar{a}_{i,j,k} = a_1 u_{i,j,k}^{(p)} + \frac{c}{6}u_{i+1,j,k}^{(p)} + ru_{i+1,j,k}^{(p+1/6)}.$$

If we take our approximations as sweeps parallel to the yz-plane the u values are then evaluated at points lying on planes which are parallel to the yz-plane and on each of these planes, the points are *re-ordered row-wise* (parallel to the y-axis, as in Figure 4.7.2), such that

$$(G_3 + G_4)\underline{u}_{[xy]} = (G_1 + G_2)\underline{u}_{[yz]}, \quad (4.7.29)$$

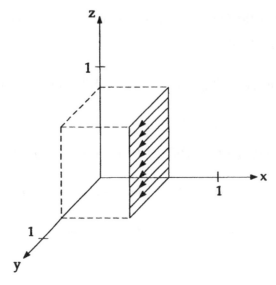

Figure 4.7.2

where

$$\underline{u}_{[yz]} = (u_{1[yz]}, u_{2[yz]}, \ldots, u_{m[yz]})^T,$$

where

$$u_{i[yz]} = (u_{i,1,1}, u_{i,2,1}, \ldots, u_{i,m,1}, u_{i,1,2}, u_{i,2,2}, \ldots, u_{i,m,2}, \ldots, u_{i,m})^T, \quad \text{for } i = 1, 2, \ldots, m.$$

With the above reordering of points we are able to derive the computational formulae for the third and fourth intermediate levels as follows:

(iii) *At the third intermediate level (the $(p+1/2)^{th}$ iterate)*: The third equation of (4.7.9) is transformed to

$$(G_1 + rI)\underline{u}_{[yz]}^{(p+1/2)} = G_1 \underline{u}_{[yz]}^{(p)} + r\underline{u}_{[yz]}^{(p+1/3)},$$

or,

$$\underline{u}_{[yz]}^{(p+1/2)} = (G_1 + rI)^{-1}[G_1 \underline{u}_{[yz]}^{(p)} + r\underline{u}_{[yz]}^{(p+1/3)}], \tag{4.7.30}$$

which leads to

$$u_{k[yz]}^{(p+1/2)} = \begin{cases} \bar{C}_1^{-1}[\hat{\bar{C}}_1 u_{i[yz]}^{(p)} + r u_{i[yz]}^{(p+1/3)}] & \text{for } k = 1, 3, \ldots, m, \tag{4.7.31a} \\ \bar{C}_2^{-1}[\hat{\bar{C}}_2 u_{i[yz]}^{(p)} + r u_{i[yz]}^{(p+1/3)}] & \text{for } k = 2, 4, \ldots, m-1. \tag{4.7.31b} \end{cases}$$

Hence we find that:

(a) for equation (4.7.31a):

$$u_{i,1,k}^{(p+1/2)} = \left(\frac{c}{6} u_{i,1,k}^{(p)} + r u_{i,1,k}^{(p+1/3)}\right) \Big/ r_1, \quad \text{for } k = 1, 3, \ldots, m; \; k = 1, 3, \ldots, m,$$

$$u_{i,j,k}^{(p+1/2)} = (r_1 b_{i,j,k} - a_1 \bar{b}_{i,j,k})/\Delta, \quad i = 1, 3, \ldots, m; \; k = 1, 3, \ldots, m;$$

$$u_{i,j+1,k}^{(p+1/2)} = (-a_1 b_{i,j,k} + r_1 \bar{b}_{i,j,k})/\Delta, \quad j = 2, 4, \ldots, m-1, \tag{4.7.32a}$$

$$u_{i,m,k}^{(p+1/2)} = \left(\frac{c}{6}u_{i,m,k}^{(p)} + ru_{i,m,k}^{(p+1/3)}\right)\bigg/r_1, \quad \text{for } k = 1, 3, \ldots, m; \; j = 2, 4, \ldots, m-1,$$

$$u_{i,j,k}^{(p+1/2)} = (r_1 b_{i,j,k} - a_1 \bar{b}_{i,j,k})/\Delta, \qquad i = 1, 3, \ldots, m; \; k = 2, 4, \ldots, m-1,$$

$$u_{i,j+1,k}^{(p+1/2)} = (-a_1 g_{i,j,k} + r_1 \bar{b}_{i,j,k})/\Delta, \quad j = 1, 3, \ldots, m-2.$$

and

(b) for equation (4.7.31b):

$$u_{i,m,k}^{(p+1/2)} = \left(\frac{c}{6}u_{i,m,k}^{(p)} + ru_{i,m,k}^{(p+1/3)}\right)\bigg/r_1, \quad \text{for } i = 2, 4, \ldots, m-1; \; k = 1, 3, \ldots, m,$$

$$u_{i,j,k}^{(p+1/2)} = (r_1 b_{i,j,k} - a_1 \bar{b}_{i,j,k})/\Delta, \qquad i = 2, 4, \ldots, m-1; \; k = 1, 3, \ldots, m;$$

$$u_{i,j+1,k}^{(p+1/2)} = (-a_1 b_{i,j,k} + r_1 \bar{b}_{i,j,k})/\Delta, \quad j = 1, 3, \ldots, m-2, \tag{4.7.32b}$$

$$u_{i,1,k}^{(p+1/2)} = \left(\frac{c}{6}u_{i,1,k}^{(p)} + ru_{i,1,k}^{(p+1/3)}\right)\bigg/r_1, \quad \text{for } i = 2, 4, \ldots, m-1; \; k = 2, 4, \ldots, m-1,$$

$$u_{i,j,k}^{(p+1/2)} = (r_1 b_{i,j,k} - a_1 \bar{b}_{i,j,k})/\Delta, \qquad i = 2, 4, \ldots, m-1; \; k = 1, 3, \ldots, m;$$

$$u_{i,j+1,k}^{(p+1/2)} = (-a_1 g_{i,j,k} + r_1 \bar{b}_{i,j,k})/\Delta, \quad j = 2, 4, \ldots, m-1,$$

where

$$b_{i,j,k} = \frac{c}{6}u_{i,j,k}^{(p)} + a_1 u_{i,j+1,k}^{(p)} + ru_{i,j,k}^{(p+1/3)}$$

and

$$\bar{b}_{i,j,k} = a_1 u_{i,j,k}^{(p)} + \frac{c}{6}u_{i,j+1,k}^{(p)} + ru_{i,j+1,k}^{(p+1/3)}.$$

(iv) *At the fourth intermediate level (the $(p+2/3)^{th}$ iterate):* The fourth equation of (4.7.9) is transformed to

$$(G_2 + rI)\underline{u}_{[yz]}^{(p+2/3)} = G_2 \underline{u}_{[yz]}^{(p)} + r\underline{u}_{[yz]}^{(p+1/2)},$$

or

$$\underline{u}_{[yz]}^{(p+2/3)} = (G_2 + rI)^{-1}[G_2 \underline{u}_{[yz]}^{(p)} + r\underline{u}_{[yz]}^{(p+1/2)}], \tag{4.7.33}$$

which leads to

$$u_{k[yz]}^{(p+2/3)} = \begin{cases} \bar{C}_2^{-1}[\hat{\bar{C}}_2 u_{i[yz]}^{(p)} + ru_{i[yz]}^{(p+1/2)}] & \text{for } i = 1, 3, \ldots, m, \tag{4.7.34a} \\ \bar{C}_1^{-1}[\hat{\bar{C}}_1 u_{i[yz]}^{(p)} + ru_{i[yz]}^{(p+1/2)}] & \text{for } i = 2, 4, \ldots, m-1. \tag{4.7.34b} \end{cases}$$

Therefore we obtain:

(a) for equation (4.7.34a):

$$u_{i,m,k}^{(p+2/3)} = \left(\frac{c}{6}u_{i,m,k}^{(p)} + ru_{i,m,k}^{(p+1/2)}\right)\bigg/r_1, \quad \text{for } k = 1, 3, \ldots, m; \; k = 1, 3, \ldots, m,$$

$$u_{i,j,k}^{(p+2/3)} = (r_1 c_{i,j,k} - a_1 \bar{c}_{i,j,k})/\Delta, \qquad i = 1, 3, \ldots, m; \; k = 1, 3, \ldots, m;$$

$$u_{i,j+1,k}^{(p+2/3)} = (-a_1 c_{i,j,k} + r_1 \bar{c}_{i,j,k})/\Delta, \quad j = 1, 3, \ldots, m-2, \tag{4.7.35a}$$

$$u_{i,1,k}^{(p+2/3)} = \left(\frac{c}{6}u_{i,1,k}^{(p)} + ru_{i,1,k}^{(p+1/2)}\right)\Big/r_1, \quad \text{for } i = 1,3,\dots,m; \ k = 2,4,\dots,m-1,$$

$$u_{i,j,k}^{(p+2/3)} = (r_1 c_{i,j,k} - a_1 \bar{c}_{i,j,k})/\Delta, \qquad i = 1,3,\dots,m; \ k = 2,4,\dots,m-1,$$

$$u_{i,j+1,k}^{(p+2/3)} = (-a_1 c_{i,j,k} + r_1 \bar{c}_{i,j,k})/\Delta, \qquad j = 2,4,\dots,m-1;$$

and

(b) for equation (4.7.35b):

$$u_{i,1,k}^{(p+2/3)} = \left(\frac{c}{6}u_{i,1,k}^{(p)} + ru_{i,1,k}^{(p+1/2)}\right)\Big/r_1, \quad \text{for } i = 2,4,\dots,m-1; \ k = 1,3,\dots,m,$$

$$u_{i,j,k}^{(p+2/3)} = (r_1 c_{i,j,k} - a_1 \bar{c}_{i,j,k})/\Delta, \qquad i = 2,4,\dots,m-1; \ k = 1,3,\dots,m;$$

$$u_{i,j+1,k}^{(p+2/3)} = (-a_1 c_{i,j,k} + r_1 \bar{c}_{i,j,k})/\Delta, \qquad j = 2,4,\dots,m-1, \hspace{2cm} (4.7.35\text{b})$$

$$u_{i,m,k}^{(p+2/3)} = \left(\frac{c}{6}u_{i,m,k}^{(p)} + ru_{i,m,k}^{(p+1/2)}\right)\Big/r_1, \quad \text{for } i = 2,4,\dots,m-1; \ k = 2,4,\dots,m-1,$$

$$u_{i,j,k}^{(p+2/3)} = (r_1 c_{i,j,k} - a_1 \bar{c}_{i,j,k})/\Delta, \qquad i = 2,4,\dots,m-1; \ k = 2,4,\dots,m-1;$$

$$u_{i,j+1,k}^{(p+2/3)} = (-a_1 c_{i,j,k} + r_1 \bar{c}_{i,j,k})/\Delta, \qquad j = 1,3,\dots,m-2,$$

where

$$c_{i,j,k} = \frac{c}{6}u_{i,j,k}^{(p)} + a_1 u_{i,j+1,k}^{(p)} + ru_{i,j,k}^{(p+1/2)}$$

and

$$\bar{c}_{i,j,k} = a_1 u_{i,j,k}^{(p)} + \frac{c}{6}u_{i,j+1,k}^{(p)} + ru_{i,j+1,k}^{(p+1/2)}.$$

To determine the AGE equations at the fifth and sixth intermediate levels, it is necessary that we consider our approximations as sweeps parallel to the *xz*-plane and then evaluate the *u* value at points lying on each of these planes as illustrated in Figure 4.7.3.

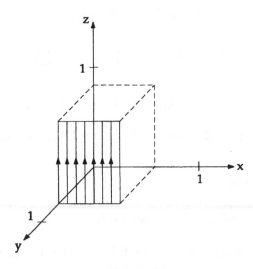

Figure 4.7.3

In this case, the points are *reordered column-wise* (parallel to the z-axis) such that,

$$(G_5 + G_6)\underline{u}_{[xy]} = (G_1 + G_2)\underline{u}_{[xz]} \tag{4.7.36}$$

where

$$\underline{u}_{[xz]} = (u_{1[xz]}, u_{2[xz]}, \ldots, u_{m[xz]})^T,$$

with

$$u_{j[xz]} = (u_{1,j,1}, u_{1,j,2}, \ldots, u_{1,j,m}, u_{2,j,1}, u_{2,j,2}, \ldots, u_{2,j,m}, \ldots, u_{m,j,1}, u_{m,j,2}, \ldots, u_{m,j,m})^T,$$

$$\text{for } j = 1, 2, \ldots, m.$$

With the above reordering of points, the fifth and sixth equations are transformed to

$$(G_1 + rI)\underline{u}_{[xz]}^{(p+5/6)} = G_1\underline{u}_{[xz]}^{(p)} + r\underline{u}_{[xz]}^{(2/3)} \tag{4.7.37}$$

and

$$(G_2 + rI)\underline{u}_{[xz]}^{(p+1)} = G_2\underline{u}_{[xz]}^{(p)} + r\underline{u}_{[xz]}^{(p+5/6)}, \tag{4.7.38}$$

respectively. We now derive the computational formulae at the fifth and sixth intermediate levels.

(v) *At the fifth intermediate level (the* $(p+5/6)^{th}$ *iterate)*: From the equation (4.7.37) we have,

$$\underline{u}_{[xz]}^{(p+5/6)} = (G_1 + rI)^{-1}[G_1\underline{u}_{[xz]}^{(p)} + r\underline{u}_{[xz]}^{(2/3)}, \tag{4.7.39}$$

which leads to

$$u_{j[xz]}^{(p+5/6)} = \begin{cases} \bar{C}_1^{-1}[\hat{\bar{C}}_1 u_{j[xz]}^{(p)} + r u_{j[xz]}^{(p+2/3)}] & \text{for } k = 1, 3, \ldots, m, \tag{4.7.40a} \\ \bar{C}_2^{-1}[\hat{\bar{C}}_2 u_{j[xz]}^{(p)} + r u_{j[xz]}^{(p+2/3)}] & \text{for } k = 2, 4, \ldots, m-1. \tag{4.7.40b} \end{cases}$$

Therefore, we obtain:

(a) for equation (4.7.40a):

$$u_{i,j,1}^{(p+5/6)} = \left(\frac{c}{6}u_{i,j,1}^{(p)} + ru_{i,j,1}^{(p+2/3)}\right)\bigg/r_1, \quad \text{for } j = 1, 3, \ldots, m; \ i = 1, 3, \ldots, m,$$

$$u_{i,j,k}^{(p+5/6)} = (r_1 d_{i,j,k} - a_1 \bar{d}_{i,j,k})/\Delta, \quad j = 1, 3, \ldots, m; \ i = 1, 3, \ldots, m;$$

$$u_{i,j+1,k}^{(p+5/6)} = (-a_1 d_{i,j,k} + r_1 \bar{d}_{i,j,k})/\Delta, \quad k = 2, 4, \ldots, m-1, \tag{4.7.41a}$$

$$u_{i,j,m}^{(p+5/6)} = \left(\frac{c}{6}u_{i,j,m}^{(p)} + ru_{i,j,m}^{(p+2/3)}\right)\bigg/r_1, \quad \text{for } j = 1, 3, \ldots, m; \ i = 2, 4, \ldots, m-1,$$

$$u_{i,j,k}^{(p+5/6)} = (r_1 d_{i,j,k} - a_1 \bar{d}_{i,j,k})/\Delta, \quad j = 1, 3, \ldots, m; \ i = 2, 4, \ldots, m-1,$$

$$u_{i,j+1,k}^{(p+5/6)} = (-a_1 d_{i,j,k} + r_1 \bar{d}_{i,j,k})/\Delta, \quad k = 1, 3, \ldots, m-1;$$

and

(b) for equation (4.7.40b):

$$u_{i,j,m}^{(p+5/6)} = \left(\frac{c}{6}u_{i,j,m}^{(p)} + ru_{i,j,m}^{(p+2/3)}\right)\bigg/r_1, \quad \text{for } j = 2, 4, \ldots, m-1; \ i = 1, 3, \ldots, m,$$

$$u_{i,j,k}^{(p+5/6)} = (r_1 d_{i,j,k} - a_1 \bar{d}_{i,j,k})/\Delta, \quad j = 2, 4, \ldots, m-1; \ i = 1, 3, \ldots, m;$$

$$u_{i,j+1,k}^{(p+5/6)} = (-a_1 d_{i,j,k} + r_1 \bar{d}_{i,j,k})/\Delta, \quad k = 1, 3, \ldots, m-2, \tag{4.7.41b}$$

$$u_{i,j,1}^{(p+5/6)} = \left(\frac{c}{6}u_{i,j,1}^{(p)} + ru_{i,j,1}^{(p+2/3)}\right)\Big/r_1, \quad \text{for } j = 2, 4, \ldots, m-1; \ i = 2, 4, \ldots, m-1,$$

$$u_{i,j,k}^{(p+5/6)} = (r_1 d_{i,j,k} - a_1 \bar{d}_{i,j,k})/\Delta, \qquad j = 2, 4, \ldots, m-1; \ i = 2, 4, \ldots, m-1;$$

$$u_{i,j,k+1}^{(p+5/6)} = (-a_1 d_{i,j,k} + r_1 \bar{d}_{i,j,k})/\Delta, \qquad k = 2, 4, \ldots, m-1,$$

where

$$d_{i,j,k} = \frac{c}{6}u_{i,j,k}^{(p)} + a_1 u_{i,j,k+1}^{(p)} + ru_{i,j,k}^{(p+2/3)}$$

and

$$\bar{d}_{i,j,k} = a_1 u_{i,j,k}^{(p)} + \frac{c}{6}u_{i,j,k+1}^{(p)} + ru_{i,j,k+1}^{(p+2/3)}.$$

(vi) *At the sixth intermediate level (the $(p+1)^{th}$ level)*: From equation (4.7.38), we have

$$\underline{u}_{[xz]}^{(p+1)} = (G_2 + rI)^{-1}[G_2 \underline{u}_{[xz]}^{(p)} + r\underline{u}_{[xz]}^{(5/6)}], \tag{4.7.42}$$

which leads to

$$u_{j[xz]}^{(p+1)} = \begin{cases} \bar{C}_2^{-1}[\hat{\bar{C}}_2 u_{j[xz]}^{(p)} + ru_{j[xz]}^{(p+5/6)}] & \text{for } j = 1, 3, \ldots, m, \tag{4.7.43a} \\ \bar{C}_1^{-1}[\hat{\bar{C}}_1 u_{j[xz]}^{(p)} + ru_{j[xz]}^{(p+5/6)}] & \text{for } j = 2, 4, \ldots, m-1. \tag{4.7.43b} \end{cases}$$

Therefore, we obtain:

(a) for equation (4.7.43a):

$$u_{i,j,m}^{(p+1)} = \left(\frac{c}{6}u_{i,j,m}^{(p)} + ru_{i,j,m}^{(p+5/6)}\right)\Big/r_1, \quad \text{for } j = 1, 3, \ldots, m; \ i = 1, 3, \ldots, m,$$

$$u_{i,j,k}^{(p+1)} = (r_1 e_{i,j,k} - a_1 \bar{e}_{i,j,k})/\Delta, \qquad j = 1, 3, \ldots, m; \ i = 1, 3, \ldots, m;$$

$$u_{i,j+1,k}^{(p+1)} = (-a_1 e_{i,j,k} + r_1 \bar{e}_{i,j,k})/\Delta, \quad k = 2, 4, \ldots, m-2, \tag{4.7.44a}$$

$$u_{i,j,1}^{(p+1)} = \left(\frac{c}{6}u_{i,j,1}^{(p)} + ru_{i,j,1}^{(p+5/6)}\right)\Big/r_1, \quad \text{for } j = 1, 3, \ldots, m; \ i = 2, 4, \ldots, m-1,$$

$$u_{i,j,k}^{(p+1)} = (r_1 e_{i,j,k} - a_1 \bar{e}_{i,j,k})/\Delta, \qquad j = 1, 3, \ldots, m; \ i = 2, 4, \ldots, m-1,$$

$$u_{i,j+1,k}^{(p+1)} = (-a_1 e_{i,j,k} + r_1 \bar{e}_{i,j,k})/\Delta, \quad k = 1, 3, \ldots, m-1;$$

and

(b) for equation (4.7.43b):

$$u_{i,j,1}^{(p+1)} = \left(\frac{c}{6}u_{i,j,1}^{(p)} + ru_{i,j,1}^{(p+5/6)}\right)\Big/r_1, \quad \text{for } j = 2, 4, \ldots, m-1, \ i = 1, 3, \ldots, m,$$

$$u_{i,j,k}^{(p+1)} = (r_1 e_{i,j,k} - a_1 \bar{e}_{i,j,k})/\Delta, \qquad j = 2, 4, \ldots, m-1; \ i = 1, 3, \ldots, m;$$

$$u_{i,j,k+1}^{(p+1)} = (-a_1 e_{i,j,k} + r_1 \bar{e}_{i,j,k})/\Delta, \quad k = 2, 4, \ldots, m-1, \tag{4.7.44b}$$

$$u_{i,j,m}^{(p+1)} = \left(\frac{c}{6}u_{i,j,m}^{(p)} + ru_{i,j,m}^{(p+5/6)}\right)\bigg/r_1, \quad \text{for } j = 2,4,\dots,m-1; \; i = 2,4,\dots,m-1,$$

$$u_{i,j,k}^{(p+1)} = (r_1 e_{i,j,k} - a_1 \bar{e}_{i,j,k})/\Delta, \qquad j = 2,4,\dots,m-1; \; i = 2,4,\dots,m-1;$$

$$u_{i,j,k+1}^{(p+1)} = (-a_1 e_{i,j,k} + r_1 \bar{e}_{i,j,k})/\Delta, \quad k = 1,3,\dots,m-2,$$

where

$$e_{i,j,k} = \frac{c}{6}u_{i,j,k}^{(p)} + a_1 u_{i,j,k+1}^{(p)} + ru_{i,j,k}^{(p+5/6)}$$

and

$$\bar{e}_{i,j,k} = a_1 u_{i,j,k}^{(p)} + \frac{c}{6}u_{i,j,k+1}^{(p)} + ru_{i,j,k+1}^{(p+5/6)}.$$

Thus, we see that the AGE algorithm corresponds to sweeping through the mesh lying on planes (where the appropriate reordering of the points are done) parallel to the xy, yz and xz-planes. This process involves tridiagonal systems which in turn entails at each stage the solution of (2×2) block systems. The iterative procedure is continued until convergence is reached, that is, when the requirement

$$|u_{i,j,k}^{(p+1)} - u_{i,j,k}^{(p)}| \leq \varepsilon,$$

is met, where ε is the convergence criterion. This then yields the solution at the $(N+1)^{st}$ time step.

4.8. The AGE Method for Solving Higher-Order Partial Differential Equations

4.8.1. Introduction

Consider the fourth-order parabolic partial differential equation,

$$\frac{\partial^2 U}{\partial t^2} + \frac{\partial^4 U}{\partial x^4} = 0, \quad 0 \leq x \leq 1, \; t > 0, \tag{4.8.1a}$$

subject to the initial conditions

$$U(x,0) = g_0(x)$$

and

$$\frac{\partial U}{\partial t}(x,0) = g_1(x), \quad \text{for } 0 \leq x \leq 1, \tag{4.8.1b}$$

and with boundary conditions at $x = 0$ and 1 of the form

$$U(0,t) = f_0(t), \qquad U(1,t) = f_1(t),$$

and

$$\frac{\partial^2 U}{\partial x^2}(0, t) = p_0(t), \quad \frac{\partial^2 U}{\partial x^2}(1, t) = p_1(t).$$ (4.8.1c)

Various finite difference approximations for the numerical solution of (4.8.1) have been proposed by Crandall (1955), Conte and Royster (1956), Conte (1954) and Albrecht (1957). With the introduction of two new variables Φ and Ψ defined by

$$\Phi = \frac{\partial U}{\partial t} \quad \text{and} \quad \Psi = \frac{\partial^2 U}{\partial x^2},$$ (4.8.2)

equation (4.8.1a) can now be written as two simultaneous partial equations of the form

$$\frac{\partial \Phi}{\partial t} = -\frac{\partial^2 \Psi}{\partial x^2} \quad \text{and} \quad \frac{\partial \Psi}{\partial t} = \frac{\partial^2 \Phi}{\partial x^2}$$ (4.8.3)

Richtmyer, (1967) or as the second-order system

$$\frac{\partial \Omega}{\partial t} = A \frac{\partial^2 \Omega}{\partial x^2},$$ (4.8.4)

where

$$\Omega = \begin{bmatrix} \phi \\ \psi \end{bmatrix} \quad \text{and} \quad A = \begin{bmatrix} 0 & -1 \\ 1 & 0 \end{bmatrix}.$$

Evans (1965) derived finite difference methods for the numerical solution of equation (4.8.4), while Fairweather and Gourlay (1966) derived explicit and implicit finite difference methods for the numerical solution of equation (4.8.4) based on the semi-explicit method of Lees (1966) and the high accuracy method of Douglas (1961) respectively.

In this paper we follow the Conte scheme (1954) where a stable implicit finite difference approximation is presented, the scheme is based on the 11 points indicated in Figure 4.8.1.

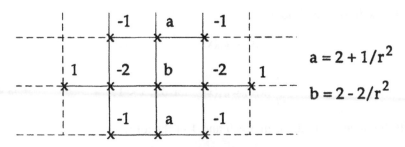

Figure 4.8.1 The 11 points scheme.

If we let $u_{i,k} = u(i\Delta x, k\Delta t)$, with Δx and Δt the grid spacings on the rectangular network, the finite difference equations at the point (i, k) on the mesh can be represented as follows:

$$u_{i+2,k} - (u_{i+1,k+1} + 2u_{i+1,k} + u_{i+1,k-1}) + 2(u_{i,k+1} + u_{i,k} + u_{i,k-1})$$

$$- (u_{i-1,k+1} + 2u_{i-1,k} + u_{i-1,k-1}) + u_{i-2,k} + (u_{i,k+1} - 2u_{i,k} + u_{i,k-1})/r^2 = 0,$$

$$(4.8.5)$$

or

$$-u_{i-1,k+1} + (2 + 1/r^2)u_{i,k+1} - u_{i+1,k+1}$$

$$= -u_{i+2,k} + 2u_{i+1,k} - 2(1 - 1/r^2)u_{i,k} + 2u_{i-1,k} - u_{i-2,k}$$

$$+ u_{i+1,k-1} - (2 + 1/r^2)u_{i,k-1} + u_{i-1,k-1},$$

$$(4.8.6)$$

where $r = \Delta t/\Delta x^2$. This scheme is unconditionally stable, and has local truncation error of $0(\Delta x^2)$. However, it is a three level scheme and so requires starting values on $t = \Delta t$ as well as on $t = 0$ which are usually specified in the initial conditions.

4.8.2. The AGE Method

Equation (4.8.6) can be expressed in matrix form as

$$A\underline{u}^{[k+1]} = B\underline{u}^{[k]} + C\underline{u}^{[k-1]} + D,$$

$$(4.8.7a)$$

i.e.,

$$A\underline{u}^{[k+1]} = \underline{b},$$

$$(4.8.7b)$$

where $\underline{b} = B\underline{u}^{[k]} + C\underline{u}^{[k-1]} + D, \underline{u}^{[k]}$ and $\underline{u}^{[k-1]}$ are the known $\underline{u}$ values at time levels k and $k-1$, respectively, $\underline{u} = (u_1, u_2, \ldots, u_n)^T$ and the coefficient matrix A is tridiagonal. This system can be solved using the Alternating Group Explicit (AGE) iterative method.

The AGE method can be applied as follows: Assume that the matrix A is of the form

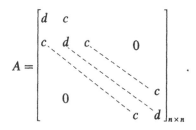

Split the matrix A into the sum of two matrices G_1 and G_2 such that

$$A = G_1 + G_2,$$

$$(4.8.8)$$

where, if n is even, we have

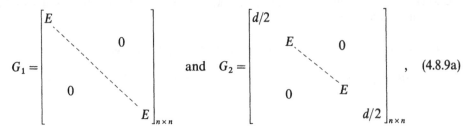

$$G_1 = \begin{bmatrix} E & & & \\ & \ddots & & 0 \\ & & \ddots & \\ 0 & & & \ddots \\ & & & & E \end{bmatrix}_{n \times n} \quad \text{and} \quad G_2 = \begin{bmatrix} d/2 & & & \\ & E & & 0 \\ & & \ddots & \\ 0 & & & E \\ & & & & d/2 \end{bmatrix}_{n \times n} , \qquad (4.8.9a)$$

and if n is odd then,

$$G_1 = \begin{bmatrix} d/2 & & & \\ & E & & 0 \\ & & \ddots & \\ 0 & & & \ddots \\ & & & & E \end{bmatrix}_{n \times n} \quad \text{and} \quad G_2 = \begin{bmatrix} E & & & \\ & \ddots & & 0 \\ & & \ddots & \\ 0 & & & E \\ & & & & d/2 \end{bmatrix}_{n \times n} , \qquad (4.8.9b)$$

where

$$E = \begin{bmatrix} d/2 & c \\ c & d/2 \end{bmatrix}.$$

G_1 and G_2 satisfies the condition that $G_1 + \rho I$ and $G_2 + \rho I$ are non-singular for any $\rho > 0$. By using equation (4.8.8) the matrix equation (4.8.7b) can now be written in the form

$$(G_1 + G_2)\underline{u}^{[k+1]} = \underline{b}. \qquad (4.8.10)$$

The AGE method is then defined by the equations

$$(G_1 + \rho I)\underline{u}^{(p+1/2)} = \underline{b} - (G_2 - \rho I)\underline{u}^{(p)}, \qquad (4.8.11a)$$

$$(G_2 + \rho I)\underline{u}^{(p+1)} = \underline{b} - (G_1 - \rho I)\underline{u}^{(p+1/2)}, \qquad (4.8.11b)$$

where ρ is an acceleration parameter and p is the iteration counter. It has been shown that the optimal parameter ρ is dependent on the upper and lower bounds for the eigenvalues of G_1 and G_2. Equations (4.8.11) can be written in explicit form as

$$\underline{u}^{(p+1/2)} = (G_1 + \rho I)^{-1}[\underline{b} - (G_2 - \rho I)\underline{u}^{(p)}], \qquad (4.8.12a)$$

$$\underline{u}^{(p+1)} = (G_2 + \rho I)^{-1}[\underline{b} - (G_1 - \rho I)\underline{u}^{(p+1/2)}], \qquad (4.8.12b)$$

with $(G_1 + \rho I)$ and $(G_2 + \rho I)$ easily invertible (2×2) matrices. The solution is obtained by using (4.8.12a) and (4.8.12b) in alternate sweeps until a suitable convergence to a specified level of accuracy ε is achieved.

4.8.3. Two-Dimensional Case

The AGE fractional scheme for the solution of the equation,

$$\frac{\partial^2 U}{\partial t^2} + \nabla^4 U = 0 \quad \text{where} \quad \Delta^2 \equiv \frac{\partial^2}{\partial x^2} + \frac{\partial^2}{\partial y^2}, \qquad (4.8.13)$$

in the region $R = R \times [0 < t \le T]$, where $R = \{(x, y) : 0 \le x \le L, \ 0 \le y \le M\}$, with appropriate initial and boundary conditions is now considered.

Consider equation (4.8.7) and let $\Delta x = L/(m+1)$, $\Delta y = M/(n+1)$ and Δt denotes the mesh size for x, y and t respectively; $x_i = i\Delta x$, $i = \Delta x$. $i = 0, 1, \dots, m+1$; $y_j = j\Delta y$, $j = 0, 1, \dots, n+1$; and $t_k = k\Delta t$, $k = 0, 1, \dots, T/\Delta t$. For simplicity of presentation, we assume that m and n are chosen so that $\Delta x = \Delta y = h$ and $\delta t = 1$, then the ratio $r = 1/h^2$ is the mesh ratio. The scheme is based on the 23 point computational molecule indicated in Figure 4.8.2.

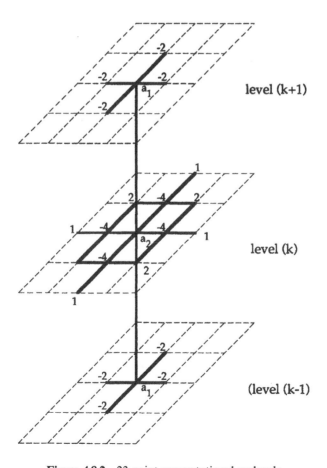

Figure 4.8.2 23 point computational molecule.

Letting $u_{i,j,k} = u(i\Delta x, j\Delta y, k\Delta t)$ the finite difference representation is

$$-2(u_{i-1,j,k+1} + u_{i+1,j,k+1} + u_{i,j-1,k+1} + u_{i,j+1,k+1}) + (8 + 1/r^2)u_{i,j,k+1}$$

$$= -(u_{i-2,j,k} + u_{i+2,j,k} + u_{i,j-2,k} + u_{i,j+2,k})$$

$$-2(u_{i-1,j-1,k} + u_{i+1,j-1,k} + u_{i-1,j+1,k} + u_{i+1,j+1,k})$$

$$+4(u_{i-1,j,k} + u_{i+1,j,k} + u_{i,j-1,k} + u_{i,j+1,k}) - (4 - 2/r^2)u_{i,j,k}$$

$$-(8 + 1/r^2)u_{i,j,k-1} + 2(u_{i-1,j,k-1} + u_{i+1,j,k-1} + u_{i,j-1,k-1} + u_{i,j+1,k-1}),$$

$$(4.8.14)$$

with local truncation error of $0(h^2 + \ell^2)$.

Equation (4.8.14) can be expressed in the matrix form (4.8.7a), with $\underline{u} = (\underline{u}_1, \underline{u}_2, \ldots, \underline{u}_n)^T$ and $\underline{u}_j = (u_{1j}, u_{2j}, \ldots, u_{mj})^T$, $j = 1, 2, \ldots, n$.

When the $m \times n$ mesh points on the rectangular grid system are ordered row-wise then the coefficient matrix A has the following block tridiagonal form:

$$A = \begin{bmatrix} A_1 & A_2 & & & \\ A_2 & A_1 & A_2 & & 0 \\ & & \ddots & & \\ 0 & & A_2 & A_1 & A_2 \\ & & & A_2 & A_1 \end{bmatrix}_{mn \times mn}, \text{ with } A_1 = \begin{bmatrix} a_1 & -2 & & & \\ -2 & a_1 & -2 & & 0 \\ & & \ddots & & \\ 0 & & & -2 & a_1 \end{bmatrix}_{m \times m}, \quad (4.8.15)$$

and $A_2 = \text{diag}(-2)$;

$$B = \begin{bmatrix} B_1 & B_2 & B_3 & & & \\ B_2 & B_1 & B_2 & B_3 & & 0 \\ B_3 & B_2 & B_1 & B_2 & B_3 & \\ & & \ddots & & \\ 0 & & & B_3 & B_2 & B_1 \end{bmatrix}_{mn \times mn}, \quad (4.8.16)$$

with

$$B_1 = \begin{bmatrix} -a_2 & 4 & -1 & & & \\ 4 & -a_2 & 4 & -1 & & \\ -1 & 4 & -a_2 & 4 & -1 & 0 \\ & & \ddots & & \\ 0 & & & -1 & 4 & -a_2 \end{bmatrix}_{m \times m}, \quad B_2 = \begin{bmatrix} 4 & -2 & & & \\ -2 & 4 & -2 & & 0 \\ & & \ddots & & \\ 0 & & & -2 & 4 \end{bmatrix}_{m \times m}, \quad (4.8.17)$$

$B_3 = \text{diag}(-1)$, $a_1 = 8 + 1/r^2$, and $a_2 = 4 - 2/r^2$.

The matrices $C = -A$ and D consists of boundary and initial conditions.

In the AGE method we split the matrix A into the form

$$A = G_1 + G_2 + G_2 + G_3 + G_4, \qquad (4.8.18)$$

where

$$G_1 + G_2 = \begin{bmatrix} A_3 & & & \\ & A_3 & & 0 \\ & & \ddots & \\ 0 & & & A_3 \end{bmatrix}_{mn \times mn} \quad \text{and} \quad G_3 + G_4 = \begin{bmatrix} A_4 & A_2 & & & \\ A_2 & A_4 & A_2 & & 0 \\ & A_2 & A_4 & \ddots & \\ & & \ddots & \ddots & \\ 0 & & & A_2 & A_4 \end{bmatrix}_{mn \times mn},$$

$$A_3 = \begin{bmatrix} a_1/2 & -2 & & & \\ -2 & a_1/2 & -2 & & 0 \\ & -2 & \ddots & \ddots & \\ & & \ddots & \ddots & -2 \\ 0 & & & -2 & a_1/2 \end{bmatrix}_{m \times m} \quad \text{and} \quad A_4 = \mathrm{diag}(a_1/2) \qquad (4.8.19)$$

If we assume that the order of the matrix A is even, we find that,

$$G_1 = \begin{bmatrix} E & & & \\ & E & & 0 \\ & & \ddots & \\ 0 & & & E \end{bmatrix}_{mn \times mn} \quad \text{and} \quad G_2 = \begin{bmatrix} F & & & \\ & F & & 0 \\ & & \ddots & \\ 0 & & & F \end{bmatrix}_{mn \times mn}, \qquad (4.8.20)$$

where

$$E = \begin{bmatrix} c & -2 & & & & \\ -2 & c & & & 0 & \\ & & \ddots & & & \\ & & & \ddots & & \\ 0 & & & & c & -2 \\ & & & & -2 & c \end{bmatrix}_{m \times m} \quad \text{and} \quad F = \begin{bmatrix} c & & & & & \\ & c & -2 & & 0 & \\ & -2 & c & & & \\ & & & \ddots & & \\ & & & & \ddots & \\ 0 & & & & & c \end{bmatrix}_{m \times m},$$

$$(4.8.21)$$

with $c = a_1/4$.

The Douglas formula for the AGE fractional scheme then takes the form

$$(G_1 + \rho I)\underline{u}_{(r)}^{(p+1/4)} = (\rho I - G_1 - 2G_2 - 2G_3 - 2G_4)\underline{u}_{(r)}^{(p)} + 2B\underline{u}_{(r)}^{[k]} + 2C_{(r)}^{[k-1]} + 2D,$$

$$(G_2 + \rho I)\underline{u}_{(r)}^{(p+1/2)} = G_2\underline{u}_{(r)}^{(p)} + \rho\underline{u}_{(r)}^{(p+1/4)},$$

$$(G_3 + \rho I)\underline{u}_{(r)}^{(p+3/4)} = G_3\underline{u}_{(r)}^{(p)} + \rho\underline{u}_{(r)}^{(p+1/2)}, \qquad (4.8.22)$$

$$(G_4 + \rho I)\underline{u}_{(r)}^{(p+1)} = G_4\underline{u}_{(r)}^{(p)} + \rho\underline{u}_{(r)}^{(p+3/4)},$$

where ρ is the iteration parameter and p is the iteration counter.

From (4.8.18), since $A = G_1 + G_2 + G_3 + G_4$, the first equation of (4.8.22) can be expressed as

$$(G_1 + \rho I)\underline{u}_{(r)}^{(p+1/4)} = (\rho I - G_1 - 2A)\underline{u}_{(r)}^{(p)} + 2B\underline{u}_{(r)}^{[k]} + 2C_{(r)}^{[k-1]} + 2D,$$

and the system (4.8.22) simplifies to the explicit form

$$\underline{u}_{(r)}^{(p+1/4)} = (G_1 + \rho I)^{-1}[(\rho I + G_1 - 2A)\underline{u}_{(r)}^{(p)} + 2B\underline{u}_{(r)}^{[k]} + 2C_{(r)}^{[k-1]} + 2D],$$

$$\underline{u}_{(r)}^{(p+1/2)} = (G_2 + \rho I)^{-1}[G_2\underline{u}_{(r)}^{(p)} + \rho\underline{u}_{(r)}^{(p+1/4)}], \qquad (4.8.23)$$

$$\underline{u}_{(r)}^{(p+3/4)} = (G_3 + \rho I)^{-1}[G_3\underline{u}_{(r)}^{(p)} + \rho\underline{u}_{(r)}^{(p+1/2)}],$$

$$\underline{u}_{(r)}^{(p+1)} = (G_4 + \rho I)^{-1}[G_4\underline{u}_{(r)}^{(p)} + \rho\underline{u}_{(r)}^{(p+3/4)}].$$

Notice that at the third intermediate level, the $(p + 3/4)^{th}$ iterate, if we order the mesh points column-wise parallel to the y-axis, we have

$$\underline{u}_{(c)} = (\underline{u}_1, \underline{u}_2, \ldots, \underline{u}_m)^T, \quad \text{with } \underline{u}_i - (u_{i1}, u_{i2}, \ldots, u_{in})^T \quad \text{for } i = 1, 2, \ldots, m;$$

we also find that

$$(G_3 + G_4)\underline{u}_{(r)} + (\hat{G}_1 + \hat{G}_2)\underline{u}_{(c)} \qquad (4.8.24)$$

and

$$G_3\underline{u}_{(r)} = \hat{G}_1\underline{u}_{(c)}, \qquad G_4\underline{u}_{(r)} = \hat{G}_2\underline{u}_{(c)}, \qquad (4.8.25)$$

hence, the third equation of (4.8.23) can be transformed to

$$\underline{u}_{(c)}^{(p+3/4)} = (\hat{G}_1 + \rho I)^{-1}[\hat{G}_1\underline{u}_{(c)}^{(p)} + \rho\underline{u}_{(c)}^{(p+1/2)}]. \qquad (4.8.26)$$

At the fourth intermediate level, the $(p + 1)^{th}$ iterate, by virtue of (4.8.25) the last equation of (4.8.23) can be transformed to

$$\underline{u}_{(c)}^{(p+1)} = (\hat{G}_2 + \rho I)^{-1}[\hat{G}_2\underline{u}_{(c)}^{(p)} + \rho\underline{u}_{(c)}^{(p+3/4)}], \qquad (4.8.27)$$

where

$$\hat{G}_1 = \begin{bmatrix} P & & & 0 \\ & \ddots & & \\ & & \ddots & \\ 0 & & & P \end{bmatrix}_{mn \times mn} \quad \text{and} \quad \hat{G}_2 = \begin{bmatrix} Q & & & 0 \\ & \ddots & & \\ & & \ddots & \\ 0 & & & Q \end{bmatrix}_{mn \times mn}, \qquad (4.8.28)$$

and the matrices P and Q are exactly the same forms as E and F of (4.8.21) but of order $(n \times n)$.

The AGE scheme corresponds to sweeping through the mesh parallel to the coordinate x and y axes involving at each stage the solution of 2×2 block systems. The iterative procedure is continued until convergence is achieved.

4.8.4. Stability Analysis of the AGE Scheme

We now examine the stability of (4.8.14), and assume that $Z_{m,n,k}$ is the error at the grid point $X = mh$, $Y = nh$, $T = k\ell$. To investigate the propagation of the error as t increases, it is necessary to find a solution of the difference equation (4.8.14) which reduces to $e^{i\beta x} e^{i\gamma y}$ when $t = 0$. Let such a solution be,

$$Z_{m,n,k} = e^{\alpha k \ell} e^{i\beta mh} e^{i\gamma nh} = \xi^k e^{i\beta mh} e^{i\gamma nh}, \qquad (4.8.29)$$

where $\xi = e^{\alpha \ell}$, β and γ are arbitrary real numbers and $\alpha = \alpha(\beta, \gamma)$ is in general complex.

Since $Z_{m,n,k}$ satisfies the difference equation (4.8.14), then by substituting $Z_{m,n,k}$ from (4.8.29) and by the cancellation of the term $\xi^{k-1} e^{i\beta mh} e^{i\gamma\gamma nh}$ we get the quadratic equation in ξ,

$$\xi^2 [2(e^{-i\beta h} + e^{i\beta h} + e^{-i\gamma h} + e^{i\gamma h}) - (8 + 1/r^2)]$$

$$+ [2(e^{-i\beta h} + e^{i\beta h} + e^{-i\gamma h} + e^{i\gamma h}) - (8 + 1/r^2)]$$

$$= \xi [e^{-2i\beta h} + e^{2i\beta h} + e^{-2i\gamma h} + e^{2i\gamma h} + 2(e^{-i\gamma h} + e^{i\gamma h})$$

$$(e^{-i\beta h} + e^{i\beta h}) - 4(e^{i-\beta h} + e^{i\beta h} + e^{i-\gamma h} + e^{i\gamma h}) + (4 - 2/r^2)]. \qquad (4.8.30)$$

After simplification, we obtain,

$$\lambda = \frac{\xi^2 + 1}{\xi} = \frac{2[1 - 8r^2(\sin^2(\beta h/2) + \sin^2(2\gamma h/2))^2 + 8r^2(\sin^2(\beta h/2) + \sin^2(2\gamma h/2))]}{1 + 8r^2(\sin^2(\beta h/2) + \sin^2(\gamma h/2))}.$$
$$(4.8.31)$$

For each value of λ we obtain two solutions ξ_1^k and ξ_2^k, where ξ_1 and ξ_2 are solutions of the characteristic equation $\xi^2 - \lambda\xi + 1 = 0$.

The difference equation is said to be stable if ξ_1 and ξ_2 are less than or equal to one in magnitude. The condition that these roots be less than or equal to one in magnitude is simply that $|\xi/2| \le 1$ for all values of βh and γh. Hence, for stability, we have

$$-1 \le \frac{1 - 8r^2(\sin^2(\beta h/2) + \sin^2(\gamma h/2))^2 + 8r^2(\sin^2(\beta h/2) + \sin^2(\gamma h/2))}{1 + 8r^2(\sin^2(\beta h/2) + \sin^2(\gamma h/2))} \le 1, \quad (4.8.32)$$

for all $\beta, \gamma \in R$.

This leads to two inequalities which are trivially satisfied for all $r > 0$. Thus, the finite difference approximation (4.8.14) is stable for all values of r.

4.8.5. Numerical Experiments

The scheme outlined in Section 4.8.3 is now used to solve the fourth-order parabolic partial differential equation,

$$\frac{\partial^2 U}{\partial t^2} + \nabla^4 U = (4\pi^4 - \pi^2)\sin \pi x \, \sin \pi y \, \cos \pi t, \qquad (4.8.33a)$$

where

$$\nabla^4 \equiv \left(\frac{\partial^2}{\partial x^2} + \frac{\partial^2}{\partial y^2}\right)^2 \equiv \frac{\partial^4}{\partial x^4} + 2\frac{\partial^4}{\partial x^2 \partial y^2} + \frac{\partial^4}{\partial y^4},$$

subject to the initial conditions

$$U(x, y, 0) = \sin \pi x \, \sin \pi y, \quad 0 \le x, y \le 1, \qquad (4.8.33b)$$

$$\frac{\partial U}{\partial t}(x, y, 0) = 0,$$

and the boundary conditions

$$U(0, y, t) = U(1, y, t) = \frac{\partial^2 U}{\partial x^2}(0, y, t) = \frac{\partial^2 U}{\partial t}(1, y, t) = 0, \quad 0 < y < 1 \text{ and } t \ge 0,$$

$$U(x, 0, t) = U(x, 1, t) = 0, \quad 0 < x < 1 \text{ and } t \ge 0,$$

$$\frac{\partial^2 U}{\partial y^2}(x, 0, t) = \frac{\partial^2 U}{\partial y^2}(x, 1, t) = 0, \quad 0 < x < 1. \qquad (4.8.33c)$$

The exact solution for this problem is

$$U(x, y, t) = \sin \pi x \, \sin \pi y \, \cos \pi t.$$

The numerical results are displayed in Tables 4.8.1–4.8.4 for different mesh sizes and different values of x and y and the program was executed for 30 time steps. The convergence test used for the AGE scheme at each time level was the absolute test with $\varepsilon = 10^{-6}$, where ε is the convergence criterion.

Table 4.8.1 Solution and numerical approximations for $\Delta x = \Delta y = 1/51$, $\Delta t = 0.0002$, $t = 0.006$, $r = 0.5202$, $\rho = 2.67$, $x = 17/51$

y	1/51	5/51	10/51	15/51	20/51	25/51
Solution	0.053304	0.262491	0.500278	0.690980	0.816651	0.865461
AGE-D	0.053307	0.262505	0.500302	0.691014	0.816690	0.865502

Table 4.8.2 Solution and numerical approximations for $\Delta x = \Delta y = 1/21$, $\Delta t = 0.0008$, $t = 0.024$, $r = 0.3528$, $\rho = 6.1$, $x = 7/21$

y	1/21	3/21	5/21	7/21	9/21	10/21
Solution	0.128708	0.374687	0.587373	0.747869	0.841914	0.861150
AGE-D	0.128768	0.374861	0.587645	0.748214	0.842301	0.861546

Table 4.8.3 Solution and numerical approximations for $\Delta x = \Delta y = 1/41$, $\Delta t = 0.0008$, $t = 0.024$, $r = 1.3448$, $\rho = 1.46$, $x = 5/41$

y	1/41	5/41	10/41	13/41	15/41	20/41
Solution	0.028534	0.139342	0.258480	0.312878	0.340140	0.372481
AGE-D	0.028584	0.139561	0.258791	0.313185	0.340427	0.372679

Table 4.8.4 Solution and numerical approximations for $\Delta x = \Delta y = 1/25$, $\Delta t = 0.001$, $t = 0.03$, $r = 0.625$, $\rho = 2.15$, $x = 5/25$

y	1/25	3/25	5/25	7/25	9/25	12/25
Solution	0.073342	0.215418	0.343958	0.450886	0.529484	0.584022
AGE-D	0.073409	0.215602	0.344227	0.451205	0.529818	0.584329

From the results obtained it can be noted that the AGE method performed quite well, is extremely competitive and the convergence is very rapid if a good previous approximation is used as a starting guess, i.e., the previous time step in parabolic problems.

Also, the AGE algorithm is an ideal algorithm for use on parallel computers as it possesses separate and independent tasks, i.e., (2×2) groups which can be executed concurrently.

5. The Alternating Group Explicit (AGE) Method for Boundary Value Problems

5.1. Introduction

In this chapter, we shall investigate the solution of the one-dimensional two-point boundary value problem using the alternating group explicit (AGE) iterative method. In particular the two-point group method is analysed in order to determine the optimum parameter r which maximises the convergence rate of the method.

Consider then the second order ordinary differential equation

$$-\frac{d^2U}{dx^2} + q(x)U = f(x),\qquad(5.1.1)$$

over the line segment $a < x < b$, subject to the boundary conditions at the points a and b, i.e.,

$$U(a) = \alpha, \quad U(b) = \beta.\qquad(5.1.2)$$

Here α and β are given real constants, and $q(x)$ and $f(x)$ are given real continuous functions on $a \le x < b$, with

$$q(x) \ge 0.$$

The finite difference approximations to (5.1.1) can be derived by using the method based on finite Taylor's series expansions of the solution $U(x)$ to (5.1.1). The finite difference replacement of (5.1.1) is given by

$$\frac{(2u_i - u_{i+1} - u_{i-1})}{h^2} + q_i u_i = f_i, \quad 1 \le i \le m,\qquad(5.1.3)$$

with a local truncation error of $(h^2/12)d^4U_i/dx^4 + \cdots$ where u_i denote the function satisfying the difference equation at the mesh point $(x_i = a + ih)$ of the discrete problem and h is the mesh spacing given by

$$h = \frac{(b-a)}{(m+1)}.\qquad(5.1.4)$$

In matrix notation (5.1.3) can be written in the form

$$A\underline{u} = \underline{b},\qquad(5.1.5)$$

where A is a real $(m \times m)$ matrix, $\underline{u}$ is the discrete approximation to the solution $U(x)$ of (5.1.1–5.1.2) and b is a column vector, i.e.,

$$A = \begin{bmatrix} 2+q_1h^2 & -1 & & & \\ -1 & 2+q_2h^2 & -1 & & 0 \\ & & \ddots & & \\ & & & & \\ 0 & & -1 & 2+q_{m-1}h^2 & -1 \\ & & & -1 & 2+q_mh^2 \end{bmatrix}, \qquad (5.1.6a)$$

$$\underline{u} = (u_1, u_2, \ldots, u_m)^T, \qquad (5.1.6b)$$

and

$$\underline{b} = h^2(f_1 + \alpha/h^2, f_2, \ldots, f_{m-1}, f_m + \beta/h^2)^T. \qquad (5.1.6c)$$

The tridiagonal matrix A is real and symmetric, and since $q(x) \geq 0$ then it is also diagonally dominant with positive diagonal entries.

5.2. Alternating Group Explicit (AGE) Method

5.2.1. Development of the Method

Consider (5.1.1), where the finite difference replacement of this equation has been obtained and is given by (5.1.3), which in matrix notation has the form (5.1.5).

We now consider a class of methods for solving (5.1.5) which is based on the "splitting" of the matrix A into the sum of three matrices

$$A = G_1 + G_2 + \Sigma, \qquad (5.2.1)$$

where Σ is a non-negative diagonal matrix and where G_1, G_2 and Σ satisfy the conditions:

(i) $G_1 + \theta\Sigma + rI$ and $G_2 + \theta\Sigma + rI$ are non-singular for any $\theta \geq 0, r > 0$.
(ii) For any vectors $\underline{v}_1$ and $\underline{v}_2$ and for any constants $\theta \geq 0$ and $r > 0$ it is easier to solve the system explicitly, i.e.,

$$z_1 = G_1^{-1}\underline{v}_1 \quad \text{and} \quad z_2 = G_2^{-1}\underline{v}_2,$$

for z_1 and z_2, respectively.

We shall be concerned here with the situation where G_1 and G_2 are either small 2×2 block systems or can be made so by a suitable permutation of their rows and corresponding columns. This procedure is easier in the sense that the work required is much less than would be required to solve the original system (5.1.5) directly.

From the above discussion, we have G_1, G_2 and Σ given by

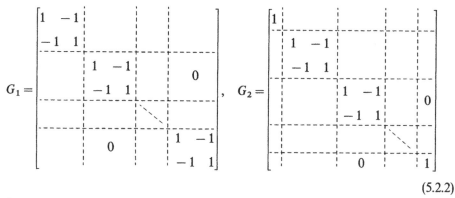

$$(5.2.2)$$

and

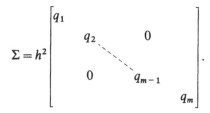

By using (5.2.1) we can write the matrix equation (5.1.5) in the form

$$(G_1 + G_2 + \Sigma)\underline{u} = \underline{b}. \tag{5.2.3}$$

Let us consider two equivalent forms

$$(G_1 + \theta\Sigma + rI)\underline{u} = \underline{b} - (G_2 + (1 - \theta)\Sigma - rI)\underline{u} \tag{5.2.4a}$$

and

$$(G_2 + \theta\Sigma + r'I)\underline{u} = \underline{b} - (G_1 + (1 - \theta)\Sigma - r'I)\underline{u}. \tag{5.2.4b}$$

Similar to the Peaceman–Rachford method [1955], we select positive iteration parameters r and r' and determine $\underline{u}^{(k+1/2)}$ and $\underline{u}^{(k+1)}$, respectively, by

$$\underline{u}^{(k+1/2)} = (G_1 + \theta\Sigma + rI)^{-1}[\underline{b} - (G_2 + (1-\theta)\Sigma = rI)\underline{u}^{(k)}], \tag{5.2.5a}$$

and

$$\underline{u}^{(k+1)} = (G_2 + \hat{\theta}\Sigma + r'I)^{-1}[\underline{b} - (G_1 + (1-\hat{\theta})\Sigma - r'I)\underline{u}^{(k+1/2)}], \tag{5.2.5b}$$

where $\underline{u}^{(0)}$ is an arbitrary initial vector approximation of the unique solution of (5.1.5) and $\underline{u}^{(k)}$ is the kth iterate to the solution of (5.1.5).

For simplicity, we shall consider here the special case where

$$\theta = \hat{\theta} = 1/2, \quad r = r', \tag{5.2.6}$$

and we let

$$\bar{G}_1 = G_1 + \Sigma/2, \quad \bar{G}_2 = G_2 + \Sigma/2. \tag{5.2.7}$$

Evidently, $\bar{G}_1$ and $\bar{G}_2$ also satisfy the following conditions:

(i) $\bar{G}_1 + rI$ and $\bar{G}_2 + rI$ are non-singular for any $r > 0$;
(ii) for any vectors $\underline{v}_1$ and $\underline{v}_2$ and for any $r > 0$ it is easier to solve the systems explicitly:

$$(\bar{G}_1 + rI)\underline{z}_1 = \underline{v}_1, \quad (\bar{G}_2 + rI)\underline{z}_2 = \underline{v}_2. \tag{5.2.8}$$

Thus, (5.2.3) becomes

$$(\bar{G}_1 + \bar{G}_2)\underline{u} = \underline{b}, \tag{5.2.9}$$

and (5.2.4a–5.2.4b) become, respectively,

$$\underline{u}^{(k+1/2)} = (\bar{G}_1 + rI)^{-1} [\underline{b} - (\bar{G}_2 - rI)\underline{u}^{(k)}], \tag{5.2.10a}$$

$$\underline{u}^{(k+1)} = (\bar{G}_2 + rI)^{-1} [\underline{b} - (\bar{G}_1 - rI)\underline{u}^{(k+1/2)}]. \tag{5.2.10b}$$

If we combine the above two equations into the form

$$\underline{u}^{(k+1)} = T_r \underline{u}^{(k)} + \underline{b}_r, \tag{5.2.11}$$

where

$$\underline{b}_r = (\bar{G}_2 + rI)^{-1} \{ I - (\bar{G}_1 - rI)(\bar{G}_1 + rI)^{-1} \} \underline{b} \tag{5.2.12a}$$

and

$$T_r = (\bar{G}_2 + rI)^{-1} (\bar{G}_1 - rI)(\bar{G}_1 + rI)^{-1} (\bar{G}_2 - rI). \tag{5.2.12b}$$

The matrix T_r is called the AGE iteration matrix.

Now to analyse the convergence of the AGE method, we assume that U is the true solution of (5.1.5) then

$$(\bar{G}_1 + \bar{G}_2) \underline{U} = \underline{b}, \tag{5.2.13}$$

and hence we have

$$(\bar{G}_1 + rI) \underline{U} = \underline{b} - (\bar{G}_2 - rI) \underline{U}. \tag{5.2.14}$$

If $\underline{e}^{(k)} = \underline{u}^{(k)} - U$ is the error vector associated with the vector iterate $\underline{u}^{(k)}$, therefore from (5.2.10a) and (5.2.14) we have

$$(\bar{G}_1 + rI) \underline{e}^{(k+1/2)} = -(\bar{G}_2 - rI) \underline{e}^{(k)}. \tag{5.2.15a}$$

Similarly

$$(\bar{G}_2 + rI) \underline{e}^{(k+1)} = -(\bar{G}_1 - rI) \underline{e}^{(k+1/2)}, \tag{5.2.15b}$$

and hence

$$\underline{e}^{(k+1)} = T_r \underline{e}^{(k)}, \tag{5.2.16}$$

where T_r is given in (5.2.12b).

To examine the convergence properties of T_r, we first prove the following theorem.

Theorem 5.2.1.
If $\bar{G}_1$ and $\bar{G}_2$ are real positive definite matrices and if $r > 0$, then $\rho(T_r) < 1$.

Proof. If we define the matrix $\tilde{T}_r$ by applying the similarity transformation as

$$\tilde{T}_r = (\bar{G}_2 + rI) T_r (\bar{G}_2 + rI)^{-1}$$

$$= (\bar{G}_1 - rI)(\bar{G}_1 + rI)^{-1}(\bar{G}_2 - rI)(\bar{G}_2 + rI)^{-1}, \tag{5.2.17}$$

then it is evident that T_r is similar to $\tilde{T}_r$, and hence from the properties of the matrix norms we have

$$\|\tilde{T}_r\| \leq \|(\bar{G}_1 - rI)(\bar{G}_1 + rI)^{-1}\| \cdot \|(\bar{G}_2 - rI)(\bar{G}_2 + rI)^{-1}\|. \tag{5.2.18}$$

But since $\bar{G}_1$ and $\bar{G}_2$ are symmetric and since $(\bar{G}_1 - rI)$ commutes with $(\bar{G}_1 + rI)^{-1}$ we have

$$\|(\bar{G}_1 - rI)(\bar{G}_1 + rI)^{-1}\| = \rho\{(\bar{G}_1 - rI)(\bar{G}_1 + rI)^{-1}\}$$

$$= \max_{\lambda} |(\lambda - r)/(\lambda + r)|, \tag{5.2.19}$$

where λ ranges over all eigenvalues of $\bar{G}_1$. But since $\bar{G}_1$ is positive definite, its eigenvalues are positive. Therefore

$$\|(\bar{G}_1 - rI)(\bar{G}_1 + rI)^{-1}\| < 1. \tag{5.2.20}$$

The same argument applied to the corresponding matrix product with $\bar{G}_2$ shows that $\|(\bar{G}_2 - rI)(\bar{G}_2 + rI)^{-1}\| < 1$, and we therefore conclude that

$$\rho(T_r) = \rho(\tilde{T}_r) \leq \|\tilde{T}_r\| < 1; \tag{5.2.21}$$

hence, the convergence follows.

Let us now assume that $\bar{G}_1$ and $\bar{G}_2$ are real positive definite matrices and that the eigenvalues λ of $\bar{G}_1$ and μ of $\bar{G}_2$ lie in the ranges

$$0 < a \leq \lambda \leq b, \quad 0 < a \leq \mu \leq b. \tag{5.2.22}$$

Evidently, if $r > 0$ we have

$$\rho(T_r) \leq \rho((\bar{G}_1 - rI)(\bar{G}_1 + rI)^{-1})\rho((\bar{G}_2 - rI)(\bar{G}_2 + rI)^{-1})$$

$$= (\max_{a \leq \lambda \leq b} |(\lambda - r)/(\lambda + r)|)(\max_{a \leq \mu \leq b} |(\mu - r)/(\mu + r)|)$$

$$= [\max_{a \leq \gamma \leq b} |(\gamma - r)/(\gamma + r)|^2] = \phi(a, b; r). \tag{5.2.23}$$

Since $(\gamma - r)/(\gamma + r)$ is an increasing function of γ we have

$$\max_{a \leq \gamma \leq b} |(\gamma - r)/(\gamma + r)| = \max(|(\gamma - \alpha)/(\gamma + \alpha)|, |(\gamma - \beta)/(\gamma + \beta)|). \tag{5.2.24}$$

When $r = \sqrt{ab}$, then

$$|(a - r)/(a + r)| = |(b - r)/(b + r)| = (\sqrt{b} - \sqrt{a})/(\sqrt{b} + \sqrt{a}). \tag{5.2.25}$$

Moreover, if $0 \leq r \leq \sqrt{ab}$, we have

$$|(b - r)/(b + r)| - (\sqrt{b} - \sqrt{a})/(\sqrt{b} + \sqrt{a})$$

$$= 2\sqrt{b}(\sqrt{ab} - r)/(b + r)(\sqrt{b} + \sqrt{a}) > 0, \tag{5.2.26}$$

and if $\sqrt{ab} < r$, then

$$|(a-r)/(a+r)| - (\sqrt{b}-\sqrt{a})/(\sqrt{b}+\sqrt{a})$$

$$= 2\sqrt{b}(r-\sqrt{ab})/(r+a)(\sqrt{b}+\sqrt{a}) > 0. \qquad (5.2.27)$$

Therefore, $\phi(a,b;r)$ is minimised when $r = \sqrt{ab}$ and

$$\rho(T_{r=\sqrt{ab}}) \le \phi(a,b;r=\sqrt{ab}) = [(\sqrt{b}-\sqrt{a})/(\sqrt{b}+\sqrt{a})]^2. \qquad (5.2.28)$$

Thus, $r = \sqrt{ab}$ is optimum in the sense that the bound $\phi(a,b;r)$ for $\rho(T_r)$ is minimised.

The convergence of the AGE method may be accelerated if we allow r to vary cyclically from iteration to iteration. This rapid convergence can be proved to hold for an appropriate choice of the iteration parameters in the commutative case, i.e., for the periodic problem with $h^{-1} = 5$, in which case the two matrices G_1 and G_2 have the form

$$G_1 = \begin{bmatrix} 1 & -1 & 0 & 0 \\ -1 & 1 & 0 & 0 \\ 0 & 0 & 1 & -1 \\ 0 & 0 & -1 & 1 \end{bmatrix} \quad \text{and} \quad G_2 = \begin{bmatrix} 1 & 0 & 0 & -1 \\ 0 & 1 & -1 & 0 \\ 0 & -1 & 1 & 0 \\ -1 & 0 & 0 & 1 \end{bmatrix},$$

such that

$$G_1 G_2 = \begin{bmatrix} 1 & -1 & 1 & -1 \\ -1 & 1 & -1 & 1 \\ 1 & -1 & 1 & -1 \\ -1 & 1 & -1 & 1 \end{bmatrix} = G_2 G_1.$$

Hence a set of optimum iteration parameters can be established. However for the non-commutative case no optimum sets of iteration (acceleration) parameters have yet been found, even for the classical ADI methods. One would expect that by using sets of parameters (i.e., Jordan–Wachspress) corresponding to the commutative case, the rates of convergence would not be greatly affected by the non-commutative feature.

The AGE method can be easily extended to obtain the solution of block tridiagonal linear systems derived from the discretisation of multi-dimensional elliptic boundary value problems, i.e., the Laplace model problem $\nabla^2 \phi = 0$ in a square region with Dirichlet boundary conditions. The coefficient matrix A has the block structure

$$A = \begin{bmatrix} D_1 & -I & & & \\ -I & D_2 & -I & & 0 \\ & & \ddots & & \\ 0 & & -I & D_{n-1} & -I \\ & & & -I & D_n \end{bmatrix}, \qquad (5.2.29)$$

where

$$D_i = \begin{bmatrix} 4 & -1 & & & & \\ -1 & 4 & -1 & & 0 & \\ & & \ddots & & & \\ & 0 & & -1 & 4 & -1 \\ & & & & -1 & 4 \end{bmatrix}, \quad i = 1, 2, \ldots, n. \quad (5.2.30)$$

In a similar technique we split the matrix A into the sum of two submatrices

$$A = G_1 + G_2, \quad (5.2.31)$$

where

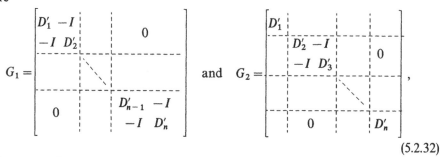

$$\quad (5.2.32)$$

if n is even and

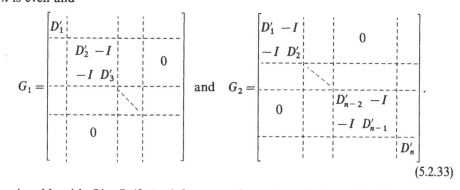

$$\quad (5.2.33)$$

If n is odd, with $D_i' = D_i/2, i = 1, 2, \ldots, n$, where G_1 and G_2 satisfy the condition $G_1 + rI$ and $G_2 + rI$ are non-singular for any $r > 0$ and are easily solvable as before. The solution procedure using the AGE algorithm has been discussed fully in Section 5.8.

5.3. Numerical Example

Consider the linear problem

$$U_1' = U_2, \quad (5.3.1a)$$

$$U_2' = 400(U_1 + \cos^2(\pi x)) + 2\pi^2 \cos(2\pi x), \quad (5.3.1b)$$

subject to the boundary conditions

$$U_1(0) = U_1(1) = 0.$$ (5.3.1c)

The exact solution for this problem is given by

$$U_1(x) = (e^{-20} \cdot e^{20x} + e^{-20x})/(1 + r^{-20}) - \cos^2(\pi x),$$

$$U_2(x) = (20e^{-20} \cdot e^{20x} - 20e^{-20x})/(1 + e^{-20}) + \pi \sin(2\pi x).$$ (5.3.2)

From (5.3.1a) and (5.3.1b) we have

$$U_1'' = 400(U_1 + \cos^2(\pi x)) + 2\pi^2 \cos(2\pi x).$$ (5.3.3)

By following the finite difference procedure of Section 5.1, equation (5.3.3) can be approximated to obtain the linear difference equation in u_i, i.e.,

$$(u_{i-1} - 2u_i + u_{i+1})/h^2 = 400[u_i + \cos^2(\pi x)] + 2\pi^2 \cos(2\pi x_i), \quad i = 1, 2, \ldots, m.$$

This equation can be simplified to the form

$$-u_{i-1} + (2 + 400h^2)u_i - u_{i+1}$$
$$= -2h^2[200\cos^2(\pi x_i) + \pi^2 \cos(2\pi x_i)], \quad i = 1, 2, \ldots, m.$$ (5.3.4)

The boundary conditions are replaced by

$$u_0 = 0 \quad \text{and} \quad u_{m+1} = 0,$$ (5.3.5)

where $h = 1/(m+1)$.

The linear system (5.3.4) can be represented in matrix notation as

$$A\underline{u} = (\bar{G}_1 + \bar{G}_2)\underline{u} = \underline{b}.$$ (5.3.6)

The vector u is defined in the usual way, and b is given by

$$\underline{b} = (c_1, c_2, \ldots, c_{m-1}, c_m)^T,$$ (5.3.7a)

where

$$c_i = -2h^2[200\cos^2(\pi x_i) + \pi^2 \cos(2\pi x_i)], \quad i = 1, 2, \ldots, m$$ (5.3.7b)

and $\bar{G}_1$ and $\bar{G}_2$ are given by

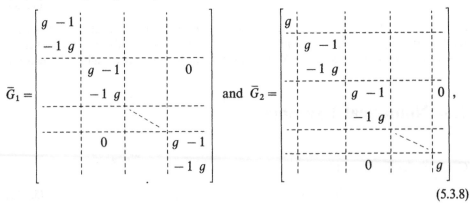

(5.3.8)

with $g = 1 + 200h^2$.

Hence by applying the AGE method of Section 5.2.1, we can determine $\underline{u}^{(k+1/2)}$ and $\underline{u}^{(k+1)}$ successively from (5.2.10). It is obvious that the (2×2) submatrices of $(\bar{G}_1 + rI)$ and $(\bar{G}_2 + rI)$ are of the form

$$\hat{G} = \begin{bmatrix} \alpha & -1 \\ -1 & \alpha \end{bmatrix},$$

where $\alpha = g + r$, and the inverse of $\hat{G}$ is given by

$$\hat{G}^{-1} = d \begin{bmatrix} \alpha & 1 \\ 1 & \alpha \end{bmatrix}, \quad \text{where} \quad d = 1/(\alpha^2 - 1). \tag{5.3.9}$$

Hence the vector $\underline{u}^{(k+1)}$ can be determined from $\underline{u}^{(k)}$ in two steps, we first determine $\underline{u}^{(k+1/2)}$ as follows:

$$\begin{bmatrix} u_1 \\ u_2 \\ u_3 \\ u_4 \\ \vdots \\ u_{m-1} \\ u_m \end{bmatrix}^{(k+1/2)} = d \begin{bmatrix} \alpha & 1 & & & & & \\ 1 & \alpha & & & & & \\ & & \alpha & 1 & & & \\ & & 1 & \alpha & & 0 & \\ & & & & \ddots & & \\ & 0 & & & & \alpha & 1 \\ & & & & & 1 & \alpha \end{bmatrix} \begin{bmatrix} c_1 - \beta u_1 \\ c_2 - \beta u_2 + u_3 \\ c_3 + u_2 - \beta u_3 \\ c_4 - \beta u_4 + u_5 \\ \vdots \\ c_{m-1} + u_{m-2} - \beta u_{m-1} \\ c_m - \beta u_m \end{bmatrix}^{(k)},$$

$$\tag{5.3.10a}$$

and by using the values of $\underline{u}^{(k+1/2)}$ we determine $\underline{u}^{(k+1)}$,

$$\begin{bmatrix} u_1 \\ u_2 \\ u_3 \\ \vdots \\ u_{m-2} \\ u_{m-1} \\ u_m \end{bmatrix}^{(k+1)} = d \begin{bmatrix} 1/d\alpha & & & & & & \\ & \alpha & 1 & & & 0 & \\ & 1 & \alpha & & & & \\ & & & \ddots & & & \\ & & & & \alpha & 1 & \\ & 0 & & & 1 & \alpha & \\ & & & & & & 1/d\alpha \end{bmatrix} \begin{bmatrix} c_1 - \beta u_1 + u_2 \\ c_2 + u_1 - \beta u_2 \\ c_3 - \beta u_3 + u_4 \\ \vdots \\ c_{m-2} + u_{m-3} - \beta u_{m-2} \\ c_{m-1} - \beta u_{m-1} + u_m \\ c_m + u_{m-1} - \beta u_m \end{bmatrix}^{(k+1/2)},$$

$$\tag{5.3.10b}$$

where $\beta = g - r$.

By carrying out the necessary algebra, (5.3.10a,b) can then be written in explicit form. Hence $\underline{u}^{(k+1/2)}$ can be determined explicitly from the computational molecule shown in Figure 5.3.1 and the equations

$$u_i^{(k+1/2)} = (a\alpha u_{i-1}^{(k)} - \alpha\beta u_i^{(k)} - \beta u_{i+1}^{(k)} + b u_{i+2}^{(k)} + \alpha c_i + c_{i+1})d, \qquad (5.3.11a)$$

$$u_{i+1}^{(k+1/2)} = (a u_{i-1}^{(k)} - \beta u_i^{(k)} - \alpha\beta u_{i+1}^{(k)} + b\alpha u_{i+2}^{(k)} + c_i + \alpha c_{i+1})d, \qquad (5.3.11b)$$

for $i = 1, 3, 5, \ldots, m-1$, with $a = 0$ if $i = 1$ and $b = 0$ if $i = m-1$ and $a = b = 1$ otherwise.

Similarly, the computational molecule for the $(k+1)^{th}$ sweep is given in Figure 5.3.2 and the equations

$$u_i^{(k+1)} = (a\alpha u_{i-1}^{(k+1/2)} - \alpha\beta u_i^{(k+1/2)} - \beta u_{i+1}^{(k+1/2)} + b u_{i+2}^{(k)} + \alpha c_i + c_{i+1})d, \qquad (5.3.12a)$$

$$u_{i+1}^{(k+1)} = (a u_{i-1}^{(k+1/2)} - \beta u_i^{(k+1/2)} - \alpha\beta u_{i+1}^{(k+1/2)} + b\alpha u_{i+2}^{(k+1/2)} + c_i + \alpha c_{i+1})d, \qquad (5.3.12b)$$

for $i = 2, 4, 6, \ldots, m-2$ with $i = 0$, $a = 0$ and $b = 1/\alpha^2 d$ in (5.3.12b) and $i = m$, $b = 0$ and $a = 1/\alpha^2 d$ in (5.3.12a) and $a = b = 1$ otherwise.

Table 5.3.1 shows the results obtained by solving the linear problem by the AGE method with the minimum number of iterations tabulated against h^{-1}; clearly the results support the linear theory developed in Section 5.2. The convergence test used is the average test with the maximum error $\varepsilon = 5 \times 10^{-6}$.

It should be noticed that for non-linear problems, the application of the AGE iterative method is of immense practical importance, as even the application of the

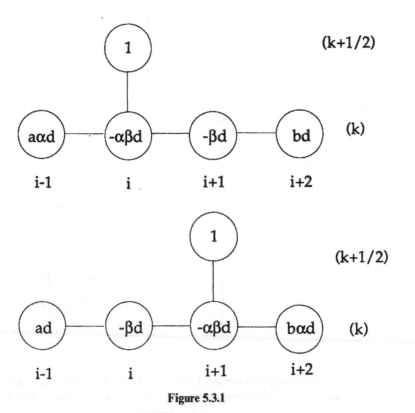

Figure 5.3.1

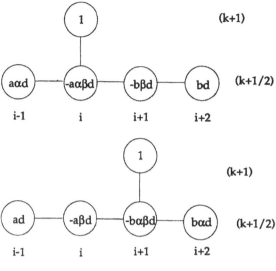

Figure 5.3.2

Table 5.3.1

| h^{-1} | AGE | |
	r	k (*no. of iterations*)
15	1.3 – 1.72	6
31	0.52 – 0.75	11
45	0.45 – 0.48	15
61	0.335 – 0.35	20
75	0.17 – 0.3	25
91	0.14 – 0.17	30
101	0.125 – 0.14	33

direct method of solution (Gaussian elimination) has to be applied iteratively for such problems.

Also, for parallel computers, it is well known that the direct method (i.e., Gaussian Elimination) consists of non-linear recurrences that must be evaluated sequentially, so there is little parallelism in the direct algorithm and therefore is not an ideal algorithm for use on parallel computers. However, the AGE method is suitable for parallel computers as it possesses separate and independent tasks, i.e., (2×2) groups which can be executed concurrently.

The Computational Complexity
In terms of arithmetic calculations, we will now estimate the amount of operations required to solve the problem by applying the AGE method. For the linear problem, $\underline{u}^{(k+1/2)}$ and $\underline{u}^{(k+1)}$ can be obtained from (5.3.10a) and (5.3.10b), respectively. Hence

we can show that the number of operations required to solve this problem by the AGE iterative method is

$(6m-7)$ multiplications $+ (4m-1\frac{1}{2})$ additions $+$ right-hand side (RHS) unit per iteration.

(b) The Modified AGE Method
Consider the linear system

$$A\underline{u} = \underline{b}, \tag{5.3.13}$$

where u and b are m-dimensional vectors and A is a real $m \times m$ matrix.

The alternating group explicit (AGE) method consists of splitting the matrix A into the form $A = \bar{G}_1 + \bar{G}_2$, selecting a positive iteration parameter r and a guess $\underline{u}^{(0)}$ of the solution, and carrying out the iterative scheme defined by

$$(\bar{G}_1 + rI)\underline{u}^{(k+1/2)} = \underline{b} - (\bar{G}_2 - rI)\underline{u}^{(k)},$$

$$(\bar{G}_2 + rI)\underline{u}^{(k+1)} = \underline{b} - (\bar{G}_1 - rI)\underline{u}^{(k+1/2)}. \tag{5.3.14}$$

In this paper we consider the modified AGE (MAGE) scheme

$$(\bar{G}_1 + rI)\underline{u}^{(k+1/2)} = \underline{b}_1 - (\bar{G}_1 - rI)\underline{u}^{(k)},$$

$$(\bar{G}_2 + rI)\underline{u}^{(k+1)} = \underline{b}_2 - (\bar{G}_2 - rI)\underline{u}^{(k+1/2)}, \tag{5.3.15}$$

where $\underline{b} = \underline{b}_1 + \underline{b}_2$ is written in an arbitrary manner. The solution vector is then $\underline{u} = \underline{y} + \underline{z}$, where $\lim \underline{u}^{(k)} = \underline{y}$ and $\lim \underline{u}^{(k+1/2)} = \underline{z}$, (Kellogg, 1964).

Consider the MAGE scheme given by

$$(\bar{G}_1 + rI)\underline{u}^{(k+1/2)} = \underline{b}_1 - (\bar{G}_1 - rI)\underline{u}^{(k)}, \tag{5.3.16a}$$

$$(\bar{G}_2 + rI)\underline{u}^{(k+1)} = \underline{b}_2 - (\bar{G}_2 - rI)\underline{u}^{(k+1/2)}, \tag{5.3.16b}$$

where $\underline{b} = \underline{b}_1 + \underline{b}_2$ is written in an arbitrary manner. Combine the above two equations into the form

$$\underline{u}^{(k+1)} = \tilde{T}_r \underline{u}^{(k)} + \tilde{\underline{b}}_r, \tag{5.3.17}$$

where

$$\tilde{\underline{b}}_r = (\bar{G}_2 + rI)^{-1}\underline{b}_2 - (\bar{G}_2 + rI)^{-1} \times (\bar{G}_2 - rI)(\bar{G}_1 + rI)^{-1}\underline{b}_1 \tag{5.3.18}$$

and

$$\tilde{T}_r = (\bar{G}_2 + rI)^{-1}(\bar{G}_2 - rI) \times (\bar{G}_1 + rI)^{-1}(\bar{G}_1 - rI). \tag{5.3.19}$$

The matrix $\tilde{T}_r$ is called the MAGE matrix.

To prove the convergence of the method, we state the following theorem of Young.

Theorem 5.3.1
The iterative method (5.3.17) is convergent if and only if $\rho(\tilde{T}_r) \leq 1$.

From the properties of the matrix norms we have

$$\|\tilde{T}_r\| \leq \|(\bar{G}_2 + rI)^{-1}(\bar{G}_2 - rI)\| \cdot \|(\bar{G}_1 + rI)^{-1}(\bar{G}_1 - rI)\|. \qquad (5.3.20)$$

But since $\bar{G}_1$ and $\bar{G}_2$ are symmetric and $(\bar{G}_2 + rI)^{-1}$ commutes with $(\bar{G}_2 - rI)$ we have

$$\|(\bar{G}_2 + rI)^{-1}(\bar{G}_2 - rI)\| = \rho((\bar{G}_2 + rI)^{-1}(\bar{G}_2 - rI))$$

$$= \max_{\lambda} |(\lambda - r)/(\lambda + r)|, \qquad (5.3.21)$$

where λ ranges over all eigenvalues of $\bar{G}_2$. But since $\bar{G}_2$ is positive definite, its eigenvalues are positive. Therefore

$$\|(\bar{G}_2 + rI)^{-1}(\bar{G}_2 - rI)\| < 1. \qquad (5.3.22)$$

The same argument applied to the corresponding matrix product with $\bar{G}_1$ shows that $\|(\bar{G}_1 + rI)^{-1}(\bar{G}_1 - rI)\| < 1$, and we therefore conclude that

$$\rho(\tilde{T}_r) \leq \|\tilde{T}_r\| < 1. \qquad (5.3.23)$$

Hence, the convergence of the method follows.

5.4. The AGE Method for Two Dimensional Problems

Consider the two-dimensional elliptic PDE

$$-a_1 \frac{\partial^2 U}{\partial x^2} - a_2 \frac{\partial^2 U}{\partial y^2} + cu = g(x, y) \qquad (5.4.1)$$

(a_1, a_2, c constants $a_1, a_2 > 0$, $c \geq 0$) over the rectangle R, with vertices $(0,0)$, (ℓ_x, ℓ_y), with U specified on the boundary ∂R of R. For the solution of (5.4.1) a uniform grid of mesh spacings h_x and h_y in the x- and y-directions is imposed on R, so that $N_x = \ell_x/h_x$, $N_y = \ell_y/h_y$ are odd integers (≥ 5). The internal nodes are taken in blocks of four, the 4-point blocks are considered in their natural ordering and also the natural ordering of the four points within each block is adopted. Whenever necessary and for illustrative purposes we shall refer to Figure 5.4.1 below, which corresponds to $N_x = 9$ and $N_y = 7$. When the classical 5-point stencil is used, equation (5.4.1) can be approximated at each internal node by a 5-point difference equation. The totality of

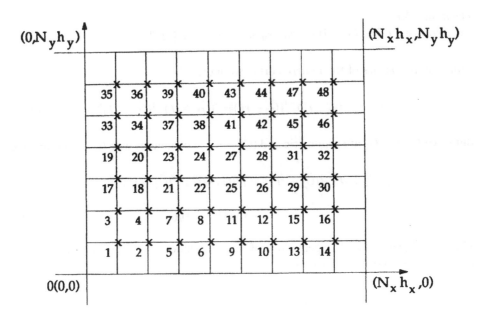

Figure 5.4.1

these $(N_x - 1)(N_y - 1)$ difference equations, where the boundary conditions are suitably incorporated, produces a linear system of the form

$$A\underline{u} = \underline{b}. \tag{5.4.2}$$

In the case of Figure 5.4.1 the matrix A is of the form

$$
A = \begin{bmatrix}
B & C & & & E & & & & & & & \\
C^T & B & C & & & E & & & & & & \\
& C^T & B & C & & & E & & & & & \\
& & C^T & B & C & & & E & & & & \\
\hline
E^T & & & C^T & B & C & & & E & & & \\
& E^T & & & C^T & B & C & & & E & & \\
& & E^T & & & C^T & B & C & & & E & \\
& & & E^T & & & C^T & B & C & & & E \\
\hline
& & & & E^T & & & C^T & B & C & & \\
& & & & & E^T & & & C^T & B & C & \\
& & & & & & E^T & & & C^T & B & C \\
& & & & & & & E^T & & & C^T & B
\end{bmatrix}, \tag{5.4.3}
$$

where

$$B = \begin{bmatrix} d & -e & -f & 0 \\ -e & d & 0 & -f \\ -f & 0 & d & -e \\ 0 & -f & -e & d \end{bmatrix}, \quad C = \begin{bmatrix} 0 & 0 & 0 & 0 \\ -e & 0 & 0 & 0 \\ 0 & 0 & 0 & 0 \\ 0 & 0 & -e & 0 \end{bmatrix},$$

$$E = \begin{bmatrix} 0 & 0 & 0 & 0 \\ 0 & 0 & 0 & 0 \\ -f & 0 & 0 & 0 \\ 0 & -f & 0 & 0 \end{bmatrix} \tag{5.4.4}$$

and

$$d = \frac{2a_1}{h_x^2} + \frac{2a_2}{h_y^2} + C, \quad e = \frac{a_1}{h_x^2}, \quad f = \frac{a_2}{h_y^2}. \tag{5.4.5}$$

The matrix A in (5.4.2), given through (5.4.3 – 5.4.5) and known to be positive definite is split as follows

$$A = F + G, \tag{5.4.6}$$

where

$$F = \text{diag}(B_1, B_1, \ldots, B_1), \quad G = A - F \tag{5.4.7}$$

and

$$B_1 = \begin{bmatrix} d_1 & -e & -f & 0 \\ -e & d_1 & 0 & -f \\ -f & 0 & d_1 & -e \\ 0 & -f & -e & d_1 \end{bmatrix}, \quad d_1 = \frac{a_1}{h_x^2} + \frac{a_2}{h_y^2} + c_1, \quad c_1 \in [0, c]. \tag{5.4.8}$$

In the next section we consider an alternative strategy for the solution of (5.4.2) besides the well known SOR, SLOR and ADI iterative methods.

5.5. Alternating Group Explicit (AGE) Iterative Schemes

A non-stationary biparametric analogue of the Alternating Direction Implicit (ADI) methods introduced by Peaceman and Rachford (1955) and Douglas (1955) (see also Varga, 1962 and Young, 1971) is now applied to solve (5.4.2). When (5.4.6) is taken into account, we shall call the strategy the Alternating Group Explicit (AGE) method.

The corresponding iterative scheme we propose for the solution of (5.4.2), in view of (5.4.3–5.4.8) is

$$(r_1^{(m)}I + F)\underline{u}^{(m+1/2)} = (r_1^{(m)}I - G)\underline{u}^{(m)} + \underline{b},$$

$$(r_2^{(m)}I + F)\underline{u}^{(m+1)} = (r_2^{(m)}I - G)\underline{u}^{(m+1/2)} + \underline{b}, \quad m = 0, 1, 2, \ldots, \tag{5.5.1}$$

where in (5.5.1) $u^{(0)}$ is arbitrary, $u^{(m+1/2)}$ an intermediate approximation to $u^{(m+1)}$ and $\{r_1^{(m)}\}, \{r_2^{(m)}\}, \ m = 0, 1, 2, \ldots$ two sequences of positive acceleration (iteration) parameters which will be determined in two specific stationary cases. The two-step iterative method introduced in (5.5.1) called Alternating Group Explicit (AGE) has the following characteristics. The matrix coefficient in the LHS of the first step is a direct sum of 4×4 block matrices corresponding to the four points of each group of Figure 5.4.1 taken (the groups) along group lines in x-direction. Each of these 4×4 blocks is the sum of the matrix B_1 of (5.4.8) and of $r_1^{(m)}I_4 \mid m = 0, 1, 2, \ldots$. Therefore its inverse can be found once, at the beginning of each step explicitly. On the other hand the coefficient in the LHS of the second step can be transformed by a permutation transformation, as is shown below, into a direct sum of 4×4 blocks, the inverses of which are readily available in an analogous way as before.

Before we proceed with the analysis of the AGE method, we determine the eigenvalues of the matrices F and G defined in (5.4.7–5.4.8). The eigenvalues of F are those of B_1 in (5.4.8) with a multiplicity $n_x n_y$ each, where

$$n_x = (N_x - 1)/2, \quad n_y = (N_y - 1)/2. \tag{5.5.2}$$

In view of (5.4.5) and (5.4.8) the eigenvalues of B_1 are, in turn, those of

$$V = \begin{bmatrix} e+f & -e & -f & 0 \\ -e & e+f & 0 & -f \\ -f & 0 & e+f & -e \\ 0 & -f & -r & e+f \end{bmatrix} \tag{5.5.3}$$

increased by c_1. The eigenvalues of V are easily determined to be the numbers 0, $2e$, $2f$, $2(e+f)$, so the eigenvalues of F are c_1, $2e + c_1$, $2f + c_1$, $2(e+f) + c_1$ each with a multiplicity $n_x n_y$.

We now observe that there always exists a permutation matrix P, which can be constructed analytically, such that the matrix $\tilde{G}$ given by

$$\tilde{G} = PGP^T \tag{5.5.4}$$

is the direct sum of 4×4 blocks. That at least one permutation matrix P exists such that (5.5.4) is valid can be easily proved by simply reordering the internal nodes of

R in the following way. First the four corner nodes are numbered in their natural ordering from 1 to 4. Then the remaining nodes on the two external rows followed by the nodes on the two external columns are numbered in their natural ordering from 5 to 2 $(N_x + N_y - 4)$. Finally the remaining nodes in groups of four (natural ordering for the groups with natural ordering of the nodes within each group) are numbered from $2(N_x + N_y - 4) + 1$ to $(N_x - 1)(N_y - 1)$ (see Figure 5.5.1 which corresponds to the new numbering of the nodes).

Now from the previous description of the new numbering the permutation matrix P in (5.5.4) can be constructed analytically. In what follows we give in Table 5.5.1 the row and column (col.) numbers in which the unit elements of P appear.

Table 5.5.1 Positions of P of (5.5.4) with a Unit Element

Row	Col.	Row	Col.	Row	Col.	Row	Col.	Row	Col.	Row	Col.
1	1	9	10	17	3	25	4	33	12	41	24
2	14	10	13	18	17	26	7	34	15	42	27
3	35	11	36	19	19	27	18	35	26	43	38
4	48	12	39	20	33	28	21	36	29	44	41
5	2	13	40	21	16	29	8	37	20	45	28
6	5	14	43	22	30	30	11	38	23	46	31
7	6	15	44	23	32	31	22	39	34	47	42
8	9	16	47	24	46	32	25	40	37	48	45

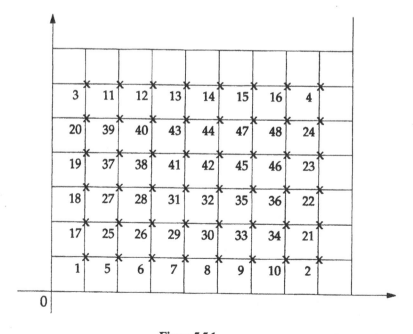

Figure 5.5.1

Thus $\tilde{G}$ is a block diagonal matrix consisting of 4×4 blocks of the following four types. The first block is the diagonal matrix of order 4

$$V_1 = \begin{bmatrix} d-d_1 & & & \\ & d-d_1 & & \mathbf{0} \\ \mathbf{0} & & d-d_1 & \\ & & & d-d_1 \end{bmatrix}. \tag{5.5.5}$$

The next $(n_x - 1)$ 4×4 blocks are of the form

$$V_2 = \begin{bmatrix} d-d_1 & -e & & \\ -e & d-d_1 & & \mathbf{0} \\ \mathbf{0} & -e & d-d_1 & -e \\ & & -e & d-d_1 \end{bmatrix}, \tag{5.5.6}$$

followed by another $(n_y - 1)$ 4×4 blocks of the type

$$V_3 = \begin{bmatrix} d-d_1 & -f & & \\ -f & d-d_1 & & \mathbf{0} \\ \mathbf{0} & -f & d-d_1 & -f \\ & & -e & d-d_1 \end{bmatrix}. \tag{5.5.7}$$

Finally the last $(n_x - 1)(n_y - 1)$ 4×4 blocks have the form

$$V_4 = \begin{bmatrix} d-d_1 & -e & -f & 0 \\ -e & d-d_1 & 0 & -f \\ -f & 0 & d-d_1 & -e \\ 0 & -f & -e & d-d_1 \end{bmatrix}. \tag{5.5.8}$$

Now V_1 has the eigenvalues $e+f+c_2$ $(c_2 = c - c_1)$ of multiplicity four, U_2 the eigenvalues $f+c_2$ and $2e+f+c_2$ of multiplicity two each, U_3 the eigenvalues $e+c_2$ and $e+2f+c_2$ of multiplicity two each and U_4 the eigenvalues c_2, $2e+c_2$, $2f+c_2$ and $2(e+f)+c_2$. The matrix G possesses the following eigenvalues: $e+f+c_2$ (multiplicity four), $f+c_2$ and $2e+f+c_2$ of multiplicity $2(n_x-1)$ each, $e+c_2$ and $e+2f+c_2$ of multiplicity $2(n_y-1)$ each and finally c_2, $2e+c_2$, $2f+c_2$ and $2(e+f)+c_2$ of multiplicity $(n_x-1)(n_y-1)$ each.

We note that the matrices F and G from (5.4.7) and (5.4.3) are real symmetric matrices, do *not* commute. The latter observation does not enable us to determine optimum sequences for the two sets of the acceleration parameters of the Jordan–Wachspress's type (see Wachspress, 1966 and also Young, 1971), where the commutativity of the operators F and G is assumed. So, for the moment, we turn our attention to the stationary biparametric and monoparametric cases of (5.5.1), before we discuss the corresponding non-stationary monoparametric one.

(1) *Stationary biparametric AGE method* It is $r_1 \equiv r_1^{(0)} = r_1^{(1)} = \cdots$ and $r_2 \equiv r_2^{(0)} = r_2^{(1)} = \cdots$ so that (5.5.1) becomes

$$(r_1 I + F)\underline{u}^{(m+1/2)} = (r_1 I - G)\underline{u}^{(m)} + \underline{b},$$

$$(r_2 I + F)\underline{u}^{(m+1)} = (r_2 I - G)\underline{u}^{(m+1/2)} + \underline{b}, \quad m = 0, 1, 2, \ldots, \tag{5.5.9}$$

and on elimination of $u^{(m+1/2)}$ from (5.5.9) we obtain

$$\underline{u}^{(m+1)} = (r_2 I + G)^{-1}(r_2 I - F)(r_1 I + F)^{-1}(r_1 I - G)\underline{u}^{(m)}$$

$$+ (r_1 + r_2)(r_2 I + G)^{-1}(r_1 I + F)^{-1}\underline{b}, \quad m = 0, 1, 2, \ldots. \tag{5.5.10}$$

The iteration matrix T_{r_1,r_2} of scheme (5.5.10) is similar to $\tilde{T}_{r_1,r_2}$ where

$$\tilde{T}_{r_1,r_2} = (r_r I + G)T_{r_1,r_2}(r_2 I + G)^{-1}$$

$$= (r_2 I - F)(r_1 I + F)^{-1}(r_1 I - G)(r_2 I + G)^{-1}. \tag{5.5.11}$$

Taking 2-norms in (5.5.11) we successively obtain

$$\rho(T_{r_1,r_2}) = \rho(\tilde{T}_{r_1,r_2}) \leq \| \tilde{T}_{r_1,r_2} \|_2$$

$$\leq \| (r_2 I - F)(r_1 + F)^{-1} \|_2 \, \| (r_1 I - G)(r_2 I + G)^{-1} \|_2$$

$$= \rho\big((r_2 I - F)(r_1 I + F)^{-1}\big)\rho\big((r_1 I - G)(r_2 I + G)^{-1}\big)$$

$$= \max_{f_i} \left| \frac{r_2 - f_i}{r_1 + f_i} \right| \max_{g_i} \left| \frac{r_1 - g_i}{r_2 + g_i} \right|,$$

or equivalently

$$\rho(T_{r_1,r_2}) \leq \max_{f_i,g_i} \left| \frac{(r_2 - f_i)(r_1 - g_i)}{(r_1 + f_i)(r_2 + g_i)} \right|, \tag{5.5.12}$$

where f_i and g_i are the eigenvalues of the matrices F and G, respectively. From the previous analysis, however, we have that the intervals in which the eigenvalues f_i and g_i vary are the following:

$$c_1 \leq f_i \leq 2(e + f) + c_1, \quad c_2 \leq g_i \leq 2(e + f) + c_2, \tag{5.5.13}$$

with c_1, e, f being given by (5.4.5) and (5.4.8) and $c_2 = c - c_1$. Thus our optimisation problem turns out to be that of the minimisation of the RHS of (5.5.12) under restrictions (5.5.13). This is always possible, except when $c = 0$ (Laplace's equation), and the optimum parameters can be obtained directly from the corresponding analysis by Avdelas and Hadjidimos (1979) in the commutative case, which still holds, where, however, the optimum parameters are not given in a closed form and the case $c_1 c_2 = 0$, with $c > 0$ is not covered. If we put

$$\mu_1 = c_1, \quad v_1 = 2(e+f)+c_1, \quad \mu_2 = c_2, \quad v_2 = 2(e+f)+c_2, \tag{5.5.14}$$

the optimum parameters are given by the expressions

$$r_1 = K/(\mu_1 v_1 - \mu_2 v_2 + \alpha \beta \gamma \delta),$$

$$r_2 = K/(\mu_2 v_2 - \mu_1 v_1 + \alpha \beta \gamma \delta), \tag{5.5.15}$$

$$M = (1-L)/(1+L),$$

where

$$K = \mu_1 v_1 (\mu_2 + v_2) + \mu_2 v_2 (\mu_1 + v_1),$$

$$\alpha = (\mu_1 + \mu_2)^{1/2}, \quad \beta = (\mu_1 + v_2)^{1/2}, \quad \gamma = (\mu_2 + v_1)^{1/2}, \quad \delta = (v_1 + v_2)^{1/2}, \tag{5.5.16}$$

$$L = (\alpha \delta)/(\beta \gamma),$$

with M being the optimum spectral radius of scheme (5.5.9).

For $c > 0$ it can be proved that the optimum M given in (5.5.15) is independent of c_1 and c_2. More specifically:

Theorem 5.5.1.
For $c = c_1 + c_2 > 0$, $c_1, c_2 \geq 0$ the value of M for the optimum stationary biparametric AGE method is independent of how c is split into c_1 and c_2.

Proof. This follows immediately since the numbers α, β, γ and δ do not depend on the separation of c into c_1 and c_2. Indeed, after some manipulation by using (5.5.15), (5.5.16) and (5.5.14) it can be found that

$$M = \frac{1 - (c(4(e+f)+c))^{1/2}(2(e+f)+c)}{1 + (c(4(e+f)+c))^{1/2}(2(e+f)+c)}.$$

(2) *Stationary monoparametric AGE method* This time it is $r = r_1^{(0)} = r_2^{(0)} = r_1^{(1)} = r_2^{(1)} = \cdots$ so that one can obtain the analogues of (5.5.9 – 5.5.11) with the only difference being that $r_1 = r_2 = r$. Except for the case of Laplace's equation ($c = 0$), when an optimum r cannot be found (as in the stationary biparametric case), in all other cases the optimum parameters can be obtained directly from the theory developed in Young (1971). More specifically, and if $c_1 \leq c_2$, it is

$$r = \left(c_2 (2(e+f)+c_2) \right)^{1/2},$$

$$M = \frac{\left((c_2(2(e+f)+c_2))^{1/2} - c_1 \right) \left((2(e+f)+d_2)^{1/2} - c_2^{1/2} \right)}{\left((c_2(2(e+f)+c_2))^{1/2} + c_1 \right) \left((2(e+f)+d_2)^{1/2} + c_2^{1/2} \right)}. \tag{5.5.17}$$

In case $c_2 \leq c_1$ one obtains again formulae analogous to (5.5.17) with the roles of c_1 and c_2 being interchanged. In what follows we prove that a minimum value for M in (5.5.17) when $c_1 \leq c_2$, is achieved for $c_1 = c_2 = c/2$. (It is understood that with the same reasoning one can also prove that in the case $c_2 \leq c_1$ the corresponding minimum for M will take place again for $c_1 = c_2 = c/2$.)

Theorem 5.5.2.
For $c = c_1 + c_2, c_1, c_2 \geq 0$, and $c_2 \geq c_1$ the least value for M in (5.5.17) is achieved for $c_1 = c_2 = c/2$.

Proof. In view of (5.5.17), M is written successively as follows:

$$M = \frac{(r - c_1)(r - c_2)}{(r + c_1)(r + c_2)}$$

$$= 1 - 2rc/[r^2 + rc + c_2(c - c_2)]$$

$$= 1 - 2rc/[\alpha c_2 + rc + cc_2]$$

$$= 1 - 2c(r/c_2)/[\alpha + c(r/c_2) + c],$$

where $\alpha = 2(e + f)$. Thus, M is a decreasing function of (r/c_2), but (r/c_2) is a decreasing function of c_2, so $\partial M/\partial c > 0$. Therefore, M increases w.r.t c_2 and its minimum is achieved for $c_2 = c/2 (= c_1)$ which proves the theorem.

Note: For $c_1 = c_2 = c/2$ (5.5.17) becomes

$$r = 1/2(c(4(e + f) + c))^{1/2},$$

$$M = \frac{(1 - (c/(4(e + f) + c))^{1/2})^2}{(1 + (c/(4(e + f) + c))^{1/2})^2}, \qquad (5.5.17')$$

and it can be proved that the value for M above is identical with that of Theorem 5.5.1. This is something which was expected since the ranges for the eigenvalues f_i and g_i are the same, namely $[c/2, 2(e - f) + c/2]$.

Although for Laplace's equation optimum results cannot be obtained, it can be proved, however, that for any choice of $r > 0$ the corresponding AGE scheme converges more specifically,

Theorem 5.5.3.
The stationary monoparametric AGE scheme (5.5.9) $(r = r_1 = r_2)$ for Laplace's equation converges for any choice of the acceleration parameter $r > 0$.

Proof. For the iteration matrix $T_{r,r}$ of scheme (5.5.9), it is shown that

$$T_{r,r} = (rI + G)^{-1}(rI - F)(rI + F)^{-1}(rI - G),$$

where now the matrices F and G are real symmetric semipositive definite with their sum $F + G = A$ being a real symmetric positive definite matrix. From Theorem 2.9 of Young (1971) with $A^{1/2}$ the square root of A (which is again real, symmetric, positive definite) we have that

$$\rho(T_{r,r}) = \rho(A^{1/2}T_{r,r}A^{-1/2}) \leq \|A^{1/2}T_{r,r}A^{-1/2}\|_2 = \|T_{r,r}\|A^{1/2} < 1, \qquad (5.5.18)$$

iff $FG + GF + 2r^2I$ is positive definite. The latter, in view of the observation made in Young (1971) immediately after the aforementioned theorem, is equivalent to $FG + rI$ being positive real. However this is true since the product FG of two real symmetric semipositive definite matrices is a real matrix with non-negative eigenvalues so that $FG + rI$, with $r > 0$, is real positive, which proves the present theorem.

(3) *Non-stationary monoparametric AGE method*   We have $r^{(0)} = r_1^{(0)} = r_2^{(0)}$, $r^{(1)} = r_1^{(1)} = r_2^{(1)}, \ldots, r^{(k-1)} = r_1^{(k-1)} = r_2^{(k-1)}$ if a set of k positive acceleration parameters is used in a cyclic way. It is readily found out that the iteration matrix $T_{r^{(0)}r^{(1)} \ldots r^{(k-1)}}$ of scheme (5.5.1) after a complete cycle of k iterations starting with $r^{(0)}$ is given by

$$T_{r^{(0)}r^{(1)} \ldots r^{(k-1)}} = T_{r^{(k-1)},r^{(k-1)}} \cdots T_{r^{(1)}r^{(1)}} T_{r^{(0)}r^{(0)}}. \qquad (5.5.19)$$

For the corresponding AGE scheme the following theorem holds.

Theorem 5.5.4.
The non-stationary monoparametric scheme (5.5.1) for $c \geq 0$ converges for any choice of the k positive acceleration parameters $r^{(0)}$, $r^{(1)}$, $\ldots$, $r^{(k-1)}$.

Proof. For $c > 0$ the assertion of the theorem is an intermediate consequence of the Corollary 2.10 of Young [1971, p. 502]. For $c = 0$ (Laplace's equation) use of Theorem 5.5.3 can be made. More specifically from (5.5.19) it can be obtained

$$\rho(T_{r^{(0)}r^{(1)} \ldots r^{(k-1)}}) \leq \|A^{1/2} T_{r^{(0)}r^{(1)} \ldots r^{(k-1)}} A^{-1/2}\|_2$$

$$\leq \| T_{r^{(k-1)},r^{(k-1)}} \| A^{1/2} \cdots \| T_{r^{(1)}r^{(1)}} \| A^{1/2} \| T_{r^{(0)}r^{(0)}} \| A^{1/2} \leq 1,$$

since from (5.5.18) each factor is strictly less than 1.

Unfortunately, even in the case $c > 0$ an optimum set of the k iteration parameters cannot be found because the matrices F and G do not commute. However, if one can exploit the result of Widlund (1966), where in a similar non-commutative case, the set of parameters obtained by Peaceman and Rachford (1955) works equally well, then our problem will have been solved. However, it has not been proved that all the lemmas and theorems on which Widlund based his analysis hold in our case.

5.6. Comparison of the AGE, the Classical ADI and the Point SOR Methods and Conclusions

Assume that for the numerical solution of (5.4.1) in R the same spacings, formulae, etc. as in Section 5.4 are used except that this time a natural ordering of the nodes in R is adopted. The linear system to be solved is again of the general form (5.4.2) namely, $A_1 \underline{u}_1 = \underline{b}_1$, where $A_1 = P_1 A P_1^T$, $\underline{u}_1 = P_1 \underline{u}$ and $\underline{b}_1 = P_1 \underline{b}$, with P_1 being a suitable permutation matrix. A_1 now is given by the expression

$$A_1 = H + V \qquad (5.6.1)$$

and

$$H = I_{N_y-1} \otimes (U_{N_x-1} + c_1 I_{N_x-1}), \quad V = (U_{N_y-1} + c_1 I_{N_y-1}) \otimes I_{N_x-1}, \qquad (5.6.2)$$

where U_{N_x-1}, U_{N_y-1} are the tridiagonal matrices $\mathrm{diag}(-e, d_1 - e)$ and $\mathrm{diag}(-f, d - d_1, -f)$ of orders $N_x - 1$ and $N_y - 1, I_{N_x-1}, I_{N_y-1}$ the unit matrices of the corresponding orders and $\otimes$ a tensor product. (The definitions for d, e, f, d_1, etc., are given in (5.5.5) and (5.5.8).) The classical stationary biparametric ADI (Peaceman–

Rachford) scheme for the solution of the corresponding system (5.5.2), with A being given by (5.6.1) and (5.6.2) is the following:

$$(\tilde{r}_1 I + H)\underline{u}_1^{(m+1/2)} = (\tilde{r}_1 I - V)\underline{u}_1^{(m)} + \underline{b}_1,$$

$$(\tilde{r}_2 I + V)\underline{u}_1^{(m+1)} = (\tilde{r}_2 I - H)\underline{u}_1^{(m+1/2)} + \underline{b}_1. \qquad (5.6.3)$$

Since the eigenvalues of H and V are $4e\sin^2(k\pi/(2N_x)) + c_1$, $k = 1(1)N_x - 1$ with multiplicity $N_y - 1$ each and $4f\sin^2(k\pi/(2N_y)) + c_2$ with multiplicity $N_x - 1$ each, the optimum parameters can be found by using formulae corresponding to (5.5.14–5.5.16). Notably we shall have

$$\tilde{M} = (1 - \tilde{L})/(1 + \tilde{L}) \qquad (5.6.4a)$$

with

$$\tilde{L} = \frac{(4e\sin^2(\pi/2N_x) + 4f\sin^2(\pi/2N_y) + c) \times (4e\cos^2(\pi/2N_x) + 4f\cos^2(\pi/2N_y) + c))^{1/2}}{(4e\sin^2(\pi/2N_x) + 4f\sin^2(\pi/2N_y) + c) \times (4f\cos^2(\pi/2N_y) + 4e\cos^2(\pi/2N_x) + c))^{1/2}}.$$

$$(5.6.4b)$$

As can be seen from (5.6.4a–5.6.4b), besides the fact that (5.6.3) always converges (even for $c = 0$) it converges faster than the corresponding AGE scheme because for all cases $\tilde{M} < M$.

For the natural ordering of the nodes the matrix A_1, used in the previous paragraph and defined by (5.6.1–5.6.2), is weakly 2-cyclic consistently ordered. If we now consider a red/black ordering of the entire set of the $(N_x - 1)(N_y - 1)$ nodes of R the corresponding linear system (5.4.2) will be of the form $A_2 \underline{u}_2 = \underline{b}_2$ with $A_2 = P_2 A P_2^T$, $\underline{u}_2 = P_2 \underline{u}$, $\underline{b}_2 = P_2 \underline{b}$ and P_2 a suitable permutation matrix. Obviously the matrix A_2 is weakly 2-cyclic consistently ordered. On the other hand it is apparent that A_1 and A_2 are similar matrices. So also will be their associated Jacobi matrices B_1 and B_2. Consequently the optimum spectral radii of the point SOR methods corresponding to the A_1 and A_2 matrices will be identical. In addition they will be given by the expression $\tilde{M}$ of (5.6.4a–5.6.4b) for the corresponding model problem. Therefore for large N_x, N_y the AGE method introduced in this paper is asymptotically as fast as the ADI and the point SOR methods ($c > 0$) with the latter methods being based on either a natural or a red/black ordering of the nodes in R.

In order to simplify the analysis we shall assume that in (5.4.1) $a_1 = a_2 = 1$, R is a square so that $\ell_x = \ell_y$ and the spacings in the x- and y-direction are equal. In other words $h \equiv h_x = h_y$, hence $N \equiv N_x = N_y$. For this model problem we shall try to find, and therefore to compare, the number of operations involved when each one of the three methods (AGE, ADI and SOR with red/black ordering) is used for the numerical solution of the associated linear system (5.4.2). Two further points will also be taken into consideration:

(i) any inherent parallelism in the operations involved so that a parallel computer of SIMD (Single Instruction Stream-Multiple Data Stream) type may be used for the solution of a real problem; and

(ii) any symmetry in the matrices or submatrices involved so that the number of operations may be reduced.

We consider the AGE (5.5.1) and ADI (5.6.3) methods with their corresponding optimum parameters for $c_1 = c_2 = c/2$, when $r_1 = r_2$ and $\tilde{r}_1 = \tilde{r}_2$, respectively. We assume further that the operations involved:

(i) to find the inverses of the block matrices

$$\frac{1}{h^2}\begin{bmatrix} \alpha_1 & -1 \\ -1 & \alpha_1 \end{bmatrix}, \qquad \frac{1}{h^2}\begin{bmatrix} \alpha_1 & -1 & -1 & 0 \\ -1 & \alpha_1 & 0 & -1 \\ -1 & 0 & \alpha_1 & -1 \\ 0 & -1 & -1 & \alpha_1 \end{bmatrix},$$

$(\alpha_1 = 2 + (r_1 + c_1)h^2)$; and

(ii) when techniques like the ones suggested in Evans and Biggins (1982) are used to save operations, do not affect the total number of operations to be performed in the AGE method.

To calculate the RHS of the methods characterised by the equations (5.4.1) or (5.6.3) takes 2 mults. (multiplications) and 3 adds. (additions) per point per iteration if symmetry is used plus 1 add. corresponding to the components of the constant vector b.

Then, for the AGE method (eqn. 5.4.1) since the inverses of the matrices on the LHS may be obtained explicitly then 4 mults. and 3 adds. per mesh point is required to obtain $\underline{u}^{(m+1/2)}$. Finally, in order to calculate $\underline{u}^{(m+1)}$ the total number of operations required is 12 mults. $+ 14$ adds. per mesh point per iteration. However, it can be observed that the calculations of each mesh point are independent of each other and can be performed simultaneously on a SIMD computer making the total computational complexity per iteration to be only 12 mults. and 14 adds.

However, for the ADI method (eqn. 5.6.3) and after the calculation of the RHS have been completed, i.e. 4 mults. $+ 8$ adds. as above, $(N-1)$ tridiagonal systems each of order $(N-1)$ have to be solved in each half iteration. On a sequential computer this requires 5 mults. and 3 adds. or only 3 mults. $+ 2$ adds. per mesh point if the multipliers in the Gaussian elimination process are retained for subsequent use whilst it has been shown that the LU decomposition of an $(N \times N)$ tridiagonal matrix can be computed in $0(\log N)$ time by recursive doubling, using N processors. Therefore, each iteration of the ADI method will require $2N \log N$ steps thus supporting the conclusion that the AGE method is more suitable on a parallel computer.

Coming now to the point SOR method based on a red/black ordering we observe that it is also a two-step procedure in the following sense. First all $(N-1)^2/2$ new red components have to be found from all $(N-1)^2$ old red and black points and then the $(N-1)^2/2$ new black components from the $(N-1)^2/2$ old black and the $(N-1)^2/2$ new red ones. Provided that the quantities $h^2/(4 + ch^2)$, ω_b and $1 - \omega_b$ are predetermined (the optimum overrelaxation parameter ω_b is found from the expression $2/(1 + (1 - (4\cos(\pi/N)/(4 + ch^2))^2)^{1/2})$ a similar result is given, then for each new component a total of 3 mults. and 5 adds. have to be performed, if symmetry is exploited. As is seen the SOR method in question, although it has the same order in its rate of convergence the AGE method, requires much fewer operations per iteration and in addition is itself a parallel procedure suitable for an SIMD computer since the $(N-1)^2/2$ new components can be computed simultaneously.

Further numerical experiments were carried out on three problems (5.4.1), where $a_1 = a_2 = 1$, $c_1 = c_2 = c/2$, and $c = 400$, $300 \cosh(xy)$ and 0 to compare the performance of the AGE method, i.e. number of iterations and parameter ranges, with the SOR and SLOR methods. The convergence test used as the average test, i.e.,

$$\| u_i^{(k+1)} - u_i^{(k)} \| / (1 + \| u_i^{(k)} \|) < \varepsilon, \quad i = 1, 2, \ldots, (N_x - 1)(N_y - 1),$$

with $\varepsilon = 5 \times 10^{-6}$. Tables 5.6.1–5.6.3 show the results obtained and confirm the theoretical results derived.

Table 5.6.1 Comparison of SOR, SLOR and AGE Methods for $c = 400$

	SOR		SLOR		AGE	
h^{-1}	ω	k	ω	k	r	k
13	1.11	8	1.07	4	1.9 – 2.6	5
25	1.32	12	1.21	6	1.1 – 1.35	9
37	1.45	17	1.34	8	0.91 – 0.94	13
49	1.54	22	1.44	10	0.78 – 0.82	19

Table 5.6.2 Performance of the AGE Method for $c = 300 \cosh(xy)$

	AGE	
h^{-1}	r	k
13	1.54 – 2.42	6
25	1.04 – 1.16	10
37	0.84 – 0.96	16
49	0.74 – 0.76	23

Table 5.6.3 Comparision of the SOR, SLOR and AGE Methods for $c = 0$ (Laplace Equation)

	AGE		SOR	SLOR
h^{-1}	r	k	k	k
13	1.35 – 1.49	12	26	20
25	0.83 – 0.86	21	52	40
37	0.68 – 0.69	36	75	55
49	0.63	55	99	72

5.7. The Normalised Alternating Group Explicit (NAGE) Method

5.7.1. Introduction

In this section we present a computationally more efficient form of the AGE algorithm.

Suppose we have the tridiagonal linear system of equations

$$A\underline{x} = \underline{g}, \tag{5.7.1}$$

where

$$A = \begin{bmatrix} a & b & & & & \\ b & a & b & & & 0 \\ & b & a & b & & \\ & & \ddots & \ddots & \ddots & \\ & & & & & b \\ & 0 & & & b & a \end{bmatrix} \in R^{n \times n}. \tag{5.7.2}$$

Sets of such equations or a large single system of equations constitute a huge number of problems in physics and engineering.

Suppose that n is even (if n is odd, only a minor change is needed), and let

$$n = 2m, \quad C = \begin{bmatrix} a/2 & b \\ b & a/2 \end{bmatrix} \tag{5.7.3}$$

and

$$G_1 = \begin{bmatrix} C & & & \\ & C & & 0 \\ & & \ddots & \\ & 0 & & C \end{bmatrix}, \quad G_2 = \begin{bmatrix} a/2 & & & & \\ & C & & 0 & \\ & & \ddots & & \\ & 0 & & C & \\ & & & & a/2 \end{bmatrix}, \tag{5.7.4}$$

such that

$$A = G_1 + G_2. \tag{5.7.5}$$

Then, the AGE method is defined by the equations

$$(G_1 + rI)\underline{x}^{(k+1/2)} = \underline{g} - (G_2 - rI)\underline{x}^{(k)},$$

$$(G_2 + rI)\underline{x}^{(k+1)} = \underline{g} - (G_1 - rI)\underline{x}^{(k+1/2)}, \quad (k \geq 0), \tag{5.7.6}$$

where $\underline{x}^{(0)}$ is the initial guess, r is an acceleration parameter. If G_1 and G_2 are semi-positive definite and if one of them is non-singular, then the iterates $\underline{x}^{(k)}$ of (5.7.6) will converge to the exact solution of (5.7.1) for any positive parameter r. It is clear that the choice of r will affect the convergence rate of the AGE method. It has been shown that the optimal parameter r is dependent on the upper and lower bounds of the eigenvalues of G_1 and G_2. The parameter r in (5.7.6) can also be changed from iteration to iteration, which results in a faster rate of convergence.

In this section we analyse the computational complexity of the AGE method. When writing the computer program for the AGE scheme, it is usual to use the two sweep formulae, i.e., first to compute $x^{(k+1/2)}$ and then to compute $x^{(k+1)}$, which results in a total of $8n$ multiplications and $6n$ additions per iteration. It is shown that by scaling and careful organisation of the AGE formulae, the number of operations per iteration can be reduced to $5n$ multiplications and $5n$ additions or $4n$ multiplications and $5n$ additions at the expense of one extra vector storage. This results in a saving of 37.5% in multiplications and 16.6% in additions for the former and 50% in multiplications and 16.6% in additions for the latter.

5.7.2. The Computational Complexity of the AGE Method

Now we consider the practical implementation of the AGE method. We will also estimate the amount of operations required per iteration.

Let

$$C_1 = \begin{bmatrix} \alpha & b \\ b & \alpha \end{bmatrix}, \quad C_2 = \begin{bmatrix} \beta & b \\ b & \beta \end{bmatrix}, \quad \text{with} \quad \begin{aligned} \alpha &= (a/2) + r, \\ \beta &= (a/2) - r. \end{aligned} \tag{5.7.7}$$

Then

$$C_1^{-1} = \frac{1}{d} \begin{bmatrix} \alpha & -b \\ -b & \alpha \end{bmatrix} = \begin{bmatrix} s & t \\ t & s \end{bmatrix}, \tag{5.7.8}$$

where

$$s = \frac{\alpha}{d}, \quad t = -\frac{b}{d}, \quad d = \alpha^2 - b^2. \tag{5.7.9}$$

Thus, (5.7.6) can be expressed by the following explicit form:

$$\begin{bmatrix} x_1 \\ x_2 \\ \vdots \\ x_n \end{bmatrix}^{(k+1/2)} = \begin{bmatrix} C_1^{-1} & & & \\ & C_1^{-1} & & 0 \\ & & \ddots & \\ 0 & & & C_1^{-1} \end{bmatrix} \times \left(\begin{bmatrix} g_1 \\ g_2 \\ \vdots \\ g_n \end{bmatrix} - \begin{bmatrix} \beta & & & \\ & C_2 & & 0 \\ & & \ddots & \\ 0 & & & C_2 \\ & & & \beta \end{bmatrix} \begin{bmatrix} x_1 \\ x_2 \\ \vdots \\ x_n \end{bmatrix}^{(k)} \right) \tag{5.7.10a}$$

and

$$
\begin{bmatrix} x_1 \\ x_2 \\ \vdots \\ x_n \end{bmatrix}^{(k+1)} = \begin{bmatrix} \alpha^{-1} & & & \\ & C_1^{-1} & & 0 \\ & & \ddots & \\ & 0 & & C_1^{-1} \\ & & & & \alpha^{-1} \end{bmatrix}
$$

$$
\times \left\{ \begin{bmatrix} g_1 \\ g_2 \\ \vdots \\ g_n \end{bmatrix} - \begin{bmatrix} C_2 & & & \\ & C_2 & & 0 \\ & & \ddots & \\ & 0 & & C_2 \end{bmatrix} \begin{bmatrix} x_1 \\ x_2 \\ \vdots \\ x_n \end{bmatrix}^{(k+1/2)} \right\}. \qquad (5.7.10b)
$$

Since C_2 and C_1^{-1} are 2×2 matrices, it follows from (5.7.10a) and (5.7.10b) that the number of operations required to calculate $\underline{x}^{(k+1/2)}$ from $x_j^{(k)}$ or $x_j^{(k+1)}$ from $x_j^{(k+1/2)}$ is 4 multiplications and 3 additions. Thus, the number of operations required to calculate $\underline{x}^{(k+1)}$ from $\underline{x}^{(k)}$ by the AGE method is (approximately):

8n multiplications and 6n additions.

We now summarise the above in the following algorithm.

Algorithm 1
Phase I.   Preprocessing step

1.1 Choose the optimal parameter r and initial vector $\underline{x}^{(0)}(x_j^{(0)}, j = 1, 2, \ldots, n = 2m)$.

1.2 Set $\alpha = a/2 + r$, $\beta = a/2 - r$, $d = (\alpha - b)(\alpha + b)$, $s = \alpha/d, t = -b/d$.

Phase II.   The AGE iteration

For $k = 0, 1, \ldots$ until convergence do.

2.1 $u_1 = g_1 - \beta x_1^{(k)}$

$u_n = g_n - \beta x_n^{(k)}$

For $\ell = 1, 2, \ldots, m-1$ do,

$$
\begin{bmatrix} u_{2\ell} \\ u_{2\ell+1} \end{bmatrix} = \begin{bmatrix} g_{2\ell} \\ g_{2\ell+1} \end{bmatrix} - \begin{bmatrix} \beta & b \\ b & \beta \end{bmatrix} \begin{bmatrix} x_{2\ell}^{(k)} \\ x_{2\ell+1}^{(k)} \end{bmatrix}.
$$

2.2 For $\ell = 1, 2, \ldots, m$ do

$$
\begin{bmatrix} v_{2\ell-1} \\ v_{2\ell} \end{bmatrix} = \begin{bmatrix} s & t \\ t & s \end{bmatrix} \begin{bmatrix} u_{2\ell-1} \\ u_{2\ell} \end{bmatrix}.
$$

2.3 For $\ell = 1, 2, \ldots, m$ do

$$
\begin{bmatrix} q_{2\ell-1} \\ q_{2\ell} \end{bmatrix} = \begin{bmatrix} g_{2\ell-1} \\ g_{2\ell} \end{bmatrix} - \begin{bmatrix} \beta & b \\ b & \beta \end{bmatrix} \begin{bmatrix} v_{2\ell-1} \\ v_{2\ell} \end{bmatrix}.
$$

2.4 $x_1^{(k+1)} = q_1/\alpha$

$\quad x_n^{(k+1)} = q_n/\alpha.$

For $\ell = 1, 2, \ldots, m-1$ do

$$\begin{bmatrix} x_{2\ell}^{(k+1)} \\ x_{2\ell+1}^{(k+1)} \end{bmatrix} = \begin{bmatrix} s & t \\ t & s \end{bmatrix} \begin{bmatrix} q_{2\ell} \\ q_{2\ell+1} \end{bmatrix}.$$

From the above algorithm it is clear that $2n$ multiplications and $2n$ additions are needed in Steps 2.1 and 2.3, respectively, and that $2n$ multiplications and n additions are needed in Steps 2.2 and 2.4, respectively. Therefore, the total requirements for one iteration are $8n$ multiplications and $6n$ additions.

However, we should note that if $b = 1$, then the number of multiplications in each of Steps 2.1 and 2.3 are n instead of $2n$. Therefore in order to reduce the multiplication requirements we should normalise or scale the system (5.7.1) so that the off-diagonal element equals unity before the AGE method is used. This will not affect the convergence property of the AGE method. In fact, from (5.7.6) the AGE iteration can be expressed by

$$\underline{x}^{(k+1)} = T_r \underline{x}^{(k)} + \underline{z}, \tag{5.7.11}$$

where z is a suitable vector and T_r is given by

$$T_r = (G_2 + rI)^{-1}(G_1 - rI)(G_1 + rI)^{-1}(G_2 - rI).$$

The convergence property of the algorithm is determined by the matrix T_r.

If we now scale (5.7.1) by multiplying both sides by b^{-1}, then we have

$$\hat{A}\underline{x} = \hat{\underline{g}}, \tag{5.7.12}$$

where

$$\hat{A} = \hat{G}_1 + \hat{G}_2, \quad \hat{G}_i = b^{-1}G_i, \quad i = 1, 2, \quad \hat{\underline{g}} = b^{-1}\underline{g}. \tag{5.7.13}$$

Thus, the AGE method applied to (5.7.12) can be expressed by

$$\hat{\underline{x}}^{(k+1)} = \hat{T}_{\hat{r}}\hat{\underline{x}}^{(k)} + \hat{\underline{z}}, \tag{5.7.14}$$

where $\hat{\underline{z}}$ is a suitable vector and $\hat{T}_{\hat{r}}$ is given by

$$\hat{T}_{\hat{r}} = (\hat{G}_2 + \hat{r}I)^{-1}(\hat{G}_1 - \hat{r}I)(\hat{G}_1 + \hat{r}I)^{-1}(\hat{G}_2 - \hat{r}I)$$

$$= (G_2 + b\hat{r}I)^{-1}(G_1 - b\hat{r}I)(G_1 + b\hat{r}I)^{-1}(G_2 - b\hat{r}I)$$

$$= \hat{T}_{b\hat{r}}.$$

Thus, let r and $\hat{r}$ be the optimal parameters for (5.7.11) and (5.7.14), respectively, thus they satisfy the relation

$$r = b\hat{r}. \tag{5.7.15}$$

Also, the two procedures (5.7.11) and (5.7.14) will have the same convergence speed if they start with the same initial vector.

On the other hand, if we let

$$
\begin{bmatrix} s' & t' \\ t' & s' \end{bmatrix} = \begin{bmatrix} \beta & b \\ b & \beta \end{bmatrix} \begin{bmatrix} s & t \\ t & s \end{bmatrix}, \tag{5.7.16}
$$

and combine Steps 2.2 and 2.3 of Algorithm 1 into one step, i.e.,

$$
\begin{bmatrix} q_{2\ell-1} \\ q_{2\ell} \end{bmatrix} = \begin{bmatrix} g_{2\ell-1} \\ g_{2\ell} \end{bmatrix} - \begin{bmatrix} s' & t' \\ t' & s' \end{bmatrix} \begin{bmatrix} u_{2\ell-1} \\ u_{2\ell} \end{bmatrix}, \quad \ell = 1, 2, \ldots, m, \tag{5.7.17}
$$

will result in a further reduction of the multiplication requirements.

Now we summarise the above discussions in the following:

Algorithm 2
Phase I.   Preprocessing step

1.1 Choose the optimal parameter r and vector $\underline{x}^{(0)}$.
1.2 $a_1 = a/b$, $\hat{g}_i = g_i/b$, $i = 1, 2, \ldots, n$
1.3 Set

$$
\alpha = \frac{a_1}{2} + r, \ \beta = \frac{a_1}{2} - r, \ d = \alpha^2 - 1, \ t = -d^{-1}, \ s = \alpha t, \ s' = \beta s + t, \ t' = \beta t + s
$$

Phase II.   The normalised AGE iteration

For $k = 0, 1, \ldots$, until convergence do.

2.1 $u_1 = \hat{g}_1 - \beta x_1^{(k)}$

$u_n = \hat{g}_n - \beta x_n^{(k)}$

For $\ell = 1, 2, \ldots, m-1$ do,

$$
\begin{bmatrix} u_{2\ell} \\ u_{2\ell+1} \end{bmatrix} = \begin{bmatrix} \hat{g}_{2\ell} - x_{2\ell+1}^{(k)} - \beta x_{2\ell}^{(k)} \\ \hat{g}_{2\ell+1} - x_{2\ell}^{(k)} - \beta x_{2\ell+1}^{(k)} \end{bmatrix}.
$$

2.2 For $\ell = 1, 2, \ldots, m$ do,

$$
\begin{bmatrix} q_{2\ell-1} \\ q_{2\ell} \end{bmatrix} = \begin{bmatrix} \hat{g}_{2\ell-1} \\ \hat{g}_{2\ell} \end{bmatrix} - \begin{bmatrix} s' & t' \\ t' & s' \end{bmatrix} \begin{bmatrix} u_{2\ell-1} \\ u_{2\ell} \end{bmatrix}.
$$

2.3 $x_1^{(k+1)} = q_1/\alpha$

$x_n^{(k+1)} = q_n/\alpha$.

For $\ell = 1, 2, \ldots, m-1$ do,

$$
\begin{bmatrix} x_{2\ell}^{(k+1)} \\ x_{2\ell+1}^{(k+1)} \end{bmatrix} \begin{bmatrix} s & t \\ t & s \end{bmatrix} \begin{bmatrix} q_{2\ell} \\ q_{2\ell+1} \end{bmatrix}.
$$

Thus, in the above Algorithm 2, the number of operations is $2n$ additions and n multiplications for Step 2.1, $2n$ additions and $2n$ multiplications for Step 2.2 and n additions and $2n$ multiplications for Step 2.3. Therefore, the total amount of arithmetic operations per iteration for Algorithm 2 is $5n$ additions and $5n$ multiplications. (Note that once again we do not include the operations in Phase I.) Hence, compared with Algorithm 1, Algorithm 2 saves 16.6% additions and 37.5% multiplications per iteration.

Note that if we used one more vector (with n-components) storage, we can reduce the number of multiplications further to $4n$. This is because the 2×2 matrix in Step 2.3 of Algorithm 2 satisfies

$$\begin{bmatrix} s & t \\ t & s \end{bmatrix} = \frac{1}{d} \begin{bmatrix} a & -1 \\ -1 & \alpha \end{bmatrix}.$$

Therefore we can rewrite Steps 2.2 and 2.3 by the following:

2.2′ For $\ell = 1, 2, \ldots, m$ do,

$$\frac{1}{d} \begin{bmatrix} q_{2\ell-1} \\ q_{2\ell} \end{bmatrix} = \begin{bmatrix} \hat{g}_{2\ell-1}/d \\ \hat{g}_{2\ell}/d \end{bmatrix} - \begin{bmatrix} s'/d & t'/d \\ t'/d & s'/d \end{bmatrix} \begin{bmatrix} u_{2\ell-1} \\ u_{2\ell} \end{bmatrix}.$$

2.3′ $x_1^{(k+1)} = d.\alpha^{-1}.(q_1/d)$

$x_n^{(k+1)} = d.\alpha^{-1}.(q_n/d)$

For $\ell = 1, 2, \ldots, m-1$ do,

$$\begin{bmatrix} x_{2\ell}^{(k+1)} \\ x_{2\ell+1}^{(k+1)} \end{bmatrix} \begin{bmatrix} a & -1 \\ -1 & \alpha \end{bmatrix} \begin{bmatrix} \begin{bmatrix} q_{2\ell} \\ q_{2\ell+1} \end{bmatrix}/d \end{bmatrix}.$$

Now the saving is clear if we store $\hat{g}_j/d$, $j = 1, 2, \ldots, n$, beforehand. The above discussions are summarised in the following algorithm.

Algorithm 3
Phase I.   Preprocessing step

1.1 $a_1 = a/b$. Choose parameter r and vector $\underline{x}^{(0)}$.

$$\alpha = \frac{a_1}{2} + r, \quad \beta = \frac{a_1}{2} - r, \quad d = \alpha^2 - 1, \quad \alpha' = d/\alpha, \quad s'' = (\alpha\beta - 1)/d_2,$$
$$t'' = 2r/d^2, \quad \hat{g}_j = g_j/b, \quad g_j' = \hat{g}_j/d, \quad j = 1, 2, \ldots, n.$$

Phase II.   The normalised AGE iteration
For $k = 0, 1, \ldots$, until convergence do.

2.1 $u_1 = \hat{g}_1 - \beta x_1^{(k)}, \quad u_n = \hat{g}_n - \beta x_n^{(k)}$

For $\ell = 1, 2, \ldots, m-1$ do,

$$\begin{bmatrix} u_{2\ell} \\ u_{2\ell+1} \end{bmatrix} = \begin{bmatrix} \hat{g}_{2\ell} - x_{2\ell+1}^{(k)} - \beta x_{2\ell}^{(k)} \\ \hat{g}_{2\ell+1} - x_{2\ell}^{(k)} - \beta x_{2\ell+1}^{(k)} \end{bmatrix}.$$

2.2 For $\ell = 1, 2, \ldots, m$ do,

$$\begin{bmatrix} v_{2\ell-1} \\ v_{2\ell} \end{bmatrix} = \begin{bmatrix} \hat{g}'_{2\ell-1} \\ \hat{g}'_{2\ell} \end{bmatrix} - \begin{bmatrix} s'' & t'' \\ t'' & s'' \end{bmatrix} \begin{bmatrix} u_{2\ell-1} \\ u_{2\ell} \end{bmatrix}.$$

2.3 $x_1^{(k+1)} = \alpha'v_1, \quad x_n^{(k+1)} = \alpha'v_n$

For $\ell = 1, 2, \ldots, m-1$ do,

$$\begin{bmatrix} x_{2\ell}^{(k+1)} \\ x_{2\ell+1}^{(k+1)} \end{bmatrix} \begin{bmatrix} \alpha & -1 \\ -1 & \alpha \end{bmatrix} \begin{bmatrix} v_{2\ell} \\ v_{2\ell+1} \end{bmatrix}.$$

Thus, it is clear that the total amount of arithmetic operations per iteration for the above algorithm are $5n$ additions and $4n$ multiplications. Thus, compared with Algorithm 1, the saving is 16.6% in additions and 50% in multiplications. However, this is at the expense of one extra vector storage.

5.7.3. Concluding Remarks

In this note three algorithms have been presented. To date Algorithm 1 has been widely used. From these results it is recommended that Algorithm 3 or 2 should be used in order to reduce the CPU time.

5.8. Block Alternating Group Explicit (BLAGE) Methods

5.8.1. Introduction

Consider the solution of the boundary value problem

$$\frac{\partial^2 \phi}{\partial x^2} + \frac{\partial^2 \phi}{\partial y^2} = 0 \qquad\qquad (5.8.1a)$$

defined in the unit square, $0 \le x, y \le 1$, with n^2 internal mesh points in the region shown in Figure 5.8.1.

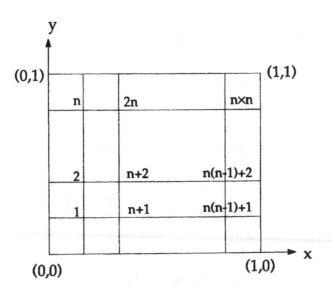

Figure 5.8.1

Previously we have shown that the central difference approximation to the partial derivatives in (5.8.1a) results in a discretised linear system of the form

$$A\underline{\phi} = \underline{b},$$ (5.8.1b)

where the coefficient matrix A has the block structure

$$A = \begin{bmatrix} D_1 & C_1 & & & & \\ A_2 & D_2 & C_2 & & & \\ & & & & 0 & \\ & & & & & \\ & 0 & & & & \\ & & A_{n-1} & D_{n-1} & C_{n-1} \\ & & & A_n & D_n \end{bmatrix},$$ (5.8.2)

with

$$A_{i+1} = C_i = -I, \quad A_1 = C_n = [0]$$ (5.8.3)

and

$$D_i = \begin{bmatrix} 4 & -1 & & & \\ -1 & 4 & -1 & & \\ & & & & 0 \\ & & & & \\ 0 & & & & \\ & & -1 & 4 & -1 \\ & & & -1 & 4 \end{bmatrix}, \quad i = 1, 2, \dots, n.$$ (5.8.4)

5.8.2. Formulation of the Method

We now follow the approach of Evans (1985) and split the matrix A into the sum of two submatrices

$$A = G_1 + G_2,$$ (5.8.5)

where

$$G_1 = \begin{bmatrix} D_1' & C_1 & & & & & \\ A_2 & D_2' & & & & & \\ & & D_3' & C_3 & & & \\ & & A_4 & D_4' & & 0 & \\ & & & & & & \\ & & & & & D_{n-1}' & C_{n-1} \\ & 0 & & & & A_n & D_n' \end{bmatrix}$$

and

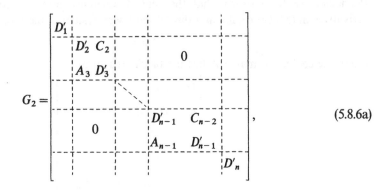

$$G_2 = \begin{bmatrix} D_1' & & & & & & \\ & D_2' & C_2 & & & & \\ & A_3 & D_3' & & & 0 & \\ & & & & & & \\ & & & & D_{n-1}' & C_{n-2} & \\ & 0 & & & A_{n-1} & D_{n-1}' & \\ & & & & & & D_n' \end{bmatrix}, \tag{5.8.6a}$$

if n is even, and

$$G_1 = \begin{bmatrix} D_1' & & & & \\ & D_2' & C_2 & & \\ & A_3 & D_3' & 0 & \\ & & & & \\ & 0 & & D_{n-1}' & C_{n-2} \\ & & & A_{n-1} & D_{n-1}' \end{bmatrix},$$

and

$$G_2 = \begin{bmatrix} D_1' & C_1 & & & & \\ A_2 & D_2' & & & & \\ & & & 0 & & \\ & & & & & \\ & 0 & & D_{n-2}' & C_{n-2} & \\ & & & A_{n-1} & D_{n-1}' & \\ & & & & & D_n' \end{bmatrix}, \tag{5.8.6b}$$

if n is odd, with $D_i' = (1/2)D_i$, $i = 1, 2, \ldots, n$, where G_1 and G_2 satisfy the condition $G_1 + rI$ and $G_2 + rI$ are non-singular for any $r > 0$.

By using (5.8.5) the matrix equation (5.8.1) can now be written in the form

$$(G_1 + G_2)\underline{\phi} = \underline{b}, \tag{5.8.7}$$

and following a strategy similar to the ADI method, $\underline{\phi}^{(k+1/2)}$ and $\underline{\phi}^{(k+1)}$ can be determined by

$$(G_1 + rI)\underline{\phi}^{(k+1/2)} = \underline{b} - (G_2 - rI)\underline{\phi}^{(k)},$$
$$(G_2 + rI)\underline{\phi}^{(k+1)} = \underline{b} - (G_1 - rI)\underline{\phi}^{(k+1/2)}, \tag{5.8.8}$$

where r is the iteration parameter.

Let $D'_i + rI = R_i$ and $D'_i - rI = P_i$, then from (5.8.8) for n is even we have

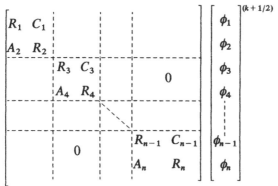

$$
= \begin{bmatrix}
b_1 - P_1\phi_1 \\
b_2 - P_2\phi_2 - C_2\phi_3 \\
b_3 - A_3\phi_2 - P_3\phi_3 \\
b_4 - P_4\phi_4 - C_4\phi_5 \\
\vdots \\
b_{n-1} - A_{n-1}\phi_{n-2} - P_{n-1}\phi_{n-1} \\
b_n - P_n\phi_n
\end{bmatrix}^{(k)}
\qquad (5.8.9a)
$$

and

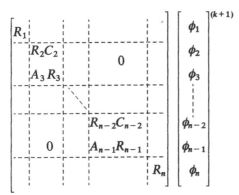

$$
= \begin{bmatrix}
b_1 - P_1\phi_1 - C_1\phi_2 \\
b_2 - A_2\phi_1 - P_2\phi_2 \\
b_3 - P_3\phi_3 - C_3\phi_4 \\
\vdots \\
b_{n-2} - A_{n-2}\phi_{n-3} - P_{n-2}\phi_{n-2} \\
b_{n-1} - P_{n-1}\phi_{n-1} - C_{n-1}\phi_n \\
b_n - A_n\phi_{n-1} - P_n\phi_n
\end{bmatrix}^{(k+1/2)}
\qquad (5.8.9b)
$$

For the model problem, we have

$$R_1 = R_2 = \cdots = R_n = \begin{bmatrix} 2+r & -1/2 & & & \\ -1/2 & 2+r & -1/2 & & 0 \\ & & \ddots & \ddots & \ddots & \\ 0 & & & -1/2 & 2+r & -1/2 \\ & & & & -1/2 & 2+r \end{bmatrix},$$

and

$$P_1 = P_2 = \cdots = P_n = \begin{bmatrix} 2-r & -1/2 & & & \\ -1/2 & 2-r & -1/2 & & 0 \\ & & \ddots & \ddots & \ddots & \\ 0 & & & -1/2 & 2-r & -1/2 \\ & & & & -1/2 & 2-r \end{bmatrix},$$

$$A_i = C_i = I, \quad i = 1, 2, \ldots, n \text{ with } A_1 = C_n = [0]. \tag{5.8.10}$$

1st Sweep

From (5.2.9a) and for $i = 1, (2), n-1$, we have

$$R\phi_i^{(k+1/2)} - \phi_{i+1}^{(k+1/2)} = z_i^{(k)}, \tag{5.8.11a}$$

$$-\phi_i^{(k+1/2)} + R\phi_{i+1}^{(k+1/2)} = z_{i+1}^{(k)}, \tag{5.8.11b}$$

where

$$z_i^{(k)} = b_i + \phi_{i-1}^{(k)} - P\phi_i^{(k)}, \tag{5.8.12a}$$

$$z_{i+1}^{(k)} = b_{i+1} - P\phi_{i+1}^{(k)} + \phi_{i+2}^{(k)}. \tag{5.8.12b}$$

From (5.8.11a),

$$\phi_{i+1}^{(k+1/2)} = R\phi_i^{(k+1/2)} - z_i^{(k)}, \tag{5.8.13}$$

hence, by substituting the value of $\phi_{i+1}^{(k+1/2)}$ from (5.8.13) in (5.8.11b)

$$-\phi_i^{(k+1/2)} + R(R\phi_i^{(k+1/2)} - z_i^{(k)}) = z_{i+1}^{(k)},$$

$$(R-I)(R+I)\phi_i^{(k+1/2)} = Rz_i^{(k)} + z_{i+1}^{(k)}. \tag{5.8.14}$$

To solve (5.2.14), we proceed in two stages, first solve

$$(R-I)\psi_i^{(k+1/2)} = Rz_i^{(k)} + z_{i+1}^{(k)} \quad \text{for } \psi^{(k+1/2)}, \tag{5.8.15a}$$

then

$$(R+I)\phi_i^{(k+1/2)} = \psi_i^{(k+1/2)} \quad \text{for } \phi_i^{(k+1/2)}, \tag{5.8.15b}$$

where both systems are tridiagonal and hence easy to solve.

Hence $\phi_{i+1}^{(k+1/2)}$ can be determined from (5.8.13).

2nd Sweep

From (5.8.9b) we have

$$R\phi_1^{(k+1)} = b_1 - P\phi_1^{(k+1/2)} + \phi_2^{(k+1/2)},$$ (5.8.16)

which can be solved and hence $\phi_1^{(k+1)}$ is obtained.

For $\phi_i^{(k+1)}$ and $\phi_{i+1}^{(k+1)}$, $i = 2, (2), n-2$ from (5.8.9b)

$$R\phi_i^{(k+1)} - \phi_{i+1}^{(k+1)} = Y_i^{(k+1/2)},$$ (5.8.17a)

$$-\phi_i^{(k+1)} + R\phi_{i+1}^{(k+1)} = Y_{i+1}^{(k+1/2)},$$ (5.8.17b)

where

$$Y_i^{(k+1/2)} = b_i + \phi_{i-1}^{(k+1/2)} - P\phi_i^{(k+1/2)},$$ (5.8.18a)

$$Y_{i+1}^{(k+1/2)} = b_{i+1} + P\phi_{i+1}^{(k+1/2)} + \phi_{i+2}^{(k+1/2)}.$$ (5.8.18b)

From (5.8.17a)

$$\phi_{i+1}^{(k+1)} = R\phi_i^{(k+1)} - Y_i^{(k+1/2)},$$ (5.8.19)

hence, from (5.8.19) and (5.8.17b) we obtain the system

$$(R-I)(R+I)\phi_i^{(k+1)} = RY_i^{(k+1/2)} + Y_{i+1}^{(k+1/2)}$$ (5.8.20)

to determine $\phi_i^{(k+1)}$, we proceed in two stages, first solve

$$(R-I)\psi_i^{(k+1)} = RY_i^{(k+1/2)} + Y_{i+1}^{(k+1/2)},$$ (5.8.21a)

then

$$(R+I)\phi_i^{(k+1)} = \psi_i^{(k+1)},$$ (5.8.21b)

and from (5.8.19), $\phi_{i+1}^{(k+1)}$ can be obtained.

Finally, $\phi_n^{(k+1)}$ is obtained by solving

$$R\phi_n^{(k+1)} = b_n + \phi_{n-1}^{(k+1/2)} - P\phi_n^{(k+1/2)}.$$ (5.8.22)

In a similar manner a solution can be obtained if n is odd.

6. Group Explicit Methods for Hyperbolic Equations

6.1. Introduction

This chapter will deal initially with the application of the GE and AGE integration schemes when used to solve the first order hyperbolic equation of the form

$$\frac{\partial U}{\partial t} = -\frac{\partial U}{\partial x}, \quad 0 \leq x \leq 1, \, t \geq 0 \tag{6.1.1}$$

and the second order hyperbolic equation, i.e., the wave equation

$$\frac{\partial^2 U}{\partial t^2} = \frac{\partial^2 U}{\partial x^2}, \quad 0 \leq x \leq 1, \, t \geq 0. \tag{6.1.2}$$

An analysis of the local truncation errors will also be performed followed by an investigation of the stability requirements of the various schemes with application to the Klein–Gordon equation.

6.2. GE Methods for the Generalised Weighted Approximation to the First Order Equation

The generalised weighted finite-difference analogue for equation (6.1.1) at the point

$$(x_j, t_{j+\theta}) = [i\Delta x, (j+\theta)\Delta t]$$

is given by

$$-\frac{1}{\Delta x} \{\theta [(1-w)u_{i+1,j+1} + (2w-1)u_{i,j+1} - wu_{i-1,j+1}] + (1-\theta)[(1-w)u_{i+1,j}$$

$$+ (2w-1)u_{ij} - wu_{i-1,j}]\} = \frac{(u_{i,j+1} - u_{ij})}{\Delta t}, \tag{6.2.1}$$

or

$$-\lambda\{\theta[(1-w)u_{i+1,j+1} + (2w-1)u_{i,j+1} - wu_{i-1,j+1}] + (1-\theta)[(1-w)u_{i+1,j}$$

$$+ (2w-1)u_{ij} - wu_{i-1,j}]\} = u_{i,j+1} - u_{ij}, \tag{6.2.2}$$

where $\lambda = \Delta t/\Delta x$, the mesh ratio and $0 < \theta < 1$.

319

With $w = 1$, this equation reduces to

$$(1 + \lambda\theta)u_{i,j+1} - \lambda\theta u_{i-1,j+1} = [1 - \lambda(1-\theta)]u_{i,j} + \lambda(1-\theta)u_{i-1,j} \qquad (6.2.3)$$

and for $w = 0$ equation (6.2.2) becomes

$$(1 - \lambda\theta)u_{i,j+1} + \lambda\theta u_{i+1,j+1} = [1 + \lambda(1-\theta)]u_{i,j} - \lambda(1-\theta)u_{i+1,j}. \qquad (6.2.4)$$

The local truncation error representations can be obtained by expanding the terms $U_{i,j+1}, U_{i-1,j+1}, U_{i-1,j}$ and U_{ij} about the point $[i\Delta x, (j+1/2)\Delta t]$ using the Taylor series. The expansion for equation (6.2.3) leads to

$$T_5 = \left(\frac{\partial U}{\partial x} + \frac{\partial U}{\partial t}\right)_{i,j+1/2} + \Delta x\left[-\frac{1}{2}\frac{\partial^2 U}{\partial x^2} - \frac{(\Delta t)^2}{16}\frac{\partial^4 U}{\partial x^2 \partial t^2}\right]_{i,j+1/2}$$

$$+ \Delta t\left[-\frac{1}{2}(1-2\theta)\frac{\partial^2 U}{\partial x\partial t} - \frac{(\Delta x)^2}{12}(1-2\theta)\frac{\partial^4 U}{\partial x^3 \partial t}\right]_{i,j+1/2}$$

$$+ (\Delta x)(\Delta t)\left[\frac{1}{4}(1-2\theta)\frac{\partial^3 U}{\partial x^2 \partial t}\right]_{i,j+1/2} + (\Delta x)^2\left[\frac{1}{6}\frac{\partial^3 U}{\partial x^3} - \frac{\Delta x}{24}\frac{\partial^4 U}{\partial x^4}\right]_{i,j+1/2}$$

$$+ (\Delta t)^2\left[\frac{1}{8}\frac{\partial^3 U}{\partial x\partial t^2} + \frac{1}{24}\frac{\partial^3 U}{\partial t^3} - \frac{\Delta t}{48}(1-2\theta)\frac{\partial^4 U}{\partial x\partial t^3}\right]_{i,j+1/2}$$

$$+ \left[\frac{(\Delta x)^4}{5!}\frac{\partial^5 U}{\partial x^5} + \frac{5(\Delta x)^3(\Delta t)}{5!2} + (1-2\theta)\frac{\partial^5 U}{\partial x^4 \partial t} + \frac{5(\Delta x)^2(\Delta t)^2}{5!2}\frac{\partial^5 U}{\partial x^3 \partial t^2}\right.$$

$$\left. + \frac{5(\Delta x)(\Delta t)^3}{5!4}(1-2\theta)\frac{\partial^5 U}{\partial x^2 \partial t^3} + \frac{5(\Delta t)^4}{5!16}\frac{\partial^5 U}{\partial x\partial t^4} + \frac{(\Delta t)^4}{5!16}\frac{\partial^5 U}{\partial t^5}\right]_{i,j+1/2} + \cdots,$$

i.e.,

$$T_5 = \Delta x\left[-\frac{1}{2}\frac{\partial^2 U}{\partial x^2} - \frac{(\Delta t)^2}{16}\frac{\partial^4 U}{\partial x^2 \partial t^2}\right]_{i,j+1/2}$$

$$+ \Delta t\left[-\frac{1}{2}(1-2\theta)\frac{\partial^2 U}{\partial x\partial t} - \frac{(\Delta x)^2}{12}(1-2\theta)\frac{\partial^4 U}{\partial x^3 \partial t}\right]_{i,j+1/2}$$

$$+ (\Delta x)(\Delta t)\left[\frac{1}{4}(1-2\theta)\frac{\partial^3 U}{\partial x^2 \partial t}\right]_{i,j+1/2} + (\Delta x)^2\left[\frac{1}{6}\frac{\partial^3 U}{\partial x^3} - \frac{\Delta x}{24}\frac{\partial^4 U}{\partial x^4}\right]_{i,j+1/2}$$

$$+ (\Delta t)^2\left[\frac{1}{8}\frac{\partial^3 U}{\partial x\partial t^2} + \frac{1}{24}\frac{\partial^3 U}{\partial t^3} - \frac{\Delta t}{48}(1-2\theta)\frac{\partial^4 U}{\partial x\partial t^3}\right]_{i,j+1/2}$$

$$+ 0\left[(\Delta x)^{\alpha_1}(\Delta t)^{\alpha_2}\right], \qquad (6.2.5)$$

with $\alpha_1 + \alpha_2 = 4$ and $0 \le \theta \le 1$. A similar expansion for the terms $U_{i,j+1}, U_{i+1,j+1}, U_{ij}$ and $U_{i+1,j}$ about the point $[i\Delta x, (j + 1/2)\Delta t]$ provides the following truncation error expression for formula (6.2.4):

$$T_6 = \left(\frac{\partial U}{\partial x} + \frac{\partial U}{\partial t}\right)_{i,j+1/2} + \Delta x \left[\frac{1}{2}\frac{\partial^2 U}{\partial x^2} + \frac{(\Delta t)^2}{16}\frac{\partial^4 U}{\partial x^2 \partial t^2}\right]_{i,j+1/2}$$

$$+ \Delta t \left[-\frac{1}{2}(1 - 2\theta)\frac{\partial^2 U}{\partial x \partial t} - \frac{1}{12}(\Delta x)^2(1 - 2\theta)\frac{\partial^4 U}{\partial x^3 \partial t}\right]_{i,j+1/2}$$

$$+ (\Delta x)(\Delta t)\left[-\frac{1}{4}(1 - 2\theta)\frac{\partial^3 U}{\partial x^2 \partial t}\right]_{i,j+1/2} + (\Delta x)^2\left[\frac{1}{6}\frac{\partial^3 U}{\partial x^3} - \frac{\Delta x}{24}\frac{\partial^4 U}{\partial x^4}\right]_{i,j+1/2}$$

$$+ (\Delta t)^2\left[\frac{1}{8}\frac{\partial^3 U}{\partial x \partial t^2} + \frac{1}{24}\frac{\partial^3 U}{\partial t^3} - \frac{\Delta t}{48}(1 - 2\theta)\frac{\partial^4 U}{\partial x \partial t^3}\right]_{i,j+1/2}$$

$$+ \left[\frac{(\Delta x)^4}{5!}\frac{\partial^5 U}{\partial x^5} + \frac{5(\Delta x)^3(\Delta t)}{5!2}(1 - 2\theta)\frac{\partial^5 U}{\partial x^4 \partial t} + \frac{5(\Delta x)^2(\Delta t)^2}{5!2}\frac{\partial^5 U}{\partial x^3 \partial t^2}\right.$$

$$\left. - \frac{5(\Delta x)(\Delta t)^3}{5!4}(1 - 2\theta)\frac{\partial^5 U}{\partial x^2 \partial t^3} + \frac{5(\Delta t)^4}{5!16}\frac{\partial^5 U}{\partial x \partial t^4} + \frac{(\Delta t)^4}{5!16}\frac{\partial^5 U}{\partial t^5}\right]_{i,j+1/2} + \cdots,$$

i.e.,

$$T_6 = \Delta x \left[\frac{1}{2}\frac{\partial^2 U}{\partial x^2} - \frac{(\Delta t)^2}{16}\frac{\partial^4 U}{\partial x^2 \partial t^2}\right]_{i,j+1/2}$$

$$+ \Delta t \left[-\frac{1}{2}(1 - 2\theta)\frac{\partial^2 U}{\partial x \partial t} - \frac{1}{12}(\Delta x)^2(1 - 2\theta)\frac{\partial^4 U}{\partial x^3 \partial t}\right]_{i,j+1/2}$$

$$+ (\Delta x)(\Delta t)\left[-\frac{1}{4}(1 - 2\theta)\frac{\partial^3 U}{\partial x^2 \partial t}\right]_{i,j+1/2} + (\Delta x)^2\left[\frac{1}{6}\frac{\partial^3 U}{\partial x^3} - \frac{\Delta x}{24}\frac{\partial^4 U}{\partial x^4}\right]_{i,j+1/2}$$

$$+ (\Delta t)^2\left[\frac{1}{8}\frac{\partial^3 U}{\partial x \partial t^2} + \frac{1}{24}\frac{\partial^3 U}{\partial t^3} - \frac{\Delta t}{48}(1 - 2\theta)\frac{\partial^4 U}{\partial x \partial t^3}\right]_{i,j+1/2}$$

$$+ 0\left[(\Delta x)^{\alpha_1}(\Delta t)^{\alpha_2}\right], \tag{6.2.6}$$

with $\alpha_1 + \alpha_2 = 4$ and $0 \le \theta \le 1$.

Now at the point $[(i - 1)\Delta x, (j + \theta)\Delta t]$, equation (6.2.4) takes the form

$$\lambda\theta u_{i,j+1} + (1 - \lambda\theta)u_{i-1,j+1} = -\lambda(1 - \theta)u_{i,j} + [1 + \lambda(1 - \theta)]u_{i-1,j}. \tag{6.2.7}$$

By coupling the equations (6.2.3) and (6.2.7) the two formulae can be written simultaneously in matrix form for the points $(i-1,j)$ and (i,j) as

$$\begin{bmatrix} -\lambda\theta & (1+\lambda\theta) \\ (1-\lambda\theta) & \lambda\theta \end{bmatrix} \begin{bmatrix} u_{i-1,j+1} \\ u_{i,j+1} \end{bmatrix} = \begin{bmatrix} \lambda(1-\theta) & 1-\lambda(1-\theta) \\ 1+\lambda(1-\theta) & -\lambda(1-\theta) \end{bmatrix} \begin{bmatrix} u_{i-1,j} \\ u_{ij} \end{bmatrix}, \quad (6.2.8)$$

i.e.,

$$A\underline{u}_{j+1} = B\underline{u}_j, \qquad (6.2.9)$$

where

$$A = \begin{bmatrix} -\lambda\theta & (1+\lambda\theta) \\ (1-\lambda\theta) & \lambda\theta \end{bmatrix}, \quad B = \begin{bmatrix} \lambda(1-\theta) & 1-\lambda(1-\theta) \\ 1+\lambda(1-\theta) & -\lambda(1-\theta) \end{bmatrix}$$

and

$$\underline{u}_j = (u_{i-1,j}, u_{i,j})^T.$$

Since the (2×2) matrix A can be easily inverted. Hence from equation (6.2.9) we have

$$\underline{u}_{j+1} = A^{-1}B\underline{u}_j, \qquad (6.2.10)$$

with

$$A^{-1} = \begin{bmatrix} -\lambda\theta & (1+\lambda\theta) \\ (1-\lambda\theta) & \lambda\theta \end{bmatrix} \quad \text{and} \quad A^{-1}B = \begin{bmatrix} (1+\lambda) & -\lambda \\ \lambda & (1-\lambda) \end{bmatrix}.$$

From equation (6.2.10) this gives rise to the following set of simple explicit equations

$$u_{i-1,j+1} = (1+\lambda)u_{i-1,j} - \lambda u_{i,j} \qquad (6.2.11)$$

and

$$u_{i,j+1} = \lambda u_{i-1,j} + (1-\lambda)u_{ij}, \qquad (6.2.12)$$

whose computational molecules are shown in Figure 6.2.1.

Now equations (6.2.11) and (6.2.12) are for adjacent points which are grouped two at a time on the mesh line. Further special formulae are needed to cope with the possibility of the existence of *ungrouped* points near the boundaries. The solution at the ungrouped point near *the right boundary* at the advanced time level $(k+1)$ can

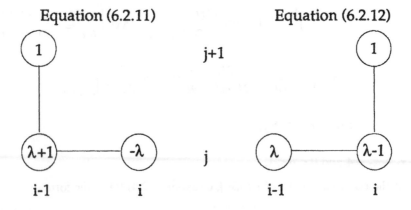

Figure 6.2.1

be computed from the equation (6.1.6) by putting $i = m - 1$. This leads to

$$u_{m-1,j+1} = \{[1 + \lambda(1 - \theta)]u_{m-1,j} - \lambda(1 - \theta)u_{m,j} - \lambda\theta u_{m,j+1}\}(1 - \lambda\theta), \quad (6.2.13)$$

where $\lambda\theta \neq 1$.

Equation (6.2.3) with $i = 1$, deals with the value of u at the ungrouped point near *the left boundary*. Thus, we have

$$u_{1,j+1} = [\lambda(1 - \theta)u_{0,j} + (1 - \lambda(1 - \theta))u_{1,j} + \lambda\theta u_{0,j+1}](1 + \lambda\theta). \quad (6.2.14)$$

Since the initial line $0 \leq x \leq 1$ is uniformly divided with a spacing or increment Δx, the manner in which the above points are grouped very much depends on whether the number m of intervals of the line segment is *even* or *odd*.

On this basis, a variety of group explicit schemes can be devised — as will presently be shown.

Even Number of Intervals
When m is even, we will have an odd number $(m - 1)$ of internal points (i.e., points that do not include the left and right boundaries whose values are given by u_0 and u_m, respectively, at every time level). Consequently, the single ungrouped point will be located near either boundary.

(i) *The GER Scheme* This refers to the group explicit with right ungrouped point (GER) scheme. It results in the consecutive application for $(1/2)(m - 2)$ times of the equations (6.2.11) and (6.2.12) for the first $(m - 1)$ points grouped two at a time. This is followed by a use of equation (6.2.13) for the final $(m - 1)^{th}$ point at every time level as shown in Figure 6.2.2. Thus, we have the following set of equations:

$$-\lambda\theta u_{i-1,j+1} + (1 + \lambda\theta)u_{i,j+1} = \lambda(1 - \theta)u_{i-1,j} + [1 - \lambda(1 - \theta)]u_{i,j},$$

$$(1 - \lambda\theta)u_{i-1,j+1} + \lambda\theta u_{i,j+1} = [1 + \lambda(1 - \theta)]u_{i-1,j} - \lambda(1 - \theta)u_{i,j},$$

$$i = 2, 4, \ldots, (m - 2),$$

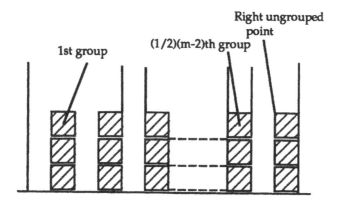

Figure 6.2.2 The GER Scheme (even number of intervals).

and

$$(1 - \lambda\theta)u_{m+1,j+1} = -\lambda\theta u_{m,j+1} - \lambda(1-\theta)u_{m,j} + [1 + \lambda(1-\theta)]u_{m-1,j}, \quad \lambda\theta \neq 1,$$

which can be written in the more compact, *implicit* matrix form as

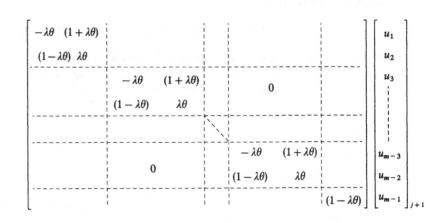

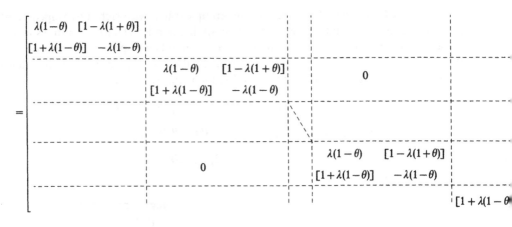

$$\times \begin{bmatrix} u_1 \\ u_2 \\ u_3 \\ \vdots \\ u_{m-3} \\ u_{m-2} \\ u_{m-1} \end{bmatrix}_j + \underline{b}_1, \tag{6.2.15}$$

where $\underline{b}_1 = [0, 0, \ldots, -\lambda(1-\theta)u_{m,j} - \lambda\theta u_{m,j+1}]^T$ which consists of known boundary

values. Now if we define

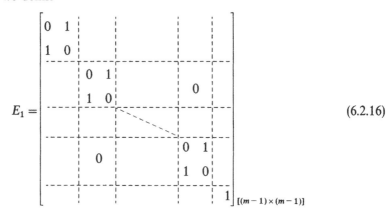

$$E_1 = \qquad (6.2.16)$$

and

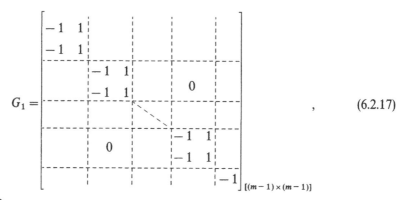

$$G_1 = \qquad , \qquad (6.2.17)$$

then we have

$$(E_1 + \lambda\theta G_1)\underline{u}_{j+1} = [E_1 - \lambda(1-\theta)G_1]\underline{u}_j + \underline{b}_1.$$

On premultiplying this equation by E_1^{-1} provides us with

$$E_1^{-1}(E_1 + \lambda\theta G_1)\underline{u}_{j+1} = E_1^{-1}[E_1 - \lambda(1-\theta)G_1]\underline{u}_j + E_1^{-1}\underline{b}_1,$$

i.e.,

$$(E_1^{-1}E_1 + \lambda\theta E_1^{-1}G_1)\underline{u}_{j+1} = [E_1^{-1}E_1 - \lambda(1-\theta)E_1^{-1}G_1]\underline{u}_j + \underline{b}_1.$$

But $E_1^{-1}E_1 = I$ and since E_1 is a permutation matrix $E_1^{-1}G_1 = G_1$ which implies that

$$(I + \lambda\theta G_1)\underline{u}_{j+1} = [I - \lambda(1-\theta)G_1]\underline{u}_j + \underline{b}_1, \qquad (6.2.18)$$

where I is the identity matrix of order $[(m-1) \times (m-1)]$. Hence, we obtain

$$\underline{u}_{j+1} = (I + \lambda\theta G_1)^{-1}[I - \lambda(1-\theta)G_1]\underline{u}_j + \underline{\hat{b}}_1, \qquad (6.2.19)$$

where

$$\underline{\hat{b}}_1 = (I + \lambda\theta G_1)^{-1}\underline{b}_1.$$

The *explicit* equation (6.2.19) is the governing equation for the computation of the GER scheme.

(ii) *The GEL Scheme* This is an abbreviation for the group explicit method with a left (GEL) ungrouped point scheme and is in fact a reverse of the GER scheme. It is obtained by the use of equation (6.1.16) for the first internal point followed by the application of equations (6.2.13) and (6.2.14) for $(1/2)(m-1)$ times for the remaining points on the mesh line. The scheme is displayed diagrammatically in Figure 6.2.3 and is determined by the following set of linear equations:

$$(1 + \lambda\theta)u_{1,j+1} = [1 - \lambda(1 - \theta)]u_{1,j} + \lambda(1 - \theta)u_{0j} + \lambda\theta u_{0,j+1}, \text{ for point 1,}$$

and

$$\left[\begin{array}{l} -\lambda\theta u_{i-1,j+1} + (1 + \lambda\theta)u_{i,j+1} = \lambda(1 - \theta)u_{i-1,j} + [1 - \lambda(1 - \theta)]u_{ij} \\[2mm] (1 - \lambda\theta)u_{i-1,j+1} + \lambda\theta u_{i,j+1} = [1 + \lambda(1 - \theta)]u_{i-1,j} - \lambda(1 - \theta)u_{ij}, \\[2mm] \qquad\qquad\qquad\qquad i = 3, 5, \ldots, m - 1; \ \lambda\theta \neq 1. \end{array} \right.$$

In *implicit matrix* form these equations can be written as

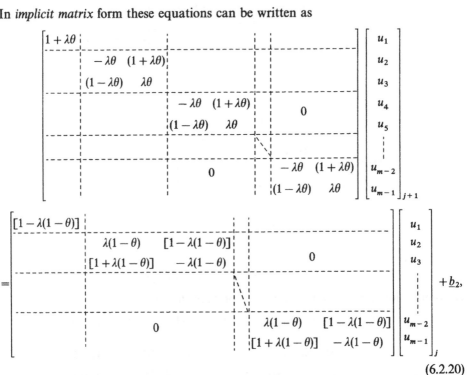

$$(6.2.20)$$

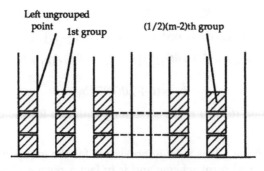

Figure 6.2.3 The GEL Scheme (even number of intervals).

where
$$\underline{b}_2 = [\lambda(1-\theta)u_{0,j} + \lambda\theta u_{0,j+1}, 0, \ldots, 0]^T.$$
If we define

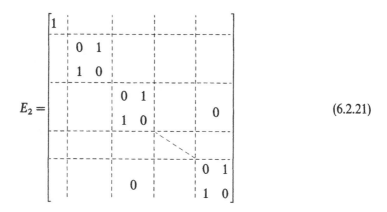

$$ E_2 = \begin{bmatrix} 1 & & & & & \\ & 0 & 1 & & & \\ & 1 & 0 & & & \\ & & & 0 & 1 & & \\ & & & 1 & 0 & & & 0 \\ & & & & & & 0 & 1 \\ & & & & 0 & & 1 & 0 \end{bmatrix} \qquad (6.2.21)$$

and

$$ G_2 = \begin{bmatrix} -1 & & & & \\ & -1 & 1 & & \\ & -1 & 1 & & \\ & & & -1 & 1 & \\ & & & -1 & 1 & & 0 \\ & & & & & & -1 & 1 \\ & & 0 & & & & -1 & 1 \end{bmatrix}, \qquad (6.2.22)$$

then the GEL scheme takes the form
$$(E_2 + \lambda\theta G_2)\underline{u}_{j+1} = [E_2 - \lambda(1-\theta)G_2]\underline{u}_j + \underline{b}_2.$$
Similarly if we premultiply this equation by E_1^{-1}, we get
$$(I + \lambda\theta G_2)\underline{u}_{j+1} = [I + \lambda(1-\theta)G_2]\underline{u}_j + \underline{b}_2, \qquad (6.2.23)$$
and this leads to the following explicit formula for the computation of the GEL scheme
$$\underline{u}_{j+1} = (I + \lambda\theta G_2)^{-1}[I - \lambda(1-\theta)G_2]\underline{u}_j + \hat{\underline{b}}_2, \qquad (6.2.24)$$
where
$$\hat{\underline{b}}_2 = (1 + \lambda\theta G_2)^{-1}\underline{b}_2.$$

(iii) *The (S)AGE Scheme* As the name suggests, this (single) alternating group explicit [(S)AGE] scheme entails the *alternate* use of the GER and the GEL formulae [i.e., equations (6.2.18) and (6.2.23)] as we march our solutions forward with respect to time as illustrated in Figure 6.2.4. Thus, the *two time level process* of the (S)AGE scheme is given by
$$(I + \lambda\theta G_1)\underline{u}_{j+1} = [I - \lambda(1-\theta)G_1]\underline{u}_j + \underline{b}_1$$
and
$$(I + \lambda\theta G_2)\underline{u}_{j+1} = [I - \lambda(1-\theta)G_2]\underline{u}_{j+1} + \underline{b}_2, \quad j = 0, 2, 4, \ldots. \qquad (6.2.25)$$

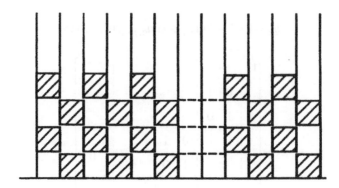

Figure 6.2.4   The (S)AGE Scheme (even number of intervals).

(iv) *The (D)AGE Scheme*   The (double) alternating group explicit scheme [(D)AGE] is a *four-time level process*. A single application of the algorithm requires the utilisation of the GER and GEL schemes for the first two time levels followed by the employment of the same formulae but now in reverse order for the second two time levels. This alternating procedure is repeated as the solutions are progressed vertically. Thus, we see that the (D)AGE scheme is basically a periodic rotation of the (S)AGE scheme with the second half cycle implemented in opposite order to complete the four time level process and is given by

$$(I + \lambda\theta G_1)\underline{u}_{j+1} = [I - \lambda(1 - \theta)G_1]\underline{u}_j + \underline{b}_1,$$

$$(I + \lambda\theta G_2)\underline{u}_{j+2} = [I - \lambda(1 - \theta)G_2]\underline{u}_{j+1} + \underline{b}_2,$$

$$(I + \lambda\theta G_2)\underline{u}_{j+3} = [I - \lambda(1 - \theta)G_2]\underline{u}_{j+2} + \underline{b}_2, \quad j = 0, 4, 8, \dots. \quad (6.2.26)$$

$$(I + \lambda\theta G_1)\underline{u}_{j+4} = [I - \lambda(1 - \theta)G_1]\underline{u}_{j+3} + \underline{b}_1,$$

The scheme is represented diagrammatically by Figure 6.2.5.

Before we proceed to establish the corresponding finite-difference analogues for the case when the line segment $0 \le x \le 1$ has an *odd* number of intervals, we observe that

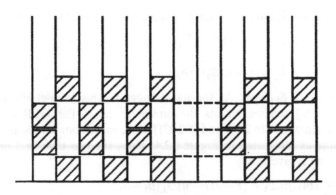

Figure 6.2.5   The (D)AGE Scheme (even number of intervals).

if we denote

$$G^{(i)} = \begin{bmatrix} -1 & 1 \\ -1 & 1 \end{bmatrix}, \quad i = 1, 2, \ldots, 1/2(m-2), \tag{6.2.27}$$

then from equations (6.2.17) and (6.2.22), we have

$$G_1 = \begin{bmatrix} G^{(1)} & & & & \\ & G^{(1)} & & 0 & \\ & & \ddots & & \\ & 0 & & G^{[(1/2)(m-2)-1]} & \\ & & & & G^{[(1/2)(m-2)]} & \\ & & & & & -1 \end{bmatrix}_{[(m-1) \times (m-1)]} \tag{6.2.28}$$

and

$$G_2 = \begin{bmatrix} 1 & & & & \\ & G^{(1)} & & & \\ & & G^{(1)} & 0 & \\ & & & \ddots & \\ & 0 & & & G^{[(1/2)(m-2)-1]} \\ & & & & & G^{[(1/2)(m-2)]} \end{bmatrix}_{[(m-1) \times (m-1)]} . \tag{6.2.29}$$

$m = $ Odd Number of Intervals

Now we will have an odd number of intervals when m is odd. Therefore, at every time level, the number of internal points is even. Accordingly, again there are two possibilities to determine the manner in which the points can be grouped on the mesh line. In the first possibility, we will have $(1/2)(m-1)$ *complete* groups of two points. In the second possibility, however, we are led to $[(m-3)/2]$ groups of two points and one point which is ungrouped adjacent to *each* boundary. Based on these observations, the following group explicit schemes can be constructed in an analogous fashion as in the *even* case.

(i) *The GEU Scheme* In this scheme, there are two points which are ungrouped, one of each adjacent to the left and right boundaries. Thus, for the *left* ungrouped point (the second point), we use equation (6.2.14) whilst the solution at the *right* ungrouped point [the $(m-1)^{th}$ point] is determined by equation (6.2.13). For the grouped points in between, we apply equations (6.2.11) and (6.2.12) in succession for $(1/2)(m-3)$ times to give the solutions at these points. This is repeated for progressive time levels and the whole procedure is known as the grouped explicit method with ungrouped (GEU) ends. Thus, the GEU method requires the solution of

$$(1 + \lambda\theta)u_{1,j+1} = [1 - \lambda(1-\theta)u_{1,j} + \lambda(1-\theta)u_{0,j} + \lambda\theta u_{0,j+1},$$

$$\left[\begin{array}{l} -\lambda\theta u_{i-1,j+1}+(1+\lambda\theta)u_{i,j+1}=\lambda(1-\theta)u_{i-1,j}+[1-\lambda(1-\theta]u_{i,j} \\ (1-\lambda\theta)u_{i-1,j+1}+\lambda\theta u_{i,j+1}=[1+\lambda(1-\theta)]u_{i-1,j}-\lambda(1-\theta)u_{i,j}, \\ \qquad\qquad\qquad\qquad\qquad i=3,5,\dots,\ m-2;\ \lambda\theta\neq1 \end{array}\right.$$

and

$$(1-\lambda\theta)u_{m-1,j+1}=-\lambda\theta u_{m,j+1}-\lambda(1-\theta)u_{m,j}+[1+\lambda(1-\theta)]u_{m-1,j},$$

which can be written in implicit matrix form as

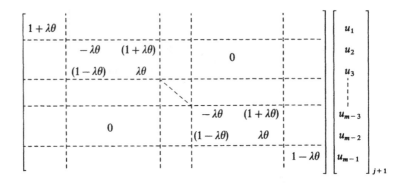

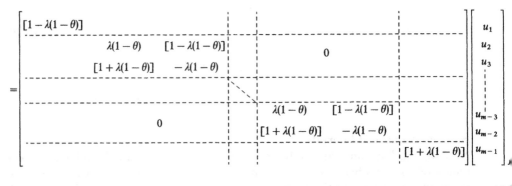

(6.2)

where $\underline{b}_3=[\lambda(1-\theta)u_{0,j}+\lambda\theta u_{0,j+1},0,\dots,0,-\lambda(1-\theta)u_{m,j}-\lambda\theta u_{m,j+1}]^T$. Now if we let

$$\hat{G}_1=\begin{bmatrix} 1 & & & & \\ G^{(1)} & & & & \\ & G^{(2)} & & 0 & \\ & & \ddots & & \\ & 0 & & G^{[(1/2)(m-3)-1]} & \\ & & & G^{[(1/2)(m-3)]} & \\ & & & & -1 \end{bmatrix} \qquad\qquad (6.2.31)$$

and

$$\hat{G}_2 = \begin{bmatrix} G^{(1)} & & & & \\ & G^{(2)} & & 0 & \\ & & \ddots & & \\ & & & G^{[(1/2)(m-3)]} & \\ 0 & & & & G^{[(1/2)(m-1)]} \end{bmatrix}, \tag{6.2.32}$$

where the (2×2) matrices $G^{(i)}$, $i = 1, 2, \ldots, (1/2)(m-1)$ are defined as in equation (6.2.25) then the GEU scheme is given by

$$(I + \lambda\theta\hat{G}_1)\underline{u}_{j+1} = [I - \lambda(1-\theta)\hat{G}_1]\underline{u}_j + \underline{b}_3, \tag{6.2.33}$$

and is described by Figure 6.2.6.

(ii) *The GEC Scheme* Alternatively, the $m-1$ internal points can be divided into a complete set of 2 point groups. This scheme, known as the group explicit complete (GEC) method is obtained by applying successively $(1/2)(m-1)$ times equation (6.2.11) and (6.2.12) for the first to $(m-1)^{th}$ point along each progressive mesh line as displayed in Figure 6.2.7. Thus, the relevant implicit equations are

$$\begin{bmatrix} -\lambda\theta u_{i-1,j+1} + (1+\lambda\theta)u_{i,j+1} = \lambda(1-\theta)u_{i-1,j} + [1-\lambda(1-\theta)]u_{i,j} \\ (1-\lambda\theta)u_{i-1,j} + \lambda\theta u_{i,j+1} = [1+\lambda(1-\theta)]u_{i-1,j} - \lambda(1-\theta)u_{i,j}, \\ i = 2, 4, \ldots, (m-1); \quad \lambda\theta \neq 1, \end{bmatrix}$$

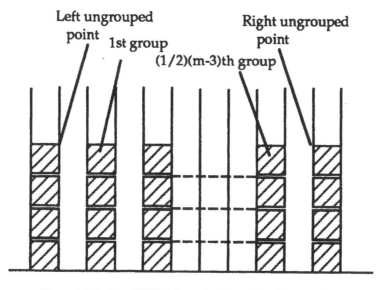

Figure 6.2.6 The GEU Scheme (odd number of intervals).

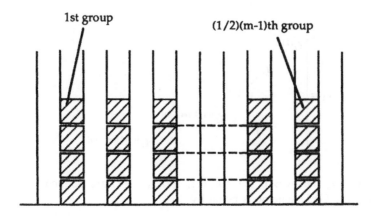

1st group   (1/2)(m-1)th group

Figure 6.2.7   The GEC Scheme (odd number of intervals).

which in matrix form can be written as

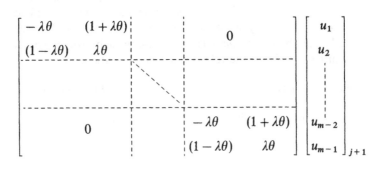

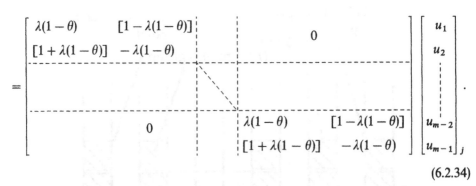

$$(6.2.34)$$

Therefore, by using equation (6.2.32), the GEC scheme can be expressed as

$$(I + \lambda\theta\hat{G}_2)\underline{u}_{j+1} = (I - \lambda(1 - \theta)\hat{G}_2)\underline{u}_j. \qquad (6.2.35)$$

(iii) *The (S)AGE and (D)AGE Scheme*   The alternating schemes corresponding to the even case are given by

$$(I + \lambda\theta\hat{G}_1)\underline{u}_{j+1} = (I - \lambda(1 - \theta)\hat{G}_1)\underline{u}_j + \underline{b}_3,$$

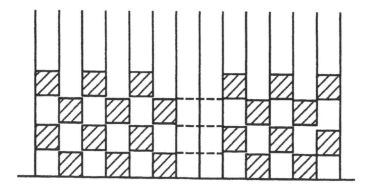

Figure 6.2.8 The (S)AGE Scheme (odd number of intervals).

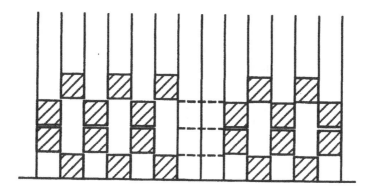

Figure 6.2.9 The (D)AGE Scheme (odd number of intervals).

$$(I + \lambda\theta\hat{G}_2)\underline{u}_{j+2} = (I - \lambda(1 - \theta)\hat{G}_2)\underline{u}_{j+1}, \tag{6.2.36}$$

for (S)AGE, Figure 6.2.8, and

$$(I + \lambda\theta\hat{G}_1)\underline{u}_{j+1} = (I - \lambda(1 - \theta)\hat{G}_1)\underline{u}_j + \underline{b}_3,$$

$$(I + \lambda\theta\hat{G}_2)\underline{u}_{j+2} = (I - \lambda(1 - \theta)\hat{G}_2)\underline{u}_{j+1},$$

$$(I + \lambda\theta\hat{G}_2)\underline{u}_{j+3} = (I - \lambda(1 - \theta)\hat{G}_2)\underline{u}_{j+2}, \tag{6.2.37}$$

$$(I + \lambda\theta\hat{G}_1)\underline{u}_{j+4} = (I - \lambda(1 - \theta)\hat{G}_1)\underline{u}_{j+3} + \underline{b}_3,$$

for (D)AGE, Figure 6.2.9.

Notice that the above GE formulae are slightly different from those obtained by Evans and Abdullah (1983) for parabolic problems in the sense that the same group $\hat{G}_i$ ($i = 1$ or 2) appears in both sides of the equations. The following conclusions may therefore be drawn:

(a) the GE schemes can be derived from the class of locally one-dimensional methods (LOD);

(b) there is no overlapping of the grouping of points. They are disjoint as shown in Figures 6.2.2–6.2.9;

(c) there is no longer a need for the commutativity of the matrices $\hat{G}_1$ and $\hat{G}_2$.

6.3. Truncation Error Analysis for the GE Methods

(i) *Truncation Error for the GER Scheme*　The set of explicit equations obtained by coupling equations (6.1.3) and (6.1.4) are

$$u_{i-1,j+1} + \lambda u_{ij} - (1+\lambda)u_{i-1,j} = 0, \tag{6.3.1}$$

$$u_{i,j+1} - \lambda u_{i-1,j} - (1-\lambda)u_{ij} = 0. \tag{6.3.2}$$

The truncation errors for any two grouped points are given for $i = 2, 4, \ldots, m-2$. By expanding by Taylor series the terms $u_{i-1,j+1}, u_{ij}, u_{i-1,j}$ in equation (6.3.1) about the point $[(i-1)\Delta x, (j+1/2)\Delta t]$, we get

$$\left(\frac{\partial U}{\partial x} + \frac{\partial U}{\partial t}\right)_{i-1,j+1/2} + \Delta x \left[\frac{1}{2}\frac{\partial^2 U}{\partial x^2} - \frac{(\Delta t)^2}{16}\frac{\partial^4 U}{\partial x^2 \partial t^2}\right]_{i-1,j+1/2}$$

$$+ \Delta t \left[-\frac{1}{2}\frac{\partial^2 U}{\partial x \partial t} - \frac{1}{12}(\Delta x)^2 \frac{\partial^4 U}{\partial x^3 \partial t}\right]_{i-1,j+1/2} + (\Delta x)(\Delta t)\left[-\frac{1}{4}\frac{\partial^3 U}{\partial x^2 \partial t}\right]_{i-1,j+1/2}$$

$$+ (\Delta x)^2 \left[\frac{1}{6}\frac{\partial^3 U}{\partial x^3} - \frac{\Delta x}{24}\frac{\partial^4 U}{\partial x^4}\right]_{i-1,j+1/2} + (\Delta t)^2 \left[\frac{1}{8}\frac{\partial^3 U}{\partial x \partial t^2} + \frac{1}{24}\frac{\partial^3 U}{\partial t^3} - \frac{\Delta t}{48}\frac{\partial^4 U}{\partial x \partial t^3}\right]_{i-1,j+1/2}$$

$$+ \left[\frac{(\Delta x)^4}{120}\frac{\partial^5 U}{\partial x^5} + \frac{(\Delta t)(\Delta x)^3}{48}\frac{\partial^5 U}{\partial x^4 \partial t} + \frac{1}{48}(\Delta x)^2(\Delta t)^2\frac{\partial^5 U}{\partial x^3 \partial t^2} + \frac{1}{96}(\Delta x)(\Delta t)^3\frac{\partial^5 U}{\partial x^2 \partial t^3}\right.$$

$$\left. + \frac{1}{384}(\Delta t)^4 \frac{\partial^5 U}{\partial x \partial t^4} + \frac{1}{1920}(\Delta t)^4 \frac{\partial^5 u}{\partial t^5}\right]_{i-1,j+1/2} + \cdots,$$

and subtracting equation (6.1.1) we obtain

$$T_{G1} = \Delta x \left[\frac{1}{2}\frac{\partial^2 U}{\partial x^2} - \frac{(\Delta t)^2}{16}\frac{\partial^4 U}{\partial x^2 \partial t^2}\right]_{i-1,j+1/2} + \Delta t \left[-\frac{1}{2}\frac{\partial^2 U}{\partial x \partial t} - \frac{1}{12}(\Delta x)^2 \frac{\partial^4 U}{\partial x^3 \partial t}\right]_{i-1,j+1/2}$$

$$+ (\Delta x)(\Delta t)\left[-\frac{1}{4}\frac{\partial^3 U}{\partial x^2 \partial t}\right]_{i-1,j+1/2} + (\Delta x)^2 \left[\frac{1}{6}\frac{\partial^3 U}{\partial x^3} - \frac{\Delta x}{24}\frac{\partial^4 U}{\partial x^4}\right]_{i-1,j+1/2}$$

$$+ (\Delta t)^2 \left[\frac{1}{8}\frac{\partial^3 U}{\partial x \partial t^2} + \frac{1}{24}\frac{\partial^3 U}{\partial t^3} - \frac{\Delta t}{48}\frac{\partial^4 U}{\partial x \partial t^3}\right]_{i-1,j+1/2} + 0[(\Delta x)^{\alpha_1}(\Delta t)^{\alpha_2}], \quad \alpha_1 + \alpha_2 = 4.$$

$$\tag{6.3.3a}$$

Similarly by expanding the terms $u_{i,j+1}, u_{i-1,j}$ and u_{ij} in equation (6.3.2) about the point $[i\Delta x, (j+1/2)\Delta t]$ leads to

$$\left(\frac{\partial U}{\partial x} + \frac{\partial U}{\partial t}\right)_{i,j+1/2} + \Delta x\left[-\frac{1}{2}\frac{\partial^2 U}{\partial x^2} + \frac{(\Delta t)^2}{16}\frac{\partial^4 U}{\partial x^2 \partial t^2}\right]_{i,j+1/2}$$

$$+ \Delta t\left[-\frac{1}{2}\frac{\partial^2 U}{\partial x \partial t} - \frac{(\Delta x)^2}{12}\frac{\partial^4 U}{\partial x^3 \partial t}\right]_{i,j+1/2} + (\Delta x)(\Delta t)\left[\frac{1}{4}\frac{\partial^3 U}{\partial x^2 \partial t}\right]_{i,j+1/2}$$

$$+ (\Delta x)^2\left[\frac{1}{6}\frac{\partial^3 U}{\partial x^3} - \frac{\Delta x}{24}\frac{\partial^4 U}{\partial x^4}\right]_{i,j+1/2} + (\Delta t)^2\left[\frac{1}{8}\frac{\partial^3 U}{\partial x \partial t^2} + \frac{1}{24}\frac{\partial^3 U}{\partial t^3} - \frac{\Delta t}{48}\frac{\partial^4 U}{\partial x \partial t^3}\right]_{i,j+1/2}$$

$$+ \left[\frac{(\Delta x)^4}{120}\frac{\partial^5 U}{\partial x^5} + \frac{(\Delta t)(\Delta x)^3}{48}\frac{\partial^5 U}{\partial x^4 \partial t} + \frac{(\Delta x)^2(\Delta t)^2}{48}\frac{\partial^5 U}{\partial x^3 \partial t^2} + \frac{(\Delta x)(\Delta t)^3}{96}\frac{\partial^5 U}{\partial x^2 \partial t^3}\right.$$

$$\left.+ \frac{1}{384}(\Delta t)^4\frac{\partial^5 U}{\partial x \partial t^4} + \frac{(\Delta t)^4}{1920}\frac{\partial^5 U}{\partial t^5}\right]_{i,j+1/2} + \cdots,$$

and subtracting equation (6.1.1) gives

$$T_{G2} = \Delta x\left[-\frac{1}{2}\frac{\partial^2 U}{\partial x^2} + \frac{(\Delta t)^2}{16}\frac{\partial^4 U}{\partial x^2 \partial t^2}\right]_{i,j+1/2} + \Delta t\left[-\frac{1}{2}\frac{\partial^2 U}{\partial x \partial t} - \frac{(\Delta x)^2}{12}\frac{\partial^4 U}{\partial x^3 \partial t}\right]_{i,j+1/2}$$

$$+ (\Delta x)(\Delta t)\left[\frac{1}{4}\frac{\partial^3 U}{\partial x^2 \partial t}\right]_{i,j+1/2} + (\Delta x)^2\left[\frac{1}{6}\frac{\partial^3 U}{\partial x^3} - \frac{\Delta x}{24}\frac{\partial^4 U}{\partial x^4}\right]_{i,j+1/2}$$

$$+ (\Delta t)^2\left[\frac{1}{8}\frac{\partial^3 U}{\partial x \partial t^2} + \frac{1}{24}\frac{\partial^3 U}{\partial t^3} - \frac{\Delta t}{48}\frac{\partial^4 U}{\partial x \partial t^3}\right]_{i,j+1/2} + 0[(\Delta x)^{\alpha_1}(\Delta t)^{\alpha_2}], \quad \alpha_1 + \alpha_2 = 4.$$

$$(6.3.3b)$$

The truncation error for the single ungrouped point near the right end is given by the truncation error incurred for equation (6.2.4). This is obtained directly by putting $i = m - 1$ in equation (6.2.6) which gives

$$T_R = \Delta x\left[\frac{1}{2}\frac{\partial^2 U}{\partial x^2} + \frac{(\Delta t)^2}{16}\frac{\partial^4 U}{\partial x^2 \partial t^2}\right]_{m-1,j+1/2}$$

$$+ \Delta t\left[-\frac{1}{2}(1-2\theta)\frac{\partial^2 U}{\partial x \partial t} - \frac{1}{12}(\Delta x)^2(1-2\theta)\frac{\partial^4 U}{\partial x^3 \partial t}\right]_{m-1,j+1/2}$$

$$+ (\Delta x)(\Delta t)\left[-\frac{1}{4}(1-2\theta)\frac{\partial^3 U}{\partial x^2 \partial t}\right]_{m-1,j+1/2} + (\Delta x)^2\left[\frac{1}{6}\frac{\partial^3 U}{\partial x^3} + \frac{\Delta x}{24}\frac{\partial^4 U}{\partial x^4}\right]_{m-1,j+1/2}$$

$$+ (\Delta t)^2\left[\frac{1}{8}\frac{\partial^3 U}{\partial x \partial t^2} + \frac{1}{24}\frac{\partial^3 U}{\partial t^3} - \frac{\Delta t}{48}(1-2\theta)\frac{\partial^4 U}{\partial x \partial t^3}\right]_{m-1,j+1/2}$$

$$+ 0[(\Delta x)^{\alpha_1}(\Delta t)^{\alpha_2}], \quad \alpha_1 + \alpha_2 = 4. \quad (6.3.3c)$$

Thus the truncation error for the overall GER strategy is $0(\Delta x) + 0(\Delta t)$.

(ii) *Truncation Error for the GEL Scheme* The truncation error for the single ungrouped point near the left boundary is given by the truncation error for equation (6.2.3). Hence with $i = 1$, the expression (6.2.5) gives

$$T_L = \Delta x \left[-\frac{1}{2} \frac{\partial^2 U}{\partial x^2} - \frac{(\Delta t)^2}{16} \frac{\partial^4 U}{\partial x^2 \partial t^2} \right]_{1, j + 1/2}$$

$$+ \Delta t \left[-\frac{1}{2}(1 - 2\theta) \frac{\partial^2 U}{\partial x \partial t} - \frac{(\Delta x)^2}{12}(1 - 2\theta) \frac{\partial^4 U}{\partial x^3 \partial t} \right]_{1, j + 1/2}$$

$$+ (\Delta x)(\Delta t) \left[\frac{1}{4}(1 - 2\theta) \frac{\partial^3 U}{\partial x^2 \partial t} \right]_{1, j + 1/2} + (\Delta x)^2 \left[\frac{1}{6} \frac{\partial^3 U}{\partial x^3} - \frac{\Delta x}{24} \frac{\partial^4 U}{\partial x^4} \right]_{1, j + 1/2}$$

$$+ (\Delta t)^2 \left[\frac{1}{8} \frac{\partial^3 U}{\partial x \partial t^2} + \frac{1}{24} \frac{\partial^3 U}{\partial t^3} - \frac{\Delta t}{48}(1 - 2\theta) \frac{\partial^4 U}{\partial x \partial t^3} \right]_{1, j + 1/2}$$

$$+ 0[(\Delta x)^{\alpha_1}(\Delta t)^{\alpha_2}]; \quad \alpha_1 + \alpha_2 = 4, \ 0 \le \theta \le 1. \tag{6.3.4}$$

We note that the truncation errors for any two ungrouped points of the GEL scheme are given by T_{G1} and T_{G2} of the equations (6.3.3a) and (6.3.3b), respectively. Thus the overall Truncation Error of the GEL scheme is $0(\Delta x) + 0(\Delta t)$.

(iii) *Truncation Error for the GEU Scheme* As indicated by Figure 6.2.6 for the case when an odd number of intervals is used, the truncation error of the scheme at the left ungrouped point is given by T_L of equation (6.3.4) whilst the error at the single ungrouped point near the right end is T_R of equation (6.3.3c). For the points in between the boundaries which are grouped two at a time, the truncation errors are given by T_{G1} and T_{G2} of equations (6.3.3a) and (6.3.3b), respectively. Similarly the overall accuracy of the GEU scheme is again $0(\Delta x) + 0(\Delta t)$.

(iv) *Truncation Error for the GEC Scheme* In this scheme, the grouping of two points at a time along each mesh line is complete as shown by Figure 6.2.7. Hence the truncation error for this scheme is given by T_{G1} and T_{G2}, respectively, for $i = 1, 3, \ldots, m - 4, m - 2$ when m is odd. Again we have the overall accuracy defined by $0(\Delta x) + 0(\Delta t)$.

(v) *Truncation Error for the S(AGE) Scheme* If we assume that the number of intervals is even, then as we know from Figure 6.2.4, this scheme entails the alternate use of the GER and the GEL schemes along successive time steps. Accordingly, its truncation error is given by the truncation errors of the GER and the GEL schemes along the alternate time level. This produces the possible effect of the cancellation of the component error terms at most internal points. A more accurate solution with this scheme is therefore expected than any of the previous GE methods. Thus the Truncation Error is of $0(\Delta t)$. A similar argument holds when m is odd.

(vi) *Truncation Error for the D(AGE) Scheme* The GER, GEL, GEL and GER, methods, in that order, are employed at each of every four time levels. By the same reasoning as above, we expect this four-step process to be of $0(\Delta t)$ and as accurate if not better than the S(AGE) method. In fact, our numerical experiment shows that the D(AGE) procedure can be more accurate than the S(AGE) scheme or any of the other GE methods.

6.4. Stability Analysis for the GE Methods

We shall now proceed to establish the stability requirement of the GE methods. From the formulae (6.2.19), (6.2.24), (6.2.33) and (6.2.35), we present below the explicit expressions for the GER, GEL, GEU and the GEC schemes:

$$\underline{u}_{j+1} = (I + \lambda\theta G_1)^{-1}[I - \lambda(1 - \theta)G_1]\underline{u}_j + \hat{\underline{b}}_1, \tag{6.4.1}$$

$$\underline{u}_{j+1} = (I + \lambda\theta G_2)^{-1}[I - \lambda(1 - \theta)G_2]\underline{u}_j + \hat{\underline{b}}_2, \tag{6.4.2}$$

$$\underline{u}_{j+1} = (I + \lambda\theta \hat{G}_1)^{-1}[I - \lambda(1 - \theta)\hat{G}_1]\underline{u}_j + \hat{\underline{b}}_3 \tag{6.4.3}$$

and

$$\underline{u}_{j+1} = (I + \lambda\theta \hat{G}_2)^{-1}[I - \lambda(1 - \theta)\hat{G}_2]\underline{u}_j. \tag{6.4.4}$$

These processes may be written in a single form as

$$\underline{u}_{j+1} = \Gamma\underline{u}_j + \hat{\underline{b}}, \tag{6.4.5}$$

where Γ, the amplification matrix, corresponds to the scheme employed and $\hat{b}$ is the relevant column vector of order $(m - 1)$ as indicated in the formulae above.

(i) *Stability of the GER Scheme* From the equations (6.4.1) and (6.4.5), the GER amplification matrix is given by

$$\Gamma_{GER} = \begin{bmatrix} (1 + \lambda) & -\lambda & & & & & \\ \lambda & (1 - \lambda) & & & & & \\ & & (1 + \lambda) & -\lambda & & & \\ & & \lambda & (1 - \lambda) & & 0 & \\ & & & & \ddots & & \\ & & & & & (1 + \lambda) & -\lambda \\ & 0 & & & & \lambda & (1 - \lambda) \\ & & & & & & 1 + \dfrac{\lambda}{(1 - \lambda\theta)} \end{bmatrix}_{[(m-1)\times(m-1)]},$$

(6.4.6)

with $\lambda\theta \neq 1$. It can be easily shown that Γ_{GER} possesses the eigenvalues 1, of multiplicity $(m - 2)$, and

$$\left[1 + \frac{\lambda}{(1 - \lambda\theta)}\right].$$

If we denote $\rho(\Gamma_{\text{GER}})$ as the spectral radius of Γ_{GER}, then for stability of the GER scheme, we require $\rho(\Gamma_{\text{GER}}) \leq 1$. This implies that

$$\left| 1 + \frac{\lambda}{(1 - \lambda\theta)} \right| \leq 1, \tag{6.4.7}$$

which gives

$$-2 \leq \frac{\lambda}{(1 - \lambda\theta)} \leq 0. \tag{6.4.8}$$

Since λ is non-negative, then $(1 - \lambda\theta) < 0$ or,

$$\lambda\theta > 1 \quad \text{and} \quad \lambda \leq -2(1 - \lambda\theta). \tag{6.4.9}$$

Different cases of θ are now treated to investigate the condition of stability of the GER scheme:

(a) For $\theta = 0$, we have

$$\left| 1 + \frac{\lambda}{(1 - \lambda\theta)} \right| = 1 + \lambda,$$

for all positive values of λ. Therefore $\rho(\Gamma_{\text{GER}}) > 1$, which shows that GER scheme is always unstable.

(b) For $0 < \theta < 1/2$, condition (6.4.8) gives

$$\lambda\theta > 1 \quad \text{and} \quad \lambda \leq \frac{2}{(2\theta - 1)}.$$

The second inequality can never be satisfied since λ is non-negative whilst $(2\theta - 1)$ is always negative. Hence, for this particular case of θ, the GER method is always unstable.

(c) For $\theta = 1/2$, we obtain

$$\left| 1 + \frac{\lambda}{(1 - \lambda\theta)} \right| = 1 + \left| 1 + \frac{\lambda}{(1 - (1/2)\lambda)} \right| > 1,$$

and as in case (a) the GER scheme is absolutely unstable.

(d) For $1/2 < \theta \leq 1$, condition (6.4.8) becomes

$$\lambda\theta > 1 \quad \text{and} \quad \lambda \geq \frac{2}{(2\theta - 1)},$$

or

$$\lambda > \frac{1}{\theta} \quad \text{and} \quad \lambda \geq \frac{1}{(\theta - \frac{1}{2})}.$$

We deduce that, the scheme is conditionally stable for

$$\lambda \geq \frac{2}{(2\theta - 1)}.$$

We conclude from cases (a)–(d) that the GER scheme is:

(1) always unstable for $0 \leq \theta \leq 1/2$,

(2) it is conditionally stable for

$$\lambda \geq \frac{2}{(2\theta - 1)},$$

when $\theta \in (1/2, 1]$.

It may therefore be summarised that none of the cases above can really be considered useful either because of their unconditional instability (when $0 \leq \theta \leq 1/2$) or due to their "inverse" conditional stability (when $1/2 < \theta < 1$ and which could lead to excessively large time steps).

(ii) *Stability of the GEL Scheme* From equations (6.4.2) and (6.4.5), the GEL amplification matrix is given by

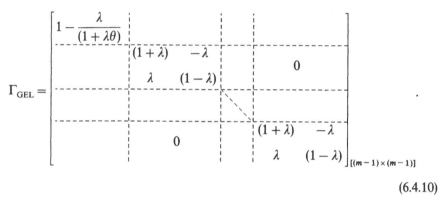

$$(6.4.10)$$

The eigenvalues of Γ_{GEL} are 1 (of multiplicity $(m-2)$) and,

$$\left[1 - \frac{\lambda}{(1 + \lambda\theta)} \right],$$

and the GEL method is stable if $\rho(\Gamma_{\text{GEL}}) \leq 1$. This requires that

$$\left| 1 - \frac{\lambda}{(1 + \lambda\theta)} \right| \leq 1 \qquad (6.4.11)$$

giving

$$0 \leq \frac{\lambda}{(1 + \lambda\theta)} \leq 2. \qquad (6.4.12)$$

Since λ is non-negative then from condition (6.4.12) we must have $(1 + \lambda\theta) > 0$. Hence,

$$\lambda \leq 2(1 + \lambda\theta). \qquad (6.4.13)$$

(a) For $\theta = 0$, we have

$$\left| 1 - \frac{\lambda}{(1 + \lambda\theta)} \right| = |1 - \lambda|.$$

In order that $\rho(\Gamma_{\text{GEL}}) \leq 1$, we must have $|1 - \lambda| \leq 1$ which is satisfied for $\lambda \leq 2$. Therefore, for this particular case of θ, the condition of stability is $\lambda \leq 2$.

(b) If $0 < \theta < 1/2$, then from condition (6.4.13) we obtain $\lambda(1 - 2\theta) \le 2$ which leads to the following condition of stability,

$$\lambda \le \frac{2}{(1 - 2\theta)}.$$

(c) For $\theta = 1/2$, we get

$$\left| 1 - \frac{\lambda}{(1 + \lambda\theta)} \right| = \left| 1 - \frac{\lambda}{(1 + (1/2)\lambda)} \right| < 1,$$

for every positive value of λ. This implies that the scheme is always stable for $\theta = 1/2$.

(d) For $1/2 < \theta \le 1$, inequality (6.4.13) leads to

$$\lambda \ge \frac{2}{(1 - 2 - \theta)}.$$

Now, the quantity $2/(1 - 2\theta)$ is always negative whilst λ is non-negative. Hence, the scheme is absolutely stable for all values of λ. From all the cases above, we conclude that the GEL scheme is:

(1) conditionally stable for $\lambda \le 2/(1 - 2\theta)$ with $0 \le \theta < 1/2$.
(2) it is absolutely stable for all values of λ when $1/2 \le \theta \le 1$.

(iii) *Stability of the GEU Scheme*   From equations (6.4.3) and (6.4.5), the GEU amplification matrix is given by

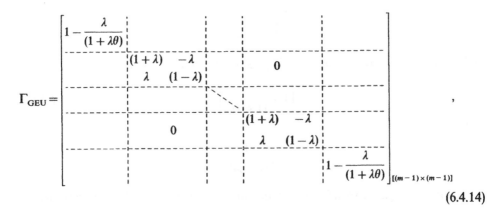

$$(6.4.14)$$

whose eigenvalues are

$$1 - \frac{\lambda}{(1 + \lambda\theta)},$$

$$1 \text{ [of multiplicity} (m - 3)],$$

and

$$1 + \frac{\lambda}{(1 - \lambda\theta)},$$

hence we can easily deduce from the conclusions drawn on the stability analysis of the GER and GEL schemes that the GEU method is conditionally stable for

$$\lambda \geq \frac{2}{(2\theta - 1)},$$

where $\theta \in (1/2, 1)$.

(iv) *Stability of the GEC Scheme* From the equations (6.4.4) and (6.4.5), the GEC amplification matrix is given by

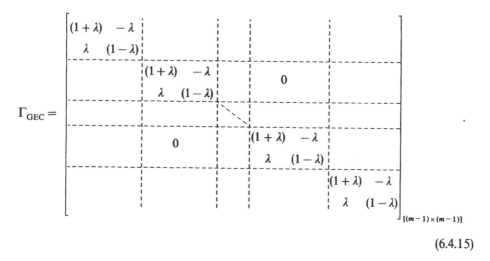

$$(6.4.15)$$

Since Γ_{GEC} has $(m-1)$ eigenvalues each equal to 1, then clearly the GEC scheme is always stable with no restrictions on λ and $\theta \in [0, 1]$.

(v) *Stability of the (S)AGE Scheme* We shall first consider the case when m is even. By means of equations (6.2.25) we obtain

$$\underline{u}_{j+2} = (I + \lambda \theta G_2)^{-1} [I - \lambda(1 - \theta)G_2]\underline{u}_{j+1} + (I + \lambda \theta G_2)^{-1} \underline{b}_2,$$

$$= (I + \lambda \theta G_2)^{-1} [I - \lambda(1 - \theta)G_2] (I + \lambda \theta G_1)^{-1}$$

$$[I - \lambda(1 - \theta)G_1]\underline{u}_j + \underline{b}_2', \qquad (6.4.16)$$

i.e.,

$$\underline{u}_{j+2} = \Gamma_{SAGE}\, \underline{u}_j + \underline{b}_2', \qquad (6.4.17)$$

where

$$\Gamma_{SAGE} = \{(I + \lambda \theta G_2)^{-1} [I - \lambda(1 - \theta)G_2]\} \{(I + \lambda \theta G_1)^{-1} [I - \lambda(1 - \theta)G_1]\}$$

$$= \Gamma_1 \Gamma_2, \qquad (6.4.18)$$

and $\underline{b}_2'$ is the appropriate column vector of order $(m - 1)$. We observe that Γ_1 and Γ_2 are exactly the amplification matrices of the GEL equation (6.4.10) and the GER equation (6.4.6) schemes, respectively. Hence by multiplying these matrices we obtain

$$\Gamma_{\text{SAGE}} = \begin{bmatrix}
a & b & & & & & & \\
c & d & -c & e & & & & \\
e & f & d & -f & & & & \\
& & c & d & -c & e & & \\
& & e & f & d & -f & & \quad 0 \\
& \quad 0 & & & & c & d & -c & e \\
& & & & & e & f & d & -f \\
& & & & & & c & d & g \\
& & & & & & e & f & h
\end{bmatrix}_{[(m-1)\times(m-1)]} , \qquad (6.4.19)$$

where

$$a = (1 + \lambda)[1 - \lambda/(1 + \lambda\theta)],$$

$$b = -\lambda[1 - \lambda/(1 + \lambda\theta)],$$

$$c = \lambda(1 + \lambda),$$

$$d = 1 - \lambda^2,$$

$$e = \lambda^2,$$

$$f = \lambda(1 - \lambda),$$

$$g = -\lambda[1 + \lambda/(1 - \lambda\theta)],$$

and

$$h = (1 - \lambda)[1 + \lambda/(1 - \lambda\theta)]. \qquad (6.4.20)$$

Note that $\text{diag}(\Gamma_{\text{SAGE}}) = (a, d, d, \ldots, d, h)$ with d occurring $(m - 3)$ times. It is difficult to evaluate directly the eigenvalues of Γ_{SAGE} in a closed form. However we know from matrix theory that if the eigenvalues of Γ_{SAGE} are denoted by μ_i, $i = 1, 2, \ldots, m - 1$, then

$$\sum_{i=1}^{m-1} \mu_i = \text{tr}(\Gamma_{\text{SAGE}}),$$

where $\text{tr}(\Gamma_{\text{SAGE}})$ is the trace of Γ_{SAGE} which is the sum of the diagonal elements of Γ_{SAGE}, i.e.,

$$\mu_1 + \mu_2 + \cdots + \mu_{m-1} = a + (m - 3)d + h.$$

Now if we insist $\rho(\Gamma_{\text{SAGE}}) \leq 1$ it follows that

$$|\mu_1 + \mu_2 + \cdots + \mu_{m-1}| = |a + (m - 3)d + h| \leq \mu_1 + \mu_2 + \cdots + \mu_{m-1}| \leq (m - 1).$$

Hence we seek the values of λ such that

$$|a + (m - 3)d + h| \leq |a| + (m - 3)|d| + |h| \leq (m - 1),$$

i.e.,

$$(1 + \lambda)\left|\left[1 - \frac{\lambda}{(1 + \lambda\theta)}\right]\right| + (m - 3)|(1 - \lambda^2)| + |(1 - \lambda)|\left|\left[1 + \frac{\lambda}{(1 - \lambda\theta)}\right]\right| \leq (m - 1).$$

Let

$$\phi(\lambda)=(1+\lambda)\left[1-\frac{\lambda}{(1+\lambda\theta)}\right]+(m-3)(1-\lambda^2)+(1-\lambda)\left[1+\frac{\lambda}{(1-\lambda\theta)}\right].$$

$\phi(\lambda)$ is non-negative and if $\lambda\leq1$ for $\theta\in[0,1]$, $\lambda\theta\neq1$ we find that $\phi(\lambda)$ will be a continuous function of λ, i.e.,

$$\phi(\lambda)=(1+\lambda)\left[1-\frac{\lambda}{(1+\lambda\theta)}\right]+(m-3)(1-\lambda^2)+(1-\lambda)\left[1+\frac{\lambda}{(1-\lambda\theta)}\right].$$

$\phi(\lambda)$ attains its greatest value of $(m-1)$ at $\lambda=0$ and $\phi(\lambda)<m-1$ in the range $0<\lambda\leq1$.

Therefore, if $\rho(\Gamma_{SAGE})\leq1$, then $\lambda\leq1$, for $\theta\in[0,1]$, $\lambda\theta\neq1$.

Now suppose that we form the sequence Γ_{SAGE}^2, Γ_{SAGE}^1,.... It is observed that the entries of the product Γ_{SAGE}^k contain combinations of powers in λ, $(1-\lambda)$ and $[1-\lambda/(1+\theta)]$. Hence if $\lambda\leq1$,

$$\lim_{k\to\infty}\Gamma_{SAGE}^k=0,$$

which implies that Γ_{SAGE} is convergent. A necessary and sufficient condition for this to be so is $\rho(\Gamma_{SAGE})<1$. We conclude that the (S)AGE scheme is stable for $\lambda\leq1$.

When m is odd, the equations constituting the (S)AGE procedure are given by equation (6.2.36). This time, however, the amplification matrix Γ_{SAGE} is the product of the amplification matrices of the GEC and GEU schemes and it takes the form

$$\Gamma_{SAGE}=\begin{bmatrix} a & -c & e \\ -b & d & -f \\ & c & d & -c & e \\ & e & f & d & -f \\ & & & c & d & -c & e \\ & & & e & f & d & -f & & 0 \\ & & & & & & \ddots \\ & & & & & & -c & d & -c & e \\ & 0 & & & & & -e & f & d & -f \\ & & & & & & & c & d & g \\ & & & & & & & e & f & h \end{bmatrix}_{[(m-1)\times(m-1)]} ,$$

(6.4.21)

The matrix has the same diagonal elements as in equation (6.4.19) and thus the (S)AGE scheme for the odd case is also conditionally stable for $\lambda\leq1$.

6.5. The AGE Iterative Method for First Order Hyperbolic Equations

Consider the linear hyperbolic equation

$$\frac{\partial U}{\partial t} + c\frac{\partial U}{\partial x} = f(x,t) \tag{6.5.1}$$

in the domain $R=\{a \leq x \leq b, \ t>0\}$, satisfying the following initial condition

$$U(x,0) = g(x), \tag{6.5.2}$$

with Dirichlet (I) and periodic (II) boundary conditions,

$$\text{Problem I:} \quad U(a,t) = q(t), \quad t>0, \tag{6.5.3a}$$

$$\text{Problem II:} \quad U(a,t) = U(b,t), \quad t>0, \tag{6.5.3b}$$

where c is a constant.

For a numerical solution of (6.5.1) we choose a uniform mesh spacing h along the x-axis, and a uniform time increment k along the t-axis. Further, if we choose

$$\left(\frac{\partial U}{\partial t}\right)_{i,j} = (u_{i,j+1} - u_{i,j})/k, \tag{6.5.4}$$

and consider the stable Crank–Nicolson implicit scheme of second order accuracy in which the partial derivative with respect to x is approximated by the average of its values at time $t=jk$ and $(j+1)k$. Then we can write

$$\left(\frac{\partial U}{\partial x}\right)_{i,j} = \frac{1}{2}\left[\frac{u_{i+1,j+1} - u_{i-1,j+1}}{2h} + \frac{u_{i+1,j} - u_{i-1,j}}{2h}\right]. \tag{6.5.5}$$

On substituting (6.5.4) and (6.5.5) into (6.5.1) and setting $s=ck/4h$, the corresponding finite difference approximation for the equation (6.5.1) becomes, on neglecting the truncation error terms,

$$-su_{i-1,j+1} + u_{i,j+1} + su_{i+1,j+1} = su_{i-1,j} + u_{i,j} - su_{i+1,j} + kf_{i,j}, \tag{6.5.6}$$

which relates the unknown values of u at the $(j+1)$ time level to the known values of u at the jth time level. If we apply the above equation (6.5.6) at each point x_i for both problems I and II on each grid line (or time level), we obtain the following linear system to be solved at each time level

$$A\underline{u}_{j+1} = \underline{d}_j, \tag{6.5.7}$$

where $\underline{d}_j$ is a known vector, and the matrix A has the form

$$A = \begin{bmatrix} 1 & s & & & & \\ -s & 1 & s & & 0 & \\ & . & . & . & & \\ 0 & & -s & 1 & s \\ & & & -s & 1 \end{bmatrix}, \quad \text{for the problem I} \qquad (6.5.8)$$

and

$$A = \begin{bmatrix} 1 & s & & & -s \\ -s & 1 & s & & 0 \\ & . & . & . & \\ 0 & & -s & 1 & s \\ s & & & -s & 1 \end{bmatrix}, \quad \text{for the problem II.} \qquad (6.5.9)$$

Thus the system (6.5.7) needs to be solved repeatedly for each time level. Fast direct method has been suggested by Hockney (1965), which is ideally suitable for a serial computer. Previously the alternating explicit group (AGE) iterative method has been used for solving (6.5.7) derived from a parabolic equation. Therefore the solution to the jth time level is a good starting vector for solving the system (6.5.7) at the $(j+1)^{th}$ time level. Another important feature for the AGE method is its complete suitability for parallel computation.

Next the convergence analysis for the AGE iterative method for the hyperbolic problem is considered and the determination of a 'good' parameter for the method given.

The alternating group explicit (AGE) method has previously been developed for parabolic problems where the constituent component of the coefficient matrix A were symmetric and positive definite.

Now we write (6.5.7) in a general form

$$A\underline{u} = \underline{d}, \qquad (6.5.10)$$

where the n by n matrix A has the form (6.5.8) or (6.5.9). Hence we assume that n is even. Let

$$C = \begin{bmatrix} 0.5 & s \\ -s & 0.5 \end{bmatrix}, \qquad (6.5.11)$$

and let A have the splitting

$$A = G_1 + G_2, \qquad (6.5.12)$$

where G_1 is given by

$$G_1 = \begin{bmatrix} C & & & 0 \\ & C & & \\ & & C & \\ 0 & & & C \end{bmatrix}, \qquad (6.5.13)$$

and G_2 is given by

$$G_2 = \begin{bmatrix} 0.5 & & & 0 \\ & C & & \\ & & C & \\ 0 & & & 0.5 \end{bmatrix}, \quad \text{for the problem I}, \qquad (6.5.14)$$

or

$$G_2 = \begin{bmatrix} 0.5 & & & -s \\ & C & & \\ & & C & \\ s & & & 0.5 \end{bmatrix}, \quad \text{for the problem II}. \qquad (6.5.15)$$

Thus the AGE method applied to (6.5.10) with A having the above splitting is given by the following:

$$(G_1 + rI)\underline{u}^{(m+1/2)} = \underline{d} - (G_2 - rI)\underline{u}^{(m)},$$

$$(G_2 + rI)\underline{u}^{(m+1)} = \underline{d} - (G_1 - rI)\underline{u}^{(m+1/2)}, \quad m \geq 0, \qquad (6.5.16)$$

or

$$\underline{y} = \underline{d} - (G_2 - rI)\underline{u}^{(m)}, \quad \cdot$$

$$(G_2 + rI)\underline{u}^{(m+1)} = \underline{d} - (G_1 - rI)(G_1 + rI)^{-1}\underline{y}, \quad m \geq 0, \qquad (6.5.17)$$

where r is the acceleration parameter. Because of the special forms of matrices G_1 and G_2, the inverses of $(G_1 + rI)$ and $(G_2 + rI)$ can be given explicitly. Therefore the AGE iterative formulae (6.5.17) or (6.5.16) can be given in an explicit form. Thus it is quite suitable for parallel computing.

It is clear the AGE iteration (6.5.17) or (6.5.16) converges to the solution $\underline{u} = A^{-1}\underline{d}$ if and only if the spectral radius $S(T)$ is less than unity and the smaller $S(T)$ is the faster the AGE method converges. Here, the matrix T is given by

$$T = (G_2 + rI)^{-1}(G_1 - rI)(G_1 + rI)^{-1}(G_2 - rI). \tag{6.5.18}$$

Now we study the convergence of the AGE method which is governed by the form of the iterative matrix T.

Since T is similar to the matrix

$$T_1 = (G_1 - rI)(G_1 + rI)^{-1}(G_2 - rI)(G_2 + rI)^{-1},$$

therefore, $S(T) = S(T_1)$. Besides, we also have

$$S(T) = S(T_1) \le \|T_1\|_2. \tag{6.5.19}$$

By the definitions of G_1 and G_2 (see 6.5.13–6.5.15), it can be shown that

$$(G_1 - rI)(G_1 + rI)^{-1} = \begin{bmatrix} D & & & 0 \\ & D & & \\ & & \ddots & \\ & & & D \\ 0 & & & D \end{bmatrix}, \quad D = (C - rI)(C + rI)^{-1} \tag{6.5.20}$$

and

$$(G_2 - rI)(G_2 + rI)^{-1} = \begin{bmatrix} \alpha & & & 0 \\ & D & & \\ & & \ddots & \\ & & & D \\ 0 & & & \alpha \end{bmatrix},$$

with $\alpha = (0.5 - r)/(0.5 + r)$ for G_2 having form (6.5.14) and that there is a permutation matrix P so that

$$PG_2P = \begin{bmatrix} C^T & & & 0 \\ & C & & \\ & & \ddots & \\ & & & C \\ 0 & & & C \end{bmatrix},$$

with C defined by (6.5.11), for G_2 having the form (6.5.15).

Hence in the latter case we have

$$P(G_2 - rI)(G_2 + rI)^{-1}P = \begin{bmatrix} D^T & & & 0 \\ & D & & \\ & & \diagdown & \\ & & & D \\ 0 & & & D \end{bmatrix}.$$

Since

$$S(T) = S(T_1 \le \|(G_1 - rI)(G_1 + rI)^{-1}\|_2 \|(G_2 - rI)(G_2 + rI)^{-1}\|_2$$

$$= \|D_2\|_2 \|P(G_2 - rI)(G_2 + rI)^{-1}P\|_2, \qquad (6.5.21)$$

therefore,

$$S(T) \le \|D\|_2 \max\{|\alpha|, \|D\|_2\} \quad \text{for } G_2 \text{ having the form } (6.5.14)$$

$$\le \|D\|_2^2 \quad \text{for } G_2 \text{ having the form } (6.5.15). \qquad (6.5.22)$$

Note that for $r > 0$

$$\|D\|_2^2 \le \|C - rI\|_2^2 \|(C + RI)^{-1}\|_2^2 = \frac{(0.5 - r)^2 + s^2}{(0.5 + r)^2 + s^2} < 1,$$

$$|\alpha| = \left|\frac{0.5 - r}{0.5 + r}\right| < 1, \qquad (6.5.23)$$

hence by (6.5.22) we have the AGE method always converges if the acceleration parameter r is positive.

Now the choice of the acceleration parameter r is crucial to the success of the AGE method. Since

$$|\alpha| \le \|D\|_2 \quad \text{for any real number } s,$$

therefore,

$$S(T) \le F(r, s) = \frac{(0.5 - r)^2 + s^2}{(0.5 + r)^2 + s^2}, \quad \text{for each case.} \qquad (6.5.24)$$

Note that

$$\frac{dF}{dr} = 2 \frac{r^2 - (0.25 + s^2)}{[(0.5 + r)^2 + s^2]^2}.$$

Hence

$$\frac{dF}{dr} = \begin{cases} < 0, & \text{if } 0 < r < r^*, \\ 0, & \text{if } r = r^*, \\ > 0, & \text{if } r > r^*. \end{cases}$$

Here r^* is given by

$$r^* = \sqrt{0.25 + s^2}. \qquad (6.5.25)$$

Thus r^* is a good choice for the accelerating parameter, and we have

$$S(T) \leq \frac{2\sqrt{0.25 + s^2} - 1}{2\sqrt{0.25 + s^2} + 1} \quad \text{for } r = r^*. \tag{6.5.26}$$

Thus, the parameter r^* is defined in terms of the parameter $s = ck/4h$ which is governed by the chosen mesh sizes and can be determined a' priòri.

Numerical results have shown that the above choice for $r = r^*$ is best. We also note that the AGE method is not too sensitive for the choice of r, i.e., there is an interval $(r_L, r_R) \ni r^*$ such that the AGE method with any r in this interval has the same convergence rate.

6.6. Group Explicit Finite Difference Approximations for Second Order Hyperbolic Equations

A natural extension is to employ group explicit finite difference procedures to second-order equations. In the process of developing these methods, we will, of course, bear in mind, as we did with first-order equations, the limitations imposed by the characteristics.

(a) Explicit methods
Let us consider the simplest of the second-order hyperbolic equations called the wave equation given by

$$\frac{\partial^2 U}{\partial x^2} = \frac{\partial^2 U}{\partial t^2}, \tag{6.6.1}$$

and the Cauchy condition

$$U(x,0) = f(x), \qquad \frac{\partial U}{\partial t}(x,0) = g(x). \tag{6.6.2}$$

As before, we take a rectangular net with constant space and time intervals given by Δx and Δt respectively and we write $u_{ij} = u(i\Delta x, j\Delta t)$, $-\infty < i < \infty$, $0 \leq j \leq \infty$. Both second partial derivatives are approximated by central difference expressions given by Chapter 2 whose truncation error is $0([\Delta x]^2 + [\Delta t]^2)$. Thus equation (6.6.1) is approximated by the explicit formula

$$u_{i,j+1} = \lambda^2(u_{i-1,j} + u_{i+1,j}) + 2(1 - \lambda^2)u_{ij} - u_{i,j-1}, \tag{6.6.3}$$

where $\lambda = \Delta t/\Delta x$. The first initial condition of (6.5.2) specifies $u_{i,0}$ on the line $t = 0$. We can use the second condition to find values on the line $t = \Delta t$ by employing a 'false' boundary and the second-order central difference formula

$$\left.\frac{\partial U}{\partial t}\right|_{i,0} = \frac{u_{i,1} - u_{i,-1}}{2\Delta t} + 0([\Delta t]^2). \tag{6.6.4}$$

Writing $g(i\Delta x) = g_i$, we have the approximation

$$u_{i,1} - u_{i,-1} = 2\Delta t \, g_i. \qquad (6.6.5)$$

From equation (6.6.3) with $j = 0$, we have

$$u_{i,1} = \lambda^2(u_{i-1,0} + u_{i+1,0}) + 2(1 - \lambda^2)u_{i,0} - u_{i,-1}.$$

Upon replacing $u_{i,-1}$ with its value, from equation (6.6.5), and solving for $u_{i,1}$ we find

$$u_{i,1} = \tfrac{1}{2}\lambda^2(f_{i-1} + f_{i+1}) + (1 - \lambda^2)f_i + \Delta t \, g_i. \qquad (6.6.6)$$

The computational molecule for equation (6.6.3) is shown in Figure 6.6.1. Superimposed on this figure are the characteristics of the wave equation, namely $t = \pm x + $ constant, whose slopes are ± 1, represented by the lines AC and BC. The solution is uniquely determined in the triangle ACB, provided the solution upto AB is known. If the absolute value of λ (i.e., the slope) exceeds 1, then equation (6.6.3) would provide a 'solution' in a region 'not reached' by the continuous solution. In such a case we would expect the result to be incorrect.

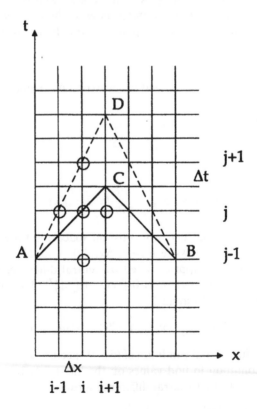

Figure 6.6.1 Comparison of the finite difference characteristics AD and DB with the true characteristics AC and BC.

If $|\lambda| \leq 1$, however, it can be shown that the method will converge under the usual assumption that certain higher derivatives exist. Convergence of the solution of equations (6.6.3) and (6.6.4) to that of the differential problem, equations (6.6.1) and (6.6.2) as $\Delta x \to 0$ and $\Delta t \to 0$, was first examined by Courant, Friedrichs, and Lewy (1928), by Lowan (1957) using operator methods and by Collatz (1960).

The proof of convergence is somewhat complicated for $\lambda > 1$ but is much simpler if the special mesh ratio condition $\lambda = 1$ holds. We shall provide the proof for the latter when the derivative initial condition is approximated by forward differences and $\Delta t = \Delta x$.

Let U_{ij} be the exact solution of the differential (wave) equation at the point (x_i, t_j) and u_{ij} the exact solution of the explicit difference equation. From the initial conditions (6.6.2) we obtain

$$g_i = (u_{i,1} - u_{i,0})/\Delta x$$

giving

$$u_{i,1} = f_i + \Delta x \, g_i. \tag{6.6.7}$$

By Taylor's expansion about the point $(i\Delta x, 0)$ we have

$$U_{i,1} = U_{i,0} + \Delta x \left(\frac{\partial U}{\partial t}\right)_{i,0} + \frac{1}{2!}(\Delta x)^2 \left(\frac{\partial^2 U}{\partial t^2}\right)_{i,\theta}, \quad (0 < \theta < 1),$$

i.e.,

$$U_{i,1} = f_i + \Delta x \, g_i + \frac{1}{2!}(\Delta x)^2 \left(\frac{\partial^2 U}{\partial t^2}\right)_{i,\theta}. \tag{6.6.8}$$

Hence from equations (6.6.7) and (6.6.8) we get the discretisation error

$$|e_{i,1}| = |U_{i,1} - u_{i,1}|$$

$$\leq \tfrac{1}{2}(\Delta x)^2 M_2, \tag{6.6.9}$$

where M_2 is the modulus of the largest value of $\partial^2 U/\partial t^2$ in the first time interval. The substitution of $u_{ij} = U_{ij} - e_{ij}$ into the finite difference equation (6.6.3) and expansion in terms of U_{ij} by Taylor's theorem gives

$$e_{i,j+1} = e_{i+1,j} + e_{i-1,j} - e_{i,j-1} + \frac{1}{24}(\Delta x)^4 \left(\frac{\partial^4 U}{\partial t^4}\right)_{i,j+\theta_1}$$

$$+ \left(\frac{\partial^4 U}{\partial t^4}\right)_{i,j+\theta_2} - \left(\frac{\partial^4 U}{\partial x^4}\right)_{i+\theta_3,j} - \left(\frac{\partial^4 U}{\partial x^4}\right)_{i+\theta_4,j},$$

where $|\theta_s| < 1$ $(s = 1, 2, 3, 4)$. Hence, if M_4 is the modulus of the largest of $\partial^4 U/\partial t^4$ and $\partial^4 U/\partial x^4$ throughout the solution domain and $|\eta| \leq 1$, then

$$e_{i,j+1} = e_{i+1,j} + e_{i-1,j} - e_{i,j-1} + \tfrac{1}{6}(\Delta x)^4 \eta M_4. \tag{6.6.10}$$

By drawing the straight line characteristics through the point $(i, j+1)$ at $\pm 45°$ to the x-axis Ox until they meet the line $j = 1$, we can mark the points within this triangle contributing terms to $e_{i,j+1}$ when working backwards using the last equation so as to express $e_{i,j+1}$ entirely in terms of the errors along $j=1$. It will be seen that there is one point at $(i,j+1)$, two points along $t=j\Delta x$, three points along $t=(j-1)\Delta x$ and following the same manner $(j+1)$ points along $t=\Delta x$. As the $(j+1)$ points along $t=\Delta x$ each contribute an error bounded by $1/2(\Delta x)^2 M_2$ as in (6.6.9) and $\Sigma_{k=1}^j k=1/2$ $j(j+1)$ points between $t=2\Delta x$ and $t=(j+1)\Delta x$ each contribute an error bounded by $1/6(\Delta x)^4 M_4$, it follows that

$$|e_{i,j+1}| \leq \tfrac{1}{2}(j+1)(\Delta x)^2 M_2 + \tfrac{1}{12}j(j+1)(\Delta x)^4 M_4. \tag{6.6.11}$$

Changing j into $(j-1)$ completes the proof. Since $j\Delta x = t$, we have

$$|e_{ij}| \leq \tfrac{1}{2}t\Delta x M_2 + \tfrac{1}{12}t^2(\Delta x)^2 M_4. \tag{6.6.12}$$

As Δx tends to zero, this error tends to zero for finite values of t. Hence convergence is established. The proof for the case $\lambda < 1$ is given by Forsythe and Wasow (1960) and Lowan (1957).

(b) Group Explicit Methods
A new strategy for solving second order hyperbolic partial differential equations using asymmetric formulae is now presented. This group explicit strategy forms some 'semi-explicit' and 'pure-explicit' algorithms which turn out to be both stable conditionally and unconditionally.

Consider the simple hyperbolic problem of the vibrating string

$$\frac{\partial^2 U}{\partial t^2} - \frac{\partial^2 U}{\partial x^2} = 0, \tag{6.6.13}$$

in the domain $R = \{(x,t): 0 \leq x \leq 1, t \geq 0\}$ satisfying the following initial conditions

$$\left.\begin{array}{c} U(x,0) = f_1(x) \\ \dfrac{\partial U}{\partial t}(x,0) = f_2(x) \end{array}\right\}, \quad 0 \leq x \leq 1, \tag{6.6.14}$$

and boundary conditions

$$\left.\begin{array}{c} U(0,t) = g_1(t) \\ U(1,t) = g_2(t) \end{array}\right\}, \quad \text{for all } t > 0. \tag{6.6.15}$$

We shall assume that these initial and boundary conditions are given with sufficient smoothness to maintain the order of accuracy of the difference scheme under consideration.

As usual, let us assume that the rectangular solution domain R is covered by a rectangular grid with grid spacing $\Delta x, \Delta t$ in the x, t directions, respectively. The grid points (x, t) are given by

$$x = x_i = i\Delta x, \quad i = 0, 1, \ldots, m, \quad m = 1/\Delta x,$$

and

$$t = t_j = j\Delta t, \quad j = 0, 1, 2, \dots .$$

The problem (6.6.13–6.6.15) can normally be solved using the classical explicit scheme derived from central difference approximations for $\partial^2 U/\partial x^2$ and $\partial^2 U/\partial t^2$ to result in the formula

$$u_{i,j+1} = 2(1 - \lambda^2)u_{i,j} + \lambda^2(u_{i-1,j} + u_{i+1,j}) - u_{i,j-1}, \tag{6.6.16}$$

$\lambda = \Delta t/\Delta x$, which is stable for $0 \le \lambda \le 1$ and with truncation error

$$T_{i,j} = (\Delta t \Delta x)^2 \left[\frac{1}{12}(\lambda^2 - 1)\frac{\partial^4 U}{\partial x^4}\bigg|_{i,j} + \frac{1}{360}(\lambda^4 - 1)\frac{\partial^6 U}{\partial x^6}\bigg|_{i,j} + \cdots \right]. \tag{6.6.17}$$

For $\lambda = 1$, the truncation error vanishes and so an exact difference representation of (6.5.13) is obtained as

$$u_{i,j+1} = u_{i-1,j} + u_{i+1,j} - u_{i,j-1}. \tag{6.6.18}$$

For the unconditionally stable scheme, normally the implicit approximation

$$\left(\frac{\partial^2 U}{\partial t^2} \right) = \frac{1}{(\Delta t)^2} \left\{ \frac{1}{4} \delta_x^2 u_{i,j+1} + \frac{1}{2} \delta_x^2 u_{i,j} + \frac{1}{4} \delta_x^2 u_{i,j-1} \right\} \tag{6.6.19}$$

is used. The truncation error is given by

$$(\Delta x)^2 \left\{ -\frac{1}{12}(4\lambda^2 + 1)\frac{\partial^4 U}{\partial x^4}\bigg|_{i,j} - \frac{1}{720}(\Delta x)^2(13\lambda^4 + 15\lambda^2 + 1)\frac{\partial^6 U}{\partial x^6}\bigg|_{i,j} + \cdots \right\}. \tag{6.6.20}$$

However the implicit schemes normally require the solution of a tridiagonal system of equations.

Besides the explicit and implicit type of methods, another class is a semi-explicit method which was proposed by Saul'yev (1964) which is formed from using an asymmetric type of approximation to the derivatives. For the equation (6.6.13), this type of approximation at the point (i,j) is given by

$$\frac{u_{i,j+1} - 2u_{i,j} + u_{i,j-1}}{\Delta t^2} - \frac{u_{i+1,j+1} - u_{i,j+1} - u_{i,j-1} + u_{i-1,j-1}}{(\Delta x)^2} = 0, \tag{6.6.21}$$

or

$$\frac{u_{i,j+1} - 2u_{i,j} + u_{i,j-1}}{\Delta t^2} - \frac{u_{i-1,j+1} - u_{i,j+1} - u_{i,j-1} + u_{i+1,j-1}}{(\Delta x)^2} = 0. \tag{6.6.22}$$

These equations in simplified form are given by

$$(1 + \lambda^2)u_{i,j+1} - \lambda^2 u_{i+1,j+1} = 2u_{i,j} - (1 + \lambda^2)u_{i,j-1} + \lambda^2 u_{i-1,j-1}, \tag{6.6.21a}$$

and

$$(1 + \lambda^2)u_{i,j+1} - \lambda^2 u_{i-1,j+1} = 2u_{i,j} - (1 + \lambda^2)u_{i,j-1} + \lambda^2 u_{i+1,j-1}, \tag{6.6.22a}$$

respectively. The 'semi-explicit property' of schemes (6.6.21a) and (6.6.22a) can be clearly seen from Figure (6.6.2), if they are implemented as an RL (right-to-left) formula and LR (left-to-right) formula, respectively.

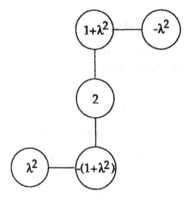

(a) RL formulae

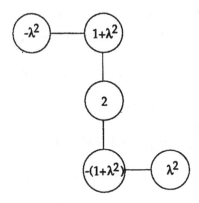

(b) LR formulae

Figure 6.6.2

These types of approximation were regarded by Saul'yev (1964) as a class of divergence approximation as their truncation errors will indicate a possible inconsistency in approximating the solution of (6.6.13). In the following we will describe how the asymmetric type of formulae are able to approximate the solution of (6.6.13).

To do this, we define a new variable $V = V(x, t)$ such that

$$V = \frac{\partial U}{\partial t}. \tag{6.6.23}$$

Therefore, the equation is now written by

$$\frac{\partial V}{\partial t} = \frac{\partial^2 U}{\partial x^2}, \tag{6.6.24}$$

with the initial conditions

$$U(x, 0) = f_1(x), \tag{6.6.25}$$

$$V(x, 0) = f_2(x),$$

and boundary conditions

$$V(0, t) = g'_1(t), \tag{6.6.26}$$

$$V(1, t) = g'_2(t)$$

in addition to the conditions (6.6.15).

The equation (6.6.24) is now in 'almost parabolic' form and hence the asymmetrical formulae can be used.

For the formulation, let us approximate the equations (6.6.23) and (6.6.24) at the point $(i, j + 1/2)$ by

$$\frac{v_{i,j+1} + v_{i,j}}{2} = \frac{u_{i,j+1} - u_{i,j}}{\Delta t}, \tag{6.6.27}$$

and

$$\frac{v_{i,j+1} - v_{i,j}}{\Delta t} = \frac{1}{(\Delta x)^2} \{u_{i+1,j+1} - u_{i,j+1} - u_{i,j} + u_{i-1,j}\}, \tag{6.6.28}$$

respectively. Whilst at the point $(i + 1, j + 1/2)$, they are approximated by

$$\frac{v_{i+1,j+1} + v_{i+1,j}}{2} = \frac{u_{i+1,j+1} - u_{i+1,j}}{\Delta t}, \tag{6.6.29}$$

and

$$\frac{v_{i+1,j+1} - v_{i+1,j}}{\Delta t} = \frac{1}{(\Delta x)^2} \{u_{i,j+1} - u_{i+1,j+1} - u_{i+1,j} + u_{i+2,j}\}, \tag{6.6.30}$$

respectively.

The equations (6.6.27–6.6.28) can be simplified to the form (as shown in Figure 6.6.3a)

$$-\lambda^2 u_{i+1,j+1} + (2 + \lambda^2) u_{i,j+1} = (2 - \lambda^2) u_{i,j} + \lambda^2 u_{i-1,j} + 2\Delta t v_{i,j}, \tag{6.6.31}$$

and the equations (6.6.29–6.6.30) are simplified to (as shown in Figure 6.6.3b)

$$-\lambda^2 u_{i,j+1} + (2 + \lambda^2) u_{i+1,j+1} = (2 - \lambda^2) u_{i+1,j} + \lambda^2 u_{i+2,j} + 2\Delta t v_{i+1,j}, \tag{6.6.32}$$

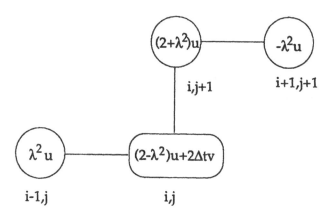

Figure 6.6.3a

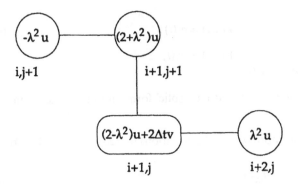

Figure 6.6.3b   Computational molecule for schemes (6.6.31) and (6.6.32).

From the above formulation the following 'semi-explicit' algorithms are possible:

(1) The equations (6.6.31) and (6.6.32) can be implemented individually as an RL-direction formulae

$$u_{i,j+1}=\frac{1}{(2+\lambda^2)}\{\lambda^2 u_{i+1,j+1}+(2-\lambda^2)u_{i,j}+\lambda^2 u_{i-1,j}+2\Delta t v_{i,j}\},\quad (6.6.33)$$

and as an LR-direction formulae

$$u_{i,j+1}=\frac{1}{(2+\lambda^2)}\{\lambda^2 u_{i-1,j+1}+(2-\lambda^2)u_{i,j}+\lambda^2 u_{i+1,j}+2\Delta t v_{i,j}\},\quad (6.6.34)$$

respectively. In both cases the solution vector v is given by

$$v_{i,j+1}=\frac{2}{\Delta t}(u_{i,j+1}-u_{i,j})-v_{i,j}.\qquad (6.6.35)$$

(2) The equations (6.6.33) and (6.6.34) are implemented alternately.

(3) At every time level, the average of the solutions of (6.6.33) and (6.6.34) is regarded as the approximate solution to the equation (6.6.13).

Besides these semi-explicit algorithms, a class of 'pure explicit' algorithms can be obtained if the equations (6.6.31) and (6.6.32) are implemented as a Group Explicit method whose implicit system is given by

$$\begin{bmatrix}2+\lambda^2 & -\lambda^2\\ -\lambda^2 & 2+\lambda^2\end{bmatrix}\begin{bmatrix}u_{i,j+1}\\ u_{i+1,j+1}\end{bmatrix}=\begin{bmatrix}2-\lambda^2 & 0\\ 0 & 2-\lambda^2\end{bmatrix}\begin{bmatrix}u_{i,j}\\ u_{i+1,j}\end{bmatrix}$$

$$+2\Delta t\begin{bmatrix}1 & 0\\ 0 & 1\end{bmatrix}\begin{bmatrix}v_{i,j+1}\\ v_{i+1,j+1}\end{bmatrix}\begin{bmatrix}\lambda^2 u_{i-1,j}\\ \lambda^2 u_{i+2,j}\end{bmatrix},\qquad (6.6.36)$$

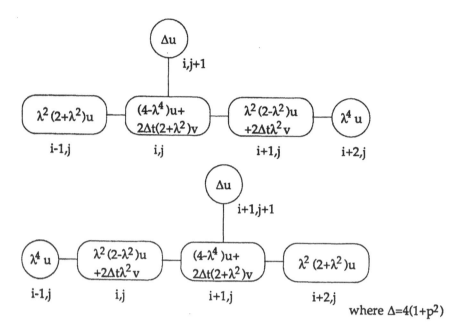

Figure 6.6.4 Computational molecule for equation (6.5.37).

and explicitly written as in Figure 6.6.4.

$$\begin{bmatrix} u_{i,j+1} \\ u_{i+1,j+1} \end{bmatrix} = \frac{1}{\Delta} \begin{bmatrix} 4 - \lambda^4 & \lambda^2(2-\lambda^2) \\ \lambda^2(2-\lambda^2) & 4-\lambda^4 \end{bmatrix} \begin{bmatrix} u_{i,j} \\ u_{i+1,j} \end{bmatrix} + 2\Delta t \begin{bmatrix} 2+\lambda^2 & \lambda^2 \\ \lambda^2 & 2+\lambda^2 \end{bmatrix} \begin{bmatrix} v_{i,j+1} \\ v_{i+1,j+1} \end{bmatrix}$$
$$+ \begin{bmatrix} \lambda^2(2+\lambda^2)u_{i-1,j} + \lambda^4 u_{i+2,j} \\ \lambda^4 u_{i-1,j} + \lambda^2(2+\lambda^2)u_{i+2,j} \end{bmatrix}, \quad (6.6.37)$$

with $\Delta = 4(1 + \lambda^2)$ for any group of 2-points.

At every time level if we assume there are an odd number of unknown points, we will always have a single ungrouped point, i.e., either $(1, j+1)$ or $(m-1, j+1)$ at each end of the range of integration.

As the value of the boundary points are known, these points can be approximated by the formulae, i.e.,

$$u_{1,j+1} = \frac{1}{(2+\lambda^2)} \{\lambda^2 u_{0,j+1} + (2-\lambda^2)u_{1,j} + \lambda^2 u_{2,j} + 2\Delta t q_{1,j}\} \quad (6.6.38)$$

and

$$u_{m-1,j+1} = \frac{1}{(2+\lambda^2)} \{\lambda^2 u_{m,j+1} + (2-\lambda^2)u_{m-1,j} + \lambda^2 u_{m-2,j} + 2\Delta t q_{m-1,j}\}. \quad (6.6.39)$$

Now the GE concept can be developed from equations (6.6.37–6.6.39).

Thus the following GE algorithms are possible:

1. Group Explicit with Left-Ungrouped Point (GEL)

Use equation (6.6.38) for the first unknown point from the left of the boundary and equation (6.6.37) for the remaining $(m-2)/2$ pairs of 2-points. This will result in the system

$$(2I + \lambda^2 G_1)\underline{u}_{j+1} = (2I - \lambda^2 G_2)\underline{u}_j + 2\Delta t \underline{v}_j + \underline{b}_{1,j}, \tag{6.6.40}$$

where

$$
G_1 =
\begin{bmatrix}
1 & & & & & & \\
 & 1 & -1 & & 0 & & \\
 & -1 & 1 & & & & \\
 & & & \ddots & & & \\
 & 0 & & & 1 & -1 & \\
 & & & & -1 & 1 &
\end{bmatrix},
\tag{6.6.41}
$$

$$
G_2 =
\begin{bmatrix}
1 & -1 & & & & \\
-1 & 1 & & 0 & & \\
 & & \ddots & & & \\
 & & & 1 & -1 & \\
 & 0 & & -1 & 1 & \\
 & & & & & 1
\end{bmatrix},
\tag{6.6.42}
$$

$$\underline{u}_j^T = [u_{1,j}, u_{2,j}, \dots, u_{m-1,j}], \qquad \underline{v}_j^T = [v_{1,j}, v_{2,j}, \dots, v_{m-1,j}]$$

and

$$\underline{b}_{1,j}^T = [-\lambda^2 u_{0,j+1}, 0, \dots, 0, \lambda^2 u_{m,j}].$$

2. Group Explicit with Right-Ungrouped Point (GER)

Use equation (6.6.37) for the first $(m-2)$ points and equation (6.6.39) for the last unknown point. This will give the system

$$(2I + \lambda^2 G_2)\underline{u}_{j+1} = (2I - \lambda^2 G_1)\underline{u}_j + 2\Delta t \underline{v}_j + \underline{b}_{2,j}, \tag{6.6.43}$$

where

$$\underline{b}_{2,j}^T = [\lambda^2 u_{0,j}, 0, \dots, 0, -\lambda^2 u_{m,j+1}].$$

3. (Single) Alternating Group Explicit (S)AGE

Use equation (6.6.40) at the $(j+1)^{th}$ time level and equation (6.6.43) at the $(j+2)^{th}$ time level, i.e.,

$$(2I + \lambda^2 G_1)\underline{u}_{j+1} = (2I - \lambda^2 G_2)\underline{u}_j + 2\Delta t \underline{v}_j + \underline{b}_{1,j},$$

$$(2I + \lambda^2 G_2)\underline{u}_{j+2} = (2I - \lambda^2 G_1)\underline{u}_{j+1} + 2\Delta t \underline{v}_{j+1} + \underline{b}_{2,j+1}. \tag{6.6.44}$$

4. (Double) Alternating Group Explicit (D)AGE

In this algorithm, the GE formulae are incorporated alternately within the four time levels with the direction reversed at the third time level. This will give the equations

$$(2I + \lambda^2 G_1)\underline{u}_{j+1} = (2I - \lambda^2 G_2)\underline{u}_j + 2\Delta t\underline{v}_j + \underline{b}_{1,j},$$

$$(2I + \lambda^2 G_2)\underline{u}_{j+2} = (2I - \lambda^2 G_1)\underline{u}_{j+1} + 2\Delta t\underline{v}_{j+1} + \underline{b}_{2,j+1},$$

$$(2I + \lambda^2 G_2)\underline{u}_{j+3} = (2I - \lambda^2 G_1)\underline{u}_{j+2} + 2\Delta t\underline{v}_{j+2} + \underline{b}_{2,j+2}, \qquad (6.6.45)$$

$$(2I + \lambda^2 G_1)\underline{u}_{j+4} = (2I - \lambda^2 G_2)\underline{u}_{j+3} + 2\Delta t\underline{v}_{j+3} + \underline{b}_{1,j+3}.$$

These are only a few of the algorithms which can be established from the original formulae (6.6.24–6.6.26).

The estimate of the truncation errors of the scheme (6.6.31) and (6.6.32) are of $0(\Delta t/\Delta x + (\Delta x)^2 + (\Delta t)^2)$ with the coefficient to $\Delta t/\Delta x$ being of opposite signs. Therefore for algorithms of alternate nature, the accuracy is approximately of $0((\Delta x)^2 + (\Delta t)^2)$.

To analyse the stability, the equations (6.6.31) and (6.6.32) are written as

$$(2I + \lambda^2 G)\underline{u}_{j+1} = (2I - \lambda^2 G^T)\underline{u}_j + 2\Delta t\underline{v}_j \qquad (6.6.46)$$

and

$$(2I + \lambda^2 G^T)\underline{u}_{j+1} = (2I - \lambda^2 G)\underline{u}_j + 2\Delta t\underline{v}_j, \qquad (6.6.47)$$

where

$$G = \begin{bmatrix} 1 & -1 & & & \\ & 1 & -1 & & 0 \\ & & \ddots & \ddots & \\ & & & 1 & -1 \\ 0 & & & & 1 \end{bmatrix}.$$

From equations (6.6.45) and (6.6.46), if the magnification error is defined as

$$\underline{e}_{j+1} = \underline{u}_{j+1} - \underline{u}_j, \qquad (6.6.48)$$

therefore an error vector relation can be given by

$$\underline{e}_{j+1} = ((2I + \lambda^2 G)^{-1}(2I - \lambda^2 G^T))^{j+1}\underline{e}_0 + 2((2I + \lambda^2 G)^{-1})^{j+1}(\underline{e}_1 - \underline{e}_0)$$

$$+ 2((2I + \lambda^2 G)^{-1})^j(\underline{e}_2 - \underline{e}_1) + \cdots + 2(2I + \lambda^2 G)^{-1}(\underline{e}_j - \underline{e}_{j-1}). \qquad (6.6.49)$$

As the difference between the individual magnification errors can be regarded as insignificant, therefore the error relation is now approximately given by

$$\underline{e}_{j+1} = \{(2I + \lambda^2 G)^{-1}(2I - \lambda^2 G^T)\}^{j+1}\underline{e}_0,$$

whose magnitude will be decaying if the magnitude of the eigenvalues

$$(2I + \lambda^2 G)^{-1}(2I - \lambda^2 G^T) \tag{6.6.50}$$

are less than or equal to unity in modulus. Since the eigenvalues of G are all equal to unity, therefore the eigenvalues of (6.6.49) are given by the relation, $(2 - \lambda^2)/(2 + \lambda^2)$ which is always less than or equal to one for all $\lambda > 0$. Hence the scheme (6.5.31) is unconditionally stable for all $\lambda > 0$.

Similarly, it can be shown that the scheme (6.6.32) is also unconditionally stable for all $\lambda > 0$.

However, even though the schemes (6.6.31) and (6.6.32) are unconditionally stable, the assumption made on the error relation (6.6.49) can only hold if the increment in the time level is taken realistically.

With the matrix method, it can easily be shown that the GER and GEL scheme are stable provided $\lambda^2 \leq 2$ and using the Lemma by Kellog the (S)AGE and (D)AGE are unconditionally stable for all $\lambda > 0$.

(c) Implicit Methods

With the expectation of gaining stability advantages, we shall now attempt to derive implicit methods for second-order equations. For the wave equation (6.6.1), the simplest implicit system is obtained by approximating $\partial^2 U/\partial t^2$, as before, by a second central difference centred at (i,j) while $\partial^2 U/\partial x^2$ is approximated by the average of two second central differences, one centred at $(i, j+1)$ and the other at $(i, j-1)$. Thus, one simple implicit approximation takes the form

$$u_{i,j+1} - 2u_{ij} + u_{i,j-1} = \frac{\lambda^2}{2}\{(u_{i+1,j+1} - 2u_{i,j+1} + u_{i-1,j+1})$$

$$+ (u_{i+1,j-1} - 2u_{i,j-1} + u_{i-1,j-1})\}. \tag{6.6.51}$$

The implicit nature of this formula is obvious by rewriting the expression to be solved on the $(j+1)$ line in terms of the values on the two preceding lines. Thus one finds the equation

$$-\lambda^2 u_{i+1,j+1} + 2(1 + \lambda^2)u_{i,j+1} - \lambda^2 u_{i-1,j+1}$$

$$= 4u_{i,j} + \lambda^2 u_{i+1,j-1} - 2(1 + \lambda^2)u_{i,j-1} + \lambda^2 u_{i-1,j-1}. \tag{6.6.52}$$

Another implicit method discussed by Richtmyer and Morton (1967) to approximate (6.6.1) is given by

$$-\frac{1}{4}\lambda^2 u_{i-1,j+1} + \left(1 + \frac{\lambda^2}{2}\right)u_{i,j+1} - \frac{\lambda^2}{4}u_{i+1,j+1}$$

$$= \frac{1}{2}\lambda^2 u_{i-1,j} + (2 - \lambda^2)u_{ij} + \frac{\lambda^2}{2}u_{i+1,j} + \frac{\lambda^2}{4}u_{i-1,j-1}$$

$$+ \left(-1 - \frac{\lambda^2}{2}\right)u_{i,j-1} + \frac{\lambda^2}{4}u_{i+1,j-1}. \tag{6.6.53}$$

By assuming that there are m mesh values to be determined, then upon writing equations (6.6.14) and (6.6.15) for each i, $i=1,2,\ldots,m$ and inserting the discretised boundary conditions, the tridiagonal form of the system becomes clear. Thus the tridiagonal algorithm of Chapter 1 may be applied to find a non-iterative solution.

Equations (6.6.3),(6.6.14) and (6.6.15) are special cases of a general three-level implicit form obtained by approximating

$$\left(\frac{\partial^2 U}{\partial t^2}\right)_{i,j} \text{ by } \frac{1}{(\Delta t)^2}\delta_t^2 u_{ij}$$

and approximating $(\partial^2 U/\partial x^2)_{ij}$ with

$$\frac{1}{(\Delta x)^2}[\alpha\delta_x^2 u_{i,j+1}+(1-2\alpha)\delta_x^2 u_{ij}+\alpha\delta_x^2 u_{i,j-1}].$$

Hence

$$\frac{1}{(\Delta t)^2}\delta_t^2 u_{ij}=\frac{1}{(\Delta x)^2}[\alpha\delta_x^2 u_{i,j+1}+(1-2\alpha)\delta_x^2 u_{ij}+\alpha\delta_x^2 u_{i,j-1}], \qquad (6.6.54)$$

where α is a weighting factor and δ^2 is the operator defined by

$$\delta_x^2 u_{ij}=u_{i+1,j}-2u_{ij}+u_{i-1,j}.$$

Note that $\alpha=0$ gives the explicit method (6.6.3), $\alpha=1/2$ gives the implicit method (6.6.14) and for $\alpha=1/4$ the implicit equation (6.6.53).

Now from equation (6.6.52) we can display the computational molecule to obtain the numerical solution of (6.6.1) as

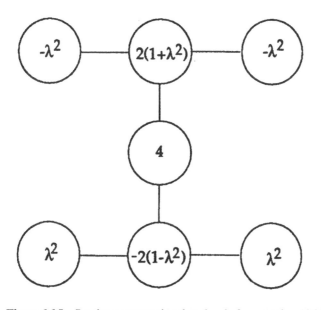

Figure 6.6.5 7-point computational molecule for equation (6.6.1).

Similarly the tridiagonal system of equations for all points $i = 1, 2, \ldots, m$ along the $(j+1)^{th}$ time level would be

$$
\begin{bmatrix}
2(1+\lambda^2) & -\lambda^2 & & & \\
-\lambda^2 & 2(1+\lambda^2) & -\lambda^2 & & \mathbf{0} \\
& & \ddots & & \\
\mathbf{0} & & & & -\lambda^2 \\
& & & -\lambda^2 & 2(1+\lambda^2)
\end{bmatrix}
\begin{bmatrix}
u_1 \\ u_2 \\ \vdots \\ u_{m-1} \\ u_m
\end{bmatrix}^{j+1}
=
\begin{bmatrix}
b_1 \\ b_2 \\ \vdots \\ b_{m-1} \\ b_m
\end{bmatrix},
\qquad (6.6.55)
$$

where

$$b_1 = 4u_{ij} + \lambda^2(u_{0,j-1} + u_{2,j-1}) - 2(1-\lambda^2)u_{1,j-1} + \lambda^2 u_{0,j+1},$$
$$b_i = 4u_{ij} + \lambda^2(u_{i-1,j-1} + u_{i+1,j-1}) - 2(1-\lambda^2)u_{i,j-1}, \quad i = 2, 3, \ldots, m-1, \quad (6.6.56)$$
$$b_m = 4u_{m.j} + \lambda^2(u_{m-1,j} + u_{m+1,j+1}) - 2(1-\lambda^2)u_{m.j-1} + \lambda^2 u_{m+1,j+1}.$$

Clearly, the linear system (6.6.55) can now be solved using the AGE iterative method as outlined in Chapter 5.

We define the matrices G_1 and G_2 for m even such that

$$
G_1 =
\begin{bmatrix}
1+\lambda^2 & -\lambda^2 & & & & & \\
-\lambda^2 & 1+\lambda^2 & & & & & \\
& & 1+\lambda^2 & -\lambda^2 & & \mathbf{0} & \\
& & -\lambda^2 & 1+\lambda^2 & & & \\
& & & & \ddots & & \\
& \mathbf{0} & & & & 1+\lambda^2 & -\lambda^2 \\
& & & & & -\lambda^2 & 1+\lambda^2
\end{bmatrix}
\qquad (6.6.57)
$$

and

$$
G_2 =
\begin{bmatrix}
1+\lambda^2 & & & & & \\
& 1+\lambda^2 & -\lambda^2 & & & \\
& -\lambda^2 & 1+\lambda^2 & & \mathbf{0} & \\
& & & \ddots & & \\
& \mathbf{0} & & & & \\
& & & & & 1+\lambda^2
\end{bmatrix},
\qquad (6.6.58)
$$

then the AGE iterative method can be applied iteratively to determine $\underline{u}^{(k+1/2)}$ and $\underline{u}^{(k+1)}$ explicitly by

$$\underline{u}^{(k+1/2)} = (G_1 + rI)^{-1}[(rI - G_2)\underline{u}^{(k)} + \underline{b}],$$
$$\underline{u}^{(k+1)} = (G_2 + rI)^{-1}[(rI - G_1)\underline{u}^{(k+1/2)} + \underline{b}], \qquad (6.6.59)$$

for $k = 0, 1, 2, \ldots$ until convergence is achieved.

6.7. The Numerical Solution of the Sine Gordon Partial Differential Equation by the AGE Method

A relativistic wave equation of a charged particle in an electromagnetic field using the recently discovered ideas of quantum theory, the Klein–Gordon equation reduces to

$$\nabla^2 U - \frac{1}{c^2}\frac{\partial^2 U}{\partial t^2} = \left(\frac{mc}{n}\right) U, \tag{6.7.1}$$

for the special case of a free particle. We may choose $c=1$ and consider the generalised non-linear form

$$\frac{\partial^2 U}{\partial t^2} - \nabla^2 U + V'(U) = 0, \tag{6.7.2}$$

for some differentiable potential function V. In particular, if $V'(U) = \sin U$ and $\partial/\partial y = \partial/\partial z = 0$, we obtain the non-linear *sine-Gordon equation* for one space dimension as

$$\frac{\partial^2 U}{\partial t^2} - \frac{\partial^2 U}{\partial x^2} + \sin U = 0. \tag{6.7.3}$$

In the following section, we shall introduce the Alternating Group Explicit (AGE) iterative method (Evans, 1985) to solve this non-linear wave equation.

Thus we seek the solution of

$$\frac{\partial^2 U}{\partial t^2} = \frac{\partial^2 U}{\partial x^2} - \sin U, \tag{6.7.4}$$

subject to the following initial conditions

$$U(x,0) = f(x),$$
$$\qquad\qquad 0 \le x \le 1, \tag{6.7.5a}$$
$$\frac{\partial U}{\partial t}(x,0) = g(x),$$

and boundary conditions

$$\frac{\partial U(0,t)}{\partial x} = h(t),$$
$$\qquad\qquad 0 \le t \le T. \tag{6.7.5b}$$
$$\frac{\partial U(1,t)}{\partial x} = k(t),$$

From the expectation of achieving stability advantages, a general weighted averaged implicit finite difference discretisation to (6.7.4) at the $(j+1)^{th}$, j^{th} and $(j-1)^{th}$ time levels is

$$-\alpha\lambda^2 u_{i-1,j+1} + (1 + 2\alpha\lambda^2)u_{i,j+1} - \alpha\lambda^2 u_{i+1,j+1}$$
$$\cong (1 - 2\alpha)\lambda^2 u_{i-1,j} + 2(1 - (1 - 2\alpha)\lambda^2)u_{i,j} + (1 - 2\alpha)\lambda^2 u_{i+1,j} + \alpha\lambda^2 u_{i-1,j-1}$$
$$-(1 + 2\alpha\lambda^2)u_{i,j-1} + \alpha\lambda^2 u_{i+1,j-1}(\Delta x^2)\sin u_{i,j}, \quad i = 1, 2, \dots, n, \tag{6.7.6}$$

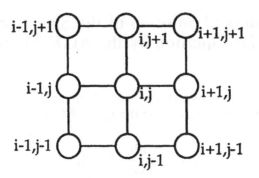

Figure 6.7.1

$0[(\Delta x)^2,(\Delta t)^2]$. Its computational molecule is given by Figure 6.7.1. The non-linear term $\sin(u)$ is centered at the point (i,j) to enable its explicit evaluation in (6.6.6). An alternative derivation is to take the average at the points $(i,j+1)$ and $(i,j-1)$ i.e.,

$$\sin(u_{ij}) \approx 1/2(\sin u_{i,j+1} + \sin u_{i,j-1}).$$

In this formulation the term $1/2 \sin u_{i,j+1}$ is implicitly evaluated on the right-hand side of (6.7.6) in each iteration.

Equation (6.7.6) gives a tridiagonal system of equations to be solved at the $(j+1)^{th}$ time level, which can be displayed in matrix form as

$$
\begin{bmatrix}
d_1 & c_1 & & & \\
a_2 & d_2 & c_2 & & \text{\Large 0} \\
 & \ddots & \ddots & \ddots & \\
\text{\Large 0} & & a_{n-1} & d_{n-1} & c_{n-1} \\
 & & & a_n & d_n
\end{bmatrix}
\begin{bmatrix}
u_1 \\ u_2 \\ \vdots \\ u_{n-1} \\ u_n
\end{bmatrix}
=
\begin{bmatrix}
b_1 \\ b_2 \\ \vdots \\ b_{n-1} \\ b_n
\end{bmatrix},
\tag{6.7.7a}
$$

or

$$A\underline{u}=\underline{b}, \tag{6.7.7b}$$

where $a_i = c_i = -\alpha\lambda^2$, $c_1 = a_n = -2\alpha\lambda^2$ and $d_i = (1 + \lambda\alpha^2)$, $i=1,2,...,n$. The vector $\underline{b}$ is defined by

$$b_1 = 2(1-(1-2\alpha)\lambda^2)u_{1,j} + 2(1-2\alpha)\lambda^2 u_{2,j} - (1+2\alpha\lambda^2)u_{1,j-1} + 2\alpha\lambda^2 u_{2,j-1}$$

$$- 2\alpha\lambda^2\Delta x h_{j+1} - 2(1-2\alpha)\lambda^2\Delta x h_j - 2\alpha\lambda^2\Delta x h_{j-1} - (\Delta x)^2\sin(u_{1,j}),$$

$$b_i = (1-2\alpha)\lambda^2 u_{i-1,j} + 2(1-(1-2\alpha)\lambda^2)u_{i,j} + (1-2\alpha)\lambda^2 u_{i+1,j} + \alpha\lambda^2 u_{i-1,j-1}$$

$$- (1+2\alpha\lambda^2)u_{i,j-1} + \alpha\lambda^2 u_{i+1,j-1} - (\Delta x)^2\sin(u_{i,j}), \quad i=2,3,...,n-1,$$

and

$$b_n = 2(1 - 2\alpha)\lambda^2 u_{n-1,j} + 2(1 - 2\alpha)\lambda^2)u_{n,j} + 2\alpha\lambda^2 u_{n-1,j-1} - (1 + 2\alpha\lambda^2)u_{n,j-1}$$
$$+ \alpha\lambda^2 \Delta x j_{j+1} + 2(1 - 2\alpha)\lambda^2 \Delta x k_j + 2\alpha\lambda^2 \Delta x k_{j-1} - (\Delta x)^2 \sin(u_{n,j}).$$

The u values on the first time level are given by the initial condition (6.6.5a). Values on the second time level are obtained by applying the forward finite difference approximation to equation (6.7.5b) at $t = 0$

$$\frac{\partial U}{\partial t}(x_{i,0}) \cong \frac{u_{i,1} - u_{i,0}}{\Delta t}$$

$$= g(x_i),$$

or,

$$u_{i,1} - u_{i,0} + \Delta t g(x_i), \tag{6.7.8}$$

which is a first-order approximation in Δt.

The solution on the third and subsequent time levels are generated iteratively by applying the AGE algorithm.

To formulate this scheme, we shall without loss of generality, assume that the space interval x is divided into an even number of sub-intervals m (odd number of points n). Then by splitting the matrix A into (Evans, 1985)

$$A = G_1 + G_2, \tag{6.7.9}$$

where

$$G_1 = \begin{bmatrix} hd_1 & & & & & \\ & hd_2 & c_2 & & & \\ & a_3 & hd_3 & & 0 & \\ \hline & & & & & \\ & 0 & & hd_{n-1} & c_{n-1} \\ & & & a_n & hd_n \end{bmatrix} \tag{6.7.10a}$$

and

$$G_2 = \begin{bmatrix} hd_1 & c_1 & & & \\ a_2 & hd_2 & & 0 & \\ \hline & & hd_{n-2} & c_{n-1} & \\ & 0 & a_{n-1} & hd_{n-1} & \\ \hline & & & & hd_n \end{bmatrix}, \tag{6.7.10b}$$

where $hd_i = d_i/2$, $i = 1, 2, \ldots, n$.

Equation (6.7.7b) can be written as

$$(G_1 + G_2)\underline{u} = \underline{b}. \tag{6.7.11}$$

Hence, the AGE iterative method in Peaceman–Rachford form can be applied to determine $\underline{u}^{(k+1/2)}$ and $\underline{u}^{(k+1)}$ implicitly by

$$(G_1 + rI)\underline{u}^{(k+1/2)} = [(rI - G_2)\underline{u}^{(k)} + \underline{b}],$$
$$(G_2 + rI)\underline{u}^{(k+1)} = [(rI - G_1)\underline{u}^{(k+1/2)} + \underline{b}], \qquad (6.7.12a)$$

or explicitly by

$$\underline{u}^{(k+1/2)} = (G_1 + rI)^{-1}[(rI - G_2)\underline{u}^{(k)} + \underline{b}],$$
$$\underline{u}^{(k+1)} = (G_2 + rI)^{-1}[(rI - G_1)\underline{u}^{(k+1/2)} + \underline{b}], \qquad (6.7.12b)$$

where the matrices $(G_1 + rI), (G_2 + rI), (rI - G_1)$ and $(rI - G_2)$ are represented by

$$(G_1 + rI) = \begin{bmatrix} w_1 & & & & \\ & w_2 & c_2 & & & \\ & a_3 & w_3 & & & \\ & & & \ddots & & \\ & & 0 & & w_{n-1} & c_{n-1} \\ & & & & a_n & w_n \end{bmatrix}, \qquad (6.7.13a)$$

$$(G_2 + rI) = \begin{bmatrix} w_1 & c_1 & & & & \\ a_2 & w_2 & & & 0 & \\ & & \ddots & & & \\ & 0 & & w_{n-2} & c_{n-1} & \\ & & & a_{n-1} & w_{n-1} & \\ & & & & & w_n \end{bmatrix}, \qquad (6.7.13b)$$

$$(rI - G_1) = \begin{bmatrix} v_1 & & & & \\ & v_2 & c_2 & & & \\ & a_3 & v_3 & & & \\ & & & \ddots & & \\ & & 0 & & v_{n-1} & c_{n-1} \\ & & & & a_n & v_n \end{bmatrix}, \qquad (6.7.13c)$$

$$(rI - G_2) = \begin{bmatrix} v_1 & c_1 & & & & \\ a_2 & v_2 & & & 0 & \\ & & \ddots & & & \\ & 0 & & v_{n-2} & c_{n-1} & \\ & & & a_{n-1} & v_{n-1} & \\ & & & & & v_n \end{bmatrix}, \qquad (6.7.13d)$$

where $w_i = hd_i + r$, $v_i = r - hd_i$, $i = 1, 2, \ldots, n$, k is now the iteration index and r is the acceleration parameter. It is clear that $(G_1 + rI)$ and $(G_2 + rI)$ are block diagonal matrices. All the diagonal elements except the first (or the last for $(G_2 + rI)$) are (2×2) sub-matrices. Therefore, $(G_1 + rI)$ and $(G_2 + rI)$ can be easily inverted by merely inverting their (2×2) block diagonal entries.

Then from equation (6.7.11b), $\underline{u}^{(k+1/2)}$ and $\underline{u}^{(k+1)}$ are given by

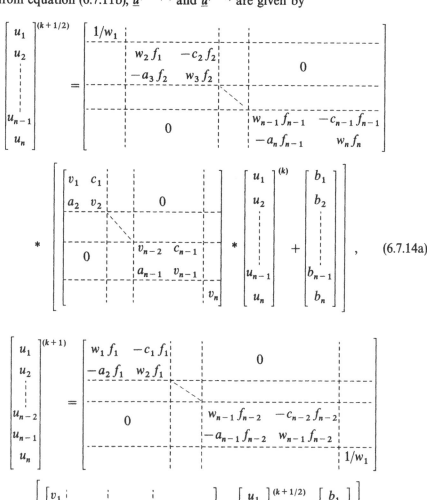

$$(6.7.14a)$$

$$(6.7.14b)$$

where $f_i = 1/(w_i^* w_{i+1} - c_i^* a_{i+1})$.

The corresponding explicit expressions for the AGE equations are obtained by carrying out the multiplications in (6.7.14a,b). Thus we have:

(i) At level $(k + 1/2)$:

$$u_1^{(k+1/2)} = (v_1 u_1^{(k)} - c_1 u_2^{(k)} + b_1)/w_1,$$

$$u_i^{(k+1/2)} = (A u_{i-1}^{(k)} + B u_i^{(k)} + C u_{i+1}^{(k)} + D u_{i+2}^{(k)} + E_i) * f_i, \tag{6.7.15}$$

and

$$u_{i+1}^{(k+1/2)} = (\tilde{A} u_{i-1}^{(k)}) + (\tilde{B} u_i^{(k)}) + (\tilde{C} u_{i+1}^{(k)}) + (\tilde{D} u_{i+2}^{(k)}) + (\tilde{E}_i) * f_i, \quad i = 2, 4, \ldots, m-1,$$

where

$$A = -a_i w_i, \qquad B = v_i w_i, \qquad C = -c_i v_{i+1}, \qquad E = b_i w_i - b_{i+1} c_i,$$

$$D = \begin{cases} 0 & \text{for } i = n-1, \\ c_i c_{i+1} & \text{otherwise}, \end{cases}$$

and

$$\tilde{A} = a_i a_{i+1}, \qquad \tilde{B} = -a_{i+1} v_i, \qquad \tilde{C} = v_{i+1} w_{i+1}, \qquad \tilde{E} = b_{i+1} w_{i+1} - a_{i+1} b_i,$$

$$\tilde{D} = \begin{cases} 0 & \text{for } i = n-1, \\ -c_{i+1} c_{i+1} & \text{otherwise}, \end{cases}$$

with the following computational molecules (Fig. 6.7.2).

(ii) At level $(k + 1)$:

$$u_i^{(k+1)} = (P u_{i-1}^{(k+1/2)} + Q u_i^{(k+1/2)} + R u_{i+1}^{(k+1/2)} + S u_{i+2}^{(k+1/2)} + T_i) * f_i,$$

$$u_{i+1}^{(k+1)} = (\tilde{P} u_{i-1}^{(k+1/2)} + \tilde{Q} u_i^{(k+1/2)} + \tilde{R} u_{i+1}^{(k+1/2)} + \tilde{S} u_{i+2}^{(k+1/2)} + \tilde{T}_i) * f_i,$$

$$i = 1, 3, \ldots, n-2, \tag{6.7.16}$$

$$u_n^{(k+1)} = (-a_n u_{n-1}^{(k+1/2)} + v_n u_n^{(k+1/2)} + b_n)/w_n,$$

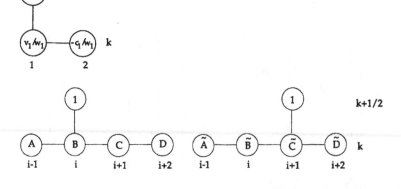

Figure 6.7.2   The AGE method at level $(k + 1/2)$.

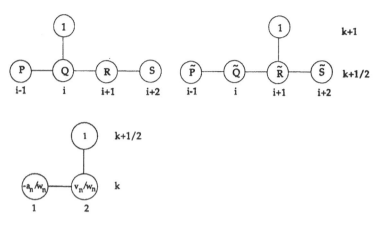

Figure 6.7.3 The AGE method at level $(k+1)$.

where

$$P = \begin{cases} 0 & \text{for } i=n, \\ -a_i w_i & \text{otherwise}, \end{cases}$$

and

$$Q = v_i w_i, \qquad R = -c_i v_{i+1}, \qquad S = c_i c_{i+1}, \qquad T_i = b_i w_i - b_{i+1} c_i,$$

$$\tilde{P} = \begin{cases} 0 & \text{for } i=n, \\ a_i a_{i+1} & \text{otherwise}, \end{cases}$$

$$\tilde{Q} = -a_{i+1} v_i, \qquad \tilde{R} = v_{i+1} w_{i+1}, \qquad \tilde{S} = -c_{i+1} w_{i+1}, \qquad \tilde{T}_i = b_{i+1} w_{i+1} - a_{i+1} b_i,$$

with its computational molecules given by Figure 6.7.3.

7. The AGE Method for Elliptic Multidimensional Partial Differential Equations

7.1. Introduction

The methods discussed in Chapter 5 are mainly concerned with the solution for the two point boundary value problem, i.e., the one dimensional problem. Now, we discuss the applicability and the viability of the methods for solving two and three dimensional problems. The scope will be limited to the investigation of the extension of the methods to solve the Helmholtz, Laplace and Poisson equations in two and three dimensions.

As shown in Chapter 5, the PR scheme had a limitation in solving the two and three dimensional problem via the AGE method. Thus, our concern will be centered on utilising the Douglas–Rachford, Douglas and Guittet's schemes. First, we outline the model for two dimensional problems, i.e., the Helmholtz equation.

The Two Dimensional Model Problem
Consider the Helmholtz equation in two dimensions

$$\frac{\partial^2 U}{\partial x^2} + \frac{\partial^2 U}{\partial y^2} - \rho U = f(x, y), \quad 0 \leq x, y \leq 1, \tag{7.1.1}$$

subject to $U = f(x, y)$ on the boundary of the unit square, $0 \leq x, y \leq 1$ with $\rho \geq 0$, and $f(x, y) \geq 0$. When $\rho = 0$ and $f(x, y) = 0$, the problem is reduced to the well known Laplace equation. If $\rho = 0$ and $f(x, y)$ is a constant but not zero, we then have a Poisson equation.

The square region is covered by a grid with sides parallel to the coordinate axes and the grid spacing is h. For $h = 1/(N + 1)$, the number of internal grid points or nodes is N^2. Figure 7.1 shows the nodes for $h = 1/6$, i.e., $N = 5$.

The coordinates of a typical internal grid point are $x_i = ih$, $y_j = jh$, where i and j are integers, and the value of u at this grid point is denoted by $u_{i,j}$. For example, from Figure 7.1, the value of $\underline{u}$ at the point $(1, 1)$ is $u_{1,1}$.

By using Taylor's theorem, we obtain

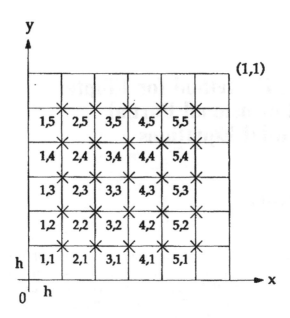

Figure 7.1 The number of nodes for $h=1/6$.

$$v_{i-1,j}=\left(v-h\frac{\partial v}{\partial x}+\frac{1}{2}h^2\frac{\partial^2 v}{\partial x^2}-\frac{1}{6}h^3\frac{\partial^3 v}{\partial x^3}+\frac{1}{24}h^4\frac{\partial^4 v}{\partial x^4}+\cdots\right)_{i,j} \qquad (7.1.2)$$

and

$$v_{i+1,j}=\left(v+h\frac{\partial v}{\partial x}+\frac{1}{2}h^2\frac{\partial^2 v}{\partial x^2}+\frac{1}{6}h^3\frac{\partial^3 v}{\partial x^3}+\frac{1}{24}h^4\frac{\partial^4 v}{\partial x^4}+\cdots\right)_{i,j}. \qquad (7.1.3)$$

By adding (7.1.2) and (7.1.3), we have

$$v_{i-1,j}+v_{i+1,j}-2v_{i,j}=\left(h^2\frac{\partial^2 v}{\partial x^2}+\frac{1}{12}h^4\frac{\partial^4 v}{\partial x^4}+\cdots\right)_{i,j}. \qquad (7.1.4)$$

Similarly

$$v_{i,j-1}+v_{i,j+1}-2v_{i,j}=\left(h^2\frac{\partial^2 v}{\partial x^2}+\frac{1}{12}h^4\frac{\partial^4 v}{\partial x^4}+\cdots\right)_{i,j}. \qquad (7.1.5)$$

Now by adding (7.1.4) and (7.1.5), we get

$$\left[h^2\left(\frac{\partial^2 v}{\partial x^2}+\frac{\partial^2 v}{\partial y^2}\right)+\frac{1}{12}h^4\left(\frac{\partial^4 v}{\partial x^4}+\frac{\partial^4 v}{\partial y^4}\right)+\cdots\right]_{i,j}, \qquad (7.1.6)$$

which leads to the five-point finite difference replacement

$$u_{i-1,j}+u_{i+1,j}-4u_{i,j}+u_{i,j-1}+u_{i,j+1}\approx 0, \qquad (7.1.7)$$

for Laplace's equation, with a local truncation error, E_{tr}, given by

$$E_{tr} = \frac{1}{12}h^4\left(\frac{\partial^4 v}{\partial x^4} + \frac{\partial^4 v}{\partial y^4}\right)_{i,j} + \cdots .$$ (7.1.8)

The principal part of the truncation error will only make sense if the derivatives of U are continuous up to order four in x and y.

Now, for the Helmholtz equation (7.1.1), we then have the five-point difference equation at the point $u_{i,j}$, $i,j = 1, 2, \ldots, N$

$$-u_{i-1,j} - u_{i+1,j} + 4gu_{i,j} - u_{i,j-1} - u_{i,j+1} = -h^2 f(x_i, y_j)$$ (7.1.9)

as an approximation to the solution of (7.1.1), where h is small.

This equation can be represented as a computational molecule in the form:

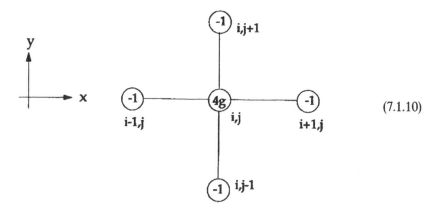

(7.1.10)

with $i,j = 1, \ldots, N$, where N is the number of points and $g = 1 + (1/4)\rho h^2$.

Our task now is to solve the linear system $A\underline{u} = \underline{b}$, derived from equation (7.1.9) for the totality of equations, $i,j = 1, 2, \ldots, N$.

From this equation we have

$$A = \begin{bmatrix} A_1 & -I & & & \\ -I & A_1 & -I & & 0 \\ & \ddots & \ddots & \ddots & \\ 0 & & -I & A_1 & -I \\ & & & -I & A_1 \end{bmatrix}_{N^2 \times N^2} \quad \text{with } A_1 = \begin{bmatrix} 4g & -1 & & & \\ -1 & 4g & -1 & & 0 \\ & \ddots & \ddots & \ddots & \\ 0 & & -1 & 4g & -1 \\ & & & -1 & 4g \end{bmatrix} .$$

(7.1.11)

The vector $\underline{u}$ and $\underline{b}$ are given by

$$\underline{u} = [u_{1,1}, \ldots, u_{N,1}; \ u_{1,2}, \ldots, u_{N,2}; \ \ldots \ ; \ u_{1,N}, \ldots, u_{N,N}]^T$$ (7.1.12)

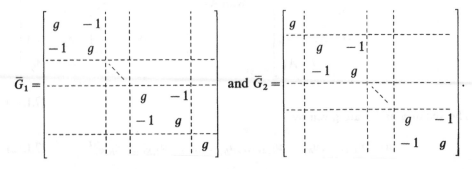

and

$$\underline{b} = [b_{1,1}, \ldots, b_{N,1}; \quad b_{1,2}, \ldots, b_{N,2}; \quad \ldots; \quad b_{1,N}, \ldots, b_{N,N}]^T, \qquad (7.1.13)$$

where the elements of vector b are as follows:

Set $b_{i,j} = -h^2 f(x_i, y_j)$, $i,j = 1, \ldots, N$. Hence, the elements next to the boundary, parallel to the x-axis are

$$b_{i,1} = b_{i,j} + u_{i,0}, \quad b_{i,N} = b_{i,N} + u_{i,N+1}, \quad i = 1, \ldots, N$$

and the elements next to the boundary, parallel to the y-axis are

$$b_{1,j} = b_{1,j} + u_{0,j}, \quad b_{N,j} = b_{N,j} + u_{N+1,j}, \quad j = 1, \ldots, N.$$

We now consider the AGE iterative method for solving $A\underline{u} = \underline{b}$, which is based on the splitting of the matrix A into

$$A = G_1 + G_2 + G_3 + G_4, \qquad (7.1.14)$$

where G_1, G_2, G_3 and G_4 are symmetric and positive definite.

These matrices can be shown to be comprised of small (2×2) block systems or can be made so by a suitable permutation on their rows and corresponding columns. As in the one dimensional case, this procedure is convenient in the sense that the work required is much less than would be required to solve $Au = b$ directly.

For the model problem, let us consider when the case N is odd. Hence for the two dimensional problem where the matrix A as in (7.1.11), we have

$$G_1 + G_2 = \begin{bmatrix} \bar{G} & & & \\ & \bar{G} & & \mathbf{0} \\ & & \ddots & \\ & & & \bar{G} \\ \mathbf{0} & & & \bar{G} \\ & & & & \bar{G} \end{bmatrix}_{N^2 \times N^2}, \text{ with } \bar{G} = \begin{bmatrix} 2g & -1 & & & \\ -1 & 2g & -1 & & \mathbf{0} \\ & \ddots & \ddots & \ddots & \\ & & -1 & 2g & -1 \\ \mathbf{0} & & & -1 & 2g \end{bmatrix},$$

and we let $\bar{G} = \bar{G}_1 + \bar{G}_2$, then

$$\bar{G}_1 = \begin{bmatrix} g & -1 & & & & \\ -1 & g & & & & \\ & & g & -1 & & \\ & & -1 & g & & \\ & & & & \ddots & \\ & & & & & g \end{bmatrix} \text{ and } \bar{G}_2 = \begin{bmatrix} g & & & & & \\ & g & -1 & & & \\ & -1 & g & & & \\ & & & \ddots & & \\ & & & & g & -1 \\ & & & & -1 & g \end{bmatrix}.$$

Also, we have

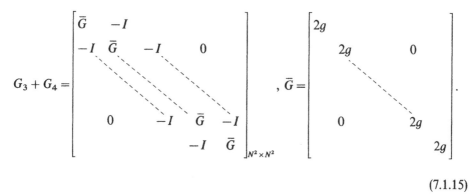

$$(7.1.15)$$

Evidently by interchanging the order of direction we have $(G_1 + G_2)$ in the form of $(G_3 + G_4)$ and vice-versa.

7.2. The Three Dimensional Model Problem

To present the model for the three dimensional problem, let us consider the Helmholtz equation in three variables x, y and z

$$\frac{\partial^2 U}{\partial x^2} + \frac{\partial^2 U}{\partial y^2} + \frac{\partial^2 U}{\partial z^2} - \sigma U = g(x, y, z), \quad (x, y, z) \in \partial R, \qquad (7.2.1)$$

bounded by

$$U(x, y, z) = f(x, y, z), \quad (x, y, z) \in \partial R, \qquad (7.2.2)$$

where ∂R is the boundary of the unit cube $0 \leq x, y, z \leq 1$.

An equally spaced three-dimensional grid is placed over the cube in such a way that it leads to N^3 internal grid points where $h = 1/(N + 1)$, with h the grid spacing. The coordinates of a grid points are

$$x_i = ih, \quad y_j = jh, \quad z_k = kh, \qquad (7.2.3)$$

where i, j, k are integers and the value of U satisfying a difference replacement of equation (7.2.1) at this grid point is given by $u_{i,j,k}$.

By following Section 7.1, the conventional seven-point finite difference replacement of equation (7.2.1), for small h is given by

$$- u_{i-1,j,k} - u_{i+1,j,k} + 6g u_{i,j,k} - u_{i,j-1,k} - u_{i,j+1,k} - u_{i,j,k-1} - u_{i,j,k+1}$$

$$= - h^2 g(x_i, y_j, z_k), \quad (7.2.4)$$

where $i, j, k = 1, \dots, N$ and $g = 1 + (1/6)\sigma h^2$.

It is quite tedious to illustrate every internal grid point for a given N, but each grid point can be represented in molecular form as

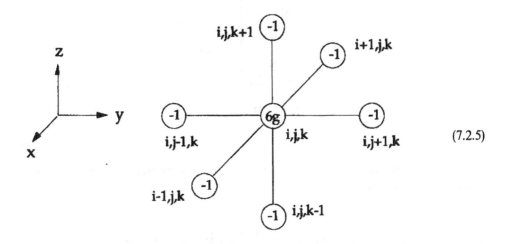

$$(7.2.5)$$

The linear system $A\underline{u} = \underline{b}$ derived from equation (7.2.4) gives

$$A = \begin{bmatrix} A_2 & -J & & & \\ -J & A_2 & -J & 0 & \\ & & & & \\ & & & & \\ 0 & & -J & A_2 & -J \\ & & & -J & A_2 \end{bmatrix}_{N^3 \times N^3}, \quad \text{with} \quad J = \begin{bmatrix} I & & & \\ & I & & 0 \\ & & & \\ & & & \\ 0 & & I & \\ & & & I \end{bmatrix}_{N^2 \times N^2},$$

$$A_2 = \begin{bmatrix} A_1 & -I & & & \\ -I & A_1 & -I & 0 & \\ & & & & \\ & & & & \\ 0 & & -I & A_1 & -I \\ & & & -I & A_1 \end{bmatrix}_{N^2 \times N^2}, \quad \text{and} \quad A_1 = \begin{bmatrix} 6g & -1 & & & \\ -1 & 6g & -1 & 0 & \\ & & & & \\ & & & & \\ 0 & & -1 & 6g & -1 \\ & & & -1 & 6g \end{bmatrix}_{N \times N},$$

$$(7.2.6)$$

the vector $\underline{u} =$

$$[u_{1,1,1}, \ldots, u_{N,1,1}; \quad u_{1,2,1}, \ldots, u_{N,2,1}; \ldots; \quad u_{1,N,1}, \ldots, u_{N,N,1}];$$

$$[u_{1,1,2}, \ldots, u_{N,1,2}; \quad u_{1,2,2}, \ldots, u_{N,2,2}; \ldots; \quad u_{1,N,2}, \ldots, u_{N,N,2}];$$

$$[u_{1,1,N}, \ldots, u_{N,1,N}; \quad u_{1,2,N}, \ldots, u_{N,2,N}; \ldots; \quad u_{1,N,N}, \ldots, u_{N,N,N}]^T, \qquad (7.2.7)$$

and the vector $\underline{b} =$

$$[b_{1,1,1}, \ldots, b_{N,1,1}; \quad b_{1,2,1}, \ldots, b_{N,2,1}; \ldots; \quad b_{1,N,1}, \ldots, b_{N,N,1}];$$

$$[b_{1,1,2}, \ldots, b_{N,1,2}; \quad b_{1,2,2}, \ldots, b_{N,2,2}; \ldots; \quad b_{1,N,2}, \ldots, b_{N,N,2}];$$

$$[b_{1,1,N}, \ldots, b_{N,1,N}; \quad b_{1,2,N}, \ldots, b_{N,2,N}; \ldots; \quad b_{1,N,N}, \ldots, b_{N,N,N}]^T, \qquad (7.2.8)$$

where the elements of vector b are as follows:

$$b_{i,j,k} = -h^2 g(x_i, y_j, z_k), \quad i, j, k = 1, \ldots, N.$$

Hence, the elements next to the boundary, parallel to the xy-plane are

$$b_{i,j,1} = b_{i,j,1} + u_{i,j,0}, \quad b_{i,j,N} = b_{i,j,N} + u_{i,j,N+1}, \qquad i, j = 1, \ldots, N.$$

The elements next to the boundary, parallel to the xz-plane are

$$b_{i,1,k} = b_{i,1,k} + u_{i,0,k}, \quad b_{i,N,k} = b_{i,N,k} + u_{i,N+1,k}, \qquad i, k = 1, \ldots, N$$

and the elements next to the boundary, parallel to the yz-plane are

$$b_{1,j,k} = b_{1,j,k} + u_{0,j,k}, \quad b_{N,j,k} = b_{N,j,k} + u_{N+1,j,k}, \qquad j, k = 1, \ldots, N.$$

By applying the AGE method, we consider a splitting of A into six submatrices, i.e.,

$$A = G_1 + G_2 + G_3 + G_4 + G_5 + G_6, \qquad (7.2.9)$$

where G_i, $i = 1, \ldots, 6$ are symmetric and positive definite.

These matrices can also be shown to be comprised of small (2×2) block systems or can be made so by a suitable permutation on their rows and corresponding column. This procedure is convenient in the sense that work required is much less than would be required to solve $A\underline{u} = \underline{b}$ directly.

Let us consider when N is odd. Thus, for the model problem where the matrix A as in (7.2.6), we have,

$$G_1 + G_2 = \begin{bmatrix} G' & & & \\ & G' & & 0 \\ & & \ddots & \\ & & & \\ 0 & & & G' \end{bmatrix}_{N^3 \times N^3}, \text{ with } G' = \begin{bmatrix} G'' & & & \\ & G'' & & 0 \\ & & \ddots & \\ & & & \\ 0 & & & G'' \end{bmatrix}_{N^2 \times N^2},$$

where

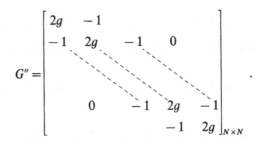

If we let $G'' = G_1'' + G_2''$, then we have

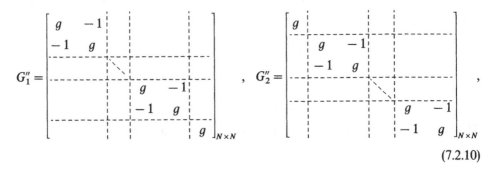

$$(7.2.10)$$

It is obvious that by interchanging the order of direction, we have $(G_1 + G_2)$ in the form of $(G_3 + G_4)$ or $(G_5 + G_6)$ and vice-versa.

With the splitting of the matrix A in both model problems, we will investigate the viability of the Douglas–Rachford, Douglas and Guittet's schemes to solve the problems by using the AGE method.

7.3. The Douglas–Rachford (DR) and Douglas Schemes

It has previously been shown that the DR and Douglas schemes for the AGE method to solve the two-point boundary-value problem, i.e., the problem in one dimension, is given by

$$(rI + G_1)\underline{u}^{(k+1/2)} = [(rI + G_1) - \omega A]\underline{u}^{(k)} + \omega \underline{b}, \qquad (7.3.1)$$

$$(rI + G_2)\underline{u}^{(k+1)} = r\underline{u}^{(k+1/2)} + G_2\underline{u}^{(k)}, \qquad (7.3.2)$$

where $A = G_1 + G_2$ and $r > 0$ is the iteration parameter. When $\omega = 1$, we have the DR scheme and for $\omega = 2$, we have the Douglas scheme. The scheme can also be extended to solve the higher dimensional problems.

Consider the two dimensional problem where the splitting of A into the four submatrices is as in (7.1.14). The AGE method in the DR and Douglas (AGE-DG-2) schemes can be presented as

$$(rI + G_1)\underline{u}^{(k+1/4)} = [(rI + G_1) - \omega A]\underline{u}^{(k)} + \omega\underline{b}, \qquad (7.3.3)$$

$$(rI + G_2)\underline{u}^{(k+1/2)} = r\underline{u}^{(k+1/4)} + G_2\underline{u}^{(k)}, \qquad (7.3.4)$$

$$(rI + G_3)\underline{u}^{(k+3/4)} = r\underline{u}^{(k+1/2)} + G_3\underline{u}^{(k)}, \qquad (7.3.5)$$

$$(rI + G_4)\underline{u}^{(k+1)} = r\underline{u}^{(k+3/4)} + G_4\underline{u}^{(k)}, \qquad (7.3.6)$$

with $\omega = 1$ for the DR scheme and $\omega = 2$ for the Douglas scheme.

This scheme corresponds to sweeping through the mesh parallel to the coordinate x and y axes. At each stage of iteration, we solve the 2×2 block systems. It is obvious that the vector $\underline{u}^{(k+1)}$ is computed in four steps for each iteration.

We now seek to analyse the convergence of the AGE-DG-2 scheme, i.e., equations (7.3.3–7.3.6). It is obvious that the AGE-DG-2 scheme is consistent. We now show that this scheme is convergent. By eliminating all the intermediate vectors of $\underline{u}$, we then have the iteration matrix as

$$T_r = I - \omega r^3 \prod_{i=4}^{1} (rI + G_1)^{-1} A. \qquad (7.3.7)$$

Let α, β, γ and δ are the respective eigenvalues of G_1, G_2, G_3 and G_4. Since G_i, $i = 1, \ldots, 4$ are symmetric and positive definite, hence all the eigenvalues are positive. By inspection, it can be shown that these eigenvalues are in the interval $[0, 2]$.

We need to show that $\mathcal{L}(T_r) < 1$.

$$\mathcal{L}(T_r) = \| T_r \|_2$$

$$= \left\| I - \omega r^3 \prod_{i=4}^{1} (rI + G_1)^{-1} A \right\|_2$$

$$= \left| 1 - \omega r^3 \left(\frac{1}{r+\alpha}\right)\left(\frac{1}{r+\beta}\right)\left(\frac{1}{r+\gamma}\right)\left(\frac{1}{r+\delta}\right)(\alpha + \beta + \gamma + \delta) \right|$$

$$= \left| 1 - \frac{2\omega r^3(\mu + v)}{(r+\mu)^2(r+v)^2} \right|, \quad \text{where } \mu = \max(\alpha, \beta), \quad v = \max(\gamma, \delta),$$

$$= \left| 1 - \frac{4\omega r^3 \lambda}{(r+\lambda)^4} \right| < 1, \quad \text{where } \lambda = \max(\mu, v), \quad \text{for } r > 0.$$

Thus, the AGE-DG-2 scheme is convergent.

In programming, we only need to consider the matrices G_1 and G_2 since after changing the direction, i.e., from row to column, equations (7.3.5) and (7.3.6) can be written as

$$(rI + G_1)\underline{u}_c^{(k+3/4)} = r\underline{u}_c^{(k+1/2)} + G_1\underline{u}_c^{(k)}, \tag{7.3.8}$$

$$(rI + G_2)\underline{u}_c^{(k+1)} = r\underline{u}_c^{(k+3/4)} + G_2\underline{u}_c^{(k)}, \tag{7.3.9}$$

where u_c stands for columnwise reordering. Hence the new set of equations for the AGE-DG-2 scheme is given as

$$(rI + G_1)\underline{u}^{(k+1/4)} = [(rI + G_1) - \omega A]\underline{u}^{(k)} + \omega\underline{b}, \tag{7.3.10}$$

$$(rI + G_2)\underline{u}^{(k+1/2)} = r\underline{u}^{(k+1/4)} + G_2\underline{u}^{(k)}, \tag{7.3.11}$$

$$(rI + G_1)\underline{u}_c^{(k+3/4)} = r\underline{u}_c^{(k+1/2)} + G_2\underline{u}_c^{(k)}, \tag{7.3.12}$$

$$(rI + G_2)\underline{u}_c^{(k+1)} = r\underline{u}_c^{(k+3/4)} + G_2\underline{u}_c^{(k)}. \tag{7.3.13}$$

The AGE-DG-2 scheme in equations (7.3.10–7.3.12) also converges and its convergence can be shown in a similar way.

It is obvious that the matrices G_1 and G_2 consist of the 2×2 block submatrices $\hat{G}$, where

$$\hat{G} = \begin{bmatrix} g & -1 \\ -1 & g \end{bmatrix}, \quad \text{with } g = 1 + \frac{1}{4}\rho h^2. \tag{7.3.14}$$

Hence, for any $r > 0$, the matrices $(rI + G_1)$ and $(rI + G_2)$ are the form of $(rI + \hat{G})$, where

$$(rI + \hat{G}) = \begin{bmatrix} \alpha & -1 \\ -1 & \alpha \end{bmatrix}, \quad \text{with } \alpha = r + g \tag{7.3.15}$$

and the inverse, $(rI + G_1)^{-1}$ and $(rI + G_2)^{-1}$ are in the form

$$(rI + \hat{G})^{-1} = d\begin{bmatrix} \alpha & 1 \\ 1 & \alpha \end{bmatrix}, \quad \text{with } d = \frac{1}{\alpha^2 - 1} \tag{7.3.16}$$

From equation (7.3.10), we set $C = (rI + G_1) - \omega A$.

Let us consider when $N = 5$. Thus, we have

$$C = \begin{bmatrix} C_1 & \omega I & & & 0 \\ \omega I & C_1 & \omega I & & \\ & \omega I & C_1 & \omega I & \\ & & \omega I & C_1 & \omega I \\ 0 & & & \omega I & C_1 \end{bmatrix}_{5^2 \times 5^2}, \quad \text{with } C_1 = \begin{bmatrix} t & s & & & 0 \\ s & t & \omega & & \\ & \omega & t & s & \\ & & s & t & \omega \\ 0 & & & \omega & t \end{bmatrix},$$

where $t = \alpha - 4g\omega$ and $s = \omega - 1$.

We now write the algorithm for general N for the AGE-DG-2 scheme using equations (7.3.10–7.3.13). In this algorithm, the number of points, N is considered odd.

Algorithm 7.3.1.
The AGE-DG-2 scheme.

Set $\qquad u_{i,j}^{(k)} = 0, \quad i, j = 0, \dots, N+1, \quad \alpha = r + g, \quad d = 1/(\alpha^2 - 1).$

$\qquad\qquad\qquad t = \alpha - 4g\omega, \quad s = \omega - 1, \quad \alpha_1 = \alpha d.$

Step 1 To compute $\underline{u}^{(k+1/4)}$. Set $i, j = 1.$

$\qquad$ while $j < N$, compute

$\qquad\qquad$ while $i \leq N - 2$, compute

$$r_1 = \omega u_{i,j-1}^{(k)} + \omega u_{i-1,j}^{(k)} + t u_{i,j}^{(k)} + s u_{i+1,j}^{(k)} + \omega u_{i,j+1}^{(k)} + \omega b_{i,j}$$

$$r_2 = \omega u_{i+1,j-1}^{(k)} + s u_{i,j}^{(k)} + t u_{i+1,j}^{(k)} + \omega u_{i+2,j}^{(k)} + \omega u_{i+1,j+1}^{(k)} + \omega b_{i+1,j}$$

$$u_{i,j}^{(k)} = \alpha_1 r_1 + r_2 d, \qquad u_{i+1,j}^{(k+1,4)} = r_1 d + \alpha_1 r_2,$$

$$i = i + 2$$

$$u_{N,j}^{(k+1/4)} = (\omega u_{N,j-1}^{(k)} + \omega u_{N-1,j}^{(k)} + t u_{N,j}^{(k)} + \omega u_{N,j+1}^{(k)} + \omega b_{N,j})/\alpha$$

$$j = j + 1.$$

Step 2 To compute $\underline{u}^{(k+1/2)}$. Set $i = 2, j = 1.$
$\qquad$ while $j \leq N$, compute

$$u_{1,j}^{(k+1/2)} = (r u_{1,j}^{(k+1/4)} + g u_{1,j}^{(k)})/\alpha$$

$\qquad$ while $i \leq N - 1$, compute

$$r_1 = r u_{i,j}^{(k+1/4)} + g u_{i,j}^{(k)} - u_{i+1,j}^{(k)}$$

$$r_2 = r u_{i+1,j}^{(k+1/4)} - u_{i,j}^{(k)} - g u_{i+1,j}^{(k)}$$

$$u_{i,j}^{(k+1/2)} = \alpha_1 r_1 + r_2 d, \qquad u_{i+1,j}^{(k+1/2)} = r_1 d + \alpha_1 r_2$$

$$i = i + 2$$

$$j = j + 1.$$

Step 3 To compute $\underline{u}^{(k+3/4)}$. Set $i, j = 1$. (Change in direction)

$\qquad$ while $i \leq N$, compute

$\qquad\qquad$ while $j \leq N - 2$, compute

$$r_1 = r u_{i,j}^{(k+1/2)} + g u_{i,j}^{(k)} - u_{i,j+1}^{(k)}$$

$$r_2 = r u_{i,j+1}^{(k+1/2)} - u_{i,j}^{(k)} + g u_{i,j+1}^{(k)}$$

$$u_{i,j}^{(k+3/4)} = \alpha_1 r_1 + r_2 d, \qquad u_{i+1,j}^{(k+3/2)} = r_1 d + \alpha_1 r_2$$

$$j = j + 2$$

$$u_{i,N}^{(k+3/4)} = (ru_{i,N}^{(k+1/2)} + gu_{i,N}^{(k)})/\alpha$$

$$i = i + 1.$$

Step 4   To compute $\underline{u}^{(k+1)}$. Set $i = 1, j = 2$.

while $i \leq N$ compute

$$u_{i,1}^{(k+1)} = (ru_{i,1}^{(k+3/4)} + gu_{i,1}^{(k)})/\alpha$$

while $j \leq N - 1$, compute

$$r_1 = ru_{i,j}^{(k+3/4)} + gu_{i,j}^{(k)} - u_{i,j+1}^{(k)}$$

$$r_2 = ru_{i,j+1}^{(k+3/4)} - u_{i,j}^{(k)} - gu_{i,j+1}^{(k)}$$

$$u_{i,j}^{(k+1)} = \alpha_1 r_1 + r_2 d, \qquad u_{i,j+1}^{(k+1)} = r_1 d + \alpha_1 r_2$$

$$j = j + 2$$

$$i = i + 1.$$

Step 5   Repeat Step 1 to Step 4 until convergence is achieved.

We now investigate the DR and Douglas scheme for the 3-dimensional problem using the AGE method, the AGE-DG-3 scheme.

Let us recall the splitting of A in (7.2.7). Then, the AGE-DG-3 scheme can be presented as

$$(rI + G_1)\underline{u}^{(k+1/6)} = [(rI + G_1) - \omega A]\underline{u}^{(k)} + \omega \underline{b}, \tag{7.3.17}$$

$$(rI + G_2)\underline{u}^{(k+1/3)} = r\underline{u}^{(k+1/6)} + G_2\underline{u}^{(k)}, \tag{7.3.18}$$

$$(rI + G_3)\underline{u}^{(k+1/2)} = r\underline{u}^{(k+1/3)} + G_3\underline{u}^{(k)}, \tag{7.3.19}$$

$$(rI + G_4)\underline{u}^{(k+2/3)} = r\underline{u}^{(k+1/2)} + G_4\underline{u}^{(k)}, \tag{7.3.20}$$

$$(rI + G_5)\underline{u}^{(k+5/6)} = r\underline{u}^{(k+2/3)} + G\underline{u}^{(k)}, \tag{7.3.21}$$

$$(rI + G_6)\underline{u}^{(k+1)} = r\underline{u}^{(k+5/6)} + G_6\underline{u}^{(k)}, \tag{7.3.22}$$

with $\omega = 1$ for the DR scheme and $\omega = 2$ for the Douglas scheme.

This scheme corresponds to sweeping through the mesh parallel to the three coordinate axes x, y and z. At each stage of the iteration, we solve the 2×2 block systems. It is obvious that the vector $u^{(k+1)}$ is computed in six steps.

We now seek to analyse the convergence of the AGE-DG-3 scheme, i.e., equations (7.3.17–7.3.22).

It is obvious that the AGE-DG-3 scheme is consistent. We now show that this scheme is convergent. By eliminating the intermediate vectors $\underline{u}$, we then have the iteration matrix as

$$T_r = I - \omega r^5 \prod_{i=6}^{1} (rI + G_i)^{-1} A. \tag{7.3.23}$$

Let $\alpha, \beta, \gamma, \tau, \sigma$ and δ be the respective eigenvalues of G_1, G_2, G_3, G_4, G_5 and G_6. Since G_i, $i = 1, \ldots, 6$ are symmetric and positive definite, hence all the eigenvalues are positive. By inspection, it can be shown that these eigenvalues are in the interval $[0,2]$.

We need to show that $\mathscr{L}(T_r) < 1$.

$$\mathscr{L}(T_r) = \|T_r\|_2$$

$$= \left\| I - \omega r^5 \prod_{i=6}^{1} (rI + G_1)^{-1} A \right\|_2$$

$$= \left| 1 - \omega r^5 \left(\frac{1}{r + \alpha} \right) \left(\frac{1}{r + \beta} \right) \left(\frac{1}{r + \gamma} \right) \left(\frac{1}{r + \tau} \right) \left(\frac{1}{r + \sigma} \right) \left(\frac{1}{r + \delta} \right) \right.$$

$$\times \left. (\alpha + \beta + \gamma + \tau + \sigma + \delta) \right|$$

$$= \left| 1 - \frac{2\omega r^5 (a + b + c)}{(r + a)^2 (r + b)^2 (r + c)^2} \right|,$$

where $a = \max(\alpha, \beta)$, $b = \max(\gamma, \tau)$ and $c = \max(\sigma, \delta)$.

If $\lambda = \max(a, b, c)$ then

$$\mathscr{L}(T_r) = \left| 1 - \frac{6\omega r^5 \lambda}{(r + \lambda)^6} \right| < 1, \quad \text{for } r > 0.$$

Thus, the AGE-DG-3 scheme is convergent.

Again, in programming we only need to consider G_1 and G_2 matrices since after reordering the direction, the new set of equations becomes

$$(rI + G_1) \underline{u}^{(k+1/6)} = [(rI + G_1) - \omega A] \underline{u}^{(k)} + \omega \underline{b}, \tag{7.3.24}$$

$$(rI + G_2) \underline{u}^{(k+1/3)} = r \underline{u}^{(k+1/6)} + G_2 \underline{u}^{(k)}, \tag{7.3.25}$$

$$(rI + G_1) \underline{u}_y^{(k+1/2)} = r \underline{u}_y^{(k+1/3)} + G_2 \underline{u}_y^{(k)}, \tag{7.3.26}$$

$$(rI + G_2) \underline{u}_y^{(k+2/3)} = r \underline{u}_y^{(k+1/2)} + G_2 \underline{u}_y^{(k)}, \tag{7.3.27}$$

$$(rI + G_1) \underline{u}_z^{(k+5/6)} = r \underline{u}_z^{(k+2/3)} + G_1 \underline{u}_z^{(k)}, \tag{7.3.28}$$

$$(rI + G_2) \underline{u}_z^{(k+1)} = r \underline{u}_z^{(k+5/6)} + G_2 \underline{u}_z^{(k)}, \tag{7.3.29}$$

where $\underline{u}_y$ stands for the y-direction and $\underline{u}_z$ stands for z-direction. The scheme (7.3.24–7.3.29) also converges and its convergence can be shown in a similar way.

It is obvious that the matrices G_1 and G_2 in the AGE-DG-3 scheme consist of the 2×2 block submatrices $\hat{G}$, where

$$\hat{G} = \begin{bmatrix} g & -1 \\ -1 & g \end{bmatrix}, \quad \text{with } g = 1 + \frac{1}{6} \sigma h^2. \tag{7.3.30}$$

Hence, for any $r > 0$, the matrices $(rI + G_1)$ and $(rI + G_2)$ are the form of $(rI + \hat{G})$, in (7.3.15), whilst the inverse, $(rI + G_1)^{-1}$ and $(rI + G_2)^{-1}$ are in the form of $(rI + \hat{G})^{-1}$ in (7.3.16).

Let us consider the case when $N = 5$. From equation (7.3.21), we set $C = (rI + G_1) - \omega A$. Thus, we have

$$
C = \begin{bmatrix}
C_1 & \omega I & & & 0 \\
\omega I & C_1 & \omega I & & \\
& \omega I & C_1 & \omega I & \\
& & \omega I & C_1 & \omega I \\
0 & & & \omega I & C_1
\end{bmatrix}_{5^3 \times 5^3} , \quad \text{for}
$$

$$
C_1 = \begin{bmatrix}
C_2 & \omega I & & & 0 \\
\omega I & C_2 & \omega I & & \\
& \omega I & C_2 & \omega I & \\
& & \omega I & C_2 & \omega I \\
0 & & & \omega I & C_2
\end{bmatrix}_{5^2 \times 5^2} , \quad \text{with} \quad
C_2 = \begin{bmatrix}
t & s & & & 0 \\
s & t & \omega & & \\
\omega & & t & s & \\
& & s & t & \omega \\
0 & & & \omega & t
\end{bmatrix} ,
$$

where $t = \alpha - 6g\omega$ and $s = \omega - 1$.

We now write the algorithm for general N for the AGE-DG-3 scheme using equations (7.3.24–7.3.29). In this algorithm, the number of points, N is considered odd.

Algorithm 7.3.2.   The AGE-DG-3 scheme.

Set $u_{i,j,k}^{(k)} = 0$, $i, j, k = 0, \ldots, N+1$, $\alpha = r + g$,

$\quad d = 1/(\alpha^2 - 1)$, $t = \alpha - 6g\omega$, $s = \omega - 1$, $\alpha_1 = \alpha d$.

Step 1   To compute $\underline{u}^{(k+1/4)}$. Set $i, j, k = 1$.
       while $k \leq N$, compute

           while $j \leq N$, compute

               while $i \leq N - 2$, compute

$$r_1 = \omega u_{i,j,k-1}^{(k)} + \omega u_{i,j-1,k}^{(k)} + \omega u_{i+1,j,k}^{(k)} + t u_{i,j,k}^{(k)}$$
$$\quad + s u_{i+1,j,k}^{(k)} + \omega u_{i,j+1,k}^{(k)} + \omega u_{i,j,k+1}^{(k)} + \omega b_{i,j,k}$$
$$r_2 = \omega u_{i+1,j,k-1}^{(k)} + \omega u_{i+1,j-1,k}^{(k)} + s u_{i,j,k}^{(k)} + t u_{i+1,j,k}^{(k)}$$
$$\quad + \omega u_{i+2,j,k}^{(k)} + \omega u_{i+1,j+1,k}^{(k)} + \omega u_{i+1,j,k}^{(k)} + \omega b_{i,j,k}$$
$$u_{i,j,k}^{(k+1/6)} = \alpha_1 r_1 + r_2 d, \qquad u_{i+1,j,k}^{(k+1/6)} = r_1 d + \alpha_1 r_2$$

$$i = i + 2$$

$$u_{N,j,k}^{(k+1/6)} = (\omega u_{N,j,k-1}^{(k)} + \omega u_{N,j-1,k}^{(k)} + \omega u_{N-1,j,k}^{(k)} + t u_{N,j,k}^{(k)}$$

$$+ \omega u_{N,j+1,k}^{(k)} + \omega u_{N,j,k+1}^{(k)} + \omega b_{N,j,k})/\alpha$$

$$j = j + 1$$

$$k = k + 1.$$

Step 2 To compute $\underline{u}^{(k+1/3)}$. Set $i = 2$, $j, k = 1$.

while $k \leq N$, compute

while $j \leq N$, compute

$$u_{1,j,k}^{(k+1/3)} = (r u_{1,j,k}^{(k+1/6)} + g u_{1,j,k}^{(k)})/\alpha$$

while $i \leq N - 1$, compute

$$r_1 = r u_{i,j,k}^{(k+1/6)} + g u_{i,j,k}^{(k)} - u_{i+1,j,k}^{(k)}$$

$$r_2 = r u_{i+1,j,k}^{(k+1/6)} - u_{i,j,k}^{(k)} + g u_{i+1,j,k}^{(k)}$$

$$u_{i,j,k}^{(k+1/3)} = \alpha_1 r_1 + r_2 d, \qquad u_{i+1,j,k}^{(k+1/3)} = r_1 d + \alpha_1 r_2$$

$$i = i + 2$$

$$j = j + 1$$

$$k = k + 1.$$

Step 3 To compute $\underline{u}^{(k+1/2)}$. Set $i, j, k = 1$. (Change in direction)

while $i \leq N$, compute
while $k \leq N$, compute

while $j \leq N - 1$, compute

$$r_1 = r u_{i,j,k}^{(k+1/3)} + g u_{i,j,k}^{(k)} - u_{i,j+1,k}^{(k)}$$

$$r_2 = r u_{i,j+1,k}^{(k+1/3)} - u_{i,j,k}^{(k)} + g u_{i,j+1,k}^{(k)}$$

$$u_{i,j,k}^{(k+1/2)} = \alpha_1 r_1 + r_2 d, \qquad u_{i,j+1,k}^{(k+1,2)} = r_1 d + \alpha_1 r_2$$

$$j = j + 2$$

$$u_{i,N,k}^{(k+1/2)} = (r u_{i,N,k}^{(k+1/3)} + g u_{i,N,k}^{(k)})/\alpha$$

$$k = k + 1$$

$$i = i + 1.$$

Step 4 To compute $\underline{u}^{(k+2/3)}$. Set $j = 2$, $i, k = 1$.

while $i \leq N$, compute

while $k \leq N$, compute

$$u_{i,1,k}^{(k+2/3)} = (r u_{i,1,k}^{(k+1/2)} + g u_{i,1,k}^{(k)})/\alpha$$

while $j \leq N-1$, compute

$$r_1 = ru_{i,j,k}^{(k+1/2)} + gu_{i,j,k}^{(k)} - u_{i,j+1,k}^{(k)}$$

$$r_2 = ru_{i,j+1,k}^{(k+1/2)} - u_{i,j,k}^{(k)} + gu_{i,j+1,k}^{(k)}$$

$$u_{i,j,k}^{(k+2/3)} = \alpha_1 r_1 + r_2 d, \qquad u_{i,j+1,k}^{(k+2/3)} = r_1 d + \alpha_1 r_2$$

$$j = j+2$$

$$k = k+1$$

$$i = i+1.$$

Step 5   To compute $\underline{u}^{(k+5/6)}$. Set $i,j,k = 1$. (Change in direction)

while $j \leq N$, compute

while $i \leq N$, compute

while $k \leq N-1$, compute

$$r_1 = ru_{i,j,k}^{(k+2/3)} + gu_{i,j,k}^{(k)} - u_{i,j,k+1}^{(k)}$$

$$r_2 = ru_{i,j,k+1}^{(k+2/3)} - u_{i,j,k}^{(k)} + gu_{i,j,k+1}^{(k)}$$

$$u_{i,j,k}^{(k+5/6)} = \alpha_1 r_1 + r_2 d, \qquad u_{i+1,j,k}^{(k+5/6)} = r_1 d + \alpha_1 r_2$$

$$k = k+2$$

$$u_{i,j,N}^{(k+5/6)} = (ru_{i,j,N}^{(k+2/3)} + gu_{i,j,N}^{(k)})/\alpha$$

$$i = i+1$$

$$j = j+1.$$

Step 6   To compute $\underline{u}^{(k+1)}$. Set $i,j = 1$, $k = 2$.

while $j \leq N$, compute

while $i \leq N$, compute

$$u_{i,j,1}^{(k+1)} = (ru_{i,j,1}^{(k+5/6)} + gu_{i,j,1}^{(k)})/\alpha$$

while $k \leq N-1$, compute

$$r_1 = ru_{i,j,k}^{(k+5/6)} + gu_{i,j,k}^{(k)} - u_{i,j+1,k+1}^{(k)}$$

$$r_2 = ru_{i,j+1,k+1}^{(k+5/6)} - u_{i,j,k}^{(k)} + gu_{i,j+1,k+1}^{(k)}$$

$$u_{i,j,k}^{(k+1)} = \alpha_1 r_1 + r_2 d, \qquad u_{i,j+1,k}^{(k+1)} = r_1 d + \alpha_1 r_2$$

$$k = k+2$$

$$i = i+1$$

$$j = j+1.$$

Step 7   Repeat Step 1 to Step 6 until convergence is achieved.

7.4. The Guittet Formula

Now Guittet's formula can be shown to successfully solve the two point boundary-value problem by using the AGE method. Guittet [1967], has shown that this scheme can be extended to solve the problem with the higher dimension.

For the two dimensional problem with the splitting of A as in (7.1.14), the AGE method in Guittet's formula, the AGE-GT-2 scheme can be presented as

$$(rI + G_1)\underline{u}^{(k+1/4)} = \left[\prod_{i=1}^{4}(rI + G_i) - \omega r^3 A\right]\underline{u}^{(k)} + \omega r^3 \underline{b}, \tag{7.4.1}$$

$$(rI + G_2)\underline{u}^{(k+1/2)} = \underline{u}^{(k+1/4)}, \tag{7.4.2}$$

$$(rI + G_3)\underline{u}^{(k+3/4)} = \underline{u}^{(k+1/2)}, \tag{7.4.3}$$

$$(rI + G_4)\underline{u}^{(k+1)} = \underline{u}^{(k+3/4)}, \tag{7.4.4}$$

where $r > 0$, r is the iteration parameter.

After eliminating the intermediate vectors u, we have the iteration matrix for the AGE-GT-2 scheme as

$$T_r = I - \omega r^3 \prod_{i=4}^{1}(rI + G_i)^{-1}A, \tag{7.4.5}$$

which is similar to the iteration matrix of the AGE-DG-2 scheme. Thus, the AGE-GT-2 scheme converges and the proof of convergence is similar to the AGE-DG-2 scheme.

In programming, the computational work can be made more easier by interchanging the rows and columns in equations (7.4.3) and (7.4.4). However, it is better to consider the matrices G_3 and G_4 as part of equation (7.4.1) as this leads to a simpler computation for $\underline{u}^{(k+1/4)}$. Thus, the new set of equations become

$$(rI + G_1)\underline{u}^{(k+1/4)} = \left[\prod_{i=1}^{4}(rI + G_i) - \omega r^3 A\right]\underline{u}^{(k)} + \omega r^3 \underline{b}, \tag{7.4.6}$$

$$(rI + G_2)\underline{u}^{(k+1/2)} = \underline{u}^{(k+1/4)} \tag{7.4.7}$$

$$(rI + G_1)\underline{u}_c^{(k+3/4)} = \underline{u}_c^{(k+1/2)}, \tag{7.4.8}$$

$$(rI + G_2)\underline{u}_c^{(k+1)} = \underline{u}_c^{(k+3/4)}, \tag{7.4.9}$$

where c denotes columnwise ordering.

The scheme also converges and its convergence can be shown in a similar way. The three dimensional problem with the splitting of A in (7.2.9), gives the AGE method in Guittet's formula, the AGE-GT-3 scheme as

$$(rI + G_1)\underline{u}^{(k+1/6)} = \left[\prod_{i=1}^{6}(rI + G_i) - \omega r^5 A\right]\underline{u}^{(k)} + \omega r^5 \underline{b}, \tag{7.4.10}$$

$$(rI + G_2)\underline{u}^{(k+1/3)} = \underline{u}^{(k+1/6)}, \tag{7.4.11}$$

$$(rI + G_3)\underline{u}^{(k+1/2)} = \underline{u}^{(k+1/3)}, \tag{7.4.12}$$

$$(rI + G_4)\underline{u}^{(k+2/3)} = \underline{u}^{(k+1/2)}, \tag{7.4.13}$$

$$(rI + G_5)\underline{u}^{(k+5/6)} = \underline{u}^{(k+2/3)}, \tag{7.4.14}$$

$$(rI + G_6)\underline{u}^{(k+1)} = \underline{u}^{(k+5/6)}, \tag{7.4.15}$$

where $r > 0$, r is the iteration parameter.

After eliminating the intermediate vectors u, we have the iteration matrix for the AGE-GT-3 scheme as

$$T_r = I - \omega r^5 \prod_{i=6}^{1} (rI + G_i)^{-1} A, \tag{7.4.16}$$

which is similar to the iteration matrix of the AGE-DG-2 scheme.

Thus, the AGE-GT-3 scheme converges and the proof of convergence is similar to the AGE-DG-3 scheme.

As in the AGE-GT-2 scheme, the computation can be made simpler by considering the interchange between the x, y, and z directions. However, the matrices G_3, G_4, G_5, and G_6 may be kept in equation (7.4.10) to ease the computational effort. Thus, we only need to consider the AGE-GT-3 scheme in the form

$$(rI + G_1)\underline{u}^{(k+1/6)} = \left[\prod_{i=1}^{6}(rI + G_i) - \omega r^5 A\right]\underline{u}^{(k)} + \omega r^5 \underline{b}, \tag{7.4.17}$$

$$(rI + G_2)\underline{u}^{(k+1/3)} = \underline{u}^{(k+1/6)}, \tag{7.4.18}$$

$$(rI + G_1)\underline{u}_y^{(k+1/2)} = \underline{u}_y^{(k+1/3)}, \tag{7.4.19}$$

$$(rI + G_2)\underline{u}_y^{(k+2/3)} = \underline{u}_y^{(k+1/2)}, \tag{7.4.20}$$

$$(rI + G_1)\underline{u}_z^{(k+5/6)} = \underline{u}_z^{(k+2/3)}, \tag{7.4.21}$$

$$(rI + G_2)\underline{u}_z^{(k+1)} = \underline{u}_z^{(k+5/6)}, \tag{7.4.22}$$

where y and z stands for the ordering in the y and z directions.

The scheme also converges and its convergence can be shown in a similar way.

The disadvantage with the AGE-GT-2 and AGE-GT-3 scheme is in the laborious calculation for obtaining the matrix

$$C_2 = \prod_{i=1}^{4}(rI + G_i) - \omega r^3 A, \tag{7.4.23}$$

for the AGE-GT-2 scheme, and

$$C_3 = \prod_{i=1}^{6}(rI + G_i) - \omega r^3 A, \tag{7.4.24}$$

for the AGE-GT-3 scheme. However, this is compensated by a simpler calculation for the remaining equations.

We now illustrate the computational algorithm to determine the matrix C_2 in (7.4.23).

Let us consider when $N = 7$. Thus

$$A = \begin{bmatrix} A_1 & -I & & & & & \\ -I & A_1 & -I & & & 0 & \\ & -I & A_1 & -I & & & \\ & & -I & A_1 & -I & & \\ & & & -I & A_1 & -I & \\ & 0 & & & -I & A_1 & -I \\ & & & & & -I & A_1 \end{bmatrix}_{7^2 \times 7^2},$$

where

$$A_1 = \begin{bmatrix} 4g & -1 & & & & & \\ -1 & 4g & -1 & & & 0 & \\ & -1 & 4g & -1 & & & \\ & & -1 & 4g & -1 & & \\ & & & -1 & 4g & -1 & \\ & 0 & & & -1 & 4g & -1 \\ & & & & & -1 & 4g \end{bmatrix}_{7^2 \times 7^2}.$$

The constituent matrices are given as follows:

$$(rI + G_1) = \begin{bmatrix} G'_1 & & & & & & \\ & G'_1 & & & 0 & & \\ & & G'_1 & & & & \\ & & & G'_1 & & & \\ & & & & G'_1 & & \\ & 0 & & & & G'_1 & \\ & & & & & & G'_1 \end{bmatrix}_{7^2 \times 7^2},$$

$$(rI + G_2) = \begin{bmatrix} G'_2 & & & & & & \\ & G'_2 & & & 0 & & \\ & & G'_2 & & & & \\ & & & G'_2 & & & \\ & & & & G'_2 & & \\ & 0 & & & & G'_2 & \\ & & & & & & G'_2 \end{bmatrix}_{7^2 \times 7^2},$$

where

$$
G'_1 = \begin{bmatrix}
\alpha & -1 & & & & & \\
-1 & \alpha & & & & & \\
& & \alpha & -1 & & & \\
& & -1 & \alpha & & & \\
& & & & \alpha & -1 & \\
& & & & -1 & \alpha & \\
& & & & & & \alpha
\end{bmatrix}, \quad
G'_2 = \begin{bmatrix}
\alpha & & & & & & \\
& \alpha & -1 & & & & \\
& -1 & \alpha & & & & \\
& & & \alpha & -1 & & \\
& & & -1 & \alpha & & \\
& & & & & \alpha & -1 \\
& & & & & -1 & \alpha
\end{bmatrix},
$$

and where $\alpha = r + g$.

Having chosen the G_3 and G_4 matrices, we can then determine $(rI + G_3)$ and $(rI + G_4)$. Let us consider the selection of the matrices G_3 and G_4 that gives the matrices $(rI + G_3)$ and $(rI + G_4)$ as

$$
(rI + G_3) = \begin{bmatrix}
G'_3 & -I & & & & & \\
-I & G'_3 & & & 0 & & \\
& & G'_3 & -I & & & \\
& & -I & G'_3 & & & \\
& & & & G'_3 & -I & \\
& 0 & & & -I & G'_3 & \\
& & & & & & G'_3
\end{bmatrix}_{7^2 \times 7^2},
$$

$$
(rI + G_4) = \begin{bmatrix}
G'_3 & & & & & & \\
& G'_3 & -I & & 0 & & \\
& -I & G'_3 & & & & \\
& & & G'_3 & -I & & \\
& & & -I & G'_3 & & \\
& 0 & & & & G'_3 & -I \\
& & & & & -I & G'_3
\end{bmatrix}_{7^2 \times 7^2},
$$

where

$$
G'_3 = \begin{bmatrix}
\alpha & & & & & & \\
& \alpha & & & 0 & & \\
& & \alpha & & & & \\
& & & \alpha & & & \\
& & & & \alpha & & \\
& 0 & & & & \alpha & \\
& & & & & & \alpha
\end{bmatrix}, \quad \text{for } \alpha = r + g.
$$

Now

$$C_2 = \prod_{i=1}^{4}(rI + G_i) - \omega r^3 A$$

$$= \begin{bmatrix} Q & P & R & & & & \\ P & Q & P & & & \mathbf{0} & \\ & P & Q & P & R & & \\ & R & P & Q & P & & \\ & & & P & Q & P & R \\ & \mathbf{0} & & R & P & Q & P \\ & & & & & P & Q \end{bmatrix}_{7^2 \times 7^2},$$

where

$$P = \omega r^3 I - G_1' G_2' G_3'$$

$$= \begin{bmatrix} x & w & -\alpha & & & & \\ w & x & w & & & \mathbf{0} & \\ & w & x & w & -\alpha & & \\ -\alpha & & w & x & w & & \\ & & & w & x & w & -\alpha \\ & \mathbf{0} & & -\alpha & & w & x & w \\ & & & & & w & x \end{bmatrix},$$

$$Q = G_1' G_2' G_3'^2 - \omega r^3 A$$

$$= \begin{bmatrix} y & x & w & & & & \\ x & y & x & & & \mathbf{0} & \\ & x & y & x & w & & \\ & w & x & y & x & & \\ & & & x & y & x & w \\ & \mathbf{0} & & & w & x & y & x \\ & & & & & x & y \end{bmatrix},$$

and

$$R = G_1' G_2'$$

$$= \begin{bmatrix} w & -\alpha & 1 & & & & \\ -\alpha & w & -\alpha & & & \mathbf{0} & \\ & -\alpha & w & -\alpha & 1 & & \\ & 1 & -\alpha & w & -\alpha & & \\ & & & -\alpha & w & -\alpha & 1 \\ & \mathbf{0} & & & 1 & -\alpha & w & -\alpha \\ & & & & & -\alpha & w \end{bmatrix},$$

with $w = \alpha^2$, $x = r^3 - \alpha^3$ and $y = \alpha^4 - 4g\omega r^3$.

We now write the algorithm for the AGE-GT-2 scheme in detail. In this algorithm, the number of points, N is assumed odd.

Algorithm 7.4.1.
The AGE-GT-2 scheme.

Set
$$u_{i,j}^{(k)} = 0, \quad i,j = 0, \ldots, N+1, \quad \alpha = r + g, \quad d = 1/(\alpha^2 - 1),$$
$$w = \alpha^2, \quad x = \omega r^3 - \alpha^3, \quad y = \alpha^4 - 4g\omega r^3 \text{ and } \alpha_1 = \alpha d.$$

Step 1   To compute $\underline{u}^{(k+1/4)}$. Set $i, j = 1$.

while $j \leq N - 2$, compute

while $r \leq N - 2$, compute

$$r_1 = wu_{i-1,j-1}^{(k)} + xu_{i,j-1}^{(k)} + wu_{i+1,j-1}^{(k)} - \alpha u_{i+2,j-1}^{(k)}$$
$$+ xu_{i-1,j}^{(k)} + yu_{i,j}^{(k)} + xu_{i+1,j}^{(k)} + wu_{i+2,j}^{(k)} + wu_{i-1,j+1}^{(k)}$$
$$+ xu_{i,j+1}^{(k)} + wu_{i+1,j+1}^{(k)} - \alpha u_{i+2,j+1}^{(k)} - \alpha u_{i-1,j+2}^{(k)}$$
$$+ wu_{i,j+2}^{(k)} - \alpha u_{i+1,j+2}^{(k)} + u_{i+2,j+2}^{(k)} + \omega r^3 b_{i,j}$$

$$r_2 = \alpha u_{i-1,j-1}^{(k)} + wu_{i,j-1}^{(k)} + xu_{i+1,j-1}^{(k)} - wu_{i+2,j-1}^{(k)}$$
$$+ wx_{i-1,j}^{(k)} + xu_{i,j}^{(k)} + yu_{i+1,j}^{(k)} + xu_{i+2,j}^{(k)} + \alpha u_{i-1,j+1}^{(k)}$$
$$+ wu_{i,j+1}^{(k)} + xu_{i+1,j+1}^{(k)} - wu_{i+2,j+1}^{(k)} - u_{i-1,j+2}^{(k)}$$
$$- \alpha u_{i,j+2}^{(k)} - wu_{i+1,j+2}^{(k)} - \alpha u_{i+2,j+2}^{(k)} + \omega r^3 b_{i+1,j}$$

$$u_{i,j}^{(k+1/4)} = \alpha_1 r_1 + r_2 d, \qquad u_{i+1,j}^{(k+1/4)} = r_1 d + \alpha_1 r_2$$
$$i = i + 2$$

$$u_{N,j}^{(k+1/4)} = (wu_{N-1,j-1}^{(k)} + xu_{N,j-1}^{(k)} + xu_{N-1,j}^{(k)} + yu_{N,j}^{(k)}$$
$$+ wu_{N-1,j+1}^{(k)} + xu_{N,j+1}^{(k)} - \alpha u_{N-1,j+2}^{(k)}$$
$$+ wu_{N,j+2}^{(k)} + \omega r^3 b_{N,j})/\alpha$$

$i = 1$.

while $i \leq N - 2$, compute

$$r_3 = -\alpha u_{i-1,j-1}^{(k)} + wu_{i,j-1}^{(k)} + \alpha u_{i+1,j-1}^{(k)} - u_{i+2,j-1}^{(k)}$$
$$+ wu_{i-1,j}^{(k)} + xu_{i,j}^{(k)} + wu_{i+1,j}^{(k)} - \alpha u_{i+2,j}^{(k)} + xu_{i-1,j+1}^{(k)}$$
$$+ yu_{i,j+1}^{(k)} + xu_{i+1,j+1}^{(k)} - wu_{i+2,j+1}^{(k)} - wu_{i-1,j+2}^{(k)}$$
$$+ xu_{i,j+2}^{(k)} - wu_{i+1,j+2}^{(k)} - \alpha u_{i+2,j+2}^{(k)} + \omega r^3 b_{i,j+1}$$

$$r_4 = -u_{i-1,j-1}^{(k)} - \alpha u_{i,j-1}^{(k)} + wu_{i+1,j-1}^{(k)} - \alpha u_{i+2,j-1}^{(k)}$$
$$- \alpha u_{i-1,j}^{(k)} + wu_{i,j}^{(k)} + xu_{i+1,j}^{(k)} + wu_{i+2,j}^{(k)} + wu_{i-1,j+1}^{(k)}$$
$$+ xu_{i,j+1}^{(k)} + yu_{i+1,j+1}^{(k)} + xu_{i+2,j+1}^{(k)} - \alpha u_{i-1,j+2}^{(k)}$$
$$+ wu_{i,j+2}^{(k)} + xu_{i+1,j+2}^{(k)} + wu_{i+2,j+2}^{(k)} + \omega r^3 b_{i+1,j+1}$$

$$u_{N,j}^{(k+1/4)} = \alpha_1 r_3 + r_4 d, \qquad u_{i+1,j}^{(k+1/4)} = r_3 d + \alpha_4 r_2$$

$$i = i + 2$$

$$u_{N,j}^{(k+1/4)} = (-\alpha u_{N-1,j-1}^{(k)} + w u_{N,j-1}^{(k)} + w u_{N-1,j}^{(k)} + x u_{N,j}^{(k)}$$

$$+ x u_{N-1,j+1}^{(k)} + y u_{N,j+1}^{(k)} + w u_{N-1,j+2}^{(k)}$$

$$+ x u_{N,j+2}^{(k)} + \omega r^3 b_{i,j+1}) / \alpha$$

$$j = j + 2$$

$$i = 1.$$

while $i \leq N - 2$, compute

$$r_5 = w u_{i-1,N-1}^{(k)} + x u_{i,N-1}^{(k)} + w u_{i+1,N-1}^{(k)} - \alpha u_{i+2,N-1}^{(k)} + x u_{i-1,N}^{(k)}$$

$$+ y u_{i,N}^{(k)} + x u_{i+1,N}^{(k)} + w u_{i+2,N}^{(k)} + \omega r^3 b_{i,N}$$

$$r_6 = -\alpha u_{i-1,N-1}^{(k)} + w u_{i,N-1}^{(k)} + x u_{i+1,N-1}^{(k)} + w u_{i+2,N-1}^{(k)} + w u_{i-1,N}^{(k)}$$

$$+ x u_{i,N}^{(k)} + y u_{i+1,N}^{(k)} + x u_{i+2,N}^{(k)} + \omega r^3 b_{i,N}$$

$$u_{N,N}^{(k+1/4)} = (w u_{N-1,N-1}^{(k)} + x u_{N,N-1}^{(k)} + y u_{N-1,N}^{(k)} + y u_{N,N}^{(k)} + \omega r^3 b_{N,N}) / \alpha$$

Step 2   To compute $\underline{u}^{(k+1/2)}$. Set $i = 2, j = 1$.

while $j \leq N$, compute

$$u_{1,j}^{(k+1/2)} = u_{1,j}^{(k+1/4)} / \alpha$$

while $i \leq N - 1$, compute

$$u_{i,j}^{(k+1/2)} = \alpha_1 u_{i,j}^{(k+1/4)} + d u_{i+1,j}^{(k+1/4)}$$

$$u_{i+1,j}^{(k+1/2)} = d u_{i,j}^{(k+1/4)} + \alpha_1 u_{i+1,j}^{(k+1/4)}$$

$$i = i + 2$$

$$j = j + 1.$$

Step 3   To compute $\underline{u}^{(k+3/4)}$. Set $i, j = 1$. (Change in direction)

while $i \leq N$, compute

while $j \leq N - 2$, compute

$$u_{i,j}^{(k+3/4)} = \alpha_1 u_{i,j}^{(k+1/2)} + d u_{i,j+1}^{(k+1/2)}$$

$$u_{i,j+1}^{(k+3/4)} = d u_{i,j}^{(k+1/2)} + \alpha_1 u_{i,j+1}^{(k+1/2)}$$

$$j = j + 2$$

$$u_{i,N}^{(k+3/4)} = u_{i,N}^{(k+1/2)} / \alpha$$

$$i = i + 1.$$

Step 4   To compute $\underline{u}^{(k+1)}$. Set $i = 1, j = 2$.

while $i \leq N$, compute

$$u_{i,1}^{(k+1)} = u_{i,1}^{(k+3/4)} / \alpha$$

while $j \leq N - 1$, compute

$$u_{i,j}^{(k+1)} = \alpha_1 u_{i,j}^{(k+3/4)} + du_{i,j+1}^{(k+3/4)}$$

$$u_{i,j+1}^{(k+1)} = du_{i,j}^{(k+3/4)} + \alpha_1 u_{i,j+1}^{(k+3/4)}$$

$$j = j + 2$$

$$i = i + 1.$$

Step 5 Repeat Step 1 to Step 4 until convergence is achieved.

Based on the Algorithm 7.4.1, more computational work will be required to derive the matrix C_3 for the AGE-GT-3 scheme. Hence, it is just sufficient to show Algorithm 7.4.1. Moreover, Mitchell (1969) has stated that, the best splitting for general ω is in the form of the AGE-DG-2 and the AGE-DG-3 schemes. The details of the amount of work needed will be shown later. Concerning ω, Guittet (1967) has also considered the choice of the parameter ω and showed that the best value of ω is 2, for the Douglas scheme which is also applicable to the AGE-DG-2 and the AGE-DG-3 schemes.

7.5. The Successive Over-relaxation (SOR) Method

For the purpose of comparison, we write the algorithm for the SOR method to solve the two and three dimensional model problems discussed in the previous sections.

a) For the two dimensional problem, equation (7.1.1).

Algorithm 7.5.1.

The SOR method for equation (7.1.1).

Set $u_{i,j}^{(k)} = 0, \quad i,j = 0, \ldots, N+1.$

Step 1 To compute $\underline{u}^{(k+1)}$

for $i = 1$ to N, compute

for $j = 1$ to N, compute

$$u_{i,j}^{(k)} = \omega\left[(b_{i,j} + u_{i-1,j}^{(k+1)} + u_{i,j-1}^{(k+1)} + u_{i+1,j}^{(k)} + u_{i,j+1}^{(k)})/4g\right] + (1-\omega)u_{i,j}^{(k)}.$$

Step 2 Repeat Step 1 until convergence is achieved.

b) For the three dimensional problem, equation (7.2.1).

Algorithm 7.5.2.

The SOR method for equation (7.2.1).

Set $u_{i,j,k}^{(k)} = 0, \quad i,j,k = 0, \ldots, N+1.$

Step 1 To compute $\underline{u}^{(k+1)}$

for $i = 1$ to N, compute
for $j = 1$ to N, compute
for $k = 1$ to N, compute

$$u_{i,j,k}^{(k+1)} = \omega[(b_{i,j,k} + u_{i-1,j,k}^{(k+1)} + u_{i,j,k}^{(k+1)} + u_{i+1,j,k-1}^{(k)}$$
$$+ u_{i+1,j,k}^{(k)} + u_{i,j+1,k}^{(k)} + u_{i,j,k+1}^{(k)})/6g] + (1 - \omega)u_{i,j,k}^{(k)}.$$

Step 2 Repeat Step 1 until convergence is achieved.

7.6. The Computational Complexity

In this section, we will be concerned with the estimation of the amount of arithmetic calculation needed in the algorithms derived in Sections 7.3 to 7.5.

In solving equation (7.1.1), i.e., the two dimensional problem, the amount of work per iteration is determined for the AGE-DG-2, AGE-GT-2 schemes and the SOR method, whilst the solution for the three dimensional problem, i.e., equation (7.2.1), the calculation is based on the AGE-DG-3 scheme and the SOR method.

Table 7.6.1 and 7.6.2 show the amount of operations for every iteration in solving equations (7.1.1) and (7.2.1) respectively.

Table 7.6.1 The Amount of Operations per Iteration for Solving Equation (7.1.1)

Method	Multiplication	Addition	Overall
AGE-DG-2	$17(N-1)^2$	$15(N-1)^2$	$32(N-1)^2$
AGE-GT-2	$13(N-1)^2$	$20(N-1)^2$	$33(N-1)^2$
SOR	$3N^2$	$6N^2$	$9N^2$

Table 7.6.2 The Amount of Operations per Iteration for Solving Equation (7.2.1)

Method	Multiplication	Addition	Overall
AGE-DG-3	$25(N-1)^3$	$23(N-1)^3$	$48(N-1)^3$
SOR	$3N^2$	$8N^2$	$11N^2$

Tables 7.6.1 and 7.6.2 show that the SOR method is far better than the AGE-DG and AGE-GT schemes in terms of the amount of computational work. However, with the limited application of the SOR method, i.e., the method only requires a single parameter, then the AGE-DG and AGE-GT schemes need further consideration. Since the computational work of the AGE-GT scheme is quite substantial, then we may consider further that the best method is the AGE-DG especially when more than 1 parameter is considered.

In the next section, the numerical results are presented to show the number of iterations needed for each method.

7.7. Experimental Results

Numerical results presented here are for the AGE-DG-2, AGE-DG-3 and AGE-GT-2 schemes and also for the SOR method for solving both the two and three dimensional problems. Five problems are considered with the first three concerned with the solution of the two dimensional problem.

Problem 4 and 5 are devoted to the solutions for problem in three dimensions. Problems 1 and 4 are the Helmholtz equation in two and three dimensions respectively. The results are presented for various values of ρ and σ. Problems 2 and 5 are the Laplace equation in two and three dimensions respectively. The results are presented for various values of ρ and σ. Problems 2 and 5 are the Laplace equation in two and three dimensions, whilst Problem 3 is a Poisson equation.

Problem 1 – The Helmholtz Equation in Two Dimensions

$$\frac{\partial^2 U}{\partial x^2} + \frac{\partial^2 U}{\partial y^2} - \rho U = 6 - \rho(2x^2 + y^2), \quad 0 \le x, y \le 1,$$

subject to the boundary conditions

$$U(x, 0) = U(x, 1) = 2x^2, \quad 0 \le x \le 1,$$

$$U(0, y) = y^2, \quad U(1, y) = 2 + y^2, \quad 0 \le y \le 1.$$

The exact solution is given by $U(x, y) = 2x^2 + y^2$.
The results are tabulated as follows:

Table 7.7.1 Problem 1 with $\rho = 0$

$\rho = 0$	SOR		AGE-DG-2 ($\omega = 2$)	
N^2	ω	iter.	r	iter.
81	1.54	25	0.95–1.04	25
361	1.74	49	0.56	49
1521	1.86	92	0.33	108
6241	1.93	178	0.19	230

Table 7.7.2 Problem 1 with $\rho = 5$

$\rho = 5$	SOR		AGE-DG-2 ($\omega = 2$)	
N^2	ω	iter.	r	iter.
81	1.49–1.50	24	1.04–1.11	22
361	1.71–1.72	46	0.59–0.61	44
1521	1.86	89	0.35	95
6241	1.92	171	0.20	205

Table 7.7.3 Problem 1 with $\rho = 20$

$\rho = 20$	SOR		AGE-DG-2 ($\omega = 2$)	
N^2	ω	*iter.*	*r*	*iter.*
81	1.40–1.25	22	1.19–1.37	18
361	1.63	42	0.71–0.72	34
1521	1.79	83	0.41	73
6241	1.89	164	0.24	158

Table 7.7.4 Problem 1 with $\rho = 40$

$\rho = 40$	SOR		AGE-DG-2 ($\omega = 2$)	
N^2	ω	*iter.*	*r*	*iter.*
81	1.31–1.34	20	1.38–1.63	15
361	1.55–1.56	40	0.79–0.84	28
1521	1.74–1.74	80	0.47	58
6241	1.85	155	0.28	127

Table 7.7.5 Problem 1 with $\rho = 70$

$\rho = 70$	SOR		AGE-DG-2 ($\omega = 2$)	
N^2	ω	*iter.*	*r*	*iter.*
81	1.23–1.26	18	1.75–1.83	12
361	1.50	34	0.91–0.99	23
1521	1.70	66	0.54	46
6241	1.83	131	0.33	101

Table 7.7.6 Problem 1 with $\rho = 100$

$\rho = 100$	SOR		AGE-DG-2 ($\omega = 2$)	
N^2	ω	*iter.*	*r*	*iter.*
81	1.21–1.23	16	1.89–2.20	11
361	1.46	30	1.01–1.12	20
1521	1.67	58	0.60–0.61	40
6241	1.81	114	0.35–0.36	87

Table 7.7.7 Problem 1 with $\rho = 200$

$\rho = 200$	SOR		AGE-DG-2 ($\omega = 2$)	
N^2	ω	*iter.*	*r*	*iter.*
81	1.12–1.20	13	2.57–3.09	9
361	1.36–1.38	23	1.29–1.46	15
1521	1.59–1.60	44	0.74–0.77	29
6241	1.76	83	0.44	61

DAVID J. EVANS

Problem 2 – The Laplace Equation in Two Dimensions

$$\frac{\partial^2 U}{\partial x^2} + \frac{\partial^2 U}{\partial y^2} = 0, \quad 0 \le x, y \le 1,$$

subject to the boundary conditions

$$U(x, 0) = U(x, 1) = \sin \pi x, \quad 0 \le x \le 1,$$

$$U(0, y) = U(1, y) = 0, \quad 0 \le y \le 1.$$

The exact solution is given by

$$U(x, y) = \operatorname{sech} \frac{\pi}{2} \cosh \pi \left(y - \frac{1}{2} \right) \sin \pi x.$$

The results are tabulated in Table 7.7.8

Table 7.7.8 Problem 2

	SOR		AGE-DG-2 ($\omega = 2$)	
N^2	ω	iter.	r	iter.
81	1.53–1.55	23	0.96–1.00	22
361	1.74	43	0.57–0.58	47
1521	1.86	85	0.34	102
6241	1.93	167	0.19	209

Problem 3 – The Poisson equation

$$\frac{\partial^2 U}{\partial x^2} + \frac{\partial^2 U}{\partial y^2} = -2, \quad 0 \le x, y \le 1,$$

subject to the boundary conditions

$$U(x, 0) = U(x, 1) = x(1 - x), \quad 0 \le x \le 1,$$
$$U(0, y) = 0, \quad U(1, y) = \sinh \pi x \sin \pi y, \quad 0 \le y \le 1.$$

The exact solution is given by

$$U(x, y) = \sinh \pi x \sin \pi y + x(1 - x).$$

The results are tabulated in Table 7.7.9.

Table 7.7.9 Problem 3

	SOR		AGE-DG-2 ($\omega = 2$)	
N^2	ω	iter.	r	iter.
81	1.54	26	0.97–1.08	27
361	1.74	52	0.58–0.59	56
1521	1.86	94	0.34	116
6241	1.93	191	0.18	228

Problem 4 – The Three Dimensional Helmholtz Problem

$$\frac{\partial^2 U}{\partial x^2} + \frac{\partial^2 U}{\partial y^2} + \frac{\partial^2 U}{\partial z^2} - \sigma U = (3 - \sigma) \cos x \cosh y \cosh z, \quad 0 \le x, y, z \le 1,$$

subject to the boundary conditions

$$U(x, y, 0) = \cosh x \cosh y, \quad 0 \le x, y \le 1,$$

$$U(x, y, 1) = 1.543 \cosh x \cosh y, \quad 0 \le x, y \le 1,$$

$$U(x, 0, z) = \cosh x \cosh z, \quad 0 \le x, z \le 1,$$

$$U(x, 1, z) = 1.543 \cosh x \cosh z, \quad 0 \le x, z \le 1,$$

$$U(0, y, z) = \cosh y \cosh z, \quad 0 \le y, z \le 1,$$

$$U(1, y, z) = 1.543 \cosh y \cosh z, \quad 0 \le y, z \le 1.$$

The exact solution is given by $U(x, y, z) = \cosh x \cosh y \cosh z$.

The results are tabulated as follows.

Table 7.7.10 Problem 4 with $\sigma = 0$

$\sigma = 0$	SOR		AGE-DG-3 ($\omega = 2$)	
N^3	ω	iter.	r	iter.
729	1.51	30	1.53–1.53	26
1331	1.57–1.59	37	1.32–1.37	33
2197	1.62	43	1.21	39
3375	1.65	49	1.11	46
4913	1.68–1.70	56	1.03–1.04	54

Table 7.7.11 Problem 4 with $\sigma = 5$

$\sigma = 5$	SOR		AGE-DG-3 ($\omega = 2$)	
N^3	ω	iter.	r	iter.
729	1.41–1.51	30	1.60–1.62	24
1331	1.53–1.56	36	1.40–1.43	30
2197	1.58–1.60	42	1.26–1.28	36
3375	1.62–1.63	48	1.15–1.19	43
4913	1.66	54	1.08	49

Table 7.7.12 Problem 4 with $\sigma = 25$

$\sigma = 25$	SOR		AGE-DG-3 ($\omega = 2$)	
N^3	ω	iter.	r	iter.
729	1.40–1.41	24	1.88–1.92	19
1331	1.47–1.48	29	1.65	23
2197	1.52–1.54	34	1.46–1.49	28
3375	1.57	38	1.33–1.37	33
4913	1.61	43	1.23–1.26	38

Table 7.7.13 Problem 4 with $\sigma = 50$

$\sigma = 50$	SOR		AGE-DG-3 ($\omega = 2$)	
N^3	ω	iter.	r	iter.
729	1.36–1.38	22	2.04–2.12	17
1331	1.44	26	1.75–1.87	21
2197	1.49–1.50	30	1.57–1.66	25
3375	1.54–1.55	34	1.43–1.50	29
4913	1.58–1.59	39	1.33–1.37	33

Table 7.7.14 Problem 4 with $\sigma = 100$

$\sigma = 100$	SOR		AGE-DG-3 ($\omega = 2$)	
N^3	ω	iter.	r	iter.
729	1.30–1.31	17	2.61–2.80	13
1331	1.37	20	2.28–2.31	15
2197	1.40–1.43	24	1.97–2.09	18
3375	1.46–1.47	27	1.76–1.90	21
4913	1.50	30	1.66–1.67	23

Table 7.7.15 Problem 4 with $\sigma = 200$

$\sigma = 200$	SOR		AGE-DG-3 ($\omega = 2$)	
N^3	ω	iter.	r	iter.
729	1.23	13	3.29–3.92	11
1331	1.27–1.31	16	2.88–3.07	12
2197	1.33–1.34	18	2.46–2.72	14
3375	1.36–1.39	21	2.18–2.44	16
4913	1.41–1.42	23	2.06–2.07	17

Problem 5– The Laplace Equation in Three Dimensions

$$\frac{\partial^2 U}{\partial x^2} + \frac{\partial^2 U}{\partial y^2} + \frac{\partial^2 U}{\partial z^2} = 0, \quad 0 \le x, y, z \le 1,$$

subject to the boundary conditions

$$U(x, y, 0) = U(x, y, 1) = 0, \quad 0 \le x, y \le 1,$$

$$U(0, y, z) = U(1, y, z) = 0, \quad 0 \le y, z \le 1,$$

$$U(x, 0, z) = 0, \quad U(x, 1, z) = \sin \pi x \sin \pi z, \quad 0 \le x, z \le 1.$$

The exact solution is given by

$$U(x, y, z) = \operatorname{sech} \frac{\pi}{\sqrt{2}} \sin \pi x \cosh \sqrt{2} \pi \left(y - \frac{1}{2} \right) \sin \pi z.$$

The results are tabulated in Table 7.7.16

Table 7.7.16 Problem 5

	SOR		AGE-DG-3 ($\omega = 2$)	
N^3	ω	*iter.*	r	*iter.*
729	1.50–1.53	27	1.42	21
2197	1.62	36	1.08–1.10	31
4913	1.69	46	0.88–0.89	40
9261	1.74	56	0.74–0.75	49

The results presented in this section are purely experimental, as no theory exists for the optimal single parameter for solving the two and three dimensional problems. The theoretical background given for the prescribed method is only concerned with the convergence of the method. However, as shown for the one dimensional problems, the AGE method again appears to be superior to the SOR method in the number of iterations when solving two and three dimensional problems governed by Dirichlet boundary conditions especially when the matrix A is strongly diagonally dominant.

Evidently, the bounds of the eigenvalues for the problems given in this section can be determined from the (2×2) block submatrices, i.e., $a = g - 1$ and $b = g + 1$, where $g = 1 + (1/4)\rho h^2$ for the two dimensional problems and $g = 1 + (1/6)\sigma h^2$ for the three dimensional problems, where ρ and σ are constants.

For the one dimensional problems with $g = 1 + (1/2)\rho h^2$, where ρ is a constant, the relation $r = \sqrt{ab}$ is satisfied for large ρ as the AGE-PR(1) scheme is theoretically derived from the ADI-PR method. It is then shown that the AGE-DG scheme gives a similar convergence as the AGE-PR(1) scheme. Thus, in order to determine the theoretical optimal r for the two and three dimensional problems we notice that $a \to 0$, hence as a guide to the experimental value of r, we should look closely to the pattern of r in the one dimensional problem when $\rho = 0$.

For a symmetric matrix, when $\rho = 0$, the value of $r \in [0,1]$, i.e., say, $r_1 \approx 0.5$. The results for the two and three dimensional problems clearly shows that when $\sigma = 0$, we obtain, $r_2 \approx 1$, and $r_3 \approx 1.5$ respectively. This assumption is based on $N = 9$ as a starting number of points. The values of r_2 and r_3 may be interpreted in many ways. The simplest is by considering the splitting of the matrices and the repetition of the same matrices used for solving the problems. Thus, this gives the value $r_2 = 2r_1$ and $r_3 = 3r_1$. The value of r for the case of larger numbers of points can then be considered to lie in the interval $[0, r_2]$ or $[0, r_3]$ as shown in Tables 7.1.7–1 to 7.1.7–16.

For a larger σ, i.e., $\sigma > 25$ and a smaller number of points, the results show that $r > b$. But, for a larger number of points, the values of r again appears to fall in the interval $[0, r_2]$ or $[0, r_3]$.

Although the AGE-DG-2 and AGE-DG-3 schemes need more computational work, we see from the simplicity of the schemes and that its parameters are easily determined, then these schemes may well be considered competitive to solve the two and three dimensional problems.

7.8. The Solution with Different Boundary Conditions

We have shown in Chapter 5 that the AGE method has been successfully applied to solve the two-point boundary-value problem with different boundary conditions. Now, with the AGE-DG-2 scheme discussed in the previous section, we will investigate the solution for problems in two dimensions governed by periodic and Neumann boundary conditions. In this section, we will consider a general problem, i.e., the Helmholtz equation in two dimensions.

Periodic Boundary Conditions
Consider the Helmholtz equation in two dimensions

$$\frac{\partial^2 U}{\partial x^2} + \frac{\partial^2 U}{\partial y^2} - \rho U = f(x,y), \quad 0 \le x,y \le 1, \tag{7.8.1}$$

in a square region $0 \le x,y \le 1$, governed by the periodic boundary conditions

$$U(x,0) = U(x,1), \quad U_y(x,0) = U_y(x,1), \quad 0 \le x \le 1, \tag{7.8.2}$$

$$U(0,y) = U(1,y), \quad U_x(0,y) = U_x(1,y), \quad 0 \le y \le 1. \tag{7.8.3}$$

The periodic boundary conditions (7.8.2) and (7.8.3) are treated in a similar fashion as in Chapter 5.

Now applying the centered finite difference approximations, for a small mesh size h, yields the conventional five-point formula

$$-u_{i,j-1} - u_{i-1,j} + 4gu_{i,j} - u_{i+1,j} - u_{i,j+1} = h^2 f(x_i, y_j), \quad i,j = 1,\ldots,N, \tag{7.8.4}$$

where $g = 1 + (1/4) \rho h^2$ and N is the number of grid points or nodes in the region.

Let us consider $h = 1/N$ and if we take $N = 6$, then the grid points can be represented as in Figure 7.8.1.

In general, for N even, equation (7.8.4) can be written in a matrix form $A\underline{u} = \underline{b}$ giving

$$A = \begin{bmatrix} A_1 & -I & & & -I \\ -I & A_1 & -I & & 0 \\ & & \ddots & & \\ & 0 & & -I & A_1 & -I \\ -I & & & & -I & A_1 \end{bmatrix}_{N^2 \times N^2} \quad \text{with } A_1 = \begin{bmatrix} 4g & -1 & & & -1 \\ -1 & 4g & -1 & & 0 \\ & & \ddots & & \\ & 0 & & -1 & 4g & -1 \\ -1 & & & & -1 & 4g \end{bmatrix}. \tag{7.8.5}$$

The vector $\underline{u}$ and $\underline{b}$ are given by

$$\underline{u} = [u_{1,1}, \ldots, u_{N,1}; \ u_{1,2}, \ldots, u_{N,2}; \ldots; \ u_{1,N}, \ldots, u_{N,N}]^T \tag{7.8.6}$$

and

$$\underline{b} = [b_{1,1}, \ldots, b_{N,1}; \ b_{1,2}, \ldots, b_{N,2}; \ldots; \ b_{1,N}, \ldots, b_{N,N}]^T, \tag{7.8.7}$$

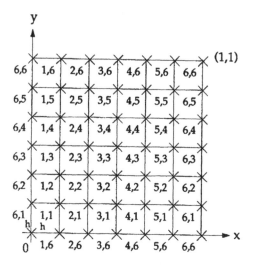

Figure 7.8.1 The number of nodes for $h = 1/6$.

where the elements of vector $\underline{b}$ are

$$b_{i,j} = h^2 f(x_i, y_j), \quad i,j = 1, \ldots, N.$$

Now, consider the splitting of the matrix A into

$$A = G_1 + G_2 + G_3 + G_4, \tag{7.8.8}$$

where G_1, G_2, G_3 and G_4 are symmetric and positive definite submatrices.

Let us consider the matrices $(G_1 + G_2)$ and $(G_3 + G_4)$ as follows:

$$G_1 + G_2 = \begin{bmatrix} A_2 & & & \\ & A_2 & & 0 \\ & & \ddots & \\ & & & A_2 \\ 0 & & & A_2 \end{bmatrix}_{N^2 \times N^2}, \text{ where } A_2 = \begin{bmatrix} 2g & -1 & & & -1 \\ -1 & 2g & -1 & & \\ & \ddots & \ddots & \ddots & 0 \\ & & -1 & 2g & -1 \\ -1 & & & -1 & 2g \end{bmatrix}_{N \times N}$$

and

$$G_3 + G_4 = \begin{bmatrix} A_3 & -I & & & -I \\ -I & A_3 & -I & & 0 \\ & \ddots & \ddots & \ddots & \\ 0 & & -I & A_3 & -I \\ -I & & & -I & A_3 \end{bmatrix}_{N^2 \times N^2}, \text{ where } A_3 = \begin{bmatrix} 2g & & & \\ & 2g & & 0 \\ & & \ddots & \\ 0 & & & 2g \\ & & & & 2g \end{bmatrix}_{N \times N}.$$

Let us consider further the matrices G_1 and G_2 for the case when N is even, i.e.,

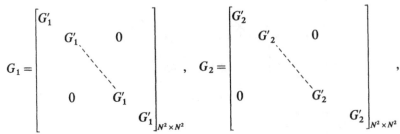

where

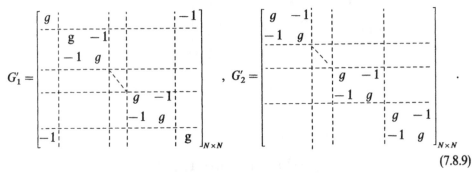

$$(7.8.9)$$

It can be shown that, after reordering the direction, i.e., from row to column, the matrices $(G_1 + G_2)$ become $(G_3 + G_4)$ and vice-versa.

Let $\alpha = r + g$, $r > 0$, where r is the iteration parameter. Thus, we have the matrices $\bar{G}_1 = (rI + G_1)$ and $\bar{G}_2 = (rI + G_2)$ which are given as

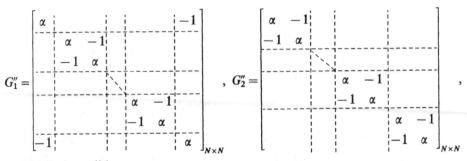

where

and easily invertible.

We now seek to analyse the convergence of the AGE-DG-2 scheme for the Poisson equation subject to periodic boundary conditions. It has been shown in Chapter 5 (for a problem subject to periodic boundary condition) that by a suitable matrix permutation P

$$G_2'' = PG_1'' P^T.$$

Thus, the eigenvalues of G_1'' and G_2'' are similar. Hence, the AGE-DG-2 scheme for solving the periodic problem is convergent.

It is obvious that, if one chooses G_2' in the form of G_1' and vice-versa, the scheme is also convergent. However, by having G_1' and G_2' as given in (7.8.9), the algorithm is found to be much shorter.

By using the AGE-DG-2 scheme in Section 7.3, the algorithm can be presented as follows. First, to obtain the matrix

$$C = [(rI + G_1)] - \omega A].$$

Suppose that $N = 6$. Then the matrix C is given as

$$C = \begin{bmatrix} C_1 & \omega I & & & & \omega I \\ \omega I & C_1 & \omega I & & 0 & \\ & \omega I & C_1 & \omega I & & \\ & & \omega I & C_1 & \omega I & \\ & 0 & & \omega I & C_1 & \omega I \\ \omega I & & & & \omega I & C_1 \end{bmatrix}_{6^2 \times 6^2}$$

with

$$C_1 = \begin{bmatrix} t & \omega & & & & s \\ \omega & t & s & & 0 & \\ & s & t & \omega & & \\ & & \omega & t & s & \\ & 0 & & s & t & \omega \\ s & & & & \omega & t \end{bmatrix}_{6 \times 6},$$

where $t = \alpha - 4\omega$ and $s = \omega - 1$.

We now write the algorithm for general N (even) in detail.

Algorithm 7.8.1.
The AGE-DG-2 scheme for equation (7.8.1).

Set $\qquad u_{1,j}^{(k)} \ i,j = 0,...,N+1, \quad \alpha = r + g, \quad d = 1/(\alpha^2 - 1),$

$$t = \alpha - 4g\omega, \quad s = \omega - 1, \quad \alpha_1 = \alpha d.$$

Step 1 To compute $\underline{u}^{(k+1/4)}$

    A. The elements $u_{1,1}^{(k+1/4)}$ to $u_{N,1}^{(k+1/4)}$. Set $i=2$

$$r_1 = tu_{1,1}^{(k)} + \omega u_{2,1}^{(k)} + su_{N,1}^{(k)} + \omega u_{1,2}^{(k)} + \omega u_{1,N}^{(k)} + \omega b_{1,1}$$

$$r_2 = su_{1,1}^{(k)} + \omega u_{N-1,1}^{(k)} + tu_{N,1}^{(k)} + \omega u_{N,2}^{(k)} + \omega u_{N,N}^{(k)} + \omega b_{N,1}$$

$$u_{1,1}^{(k+1/4)} = \alpha_1 r_1 + r_2 d, \qquad u_{N,1}^{(k+1/4)} = r_1 d + \alpha_1 r_2$$

        while $i \leq N-2$, compute

$$r_1 = \omega u_{i-1,1}^{(k)} + tu_{i,1}^{(k)} + su_{i+1,1}^{(k)} + \omega u_{i,N}^{(k)} + \omega b_{i,1}$$

$$r_2 = su_{i,1}^{(k)} + tu_{i+1,1}^{(k)} + \omega u_{i+2,1}^{(k)} + \omega u_{i+1,2}^{(k)} + \omega u_{i+1,N}^{(k)} + \omega b_{i+1,1}$$

$$u_{i,1}^{(k+1/4)} = \alpha_1 r_1 + r_2 d, \qquad u_{i+1,1}^{(k+1/4)} = r_1 d + \alpha_1 r_2$$

$$i = i+2$$

    B. The elements $u_{1,2}^{(k+1/4)}$ to $u_{N-1,1}^{(k+1/4)}$. Set $i,j = 2$

        while $j \leq N-2$, compute

$$r_1 = \omega u_{1,j-1}^{(k)} + tu_{1,j}^{(k)} + \omega u_{2,j}^{(k)} + su_{N,j}^{(k)} + \omega u_{1,j+1}^{(k)} + \omega b_{1,j}$$

$$r_2 = \omega u_{N,j-1}^{(k)} + su_{1,j}^{(k)} + \omega u_{N-1,j}^{(k)} + tu_{N,j}^{(k)} + \omega u_{N,j+1}^{(k)} + \omega b_{N,j}$$

$$u_{1,j}^{(k+1/4)} = \alpha_1 r_1 + r_2 d, \qquad u_{N,j}^{(k+1/4)} = r_1 d + \alpha_1 r_2$$

           while $i \leq N-2$, compute

$$r_1 = \omega u_{i,j-1}^{(k)} + \omega u_{i-1,j}^{(k)} + tu_{i,j}^{(k)} + su_{i+1,j}^{(k)} + \omega u_{i,j+1}^{(k)} + \omega b_{i,j}$$

$$r_2 = \omega u_{i,j-1}^{(k)} + su_{i,j}^{(k)} + tu_{i+1,j}^{(k)} + \omega u_{i+2,j}^{(k)} + \omega u_{i+1,j+1}^{(k)} + \omega u_{i+1,j+1}^{(k)}$$

$$\qquad + \omega b_{i+1,j+1}$$

$$u_{i,j}^{(k+1/4)} = \alpha_1 r_1 + r_2 d, \qquad u_{i+1,j}^{(k+1/4)} = r_1 d + \alpha_1 r_2$$

$$i = i+2$$

$$j = j+1.$$

    C. The elements $u_{1,N}^{(k+1/4)}$ to $u_{N,N}^{(k+1/4)}$. Set $i=2$

$$r_1 = \omega u_{1,1}^{(k)} + \omega u_{1,N-1}^{(k)} + tu_{1,N}^{(k)} + \omega u_{2,N}^{(k)} + su_{N,N}^{(k)} + \omega b_{1,N}$$

$$r_2 = \omega u_{N,1}^{(k)} + \omega u_{N,N-1}^{(k)} + su_{1,N}^{(k)} + \omega u_{N-1,N}^{(k)} + tu_{N,N}^{(k)} + \omega b_{N,N}$$

$$u_{1,N}^{(k+1/4)} = \alpha_1 r_1 + r_2 d, \qquad u_{N,N}^{(k+1/4)} = r_1 d + \alpha_1 r_2$$

        while $i \leq N-2$, compute

$$r_1 = \omega u_{i,1}^{(k)} + \omega u_{i,N-1}^{(k)} + tu_{i,N}^{(k)} + su_{i+1,N}^{(k)} + \omega b_{i,N}$$

$$r_2 = \omega u_{i+1,1}^{(k)} + \omega u_{i+1,N-1}^{(k)} + su_{i,N}^{(k)} + tu_{i+1,N}^{(k)} + \omega u_{i+2,N}^{(k)} + \omega b_{i+1,N}$$

$$u_{i,N}^{(k+1/4)} = \alpha_1 r_1 + r_2 d, \qquad u_{i+1,N}^{(k+1/4)} = r_1 d + \alpha_1 r_2$$

$$i = i+2$$

Step 2 To compute $\underline{u}^{(k+1/2)}$, $i = 1, j = 1$.

    while $j \leq N$, compute

        while $i \leq N-1$, compute

$$r_1 = ru_{i,j}^{(k+1/4)} + gu_{i,j}^{(k)} - u_{i+1,j}^{(k)}$$
$$r_2 = ru_{i+1,j}^{(k+1/4)} - u_{i,j}^{(k)} + gu_{i+1,j}^{(k)}$$
$$u_{i,j}^{(k+1/2)} = \alpha_1 r_1 + r_2 d, \qquad u_{i+1,j}^{(k+1/2)} = r_1 d + \alpha_1 r_2$$
$$i = i + 2$$
$$j = j + 1.$$

Step 3 To compute $\underline{u}^{(k+3/4)}$, $i = 1$, $j = 2$ (Change in direction).

while $i \leq N$, compute

$$r_1 = ru_{i,1}^{(k+1/2)} + gu_{i,1}^{(k)} - u_{i,N}^{(k)}$$
$$r_2 = ru_{1,N}^{(k+1/2)} - u_{i,1}^{(k)} + gu_{i,N}^{(k)}$$
$$u_{i,j}^{(k+3/4)} = \alpha_1 r_1 + r_2 d, \qquad u_{i,N}^{(k+3/4)} = r_1 d + \alpha_1 r_2$$

while $j \leq N - 2$, compute

$$r_1 = ru_{i,j}^{(k+1/2)} + gu_{i,j}^{(k)} - u_{i,j+1}^{(k)}$$
$$r_2 = ru_{i,j+1}^{(k+1/2)} - u_{i,j}^{(k)} + gu_{i,j+1}^{(k)}$$
$$u_{i,j}^{(k+3/4)} = \alpha_1 r_1 + r_2 d, \qquad u_{i,j+1}^{(k+3/4)} = r_1 d + \alpha_1 r_2$$
$$j = j + 2$$
$$i = i + 1.$$

Step 4 To compute $\underline{u}^{(k+1)}$, $i = 1$, $j = 1$.

while $i \leq N$, compute

while $j \leq N - 1$, compute

$$r_1 = ru_{i,j}^{(k+3/4)} + gu_{i,j}^{(k)} - u_{i,j+1}^{(k)}$$
$$r_2 = ru_{i,j+1}^{(k+3/4)} - u_{i,j}^{(k)} + gu_{i,j+1}^{(k)}$$
$$u_{i,j}^{(k+1)} = \alpha_1 r_1 + r_2 d, \qquad u_{i,N}^{(k+1)} = r_1 d + \alpha_1 r_2$$
$$j = j + 2$$
$$i = i + 1.$$

Step 5 Repeat Step 1 to Step 4 until convergence is achieved.

Neumann Boundary Conditions

We now consider the Helmholtz equation in two dimensions

$$\frac{\partial^2 U}{\partial x^2} + \frac{\partial^2 U}{\partial y^2} - \rho U = f(x,y), \quad 0 \leq x,y \leq 1, \tag{7.8.10}$$

subject to the Neumann boundary conditions

$$U_y(x,0) = g_0(x), \qquad U_y(x,1) = g_1(x), \tag{7.8.11}$$
$$U_x(0,y) = h_0(y), \qquad U_x(1,y) = h_1(y), \tag{7.8.12}$$

on the boundary of a square region $0 \leq x,y \leq 1$.

Let h be the grid spacing and if $h = 1/N$, then for $N = 4$, the grid points can be illustrated as in Figure 7.8.2.

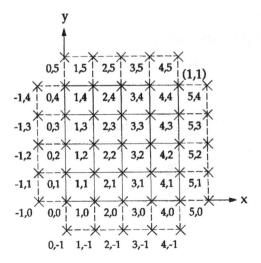

Figure 7.8.2   The number of nodes for $h = 1/4$.

Unlike the Dirichlet problem, we are now dealing with fictitious points, i.e., these are the points which lie just outside the region. These false boundaries are illustrated in Figure 7.8.2 as the points joined to the region by the dotted line. These points can be eliminated by using the centered approximation, for small h,

$$\left[\frac{\partial U}{\partial x}\right]_{i,j} = \frac{u_{i+1,j} - u_{i-1,j}}{2h} \quad \text{and} \quad \left[\frac{\partial U}{\partial y}\right]_{i,j} = \frac{u_{i,j+1} - u_{i,j-1}}{2h}, \qquad (7.8.13)$$

which gives

$$u_{i-1,j} = u_{i+1,j} - 2h\left[\frac{\partial U}{\partial x}\right]_{i,j} \quad \text{and} \quad u_{i,j-1} = u_{i,j+1} - \left[\frac{\partial U}{\partial y}\right]_{i,j}. \qquad (7.8.14)$$

For example, $u_{-1,0} = u_{1,0} - 2h(\partial U/\partial x)_{0,0}$, etc.

Obviously, we have to solve $(N+1)^2$ points including the points at the boundaries. By using the centered finite difference approximations, for small h, we have the conventional five-point formula

$$-u_{i,j-1} - u_{i-1,j} + 4gu_{i,j} - u_{i+1,j} - u_{i,j+1} = h^2 f(x_i, y_j), \quad i,j = 1, \dots, N, \qquad (7.8.15)$$

where $g = 1 + (1/4)\rho h^2$, which gives the linear system $A\underline{u} = \underline{b}$. For our model problem, the linear system derived from equation (7.8.15) gives

$$A = \begin{bmatrix} A_1 & -2I & & \\ -I & A_1 & -I & 0 \\ & & \ddots & \\ 0 & & -I & A_1 & -I \\ & & & -2I & A_1 \end{bmatrix}_{N^2 \times N^2} \quad \text{with } A_1 = \begin{bmatrix} 4g & -2 & & \\ -1 & 4g & -1 & 0 \\ & & \ddots & \\ 0 & & -1 & 4g & -1 \\ & & & -2 & 4g \end{bmatrix}_{N \times N}. \qquad (7.8.16)$$

The vector $\underline{u}$ and $\underline{b}$ are given by

$$\underline{u} = [u_{0,0}, \ldots, u_{N,0}; \ u_{0,1}, \ldots, u_{N,1}; \ldots; \ u_{0,N}, \ldots, u_{N,N}]^T, \tag{7.8.17}$$

$$\underline{b} = [b_{0,0}, \ldots, b_{N,0}; \ b_{0,1}, \ldots, b_{N,1}; \ldots; \ b_{0,N}, \ldots, b_{N,N}]^T, \tag{7.8.18}$$

where the elements of vector $\underline{b}$ are

$$b_{i,0} = h^2 f(x_i, y_0) - 2h \left[\frac{\partial U}{\partial y}\right]_{i,0}, \quad b_{i,N} - h^2 f(x_i, y_N) + 2h \left[\frac{\partial U}{\partial y}\right]_{i,N}, \quad i = 1, \ldots, N,$$

$$b_{j,0} = h^2 f(x_0, y_j) - 2h \left[\frac{\partial U}{\partial x}\right]_{0,j}, \quad b_{N,j} - h^2 f(x_N, y_j) + 2h \left[\frac{\partial U}{\partial x}\right]_{N,j}, \quad j = 1, \ldots, N,$$

$$b_{0,0} = h^2 f(x_0, y_0) - 2h \left\{\left[\frac{\partial U}{\partial x}\right]_{0,0} + \left[\frac{\partial U}{\partial y}\right]_{0,0}\right\},$$

$$b_{N,0} = h^2 f(x_N, y_0) - 2h \left\{\left[\frac{\partial U}{\partial x}\right]_{N,0} + \left[\frac{\partial U}{\partial y}\right]_{N,0}\right\},$$

$$b_{0,N} = h^2 f(x_0, y_N) - 2h \left\{\left[\frac{\partial U}{\partial x}\right]_{0,N} + \left[\frac{\partial U}{\partial y}\right]_{0,N}\right\}$$

and

$$b_{N,N} = h^2 f(x_N, y_N) - 2h \left\{\left[\frac{\partial U}{\partial x}\right]_{N,N} + \left[\frac{\partial U}{\partial y}\right]_{N,N}\right\}.$$

Evidently, for $\rho = 0$, the matrix A is singular since one of the eigenvalues is zero. Thus, we will only consider $\rho > 0$.

Now, consider the splitting of the matrix A into

$$A = G_1 + G_2 + G_3 + G_4, \tag{7.8.19}$$

where the matrices G_1, G_2, G_3 and G_4 have non-negative eigenvalues.

We now seek to analyse the convergence of the AGE-DG-2 scheme for the Helmholtz equation in two dimensions subject to Neumann boundary conditions.

Let us consider the matrices $(G_1 + G_2)$ and $(G_3 + G_4)$ as

$$G_1 + G_2 = \begin{bmatrix} G_1' & & & \\ & G_1' & & 0 \\ & & \ddots & \\ & 0 & & G_1' \\ & & & & G_1' \end{bmatrix}_{N^2 \times N^2}, \quad \text{where } G_1' = \begin{bmatrix} 2g & -2 & & & \\ -1 & 2g & -1 & & 0 \\ & \ddots & \ddots & \ddots & \\ & 0 & & -1 & 2g & -1 \\ & & & & -2 & 2g \end{bmatrix}_{N \times N},$$

$$\tag{7.8.20}$$

$$
G_3 + G_4 = \begin{bmatrix} G_3' & -2I & & \\ -I & G_3' & -I & 0 \\ & & \ddots & \ddots & \ddots \\ 0 & & -I & G_3' & -I \\ & & & -2I & G_3' \end{bmatrix}_{N^2 \times N^2}, \quad \text{where } G_3' = \begin{bmatrix} 2g & & & \\ & 2g & & 0 \\ & & \ddots & \\ 0 & & 2g & \\ & & & 2g \end{bmatrix}_{N \times N}.
$$

(7.8.21)

Now, let us consider further the case when N is odd, the matrices

$$
G_1 = \begin{bmatrix} G_1'' & & & \\ & G_1'' & & 0 \\ & & \ddots & \\ 0 & & G_1'' & \\ & & & G_1'' \end{bmatrix}_{N^2 \times N^2}, \quad \text{where } G_1'' = \begin{bmatrix} \tilde{G} & & & \\ & \hat{G} & & 0 \\ & & \ddots & \\ 0 & & \hat{G} & \\ & & & \bar{G} \end{bmatrix}_{N \times N},
$$

(7.8.22)

$$
G_2 = \begin{bmatrix} G_2'' & & & \\ & G_2'' & & 0 \\ & & \ddots & \\ 0 & & G_2'' & \\ & & & G_2'' \end{bmatrix}_{N^2 \times N^2}, \quad \text{where } G_2'' = \begin{bmatrix} 0 & & & \\ & \hat{G} & & \\ & & \ddots & \\ & & & \hat{G} \\ & & & & 0 \end{bmatrix}_{N \times N},
$$

(7.8.23)

and where

$$
\hat{G} = \begin{bmatrix} g & -1 \\ -1 & g \end{bmatrix}, \quad \tilde{G} = \begin{bmatrix} 2g & -2 \\ -1 & g \end{bmatrix}, \quad \text{and} \quad \bar{G} = \begin{bmatrix} g & -1 \\ -2 & 2g \end{bmatrix}.
$$

The matrix $(G_3 + G_4)$ can have the form of the matrix $(G_1 + G_2)$ by interchanging from row to column, and vice-versa.

From (7.8.21), the eigenvalues of the matrix G_1 are given as

$$
g - 1, \quad g + 1, \quad \tfrac{3}{2}g - \tfrac{1}{2}\sqrt{g^2 + 8}, \quad \text{and} \quad \tfrac{3}{2}g + \tfrac{1}{2}\sqrt{g^2 + 8},
$$

and from (7.8.23), we have the eigenvalues of the matrix G_2 as

$$
0, \quad g - 1 \quad \text{and} \quad g + 1.
$$

It is clear that the matrix G_2 is singular. However, for any iteration parameter, $r > 0$, the matrix $(rI + G_2)$ is non-singular, thus its inverse, $(rI + G_2)^{-1}$ exists. As shown in Chapter 5 for the one-dimensional problem, the proof for the convergence is similar. Thus, for any $\rho > 0$, the AGE-DG-2 scheme for solving the Neumann problem is convergent.

By using the AGE-DG-2 scheme, we now present the algorithm for the model problem governed by the Neumann boundary conditions.

Let us consider the case when $N = 5$. Then, the solution vector $\underline{u}$ are $u_{0,0}, \ldots, u_{5,0} ; \ldots;$ $u_{0,5}, \ldots, u_{5,5}$.

First, to obtain the matrix $C = [(rI + G_1) - \omega A]$.

For $N = 5$, the matrix C is given as

$$C = \begin{bmatrix} C_1 & s_2 I \\ \omega I & C_1 & \omega I & & 0 \\ & \omega I & C_1 & \omega I \\ & 0 & \omega I & C_1 & \omega I \\ & & & \omega I & C_1 & \omega I \\ & & & & s_2 I & C_1 \end{bmatrix}_{6^2 \times 6^2},$$

with

$$C_1 = \begin{bmatrix} t_1 & s_1 \\ s & t & \omega & & 0 \\ & \omega & t & s \\ & 0 & s & t & \omega \\ & & & \omega & t & s \\ & & & & s_1 & t_1 \end{bmatrix}_{6 \times 6},$$

where $t_1 = \alpha_1 - 4g\omega$, $t = \alpha - 4g\omega$, $s = \omega - 1$, $s_1 = 2s$ and $s_2 = 2\omega$, with $\alpha = r + g$ and $\alpha_1 = \alpha + 2g$. The matrices G_1 and G_2 are as in (7.8.22) and (7.8.23), respectively.

For any $r > 0$

$$(rI + \hat{G})^{-1} = d \begin{bmatrix} \alpha & 1 \\ 1 & \alpha \end{bmatrix}, \quad (rI + \tilde{G})^{-1} = d_1 \begin{bmatrix} \alpha & 2 \\ 1 & \alpha_1 \end{bmatrix}$$

and

$$(rI + \bar{G})^{-1} = d_1 \begin{bmatrix} \alpha_1 & 1 \\ 2 & \alpha \end{bmatrix}, \quad \text{where } d = 1/(a^2 - 1), \quad d_1 = 1/(\alpha \alpha_1 - 2).$$

We now write the algorithm for a general N (odd) in detail.

Algorithm 7.8.2.
The AGE-DG-2 scheme for equation (7.8.10).

Set $\quad u_{i,j}^{(k)} = 0, \quad i,j = 0, \ldots, N, \quad g = 1 + \tfrac{1}{4}\rho h^2, \quad \alpha = r + g, \quad \alpha_1 = r + 2g,$

$\quad d = 1/(\alpha^2 - 1), \quad d_1 = 1/(\alpha \alpha_1 - 2), \quad t = \alpha - 4g\omega, \quad t_1 = \alpha_1 - rg\omega,$

$\quad s_2 = 2\omega, \quad s = \omega - 1, \text{ and } s_1 = 2s.$

Step 1 To compute $\underline{u}^{(k+1/4)}$.

    A. The elements $u_{0,0}^{(k+1/4)}$ to $u_{N,0}^{(k+1/4)}$. Set $i=2$

$$r_1 = t_1 u_{0,0}^{(k)} + s_1 u_{1,0}^{(k)} + s_2 u_{0,1}^{(k)} + \omega b_{1,0}$$

$$r_2 = s u_{0,0}^{(k)} + t u_{1,0}^{(k)} + \omega u_{2,0}^{(k)} + s_2 u_{1,1}^{(k)} + \omega b_{1,0}$$

$$u_{0,0}^{(k+1/4)} = (\alpha r_1 + 2r_2)d_1, \qquad u_{1,0}^{(k+1/4)} = (r_1 + \alpha_1 r_2)d_1$$

        while $i \le N-3$, compute

$$r_1 = \omega u_{i-1,0}^{(k)} + t u_{i,0}^{(k)} + s u_{i+1,0}^{(k)} + s_2 u_{i,1}^{(k)} + \omega b_{i,0}$$

$$r_2 = s u_{i,0}^{(k)} + t u_{i+1,0}^{(k)} + \omega u_{i+2,0}^{(k)} + s_2 u_{i+1,1}^{(k)} + \omega b_{i+1,0}$$

$$u_{i,0}^{(k+1/4)} = (\alpha_1 r_1 + r_2)d, \qquad u_{i+1,0}^{(k+1/4)} = (r_1 + \alpha r_2)d$$

$$i = i+2$$

$$r_1 = \omega u_{N-2,0}^{(k)} + t u_{N-1,0}^{(k)} + s u_{N,0}^{(k)} + s_2 u_{N-1,1}^{(k)} + \omega b_{N-1,0}$$

$$r_2 = s_1 u_{N-1,0}^{(k)} + t_1 u_{N,0}^{(k)} + s_2 u_{N,1}^{(k)} + \omega b_{N,0}$$

$$u_{N-1,0}^{(k+1/4)} = (\alpha_1 r_1 + r_2)d_1, \qquad u_{N,0}^{(k+1/4)} = (2r_1 + \alpha r_2)d_1$$

    B. The elements $u_{0,1}^{(k+1/4)}$ to $u_{N,N-1}^{(k+1/4)}$. Set $i=2$, $j=1$.

        while $j \le N-3$, compute

$$r_1 = \omega u_{0,j-1}^{(k)} + t_1 u_{0,j}^{(k)} + s_1 u_{1,j}^{(k)} + \omega u_{0,j+1}^{(k)} + \omega b_{0,j}$$

$$r_2 = \omega u_{1,j-1}^{(k)} + s u_{0,j}^{(k)} + t u_{1,j}^{(k)} + \omega u_{2,j}^{(k)} + \omega u_{1,j+1}^{(k)} + \omega b_{1,j}$$

$$u_{0,j}^{(k+1/4)} = (\alpha_1 r_1 + r_2)d_1, \qquad u_{1,j}^{(k+1/4)} = (r_1 + \alpha r_2)d_1$$

          while $i \le N-3$, compute

$$r_1 = \omega u_{i,j-1}^{(k)} + \omega u_{i-1,j}^{(k)} + t u_{i,j}^{(k)} + s u_{i+1,j}^{(k)} + \omega u_{i,j+1}^{(k)} + \omega b_{i,j}$$

$$r_2 = \omega u_{i+1,j-1}^{(k)} + s u_{i,j}^{(k)} + t u_{i+1,j}^{(k)} + \omega u_{i+2,j}^{(k)} + \omega u_{i+1,j+1}^{(k)} + \omega b_{i+1,j}$$

$$u_{i,j}^{(k+1/4)} = (\alpha_1 r_1 + r_2)d, \qquad u_{i+1,j}^{(k+1/4)} = (r_1 + \alpha r_2)d$$

$$i = i+2$$

$$r_1 = \omega u_{N-1,j-1}^{(k)} + \omega u_{N-2,j}^{(k)} + t u_{N-1,j}^{(k)} + s u_{N,j}^{(k)} + \omega u_{N-1,j+1}^{(k)} + \omega b_{N-1,j}$$

$$r_2 = \omega u_{N,j-1}^{(k)} + s_1 u_{N-1,j}^{(k)} + t_1 u_{N,j}^{(k)} + \omega u_{N,j+1}^{(k)} + \omega b_{N,j}$$

$$u_{N-1,j}^{(k+1/4)} = (\alpha_1 r_1 + r_2)d_1, \qquad u_{N,j}^{(k+1/4)} = (r_1 + \alpha r_2)d_1$$

$$j = j+1.$$

    C. The elements $u_{0,N}^{(k+1/4)}$ to $u_{N,N}^{(k+1/4)}$. Set $i=2$,

$$r_1 = s_2 u_{0,N-1}^{(k)} + t_1 u_{0,N}^{(k)} + s_1 u_{1,N}^{(k)} + \omega b_{0,N}$$

$$r_2 = s_2 u_{1,N-1}^{(k)} + s u_{0,N}^{(k)} + t u_{1,N}^{(k)} + \omega u_{2,N}^{(k)} + \omega b_{1,N}$$

$$u_{0,N}^{(k+1/4)} = (\alpha r_1 + 2r_2)d_1, \qquad u_{1,N}^{(k+1/4)} = (r_1 + \alpha_1 r_2)d_1$$

while $i \leq N - 3$, compute

$$r_1 = s_2 u^{(k)}_{i,N-1} + \omega u^{(k)}_{i-1,N} + t u^{(k)}_{i,N} + s u^{(k)}_{i+1,N} + \omega b_{i,N}$$

$$r_2 = s_2 u^{(k)}_{i+1,N-1} + s u^{(k)}_{i,N} + \omega u^{(k)}_{i+1,N} + \omega u^{(k)}_{i+2,N} + \omega b_{i+1,N}$$

$$u^{(k+1/4)}_{i,N} = (\alpha_1 r_1 + r_2)d, \qquad u^{(k+1/4)}_{i+1,N} = (r_1 + \alpha r_2)d$$

$$i = i + 2$$

$$r_1 = s_2 u^{(k)}_{N-1,N-1} + \omega u^{(k)}_{N-2,N} + t u^{(k)}_{N-1,N} + s u^{(k)}_{N,N} + \omega b_{N-1,N}$$

$$r_2 = s_2 u^{(k)}_{N,N-1} + s_1 u^{(k)}_{N-1,N} + t_1 u^{(k)}_{N,N} + \omega u^{(k)}_{i+2,N} + \omega b_{N,N}$$

$$u^{(k+1/4)}_{N-1,N} = (\alpha_1 r_1 + r_2)d_1 \qquad u^{(k+1/4)}_{N,N} = (r_1 + \alpha r_2)d_1$$

Step 2 To compute $\underline{u}^{(k+1/2)}$, $i = 1$, $j = 0$

while $j \leq N$, compute

$$u^{(k+1/2)}_{0,j} = u^{(k+1/4)}_{0,j}$$

while $i \leq N - 2$, compute

$$r_1 = r u^{(k+1/4)}_{i,j} + g u^{(k)}_{i,j} - u^{(k)}_{i+1,j}$$

$$r_2 = r u^{(k+1/4)}_{i+1,j} - u^{(k)}_{i,j} + g u^{(k)}_{i+1,j}$$

$$u^{(k+1/2)}_{i,j} = (\alpha r_1 + r_2)d, \qquad u^{(k+1/2)}_{i+1,j} = (r_1 + \alpha r_2)d$$

$$i = i + 2$$

$$u^{(k+1/2)}_{N,j} = u^{(k+1/4)}_{N,j}$$

$$j = j + 1.$$

Step 3 To compute $\underline{u}^{(k+3/4)}$, $i = 0$, $j = 2$. (Change in direction.)

while $i \leq N$, compute

$$r_1 = r u^{(k+1/2)}_{i,0} + 2 g u^{(k)}_{i,0} - 2 u^{(k)}_{i,1}$$

$$r_2 = r u^{(k+1/2)}_{i,1} - u^{(k)}_{i,0} + g u^{(k)}_{i,1}$$

$$u^{(k+3/4)}_{i,0} = (\alpha r_1 + r_2)d_1, \qquad u^{(k+3/4)}_{i,1} = (r_1 + \alpha r_2)d_1$$

while $j \leq N - 3$, compute

$$r_1 = r u^{(k+1/2)}_{i,j} + g u^{(k)}_{i,j} - u^{(k)}_{i,j+1}$$

$$r_2 = r u^{(k+1/2)}_{i,j+1} - u^{(k)}_{i,j} + g u^{(k)}_{i,j+1}$$

$$u^{(k+3/4)}_{i,j} = (\alpha r_1 + r_2)d, \qquad u^{(k+3/4)}_{i,j+1} = (r_1 + \alpha r_2)d$$

$$j = j + 2$$

$$r_1 = r u^{(k+1/2)}_{i,N-1} + g u^{(k)}_{i,N-1} - u^{(k)}_{i,N}$$

$$r_2 = r u^{(k+1/2)}_{i,N} - 2 u^{(k)}_{i,N-1} + 2 g u^{(k)}_{i,N}$$

$$u^{(k+3/4)}_{i,N-1} = (\alpha r_1 + r_2)d_1, \qquad u^{(k+3/4)}_{i,N} = (r_1 + \alpha r_2)d_1$$

$$i = i + 1.$$

Step 4　To compute $\underline{u}^{(k+1)}$, $i=1$, $j=1$.

> while $i \leq N$, compute
>
> $u_{i,0}^{(k+1/2)} = u_{i,0}^{(k+3/4)}$
> while $j \leq N-2$, compute
>
> $r_1 = r u_{i,0}^{(k+3/4)} + g u_{i,j}^{(k)} - u_{i+1,j+1}^{(k)}$
>
> $r_2 = r u_{i,j+1}^{(k+3/4)} - u_{i,j}^{(k)} + g u_{i,j+1}^{(k)}$
>
> $u_{i,j}^{(k+1)} = (\alpha r_1 + r_2)d,\qquad u_{i,j+1}^{(k+1)} = (r_1 + \alpha r_2)d$
>
> $j = j + 2$
>
> $u_{i,N}^{(k+1)} = u_{i,N}^{(k+3/4)}$
>
> $i = i + 1.$

Step 5　Repeat Step 1 to Step 4 until convergence is achieved.

7.9. Experimental Results

Numerical results presented in this section are for the AGE-DG-2 scheme and SOR method for solving the two-dimensional problem subject to the periodic and Neumann boundary conditions discussed in Section 7.8. One problem is considered for each type of boundary condition.

Problem 1 – Periodic Boundary Conditions

$$\frac{\partial^2 U}{\partial x^2} + \frac{\partial^2 U}{\partial y^2} - \rho U = -(8\pi^2 + \rho)\sin 2\pi x \sin 2\pi y, \quad 0 \leq x, y \leq 1,$$

subject to the boundary condition

$$U(x,0) = u(x,1), \quad U_y(x,0) = U_y(x,1), \qquad 0 \leq x \leq 1,$$

$$U(0,y) = U(1,y), \quad U_x(0,y) = U_x(1,y), \qquad 0 \leq y \leq 1.$$

By applying the centered finite difference approximation for small h yields

$$- u_{i,j-1} - u_{i-1,j} + 4g u_{i,j} - u_{i+1,j} - u_{i,j+1}$$
$$= (8\pi^2 + \rho)h^2 \sin 2\pi x_i \sin 2\pi y_j, \quad i,j = 1, \ldots, N,$$

with $g = +(1/4)\rho h^2$. The exact solution is $U(x,y) = \sin 2\pi x \sin 2\pi y$.

In the linear system $A\underline{u} = \underline{b}$, we have the matrix A as in (7.8.5), the vector $\underline{u}$ as in (7.8.6) and the vector $\underline{b}$ as in (7.8.7) with its elements

$$b_{i,j} = (8\pi + \rho)h \sin 2\pi x_i \sin 2\pi y_j, \quad i,j = 1, \ldots, N.$$

The results for various ρ are tabulated in Tables 7.9.1–7.9.7.

Table 7.9.1 Problem 1 with $\rho = 0$

$\rho = 0$	SOR		AGE-DG-2 ($\omega = 2$)	
N^2	ω	iter.	r	iter.
100	1.27–1.32	30	1.75–2.07	12
400	1.54–1.58	57	1.06–1.11	24
1600	1.75	108	0.58	48
6400	1.85–1.86	202	0.21	38

Table 7.9.2 Problem 1 with $\rho = 1$

$\rho = 1$	SOR		AGE-DG-2 ($\omega = 2$)	
N^2	ω	iter.	r	iter.
100	1.31–1.32	29	1.75–2.09	12
400	1.54–1.55	56	1.06–1.12	24
1600	1.74–1.75	107	0.58–0.59	48
6400	1.85–1.86	199	0.21	37

Table 7.9.3 Problem 1 with $\rho = 5$

$\rho = 5$	SOR		AGE-DG-2 ($\omega = 2$)	
N^2	ω	iter.	r	iter.
100	1.25–1.26	32	1.74–2.16	12
400	1.51–1.53	54	1.11–1.12	23
1600	1.73–1.74	102	0.59–0.60	47
6400	1.85–1.86	190	0.18	41

Table 7.9.4 Problem 1 with $\rho = 10$

$\rho = 10$	SOR		AGE-DG-2 ($\omega = 2$)	
N^2	ω	iter.	r	iter.
100	1.22–1.27	32	1.99–2.11	11
400	1.47–1.53	52	1.09–1.17	23
1600	1.71–1.73	97	0.59–0.62	46
6400	1.85	180	0.18	40

Table 7.9.5 Problem 1 with $\rho = 20$

$\rho = 20$	SOR		AGE-DG-2 ($\omega = 2$)	
N^2	ω	iter.	r	iter.
100	1.27–1.42	26	1.95–2.28	11
400	1.48–1.48	47	1.12–1.23	22
1600	1.70–1.72	88	0.62–0.63	43
6400	1.84–1.85	164	0.22	32

Table 7.9.6 Problem 1 with $\rho = 50$

$\rho = 50$	SOR		AGE-DG-2 ($\omega = 2$)	
N^2	ω	*iter.*	r	*iter.*
100	1.28–1.31	19	2.22–2.56	10
400	1.47–1.50	37	1.25–1.33	19
1600	1.69–1.73	70	0.68–0.69	37
6400	1.84–1.85	130	0.24	32

Table 7.9.7 Problem 1 with $\rho = 100$

$\rho = 100$	SOR		AGE-DG-2 ($\omega = 2$)	
N^2	ω	*iter.*	r	*iter.*
100	1.26–1.29	15	2.60–3.00	9
400	1.42–1.58	29	1.44–1.49	16
1600	1.68–1.72	53	0.76–0.77	31
6400	1.84–1.85	98	0.21	30

Table 7.9.8 Problem 1 with $\rho = 200$

$\rho = 200$	SOR		AGE-DG-2 ($\omega = 2$)	
N^2	ω	*iter.*	r	*iter.*
100	1.10–1.29	12	3.19–3.79	8
400	1.47–1.49	20	1.72–1.77	13
1600	1.66–1.72	37	0.88–0.92	25
6400	1.82–1.85	68	0.22	25

The results presented for Problem 1 with various values of ρ in this section are based on the numerical experiments. These results show a good improvement in terms of the number if iterations for the AGE-DG-2 scheme over the SOR method which give clear evidence of the superiority of the AGE method. These gains are significant irrespective of the value of ρ. The factors which help to speed up the convergence, especially for the larger number of points are the commutativity properties between G_1 and G_2, and when the matrix A becomes strongly diagonally dominant.

Our concern here is the determination of the optimal single parameter r. To date, no such theory exists on how this parameter might be determined. Again, we will use some similarities from the solution of the one-dimensional problem governed by periodic boundary conditions. Let us consider the starting value of N as 10. This will give the grid mesh size $h = 1/10$.

It has been shown that for the one-dimensional problem, the experimental $r \in (1, b)$, where $b = 2 + (1/2)\rho h^2$, is the largest eigenvalue. The value of r, however, is greater than b when the matrix A is strongly diagonally dominant. Since we are now solving a similar but larger problem, then for some ρ, we may expect that the value of r to fall in the interval $(1, b)$, where $b = 2 + (1/4)\rho h^2$ and for larger $\rho, r > b$.

The results for Problem 1, agree with these assumptions and for $\rho < 10$, we may consider the value of $r \in (1, b)$. For $\rho > 0$, we consider $r > b$ as a guide to the experimental value of r.

With these gains, i.e., the number of iterations, the simplicity of the method and that the value of r can easily be determined, indicates that the AGE method to solve the Helmholtz equation governed by periodic boundary conditions might well be recommended over the SOR method.

Problem 2 – Neumann Boundary Conditions

$$\frac{\partial^2 U}{\partial x^2} + \frac{\partial^2 U}{\partial y^2} - \rho U = 6 - \rho(2x^2 + y^2), \quad \rho > 0,$$

subject to the boundary conditions

$$U_y(x, 0) = 0, \quad U_y(x, 1) = 2, \qquad 0 \le x \le 1,$$

$$U_x(0, y) = 0, \quad U_x(1, y) = 4, \qquad 0 \le y \le 1.$$

By applying the central finite difference approximations, for small, h yields

$$-u_{i,j-1} - u_{i-1,j} + 4g u_{i,j} - u_{i+1,j} - u_{i,j+1} = h^2 [\rho(2x_i^2 + y_j^2) - 6], \quad i, j = 1, \ldots, N,$$

where $g = 1 + (1/4)\rho h^2$, with $h = 1/N$.

The exact solution is $U(x, y) = 2x^2 + y^2$.

In the linear system $A\underline{u} = \underline{b}$, we have the matrix A as in (7.8.16), the vector $\underline{u}$ as in (7.8.17) and the vector $\underline{b}$ as (7.8.18) with its elements

$$b_{i,N} = h^2 [\rho(2x_i^2 + y_N^2) - 6] + 4h, \quad i = 0, \ldots, N,$$

$$b_{N,j} = h^2 [\rho(2x_N^2 + y_j^2) - 6] + 8h, \quad j = 0, \ldots, N,$$

and

$$b_{i,j} = h^2 [\rho(2x_i^2 + y_j^2) - 6], \qquad i, j = 0, \ldots, N-1.$$

The results for various ρ are tabulated in Tables 7.9.9–7.9.16.

Table 7.9.9 Problem 2 with $\rho = 1$

$\rho = 1$	SOR		AGE-DG-2 $(\omega = 2)$	
$(N+1)^2$	ω	iter.	r	iter.
100	1.86	77	0.09	54
400	1.93	167	0.01	49
1600	1.97	340	0.005	99
6400	1.984	567	0.002	186

Table 7.9.10 Problem 2 with $\rho = 5$

$\rho = 5$	SOR		AGE-DG-2 ($\omega = 2$)	
$(N+1)^2$	ω	*iter.*	r	*iter.*
100	1.72–1.73	41	0.35	32
400	1.85	82	0.07–0.08	48
1600	1.93	171	0.02	66
6400	1.94	349	0.01	132

Table 7.9.11 Problem 2 with $\rho = 10$

$\rho = 10$	SOR		AGE-DG-2 ($\omega = 2$)	
$(N+1)^2$	ω	*iter.*	r	*iter.*
100	1.62	32	0.50–0.54	25
400	1.80	65	0.14–0.17	36
1600	1.90	130	0.04	61
6400	1.95	235	0.02	119

Table 7.9.12 Problem 2 with $\rho = 20$

$\rho = 20$	SOR		AGE-DG-2 ($\omega = 2$)	
$(N+1)^2$	ω	*iter.*	r	*iter.*
100	1.50–1.51	26	0.72–0.80	19
400	1.74	52	0.26–0.27	29
1600	1.87	101	0.07	43
6400	1.93	204	0.04	87

Table 7.9.13 Problem 2 with $\rho = 50$

$\rho = 50$	SOR		AGE-DG-2 ($\omega = 2$)	
$(N+1)^2$	ω	*iter.*	r	*iter.*
100	1.32	20	1.20–1.29	13
400	1.60	40	0.47–0.51	22
1600	1.78	81	0.14–1.18	32
6400	1.89	161	0.07	58

Table 7.9.14 Problem 2 with $\rho = 100$

$\rho = 100$	SOR		AGE-DG-2 ($\omega = 2$)	
$(N+1)^2$	ω	*iter.*	r	*iter.*
100	1.26	16	1.80–1.86	10
400	1.52	31	0.71–0.79	17
1600	1.72	60	0.25	25
6400	1.85	116	0.11–0.12	46

Table 7.9.15 Problem 2 with $\rho = 200$

$\rho = 200$	SOR		AGE-DG-2 ($\omega = 2$)	
$(N+1)^2$	ω	iter.	r	iter.
100	1.81–1.20	12	2.31–3.49	9
400	1.39–1.41	23	1.03–1.21	13
1600	1.64	42	0.42–0.43	20
6400	1.79	80	0.16–0.17	33

Table 7.9.16 Problem 2 with $\rho = 400$

$\rho = 400$	SOR		AGE-DG-2 ($\omega = 2$)	
$(N+1)^2$	ω	iter.	r	iter.
100	1.09–1.13	9	3.60–5.80	8
400	1.27–1.29	16	1.55–1.76	10
1600	1.52–1.53	30	0.65–0.72	16
6400	1.72	56	0.21–0.27	24

The results repeat the impressive performance by the AGE method over the SOR method as shown in the one-dimensional problems governed by Neumann boundary conditions. In addition, the gains in terms of the number of iterations are significant irrespective of the value of ρ.

Our concern is again to find a suitable method to determine the optimal single parameter r, theoretically, that would become a guide to determine the value of r, experimentally. It has been shown that for the one-dimensional problems, $r \in [a, \sqrt{ab}]$ for smaller ρ, and $r \in [\sqrt{ab}, b]$ for larger values of ρ. However, this assumption is not valid for the two-dimensional problem as the relation $r = \sqrt{ab}$ no longer exists.

Instead, we will consider the relation $r = (a+b)/2$. Since we are solving similar but higher-dimensional problems, we may expect that the value of r would either be in $[a, (a+b)/2]$ for smaller values of ρ, or in $[(a+b)/2, b]$ for larger ρ. The results in Table 7.9.17 confirm this argument.

Table 7.9.17 Problem 2 the range for r

$N=9$	AGE-DG-2	The eigenvalues		
ρ	r	a	$(a+b)/2$	b
1	0.09	0.003	1.504	3.005
5	0.50–0.54	0.015	1.521	3.026
10	0.50–0.54	0.031	1.541	3.052
20	0.72–0.80	0.062	1.582	3.103
50	1.20–1.29	0.154	1.707	3.259
100	1.80–1.86	0.309	1.915	3.521
200	2.31–3.49	0.617	2.336	4.055
400	3.60–5.80	1.235	3.194	5.153

The results show that for $\rho \geq 100$, the value of r is within the interval $[a,(a+b)/2]$. For larger values of ρ, r is in $[(a+b)/2,b]$. By having this starting value, the value of r for larger numbers of points, can easily be determined, i.e., this value is in a smaller range than the previous interval.

It should be noticed that, these results do not include the case when $\rho=0$ as the matrix A will yield a zero eigenvalue. Consequently, the matrix A is singular and the computed solution will not converge to the exact solution but to the corresponding eigenvector of the matrix A. The results obtained for other values of ρ are only from one problem but it is sufficient. A given problem differs only on the right hand side vector $\underline{b}$, hence, it will only affect the number of iterations.

With these findings, the AGE method can be considered to be better than the SOR method, and be recommended to solve the Helmholtz equation in two dimensions for periodic and Neumann boundary conditions.

7.10. The Complete AGE-DG Computational Form

It has been shown in Chapter 5 that the solution of the two-point boundary-value problem with alternative forms give an improvement on the CPU time. In this section, we will investigate further the forms given in Chapter 5 to solve the two and three dimensional problems.

With the limitation of the method, our discussion in Sections 7.3 and 7.4 was mainly concerned with the AGE-DG and AGE-GT schemes. As shown in the tables of these two schemes the AGE-DG performed better in terms of computational work. Thus, we would expect more laborious work if the AGE-GT scheme is rearranged into other forms. Thus, we will confine our investigation only to the AGE-DG scheme.

Alternative Computational Form of the AGE-DG Method
Let us recall the respective Algorithms 7.3.1 and 7.3.2 for the AGE-DG-2 and AGE-DG-3 schemes. In these algorithms, we have used the intermediate variables r_1 and r_2 prior to computing the vector u and its intermediate value at each step. Our aim now is to eliminate these variables so that each equation will be in a complete computational molecule form. Having eliminated r_1 and r_2 at each step, the algorithm for the AGE-DG-2 scheme in this form, i.e., the COMP-AGE-DG-2 scheme can be written as follows.

Algorithm 7.10.1.
The COMP-AGE-DG-2 scheme.

Set $\qquad u_{i,j}^{(k)}=0, \quad i,j=0,\dots,N+1, \quad \alpha=r+g, \quad d=1/(\alpha^2-1), \quad s=w-1,$

$\qquad\qquad t=\alpha-4g\omega, \quad D=d\omega, \quad A=D\alpha, \quad B=d(\alpha t+s), \quad C=d(\alpha s+t),$

$\qquad\qquad Q=dr, \quad P=Q\alpha, \quad R=d(\alpha g-1), \quad T=d(g-\alpha).$

Step 1 $\qquad$ To compute $\underline{u}^{(k+1/4)}$, $i=1, j=1$.

$\qquad\qquad$ while $j \leq N$, compute

while $i \leq N-2$, compute

$$u_{i,j}^{(k+1/4)} = Au_{i-1,j}^{(k)} + Au_{i,j-1}^{(k)} + Au_{i,j+1}^{(k)} + Bu_{i,j}^{(k)} + Cu_{i+1,j}^{(k)}$$
$$+ Du_{i+2,j}^{(k)} + Du_{i+1,j-1}^{(k)} + Du_{i+1,j+1}^{(k)} + Ab_{i,j} + Db_{i+1,j}$$

$$u_{i+1,j}^{(k+1/4)} = Du_{i-1,j}^{(k)} + Du_{i,j-1}^{(k)} + Du_{i,j+1}^{(k)} + Cu_{i,j}^{(k)} + Bu_{i+1,j}^{(k)}$$
$$+ Au_{i+2,j}^{(k)} + Au_{i+1,j-1}^{(k)} + Au_{i+1,j+1}^{(k)} + Db_{i,j} + Ab_{i+1,j}$$

$i = i+2.$

$$u_{N,j}^{(k)} = (\omega u_{N,j-1}^{(k)} + \omega u_{N-1,j}^{(k)} + tu_{N,j}^{(k)} + \omega u_{N,j+1}^{(k)} + \omega b_{N,j})/\alpha$$

$j = j+1.$

Step 2   To compute $\underline{u}^{(k+1/2)}$, $i=2, j=1$.

while $j \leq N$, compute

$$u_{1,j}^{(k+1/2)} = (ru_{1,j}^{(k+1/4)} + gu_{1,j}^{(k)})/\alpha$$

while $i \leq N-1$, compute

$$u_{1,j}^{(k+1/2)} = Pu_{1,j}^{(k+1/4)} + Qu_{i+1,j}^{(k+1/4)} + Ru_{i,j}^{(k)} + Tu_{i+1,j}^{(k)}$$
$$u_{i+1,j}^{(k+1/2)} = Qu_{i,j}^{(k+1/4)} + Pu_{i+1,j}^{(k+1/4)} + Tu_{i,j}^{(k)} + Ru_{i+1,j}^{(k)}$$

$i = i+2.$

$j = j+1.$

Step 3   To compute $\underline{u}^{(3/4)}$, $i=1, j=1$. (Change in direction).

while $i \leq N$, compute
  while $j \leq N-2$, compute

$$u_{i,j}^{(k+3/4)} = Pu_{i,j}^{(k+1/2)} + Qu_{i,j+1}^{(k+1/2)} + Ru_{i,j}^{(k)} + Tu_{i,j+1}^{(k)}$$
$$u_{i,j+1}^{(k+3/4)} = Qu_{i,j}^{(k+1/2)} + Pu_{i,j+1}^{(k+1/2)} + Tu_{i,j}^{(k)} + Ru_{i,j+1}^{(k)}$$

$j = j+2.$

$$u_{i,N}^{(k+3/4)} = (ru_{i,N}^{(k+1/2)} + gu_{i,N}^{(k)})/\alpha$$

$i = i+1.$

Step 4   To compute $\underline{u}^{(k+1)}$, $i=1, j=2$.

while $i \leq N$, compute

$$u_{i,1}^{(k+1)} = (ru_{i,1}^{(k+3/4)} + gu_{i,1}^{(k)})/\alpha$$

while $j \leq N-1$, compute

$$u_{i,1}^{(k+1)} = Pu_{i,1}^{(k+3/4)} + Qu_{i,j+1}^{(k+3/4)} + Ru_{i,j}^{(k)} + Tu_{i,j+1}^{(k)}$$
$$u_{i,j+1}^{(k+1)} = Qu_{i,j}^{(k+3/4)} + Pu_{i,j+1}^{(k+3/4)} + Tu_{i,j}^{(k)} + Su_{i,j+1}^{(k)}$$

$j = j+2.$

$i = i+1.$

Step 5   Repeat Step 1 to Step 4 until convergence is achieved.

The computational molecule for each intermediate vector $\underline{u}$ derived from Algorithm 7.10.1, can be given as follows:

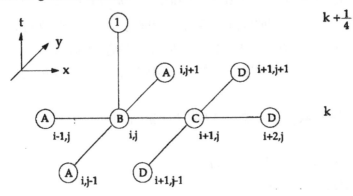

Figure 7.10.1 The computational molecule for $u_{i,j}^{(k+1/4)}$, for large N, Algorithm 7.10.1.

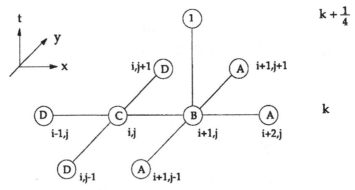

Figure 7.10.2 The computational molecule for $u_{i+1,j}^{(k+1/4)}$, for large N, Algorithm 7.10.1.

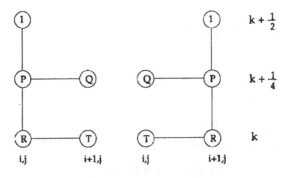

Figure 7.10.3 The computational molecule for $u_{i,j}^{(k+1/2)}$ and $u_{i,j+1}^{(k+1/2)}$ for large N, Algorithm 7.10.1.

The computation for the vectors $\underline{u}^{(3/4)}$ and $\underline{u}^{(k+1)}$ has a similar form as $\underline{u}^{(k+1/2)}$. Thus, it is sufficient to present only the molecules for this vector.

The algorithm for the AGE-DG-3 scheme in the computational form, i.e., the COMP-AGE-DG-3 scheme can be presented as follows:

Algorithm 7.10.2.
The COMP-AGE-DG-3 scheme.

Set

$$u_{i,j,k}^{(k)} = 0, \quad i,j,k = 0,\dots,N+1, \quad \alpha = r+g, \quad d = 1/(\alpha^2 - 1),$$

$$s = \omega - 1, \quad t = \alpha - 4g\omega, \quad D = d\omega, \quad A = D\alpha, \quad B = d(\alpha t + s),$$

$$C = d(\alpha s + t), \quad Q = dr, \quad P = Q\alpha, \quad R = d(\alpha g - 1), \quad T = d(g - \alpha).$$

Step 1   To compute $\underline{u}^{(k+1/6)}$. Set $i,j,k = 1$.

while $k \leq N$, compute

while $j \leq N$

while $i \leq N-2$, compute

$$u_{i,j,k}^{(k+1/6)} = Au_{i-1,j,k}^{(k)} + Au_{i,j,k-1}^{(k)} + Au_{i,j,k+1}^{(k)} + Bu_{i,j,k}^{(k)}$$
$$+ Au_{i,j-1,k}^{(k)} + Au_{i,j+1,k}^{(k)} + Du_{i+1,j,k-1}^{(k)}$$
$$+ Du_{i+1,j,k+1}^{(k)} + Cu_{i+1,j,k}^{(k)} + Du_{i+1,j-1,k}^{(k)}$$
$$+ Du_{i+1,j+1,k}^{(k)} + Du_{i+2,j,k}^{(k)} + Ab_{i,j,k} + Db_{i+1,j,k}$$

$$u_{i+1,j,k}^{(k+1/6)} = Du_{i-1,j,k}^{(k)} + Du_{i,j,k-1}^{(k)} + Du_{i,j,k+1}^{(k)} + Cu_{i,j,k}^{(k)}$$
$$+ Du_{i,j-1,k}^{(k)} + Du_{i,j+1,k}^{(k)} + Au_{i+1,j,k-1}^{(k)}$$
$$+ Au_{i+1,j,k+1}^{(k)} + Bu_{i+1,j,k}^{(k)} + Au_{i+1,j-1,k}^{(k)}$$
$$+ Au_{i+1,j+1,k}^{(k)} + Au_{i+2,j,k}^{(k)} + Db_{i,j,k} + Ab_{i+1,j,k}$$

$i = i + 2$

$$u_{N,j,k}^{(k+1/6)} = (\omega u_{N,j,k-1}^{(k)} + \omega u_{N,j-1,k}^{(k)} + \omega u_{N-1,j,k}^{(k)} + t u_{N,j,k}^{(k)}$$
$$+ \omega u_{N,j+1,k}^{(k)} + \omega u_{N,j,k+1}^{(k)} + \omega b_{N,j,k})/\alpha$$

$j = j + 1$

$k = k + 1$.

Step 2   To compute $\underline{u}^{(k+1/3)}$. Set $i = 2, j, k = 1$.

while $k \leq N$ compute

while $j \leq N$, compute

$$u_{i,j,k}^{(k+1/3)} = (r u_{1,j,k}^{(k+1/6)} + g u_{1,j,k}^{(k)})/\alpha$$

while $i \leq N-1$, compute

$$u_{i,j,k}^{(k+1/3)} = P u_{i,j,k}^{(k+1/6)} + Q u_{i+1,j,k}^{(k+1/6)} + R u_{i,j,k}^{(k)} + T u_{i+1,j,k}^{(k)}$$

$$u_{i+1,j,k}^{(k+1/3)} = Q u_{i,j,k}^{(k+1/6)} + P u_{i+1,j,k}^{(k+1/6)} + T u_{i,j,k}^{(k)} + R u_{i+1,j,k}^{(k)}$$

$i = i + 2$

$j = j + 1$

$k = k + 1$.

Step 3 To compute $\underline{u}^{(k+1/2)}$. Set $i, j, k = 1$. (Change in direction)

while $i \le N$ compute

while $k \le N$

while $j \le N - 1$, compute

$$u_{i,j,k}^{(k+1/2)} = Pu_{i,j,k}^{(k+1/3)} + Qu_{i,j+1,k}^{(k+1/3)} + Ru_{i,j,k}^{(k)} + Tu_{i,j+1,k}^{(k)}$$

$$u_{i,j+1,k}^{(k+1/2)} = Qu_{i,j,k}^{(k+1/3)} + Pu_{i,j+1,k}^{(k+1/3)} + Tu_{i,j,k}^{(k)} + Ru_{i,j+1,k}^{(k)}$$

$$j = j + 2$$

$$u_{i,N,k}^{(k+1/2)} = (ru_{i,N,k}^{(k+1/3)} + gu_{i,N,k}^{(k)})/\alpha$$

$$k = k + 1$$

$$i = i + 1.$$

Step 4 To compute $\underline{u}^{(k+2/3)}$. Set $i, k = 1, j = 2$.

while $i \le N$ compute

while $k \le N$, compute

$$u_{i,1,k}^{(k+2/3)} = (ru_{i,1,k}^{(k+1/2)} + gu_{i,1,k}^{(k)})/\alpha$$

while $j \le N - 1$, compute

$$u_{i,j,k}^{(k+2/3)} = Pu_{i,j,k}^{(k+1/2)} + Qu_{i,j+1,k}^{(k+1/2)} + Ru_{i,j,k}^{(k)} + Tu_{i,j+1,k}^{(k)}$$

$$u_{i,j+1,k}^{(k+2/3)} = Qu_{i,j,k}^{(k+1/2)} + Pu_{i,j+1,k}^{(k+1/2)} + Tu_{i,j,k}^{(k)} + Ru_{i,j+1,k}^{(k)}$$

$$j = j + 2$$

$$k = k + 1$$

$$i = i + 1.$$

Step 5 To compute $\underline{u}^{(k+5/6)}$. Set $i, j, k = 1$. (Change in direction)

while $j \le N$, compute

while $i \le N$

while $k \le N - 1$, compute

$$u_{i,j,k}^{(k+5/6)} = Pu_{i,j,k}^{(k+2/3)} + Qu_{i,j,k+1}^{(k+2/3)} + Ru_{i,j,k}^{(k)} + Tu_{i,j,k+1}^{(k)}$$

$$u_{i,j,k+1}^{(k+5/6)} = Qu_{i,j,k}^{(k+2/3)} + Pu_{i,j,k+1}^{(k+2/3)} + Tu_{i,j,k}^{(k)} + Ru_{i,j,k+1}^{(k)}$$

$$k = k + 2$$

$$u_{i,j,N}^{(k+5/6)} = (ru_{i,j,N}^{(k+2/3)} + gu_{i,j,N}^{(k)})/\alpha$$

$$i = i + 1$$

$$j = j + 1.$$

Step 6 To compute $\underline{u}^{(k+1)}$. Set $i, j = 1, k = 2$.

while $j \le N$ compute

while $i \le N$, compute

$$u_{i,j,1}^{(k+1)} = (ru_{i,j,1}^{(k+5/6)} + gu_{i,j,1}^{(k)})/\alpha$$

while $k \leq N - 1$, compute

$$u_{i,j,k}^{(k+1)} = Pu_{i,j,k}^{(k+5/6)} + Qu_{i,j,k+1}^{(k+5/6)} + Ru_{i,j,k}^{(k)} + Tu_{i,j,k+1}^{(k)}$$

$$u_{i,j,k+1}^{(k+1)} = Qu_{i,j,k}^{(k+5/6)} + Pu_{i,j,k+1}^{(k+5/6)} + Tu_{i,j,k}^{(k)} + Ru_{i,j,k+1}^{(k)}$$

$$k = k + 2$$

$$i = i + 1$$

$$j = j + 1.$$

Step 7 Repeat Step 1 to Step 6 until convergence is achieved.

The computational molecule for each intermediate vector u derived from Algorithm 7.10.2, can be given as follows:

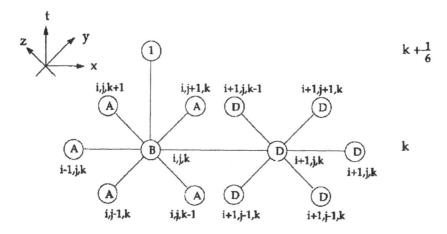

Figure 7.10.4 The computational molecule for $u_{i,j,k}^{(k+1/6)}$, for large N, Algorithm 7.10.2.

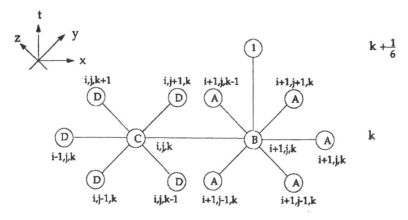

Figure 7.10.5 The computational molecule for $u_{i+1,j,k}^{(k+1/6)}$, for large N, Algorithm 7.10.2.

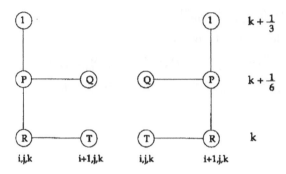

Figure 7.10.6 The computational molecule for $u_{i+1,j,k}^{(k+1/3)}$, and $u_{i,j,k}^{(k+1/3)}$ for large N Algorithm 7.10.2.

From the Algorithm 7.10.2, we find that the vectors $\underline{u}^{(k+1/2)}, \underline{u}^{(k+2/3)}, \underline{u}^{(k+5/6)}$ and $\underline{u}^{(k+1)}$ have a similar computational form as $\underline{u}^{(k+1/3)}$. Thus, it is sufficient to present only the molecules from this vector.

Before we summarise our findings, i.e., the number of operations per iteration for this scheme, let us consider the other form of presentation of the AGE-DG scheme, called the coupled AGE (CAGE) method.

The CAGE Form of the AGE-DG Method

We now consider the solution of vector u in a coupled form, where the four equations and six equations in the respective AGE-DG-2 and AGE-DG-3 schemes are made up into two and three equations. In other words, the first equation is combined with the second equation, the third with the fourth and the fifth with the sixth (in the AGE-DG-3 scheme).

Let us recall the AGE-DG-2 scheme, i.e., equations (7.3.10–7.3.13) which can be written explicitly as

$$\underline{u}^{(k+1/4)} = [I - \omega(rI + G_1)^{-1}A]\underline{u}^{(k)} + \omega(rI + G_1)^{-1}\underline{b}, \qquad (7.10.1)$$

$$\underline{u}^{(k+1/2)} = (rI + G_2)^{-1}[r\underline{u}^{(k+1/4)} + G_2\underline{u}^{(k)}], \qquad (7.10.2)$$

$$\underline{u}_c^{(k+3/4)} = (rI + G_1)^{-1}[r\underline{u}_c^{(k+1/2)} + G_1\underline{u}_c^{(k)}], \qquad (7.10.3)$$

$$\underline{u}_c^{(k+1)} = (rI + G_2)^{-1}[r\underline{u}_c^{(k+3/4)} + G_2\underline{u}_c^{(k)}]. \qquad (7.10.4)$$

In the coupled form, equations (7.10.1–7.10.4) can be explicitly combined to give a set of new equations, i.e., the CAGE-DG-2 scheme,

$$\underline{u}^{(k+1/2)} = [I - \omega r(rI + G_2)^{-1}(rI + G_1)^{-1}A]\underline{u}^{(k)}$$
$$+ \omega r(rI + G_2)^{-1}(rI + G_1)^{-1}\underline{b}, \qquad (7.10.5)$$

$$\underline{u}_c^{(k+1)} = r^2(rI + G_2)^{-1}(rI + G_1)^{-1}\underline{w}_c^* + \underline{u}_c^{(k)}, \qquad (7.10.6)$$

where $\underline{w}_c^* = \underline{u}_c^{(k+1/2)} - \underline{u}_c^{(k)}$.

Let us consider A and $(G_1 + G_2)$ as given in (7.1.11) and (7.1.15) respectively. Obviously, for $\omega = 2$, equation (7.10.5) is similar to the CAGE in the one dimensional problem, i.e., see Chapter 5, except that the matrices A, G_1 and G_2 are derived from a two dimensional problem. Thus, we may expect that the matrix

$$C_r = I - \omega r (rI + G_2)^{-1}(rI + G_1)^{-1}A, \qquad (7.10.7)$$

at least has the form of the CAGE matrix with similar forms for

$$D_r = \omega r (rI + G_2)^{-1}(rI + G_1)^{-1} \text{ and } E_r = r^2(rI + G_2)^{-1}(rI + G_1)^{-1}. \qquad (7.10.8)$$

We now write the matrices C_r, D_r and E_r for the case when $N = 7$.

From (7.1.15), for general N (odd and for any $r > 0$), we have

$$(rI + G_1)^{-1} = \begin{bmatrix} \tilde{G}_1^{-1} & & & \\ & \tilde{G}_1^{-1} & & 0 \\ & & \ddots & \\ & 0 & & \tilde{G}_1^{-1} \\ & & & & \tilde{G}_1^{-1} \end{bmatrix}_{N^2 \times N^2}, \quad \tilde{G}_1^{-1} = \begin{bmatrix} \alpha d & d & & & \\ d & \alpha d & & & \\ \hline & & \alpha d & d & \\ & & d & \alpha d & \\ \hline & & & & 1/\alpha \end{bmatrix}_{N \times N},$$

$$(7.10.9)$$

$$(rI + G_2)^{-1} = \begin{bmatrix} \tilde{G}_2^{-1} & & & \\ & \tilde{G}_2^{-1} & & 0 \\ & & \ddots & \\ & 0 & & \tilde{G}_2^{-1} \\ & & & & \tilde{G}_2^{-1} \end{bmatrix}_{N^2 \times N^2}, \quad \tilde{G}_2^{-1} = \begin{bmatrix} 1/\alpha & & & \\ \hline & \alpha d & d & \\ & d & \alpha d & \\ \hline & & & \ddots & \\ & & & \alpha d & d \\ & & & d & \alpha d \end{bmatrix}_{N \times N},$$

$$(7.10.10)$$

where $\alpha = r + g$ and $d = 1/(\alpha^2 - 1)$.

Now, we multiply the matrices $(rI + G_2)^{-1}$ and $(rI + G_1)^{-1}$ to give

$$(rI + G_2)^{-1}(rI + G_1)^{-1} = \begin{bmatrix} \tilde{G}_2^{-1} & & & \\ & \tilde{G}_2^{-1} & & 0 \\ & & \ddots & \\ & 0 & & \tilde{G}_2^{-1} \\ & & & & \tilde{G}_2^{-1} \end{bmatrix} \begin{bmatrix} \tilde{G}_1^{-1} & & & \\ & \tilde{G}_1^{-1} & & 0 \\ & & \ddots & \\ & 0 & & \tilde{G}_1^{-1} \\ & & & & \tilde{G}_1^{-1} \end{bmatrix}_{N^2 \times N^2},$$

$$= \begin{bmatrix} \tilde{G}_2^{-1}\tilde{G}_1^{-1} & & & \\ & \tilde{G}_2^{-1}\tilde{G}_1^{-1} & & 0 \\ & & \ddots & \\ & 0 & & \tilde{G}_2^{-1}\tilde{G}_1^{-1} \\ & & & & \tilde{G}_2^{-1}\tilde{G}_1^{-1} \end{bmatrix}_{N^2 \times N^2}, \qquad (7.10.11)$$

where for the case $N = 7$,

$$\tilde{G}_2^{-1}\tilde{G}_1^{-1} = \begin{bmatrix} d & S_4 & & & & & \\ S_2 & S_3 & S_2 & S_1 & & & \\ S_1 & S_2 & S_3 & S_2 & & & \\ & & S_2 & S_3 & S_2 & S_1 & \\ & & S_1 & S_2 & S_3 & S_2 & \\ & & & & S_2 & S_3 & S_4 \\ & & & & S_1 & S_2 & d \end{bmatrix}_{7\times 7} , \qquad (7.10.12)$$

where $S_1 = d^2$, $S_2 = S_1\alpha$, $S_3 = S_2\alpha$ and $S_4 = d/\alpha$. Clearly,

$$D_r = \omega r \tilde{G}_2^{-1}\tilde{G}_1^{-1} \text{ and } E_r = r^2 \tilde{G}_2^{-1}\tilde{G}_1^{-1} . \qquad (7.10.13)$$

Now, for the matrix C_r with the matrix A as in (7.1.11), we have

$$C_r = I - \omega r(rI + G_2)^{-1}(rI + G_1)^{-1}A$$

$$= \begin{bmatrix} A_r & -D_r & & & & & \\ -D_r & A_r & -D_r & & & \mathbf{0} & \\ & -D_r & A_r & -D_r & & & \\ & & -D_r & A_r & -D_r & & \\ & & & -D_r & A_r & -D_r & \\ & \mathbf{0} & & & -D_r & A_r & -D_r \\ & & & & & -D_r & A_r \end{bmatrix}_{7^2 \times 7^2} , \qquad (7.10.14)$$

where

$$A_r = \begin{bmatrix} A & B & C & & & & \\ Q & R & S & T & V & & \\ T & S & R & Q & P & & \\ & P & Q & R & S & T & V \\ & & V & T & S & R & Q & P \\ & & & & P & Q & W & X \\ & & & & V & T & Y & Z \end{bmatrix}_{7\times 7} , \qquad (7.10.15)$$

with

$$d_1 = d\omega r, \quad C = d_1/\alpha, \quad V = d_1 d, \quad A = 1 + C(1 - 4g\alpha), \quad B = C(1 - 4g),$$

$$P = V\alpha, \quad Q = P(\alpha - 4g), \quad R = 1 + 2P(1 - 2g\alpha), \quad S = V(1 + \alpha^2 - 4g\alpha),$$

$$T = V(\alpha - 4g), \quad W = 1 + C + P(1 - 4g\alpha), \quad X = C(d\alpha^3 - 4g),$$

$$Y = d_2[1 + d(1 - 4g\alpha)], \quad Z = 1 + P - 4d_1 g.$$

The computation that involves the vector $\underline{b}$ can be performed outside the iteration loop and can be assigned to a single array. We now write the algorithm for the CAGE-DG-2 scheme for general N (odd).

Algorithm 7.10.3.

The CAGE-DG-2 scheme.

Set $\qquad u_{i,j}^{(k)} = 0, i,j = 0, \ldots, N+1, \quad \alpha = r + g, \quad d = 1/(\alpha^2 - 1),$

and other coefficients as given in (7.10.11)–(7.10.14).

Step 1 To compute the vector $\underline{v}$. Set $i = 2, \ j = 1.$

while $j \leq N$, compute

$$v_{1,j} = \omega r(db_{1,j} + S_4 b_{2,j})$$

while $i \leq N - 3$, compute

$$v_{1,j} = \omega r(S_2 b_{i-1,j} + S_3 b_{i,j} + S_2 b_{i+1,j} + S_1 b_{i+2,j})$$
$$v_{i+1,j} = \omega r(S_1 b_{i-1,j} + S_2 b_{i,j} + S_3 b_{i+1,j} + S_2 b_{i+2,j})$$
$$i = i + 2$$

$$v_{N-1,j} = \omega r(S_2 b_{N-1,j} + S_3 b_{N,j} + S_4 b_{N+1,j})$$
$$v_{N,j} = \omega r(S_1 b_{N-1,j} + S_2 b_{N,j} + db_{N,j})$$
$$j = j + 1.$$

Step 2 To compute $\underline{u}^{(k+1/2)}$. Set $i = 2, \quad j = 1.$

while $j \leq N$, compute

$$
\begin{aligned}
u_{1,j}^{(k+1/2)} = {} & -du_{1,j-1}^{(k)} - S_4 u_{2,j-1}^{(k)} + A u_{1,j}^{(k)} + B u_{2,j}^{(k)} + C u_{3,j}^{(k)} \\
& - du_{1,j+1}^{(k)} - S_4 u_{2,j+1}^{(k)}
\end{aligned}
$$

while $i \leq N - 3$, compute

$$
\begin{aligned}
u_{i,j}^{(k+1/2)} = {} & -S_2 u_{i-1,j-1}^{(k)} - S_3 u_{i,j-1}^{(k)} - S_2 u_{i+1,j-1}^{(k)} \\
& - S_1 u_{i+2,j-1}^{(k)} + P u_{i-2,j}^{(k)} + Q u_{i-1,j}^{(k)} + R u_{i,j}^{(k)} \\
& + S u_{i+1,j}^{(k)} + T u_{i+2,j}^{(k)} + V u_{i+3,j}^{(k)} + S_2 u_{i-1,j+1}^{(k)} \\
& - S_3 u_{i,j+1}^{(k)} + S_2 u_{i+1,j+1}^{(k)} - S_1 u_{i+2,j+1}^{(k)} + V_{i,j}
\end{aligned}
$$

$$
\begin{aligned}
u_{i+1,j}^{(k+1/2)} = {} & -S_1 u_{i-1,j-1}^{(k)} - S_2 u_{i,j-1}^{(k)} - S_3 u_{i+1,j-1}^{(k)} \\
& - S_2 u_{i+2,j-1}^{(k)} + V u_{i-2,j}^{(k)} + T u_{i-1,j}^{(k)} + S u_{i,j}^{(k)} \\
& + R u_{i+1,j}^{(k)} + Q u_{i+2,j}^{(k)} + P u_{i+3,j}^{(k)} + S_1 u_{i-1,j+1}^{(k)} \\
& - S_2 u_{i,j+1}^{(k)} + S_3 u_{i+1,j+1}^{(k)} - S_2 u_{i+2,j+1}^{(k)} + V_{i+1,j}
\end{aligned}
$$

$$i = i + 2$$

$$
\begin{aligned}
u_{N-1,j}^{(k+1/2)} = {} & -S_2 u_{N-1,j-1}^{(k)} - S_3 u_{N,j-1}^{(k)} - S_4 u_{N+1,j-1}^{(k)} + P u_{N-2,j}^{(k)} \\
& + Q u_{N-1,j}^{(k)} + W u_{N,j}^{(k)} + X u_{N+1,j}^{(k)} - S_2 u_{N-1,j+1}^{(k)} \\
& - S_3 u_{N,j+1}^{(k)} - S_4 u_{N+1,j+1}^{(k)} + v_{N-1,j}
\end{aligned}
$$

$$u_{N,j}^{(k+1/2)} = -S_1 u_{N-1,j-1}^{(k)} - S_2 u_{N,j-1}^{(k)} - d u_{N+1,,j-1}^{(k)} + V u_{N-2,j}^{(k)}$$
$$+ T u_{N-1,j}^{(k)} + Y u_{N,j}^{(k)} + Z u_{N+1,j}^{(k)} - S_1 u_{N-1,j+1}^{(k)}$$
$$- S_2 u_{N,j+1}^{(k)} - d u_{N+1,j+1}^{(k)} + v_{N,j}$$

$j = j + 1.$

Step 3   To compute the vector $\underline{w}^*$

for $j = 1$ to N, compute

     for $i = 1$ to N, compute

         $w_{i,j}^* = u_{i,j}^{(k+1/2)} - u_{i,j}^{(k)}.$

Step 4   To compute $\underline{u}^{(k+1)}$. Set $i = 1$, $j = 2$. (Change in direction).

while $i \leq N$, compute

$$u_{i,1}^{(k+1)} = r^2 \left(d w_{i,1}^* + S_4 w_{i,2}^* \right) + u_{i,1}^{(k)}$$

while $j \leq N - 3$, compute

$$u_{i,j}^{(k+1)} = r^2 \left(S_2 w_{i,j-1}^* + S_3 w_{i,j}^* + S_2 w_{i,j+1}^* + S_1 w_{i,j+2}^* \right) + u_{i,j}^{(k)}$$
$$u_{i,j+1}^{(k+1)} = r^2 \left(S_1 w_{i,j-1}^* + S_2 w_{i,j}^* + S_3 w_{i,j+1}^* + S_2 w_{i,j+2}^* \right) + u_{i,j+1}^{(k)}$$

$j = j + 2$

$$u_{i,N-1}^{(k+1)} = r^2 \left(S_2 w_{i,N-1}^* + S_3 w_{i,N}^* + S_4 w_{i,N+1}^* \right) + u_{i,N-1}^{(k)}$$
$$u_{i,N}^{(k+1)} = r^2 \left(S_1 w_{i,N-1}^* + S_2 w_{i,N}^* + d w_{i,N}^* \right) + u_{i,N}^{(k)}$$

$i = i + 1.$

Step 5   Repeat Step 2 to Step 4 until convergence is achieved.

From Algorithm 7.10.3, it can be deduced that for large N, the computational molecules for the CAGE-DG-2 scheme are as follows:
We now extend the coupled form of the AGE-DG method to three dimensions, the CAGE-DG-3 scheme. Let us recall the AGE-DG-3 scheme, i.e., equations (7.3.24–7.3.29). In explicit form, we have

$$\underline{u}^{(k+1/6)} = [I - (rI + G_1)^{-1}A]\underline{u}^{(k)} + \omega(rI + G_1)^{-1}\underline{b}, \tag{7.10.16}$$

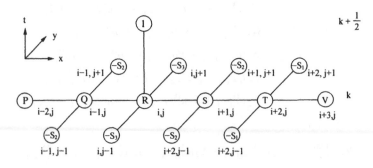

Figure 7.10.7   Computational molecules for computing $u_{i,j}^{(k+1/2)}$ for large N, Algorithm 7.10.3, the CAGE-DG-2 scheme.

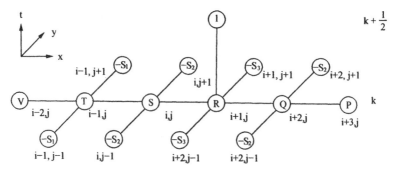

Figure 7.10.8 Computational molecules for computing $u_{i+1,j}^{(k+1/2)}$ for large N, Algorithm 7.10.3, the CAGE-DG-2 scheme.

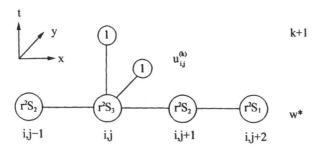

Figure 7.10.9 Computational molecules for computing $u_{i,j}^{(k+1)}$ for large N, Algorithm 7.10.3, the CAGE-DG-2 scheme.

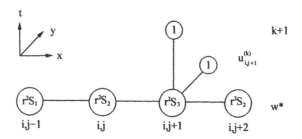

Figure 7.10.10 Computational molecules for computing $u_{i,j+1}^{(k+1)}$ for large N, Algorithm 7.10.3, the CAGE-DG-2 scheme.

$$\underline{u}^{(k+1/3)} = (rI + G_2)^{-1}[r\underline{u}^{(k+1/6)} + G_2\underline{u}^{(k)}], \qquad (7.10.17)$$

$$\underline{u}_y^{(k+1/2)} = (rI + G_1)^{-1}[r\underline{u}_y^{(k+1/3)} + G_1\underline{u}_y^{(k)}], \qquad (7.10.18)$$

$$\underline{u}_y^{(k+2/3)} = (rI + G_2)^{-1}[r\underline{u}_y^{(k+1/2)} + G_2\underline{u}_y^{(k)}], \qquad (7.10.19)$$

$$\underline{u}_z^{(k+5/6)} = (rI + G_1)^{-1}[r\underline{u}_z^{(k+2/3)} + G_1\underline{u}_z^{(k)}], \qquad (7.10.20)$$

$$\underline{u}_z^{(k+1)} = (rI + G_2)^{-1}[r\underline{u}_z^{(k+5/6)} + G_2\underline{u}_z^{(k)}]. \qquad (7.10.21)$$

In the coupled form, equations (7.10.16–7.10.21) can be combined explicitly to give a set of new equations, i.e., the CAGE-DG-3 scheme as

$$\underline{u}^{(k+1/3)} = [I - \omega r(rI + G_2)^{-1}(rI + G_1)^{-1}A]\underline{u}^{(k)}$$

$$+ \omega r(rI + G_2)^{-1}(rI + G_1)^{-1}\underline{b}, \tag{7.10.22}$$

$$\underline{u}_y^{(k+2/3)} = r^2(rI + G_2)^{-1}(rI + G_1)^{-1}\underline{w}_y^* + \underline{u}_y^{(k)}, \tag{7.10.23}$$

$$\underline{u}_z^{(k+1)} = r^2(rI + G_2)^{-1}(rI + G_1)^{-1}\underline{w}_z^{**} + \underline{u}_z^{(k)}, \tag{7.10.24}$$

where

$$\underline{w}_y^* = [\underline{u}_y^{(k+1/3)} - \underline{u}_y^{(k)}] \text{ and } \underline{w}_z^{**} = [\underline{u}_z^{(k+2/3)} - \underline{u}_z^{(k)}].$$

Thus, for the CAGE-DG-3 scheme, the computational molecules for equations (7.10.22) are expected to be similar to the ones derived in Figures 7.10.7 and 7.10.8 with the additional complexity of the z direction.

For equations (7.10.23) and (7.10.24), we would expect the computational molecules to be given as Figures 7.10.9 and 7.10.10. The molecules for computing $u_{i,j}^{(k+1/3)}$ and $u_{i+1,j}^{(k+1/3)}$ can be illustrated as follows:

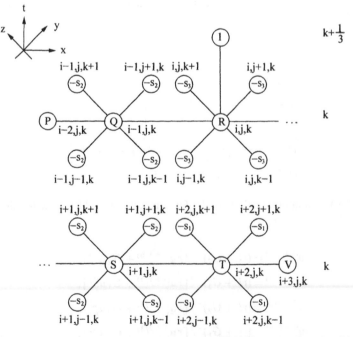

Figure 7.10.11 Computational molecules for computing $u_{i,j}^{(k+1/3)}$, for large N, the CAGE-DG-3 scheme.

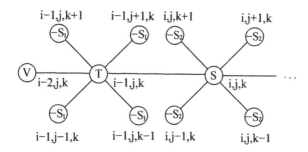

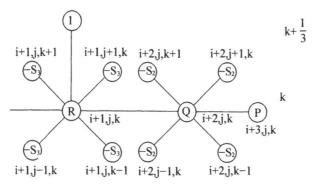

Figure 7.10.12 Computational molecules for computing $u_{i,j}^{(k+1/3)}$, for large N, the CAGE-DG-3 scheme.

The Computational Complexity

In this section, our concern is to find the best computational form which gives the fastest CPU time. In other words, we will determine the computational scheme that needs the least amount of mathematical operations in each iteration. Table 7.10.1 below summarises our findings.

Table 7.10.1 The number of operations per iteration

Method	Multiplication	Addition	Overall
AGE-DG-2	$17(N-1)^2$	$15(N-1)^2$	$32(N-1)^2$
COMP-AGE-DG-2	$16(N-1)^2$	$18(N-1)^2$	$34(N-1)^2$
CAGE-DG-2	$12(N-3)^2$	$18(N-1)^2$	$30(N-1)^2$
AGE-DG-3	$25(N-1)^3$	$23(N-1)^3$	$48(N-1)^3$
COMP-AGE-DG-3	$24(N-1)^3$	$28(N-1)^3$	$52(N-1)^3$
CAGE-DG-3	$15(N-1)^3$	$30(N-1)^3$	$45(N-1)^3$

The result indicates that there is a difference in the amount of work needed for each scheme. It appears that the COMP-AGE-DG schemes are less effective since the schemes take more work (overall) when compared with the other schemes. Although the CAGE-DG schemes give the best performance for the computational work, symbolically, these schemes require more mathematical work to obtain. Thus, it appears that the best scheme is the standard AGE-DG-2 and AGE-DG-3 schemes.

7.11. Summary

The extension of the AGE-DG scheme for solving the two- and three-dimensional elliptic problems governed by different boundary conditions was the main concern in this chapter. The results obtained not only show that the extension is viable, but also, gives further clear evidence of the superiority of the AGE method over the SOR method. Although, the SOR method needs less computational effort, the simplicity of the AGE method, makes the method as powerful as SOR.

The focal point in the early sections was to ascertain the consistency of the method and that the method converges to the exact solution when solving the problems subject to Dirichlet boundary conditions. This has been shown in detail on how the convergence is achieved. The other interesting matter is the determination of the theoretical optimal parameter, r. By applying a simple assumption, this parameter is shown to be closely related to the theoretical optimal parameters for the one-dimensional problem. As theory is not available to date, this assumption can be regarded as a good approximation for determining the experimental value of r.

8. Parallel Implementation of the Alternating Group Explicit (AGE) Method

8.1. Introduction

Systems of tridiagonal equations frequently arise in practical applications related to solving ordinary and partial differential equations by discrete numerical methods. Consider the parabolic equation

$$\frac{\partial U}{\partial t} = \frac{\partial^2 U}{\partial x^2}, \quad 0 < x < 1, \tag{8.1.1}$$

with initial and boundary conditions:

$$U(x,0) = f(x),$$

$$U(0,t) = g(t), \tag{8.1.2}$$

$$U(1,t) = q(t).$$

Let

$$\Delta x = 1/(n+1), \qquad \Delta t = M/m, \tag{8.1.3}$$

and

$$x_i = i\Delta x, \quad i = 0, 1, \ldots, n+1,$$

$$t_j = j\Delta t, \quad j = 0, 1, \ldots, m,$$

$$u_{i,j} = u(x_i, t_j),$$

where M is a positive number. Then the Crank–Nicolson implicit finite difference equation to (8.1.1) is given, for $i = 1, 2, \ldots, n$, by

$$\frac{u_{i,j+1} - u_{i,j}}{\Delta t} = \frac{1}{2}\left[\frac{u_{i+1,j+1} - 2u_{i,j+1} + u_{i-1,j+1}}{\Delta x^2} + \frac{u_{i+1,j} - 2u_{i,j} + u_{i-1,j}}{\Delta x^2}\right]$$

giving the system

$$A\underline{u}_{j+1} = B\underline{u}_j + \underline{\delta}_j,$$

435

where

$$A = \text{diag}(-1, a, -1) = \begin{bmatrix} a & -1 & & & & \\ -1 & a & -1 & & 0 & \\ & & & & & \\ & & & & & \\ & 0 & & -1 & a & -1 \\ & & & & -1 & a \end{bmatrix}, \qquad (8.1.4)$$

and

$$B = \text{diag}(1, c, 1),$$

$$\underline{u}_j^T = (u_{1,j}, u_{2,j}, \dots, u_{n,j}),$$

$$\underline{\delta}_j^T = (g_{j+1} + g_j, 0, \dots, 0, q_{j+1} + q_j).$$

Here, $g_j = g(t_j)$, $q_j = q(t_j)$, $c = (2/r) - 2$ and,

$$r = \Delta t / \Delta x^2, \qquad a = (2/r) + 2. \qquad (8.1.5)$$

Since u_0 is known from the initial condition, $\underline{u}_1$ can be found from the above linear system, then $\underline{u}_2$ and so on. Therefore we need to solve many tridiagonal systems of linear equations

$$A\underline{u} = \underline{b}, \qquad (8.1.6)$$

where the matrix A is given by (8.1.4).

The direct methods (such as Gauss elimination) for the above system (8.1.6) often lead to sequential nonlinear recurrences that have to be evaluated term by term. These algorithms are more ideally suited for serial computers.

The iterative methods of solutions are frequently used for time dependent problems. There, a solution of one system of linear equations at the jth-time level is often a very good starting vector for the system at the $(j+1)$-time level. An explicit iterative method for solving such systems was introduced by Evans (1985) and is known as the Alternating Group Explicit (AGE) method analogous to the well known ADI method (1955). The AGE method is simple conceptually and leads to a straight-forward implementation on a MIMD computer such as the Sequent Balance.

In this paper, the standard AGE iterative procedure is reorganised so that the computational work is saved by upto 25%, which is confirmed by our numerical results. Then the parallel implementation on a Sequent Balance is given. Numerical experiments show that the AGE method has a good performance on multiprocessors and a near linear speed-up is obtained. Finally, further extensions for the AGE method are presented.

8.2. Alternating Group Explicit (AGE) Method

The Standard AGE Iteration

Suppose n is even (we can usually choose n to be even). Let

$$C = \begin{bmatrix} e & -1 \\ -1 & e \end{bmatrix}, \qquad e = a/2, \tag{8.2.1}$$

then the matrix A can be split into

$$A = G_1 + G_2. \tag{8.2.2}$$

Here

$$G_1 = \text{diag}(C, \dots, C), \qquad G_2 = \text{diag}(e, C, \dots, C, e). \tag{8.2.3}$$

Thus, the AGE method for (8.1.6) can be expressed as follows:

$$(G_1 + sI)\underline{u}^{(k+1/2)} = \underline{b} - (G_2 - sI)\underline{u}^{(k)},$$
$$(G_2 + sI)\underline{u}^{(k+1)} = \underline{b} - (G_1 - sI)\underline{u}^{(k+1/2)}, \tag{8.2.4}$$

where s is the acceleration parameter. Since G_1 and G_2 are symmetric and positive definite, the AGE iterative procedure (8.2.4) converges to the solution $u = A^{-1}b$ for any positive number s. Also since G_1 and G_2 have only three different eigenvalues: $e - 1$, e, and $e + 1$, therefore by the results of the classical ADI method with one parameter we have the optimum parameter s^* given by

$$s^* = \sqrt{e^2 - 1}. \tag{8.2.5}$$

However we may change the parameter s from iteration to iteration resulting in a nonstationary method, which would result in a faster convergence rate as is the case for the classical ADI method.

Note that the matrices G_1 and G_2 are diagonal blocks with the maximum order of the sub-blocks being 2. Therefore the inverses of G_1 and G_2 can be easily given in explicit form. Let

$$C_1 = \begin{bmatrix} \alpha & -1 \\ -1 & \alpha \end{bmatrix}, \qquad C_2 = \begin{bmatrix} \beta & -1 \\ -1 & \beta \end{bmatrix}, \qquad \alpha = e + s, \qquad \beta = e - s,$$

and

$$C_1^{-1} = \frac{1}{d}\begin{bmatrix} \alpha & -1 \\ -1 & \alpha \end{bmatrix}^{-1} = \begin{bmatrix} v & w \\ w & v \end{bmatrix}, \qquad d = \alpha^2 - 1, \qquad v = \alpha/d, \qquad w = 1/d.$$

Then, the procedure (8.2.4) can be expressed by the following explicit algorithm.

Algorithm 8.1.

First sweep

$$y_1 = b_1 - \beta u_1^{(k)}, \qquad y_n = b_n - \beta u_n^{(k)},$$

$$\begin{bmatrix} y_i \\ y_{i+1} \end{bmatrix} = \begin{bmatrix} b_i \\ b_{i+1} \end{bmatrix} - \begin{bmatrix} \beta & -1 \\ -1 & \beta \end{bmatrix} \begin{bmatrix} u_i^{(k)} \\ u_{i+1}^{(k)} \end{bmatrix}, \quad i = 2, 4, \ldots, n-2,$$

$$\begin{bmatrix} u_i^{(k+1/2)} \\ u_{i+1}^{(k+1/2)} \end{bmatrix} = \begin{bmatrix} v & w \\ w & v \end{bmatrix} \begin{bmatrix} y_i \\ y_{i+1} \end{bmatrix}, \quad i = 1, 3, \ldots, n-1.$$

Second sweep

$$\begin{bmatrix} z_i \\ z_{i+1} \end{bmatrix} = \begin{bmatrix} b_i \\ b_{i+1} \end{bmatrix} - \begin{bmatrix} \beta & -1 \\ -1 & \beta \end{bmatrix} \begin{bmatrix} u_i^{(k+1/2)} \\ u_{i+1}^{(k+1/2)} \end{bmatrix}, \quad i = 1, 3, \ldots, n-1,$$

$$u_1^{(k+1)} = z_1/\alpha, \qquad u_n^{(k+1)} = z_n/\alpha,$$

$$\begin{bmatrix} u_i^{(k+1)} \\ u_{i+1}^{(k+1)} \end{bmatrix} = \begin{bmatrix} v & w \\ w & v \end{bmatrix} \begin{bmatrix} z_i \\ z_{i+1} \end{bmatrix}, \quad i = 2, 4, \ldots, n-2.$$

From the above algorithm, for each sweep $3n$ multiplications and $3n-2$ additions are needed, therefore the total complexity per iteration is

$$6n \text{ multiplications and } 6n-4 \text{ additions.}$$

Reorganisation of the AGE Procedure
Now if we let

$$y = b - (G_2 - sI)u^{(k)},$$

then $u^{(k+1)}$ produced by the AGE procedure (8.2.4) can be expressed by

$$u^{(k+1)} = (G_2 + sI)^{-1}[b - (G_1 - sI)(G_1 + sI)^{-1}y]$$
$$= (G_2 + sI)^{-1}b - (G_2 + sI)^{-1}(G_1 - sI)(G_1 + sI)^{-1}y.$$

Thus if we let

$$C_3 = d^{-1}C_2C_1^{-1} = \begin{bmatrix} v' & w' \\ w' & v' \end{bmatrix}, \qquad b' = (G_2 + sI)^{-1}b, \qquad d' = d/\alpha,$$

then, the above discussion can be summarised by the following:

First sweep

$$\begin{bmatrix} y_i \\ y_{i+1} \end{bmatrix} = \begin{bmatrix} b_i \\ b_{i+1} \end{bmatrix} - \begin{bmatrix} \beta & -1 \\ -1 & \beta \end{bmatrix} \begin{bmatrix} u_i^{(k)} \\ u_{i+1}^{(k)} \end{bmatrix}, \quad i = 2, 4, \ldots, n-2.$$

Second sweep

$$\begin{bmatrix} z_i \\ z_{i+1} \end{bmatrix} = \begin{bmatrix} v' & w' \\ w' & v' \end{bmatrix} \begin{bmatrix} y_i \\ y_{i+1} \end{bmatrix}, \quad i = 1, 3, \ldots, n-1.$$

$$u_1^{(k+1)} = b_1' - d'z_1, \qquad u_n^{(k+1)} = b_1' - d'z_n,$$

$$\begin{bmatrix} u_i^{(k+1)} \\ u_{i+1}^{(k+1)} \end{bmatrix} = \begin{bmatrix} b_i' \\ b_{i+1}' \end{bmatrix} - \begin{bmatrix} \alpha & -1 \\ -1 & \alpha \end{bmatrix} \begin{bmatrix} z_i \\ z_{i+1} \end{bmatrix}, \qquad i = 2, 4, \ldots, n-2.$$

Thus it is clear that the total amount of arithmetic operations per iteration for the above algorithm are

$4n$ multiplications and $5n - 4$ additions.

Thus compared with Algorithm 1, the saving is 33.3% in multiplications and 16.6% in additions. If the multiplication and addition operations takes roughly the same CPU time (such as on the Sequent Balance) then the total saving is 25% compared with Algorithm 8.1. However, this is at the expense of one extra vector storage (i.e., for b').

8.3. Parallel Implementations

The explicit AGE algorithm leads immediately to a simple vector or parallel form. Since $\underline{u}^{(k)}$ is known, then y_i, $i = 1, 2, \ldots, n$, in the first sweep of Algorithm 2 can be computed fully in parallel. When y_i, $i = 1, 2, \ldots, n$, are known, the unknown $\underline{u}^{(k+1)}$ in the second sweep can also be computed in parallel. This is because $\underline{u}^{(k+1)}$ only depends on y_i, $i = 1, 2, \ldots, n$. In fact each $u_i^{(k+1)}$ only depends on y_{i-1}, y_i, y_{i+1} and y_{i+2}.

Suppose we have 2 processors and that operations "$+, (-)$ and $*$" take the same CPU time. Then we divide the points into two groups. We can let the first processor P_1 update the points: $u_1, u_2, \ldots, u_{n/2}$ and the second processor P_2 update the points: $u_{n/2+1}, u_{n/2+2}, \ldots, u_n$. Thus for sequential processing we need $9n - 4$ operations; while each of processors P_1 and P_2 needs $(9n/2) - 2$ operations.

Therefore if we do not consider any other influences (i.e., overheads), the speed-up would be 2. Now suppose n processors are available, then we can arrange processor P_j be responsible for updating u_j. Thus P_1 or P_n needs 7 operations; while each of the others needs 9 operations. Therefore, the speed-up would be

$$\frac{9n-4}{9} = n - \frac{4}{9} = n\left(1 - \frac{4}{9n}\right),$$

which tends to n if n is large. However a more realistic assumption is that the number of processors p is less than n. Now suppose $n = 2*p*\ell - n_c$, where ℓ and n_c are integers with n_c satisfying $0 \le n_c \le 2p$. Therefore we divide the unknown points into p groups, where in each group there are at most $2*\ell$ points. In this case, the speed-up S_p would be given by

$$S_p = \frac{n*9-4}{2*\ell*9} = p\frac{n*9-4}{(n+n_c)*9} = p\left(\frac{1-4/(n*9)}{1+n_c/n}\right),$$

which will tend to p if n is quite large.

8.4. Numerical Results

Now we consider solving the linear system (8.1.6) with A having the form (8.1.4). Since $r = \Delta t/\Delta x^2$, if we let $\Delta t = 1/50$ then $a = 2 + 100*\Delta x^2$.

First we compare Algorithm 8.1 and Algorithm 8.2 in sequential form. Theoretically, Algorithm 8.2 will save upto 25% work compared with Algorithm 8.1. This is confirmed by our results. We let $n = 100, 200, \dots, 1000$.

Then we obtain the following table.

Table 8.4.1 CPU Times (in secs) of Algorithms 2 and 1

n	100	200	300	400	500	600	700	800	900	1000
A2	1.51	6.42	14.85	26.27	42.20	62.60	86.20	113.80	147.10	177.10
A1	1.94	8.13	19.02	34.53	55.46	80.10	111.85	148.0	186.73	233.13

In the above table, the second and third rows are the CPU times in seconds for Algorithms 8.2 and 8.1, respectively. Thus when $n = 500$, for Algorithm 8.2, the saving is $1 - 42.20/55.46 = 1 - 0.76 = 0.24 = 24\%$, compared with Algorithm 8.1.

Now we consider the parallel form. We implement the AGE iteration (Algorithm 8.2) on the Sequent Balance 8000. This parallel computer has a shared memory with 9 processors. The program was written, according to Figure 8.4.1 for any value of p (the number of processors) and an even value of n. To obtain the experimental

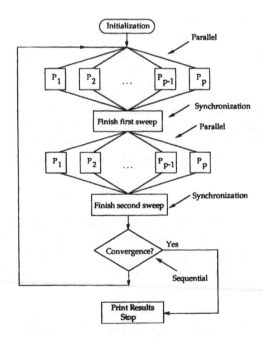

Figure 8.4.1   Flow diagram of parallel AGE algorithm.

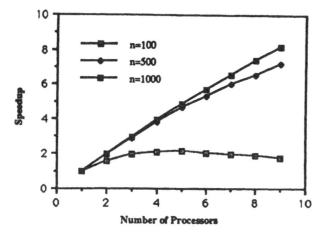

Figure 8.4.2   Diagram of speed-up vs number of processors.

we run the program for matrix A sizes 100, 200 upto 1000 and the results are listed in Figure 8.4.2.

From Figure 8.4.2 it is clear that when n is large we get an almost linear speed-up. However, when n is small, we do not get the expected speed-up as the number of processors increases. This is because of overheads such as shared memory clashing and parallel software overheads, etc., which dominate the whole computation when n is small. In general the explicit AGE method is conceptually simple and can be fully parallelised. When n is large the expected linear speed-up can be obtained.

8.5. Further Applications

We have considered the AGE methods for the special tridiagonal linear equations (8.1.6). However, the application of the AGE method is not limited to this special form. We choose the special form to simplify our discussions. Suppose we have a general tridiagonal system $A\underline{x} = \underline{b}$ with A having the form

$$A = \begin{bmatrix} a_1 & c_1 & & & & \\ b_2 & a_2 & c_2 & & 0 & \\ & & & \ddots & & \\ & 0 & & b_{n-1} & a_{n-1} & c_{n-1} \\ & & & & b_n & a_n \end{bmatrix}.$$

If we let

$$D_j = \begin{bmatrix} a_j/2 & c_j \\ b_{j+1} & a_{j+1}/2 \end{bmatrix}$$

and

$$G_1 = \text{diag}(D_1, D_3, \dots, D_{n-1}), \qquad G_2 = \text{diag}(a_1/2, D_2, D_4, \dots, D_{n-2}, a_n/2).$$

Then A has the splitting

$$A = G_1 + G_2.$$

Therefore we can establish the AGE iterative procedure (8.2.4) as before.

8.6. The Parallel AGE Fractional Method for the Elliptic Problem in Two Dimensions

The basic concepts of the AGE schemes for solving a system of equations were presented by Evans (1985). In this paper we proceed with the parallel implementation of the method.

Three strategies are investigated and implemented to solve two-dimensional boundary value problems. The strategies are mainly concerned with the way the problem to be solved is decomposed into many tasks that can be run in parallel. In the first two strategies the problem is solved by decomposing its interval into subsets and each subset is assigned to different processors which run them in parallel, whilst in the third strategy the problem is solved by decomposing its domain into partitions and each partition is assigned to different processors.

These three strategies are programmed on the SEQUENT Balance MIMD system using both the synchronous and asynchronous approach. The results from the implementations of these approaches, such as the time needed to solve the problem, number of iterations required and the 'speed-up' ratios are obtained and compared.

Using these strategies shared memory is used to hold the input, the results from the first sweep and the final output component values. These values can then be accessed by different processes. Before the process iterates on its task, it needs to read all its components first from private memory, then it releases all the values of the components for the next iteration. In the different parallel versions various mesh sizes are evaluated. The results shown are an average of many runs.

8.6.1. Problem Formulation

Consider the Dirichlet problem on the region R,

$$\frac{\partial^2 U}{\partial x^2} + \frac{\partial^2 U}{\partial y^2} = 0, \quad (x, y) \in R, \tag{8.6.1}$$

subject to the boundary conditions:

$$U(0,y) = U(1,y) = 0, \qquad 0 \le y \le 1, \tag{8.6.2a}$$

$$U(x,0) = U(x,1) = \sin \pi x, \quad 0 \le x \le 1. \tag{8.6.2b}$$

The exact solution of this problem is given by

$$U(x,y) = \operatorname{sech} \tfrac{1}{2} \pi \, \cosh(\pi(y - \tfrac{1}{2})) \sin \pi x. \tag{8.6.3}$$

By following the usual finite difference discretisation procedure equation (8.6.1) can be approximated to

$$- u_{i-1,j} - u_{i+1,j} + 4u_{i,j} - u_{i,j-1} - u_{i,j+1} = 0, \quad 1 \le i, j \le n, \tag{8.6.4}$$

and $x_i = ih$, $y_j = jh$, for $0 \le i, j \le n+1$.

We assume that R is a regular region and we order the n^2 (n is even) internal mesh points row-wise, as shown in Figure 8.6.1.

Applying (8.6.4) at each mesh point yields the system $A\underline{u} = \underline{b}$, i.e.,

$$\begin{bmatrix} B & -I & & & \\ -I & B & -I & & 0 \\ & & \ddots & & \\ 0 & & -I & B & -I \\ & & & -I & B \end{bmatrix} \begin{bmatrix} u_1 \\ \vdots \\ \\ \vdots \\ u_{n^2} \end{bmatrix} = \begin{bmatrix} \sin \pi x_1 \\ \vdots \\ \\ \vdots \\ \sin \pi x_{n^2} \end{bmatrix}, \tag{8.6.5}$$

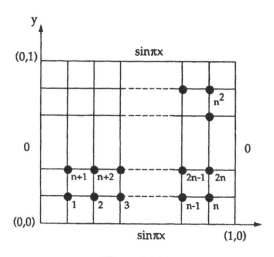

Figure 8.6.1

where

$$B = \begin{bmatrix} 4 & -1 & & & & \\ -1 & 4 & -1 & & 0 & \\ & & & & & \\ & & & -1 & 4 & -1 \\ & 0 & & & -1 & 4 \end{bmatrix}.$$

If we split A into the sum of its constituent symmetric and positive definite matrices G_1, G_2, G_3 and G_4 we have

$$A = G_1 + G_2 + G_3 + G_4, \qquad (8.6.6)$$

where G_1 and G_2 are the forward and backward differences in the x-plane and G_3 and G_4 are similar differences in the y-plane.

Then

$$G_1 + G_2 = \qquad , \qquad (8.6.7.a)$$

with $\mathrm{diag}(G_1 + G_2) = \frac{1}{2}\mathrm{diag}(A)$, and

$$G_3 + G_4 = \qquad , \qquad (8.6.7.b)$$

with $\mathrm{diag}(G_3 + G_4) = \frac{1}{2}\mathrm{diag}(A)$.

Hence

$$\mathrm{diag}(G_1) = \mathrm{diag}(G_2) = \tfrac{1}{4}\mathrm{diag}(A) \text{ and}$$

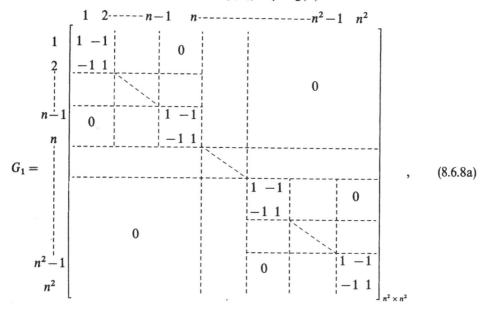

, (8.6.8a)

and

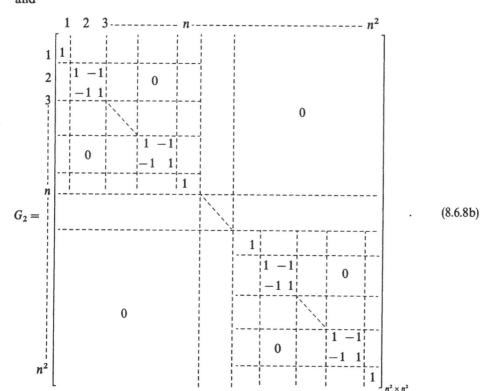

. (8.6.8b)

Also, by reordering the points column-wise, i.e., along the y-direction, then we find that G_3 and G_4 have the same structure as G_1 and G_2, respectively, i.e.,

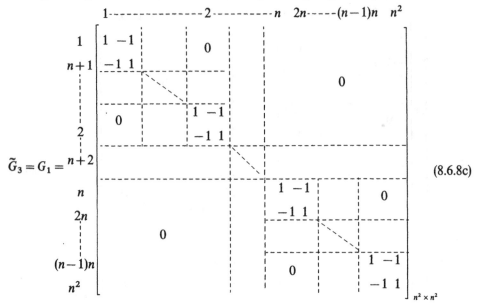

$$\tilde{G}_3 = G_1 = \tag{8.6.8c}$$

and

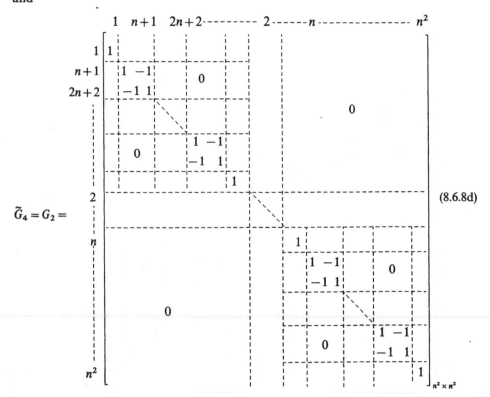

$$\tilde{G}_4 = G_2 = \tag{8.6.8d}$$

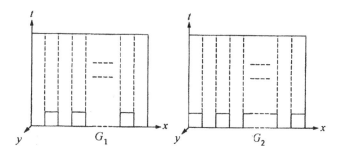

Figure 8.6.2

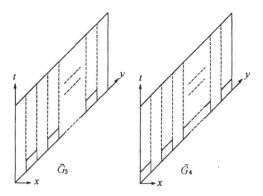

Figure 8.6.3

Figures 8.6.2 and 8.6.3 illustrate the way the points are grouped in $G_1, G_2, \tilde{G}_3$ and $\tilde{G}_4$ in the x- and y-directions, respectively.

The Operator Splitting Iterative Method

The Douglas–Rachford formula for the AGE fractional scheme takes the following form, with r a suitably chosen iteration parameter

$$(G_1+rI)\underline{u}^{(k+1/4)} = (rI-G_1-2G_2-2G_3-2G_4)\underline{u}^{(k)}+2\underline{b}, \qquad (8.6.9a)$$

$$(G_2+rI)\underline{u}^{(k+1/2)} = G_2\underline{u}^{(k)}+r\underline{u}^{(k+1/4)}, \qquad (8.6.9b)$$

$$(G_3+rI)\underline{u}^{(k+3/4)} = G_3\underline{u}^{(k)}+r\underline{u}^{(1/2)}, \qquad (8.6.9c)$$

$$(G_4+rI)\underline{u}^{(k+1)} = G_4\underline{u}^{(k)}+r\underline{u}^{(k+3/4)}. \qquad (8.6.9d)$$

We can obtain the values of $\underline{u}^{(k+1/4)}$, $\underline{u}^{(k+1/2)}$, $\underline{u}^{(k+3/4)}$ and $\underline{u}^{(k+1)}$, respectively, from

$$\underline{u}^{(k+1/4)}=(G_1+rI)^{-1}[(rI-G_1\ 2G_2-2G_3-2G_4)\underline{u}^{(k)}+2\underline{b}], \qquad (8.6.10a)$$

$$\underline{u}^{(k+1/2)}=(G_2+rI)^{-1}[G_2\underline{u}^{(k)}+r\underline{u}^{(k+1/4)}], \qquad (8.6.10b)$$

$$\underline{u}^{(k+3/4)}=(G_3+rI)^{-1}[G_3\underline{u}^{(k)}+r\underline{u}^{(k+1/2)}], \qquad (8.6.10c)$$

and

$$\underline{u}^{(k+1)} = (G_4+rI)^{-1}[G_4\underline{u}^{(k)}+r\underline{u}^{(k+3/4)}]. \qquad (8.6.10d)$$

Since the matrices (G_i+rI), $i=1,2,3,4$, are all (2×2) block submatrices, they are all

easily invertible as shown previously. To find the inverses of the above systems for
G_1, we let $w = 1 + r$, so

$$\hat{G}_1 = \begin{bmatrix} w & -1 \\ -1 & w \end{bmatrix}, \quad \text{then } \hat{G}_1 = \frac{1}{\det} \begin{bmatrix} w & 1 \\ 1 & w \end{bmatrix},$$

where $\det = (w^2 - 1)$ then

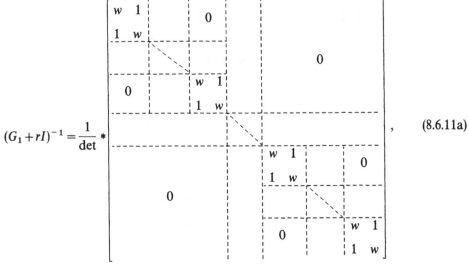

$$(G_1 + rI)^{-1} = \frac{1}{\det} * \quad , \quad (8.6.11a)$$

and

$$(G_2 + rI)^{-1} = \frac{1}{\det} * \quad .$$

(8.6.11b)

The inverses $(G_1+rI)^{-1}$ and $(G_2+rI)^{-1}$ are numbered in the order $1, 2, 3, \ldots, n$, in the x-direction. Also $(G_3+rI)^{-1}$ is similar to (8.6.11a) and $(G_4+rI)^{-1}$ is similar to (8.6.11b) but they have a different ordering, but when applied in the y-direction they become similar.

We now consider the iterative formula, equation (8.6.10) at each of the four intermediate levels:

(i) $(k+1/4)^{th}$ *time step* For the solution at the first intermediate level (the $(k+1/4)^{th}$ iterate). Equation (8.6.10a) is rearranged as

$$\underline{u}^{(k+1/4)} = (G_1+rI)^{-1}[(rI-(G_1+2G_2+2G_3+2G_4))\underline{u}^{(k)}+2\underline{b}].$$

Since the coefficients of $\underline{u}^{(k)}$, are all matrices of the same size, we can add up their elements paying attention to their direction and ordering. Hence $(G_1+2G_2+2G_3+2G_4)$ will become the matrix C:

$$C = \begin{bmatrix} F & -2I & & & & \\ -2I & F & -2I & & 0 & \\ & & & & & \\ & 0 & & -2I & F & -2I \\ & & & & -2I & F \end{bmatrix}, \qquad (8.6.12)$$

where

$$F = \begin{bmatrix} 7 & -1 & & & & \\ -1 & 7 & -2 & & & \\ & -2 & 7 & -1 & & 0 \\ & & & & & \\ & 0 & & -1 & 7 & -2 \\ & & & & -1 & 7 \end{bmatrix},$$

then, for the first step of the algorithm we have

$$\underline{u}^{(k+1/4)} = (G_1+rI)^{-1}[(rI-C)\underline{u}^{(k)}+2\underline{b}].$$

Now let $t = r-7$, where 7 is the value along the diagonal of C. Then the new coefficient matrix of $u^{(k)}$ will have the same structure as C, but will have the value t

along the diagonal. Hence

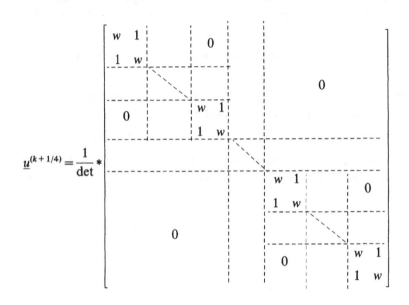

$$
*
\begin{bmatrix}
tu_1^{(k)} + u_2^{(k)} + 2u_{n+1}^{(k)} + 2\sin \pi x_1 \\
u_1^{(k)} + tu_2^{(k)} + 2u_3^{(k)} + 2u_{n+2}^{(k)} + 2\sin \pi x_2 \\
\vdots \\
2u_{n-2}^{(k)} + tu_{n-1}^{(k)} + u_n^{(k)} + 2u_{2n-1}^{(k)} + 2\sin \pi x_{n-1} \\
u_{n-1}^{(k)} + tu_n^{(k)} + 2u_{2n}^{(k)} + 2u_{n+2}^{(k)} + 2\sin \pi x_n \\
\vdots \\
2u_{n^2-2n+1}^{(k)} + tu_{n^2-n+1}^{(k)} + u_{n^2-n+2}^{(k)} + 2\sin \pi x_{n^2-n+1} \\
2u_{n^2-2n+2}^{(k)} + u_{n^2-n+1}^{(k)} + tu_{n^2-n+2}^{(k)} + 2u_{n^2-2n+3}^{(k)} + 2\sin \pi x_{n^2-n+2} \\
\vdots \\
2u_{n^2-n-2}^{(k)} + 2u_{n^2-2}^{(k)} + tu_{n^2-1}^{(k)} + tu_{n^2-1}^{(k)} + u_{n^2}^{(k)} + 2\sin \pi x_{n^2-1} \\
2u_{n^2-n-1}^{(k)} + u_{n^2-1}^{(k)} + tu_{n^2}^{(k)} + 2\sin \pi x_{n^2}
\end{bmatrix}
\qquad .(8.6.13a)
$$

By carrying out these matrix–vector operations, we obtain the values of $\underline{u}^{(k+1/4)}$ at the first intermediate level.

(ii) $(k + 1/2)^{th}$ *time step* For the solution at the second intermediate level (the $k + 1/2)^{th}$ iterate, equation (8.6.10b),

$$
\underline{u}^{(k+1/2)} = (G_2 + rI)^{-1}[G_2\underline{u}^{(k)} + r\underline{u}^{(k+1/4)}]
$$

can be written in a matrix form as

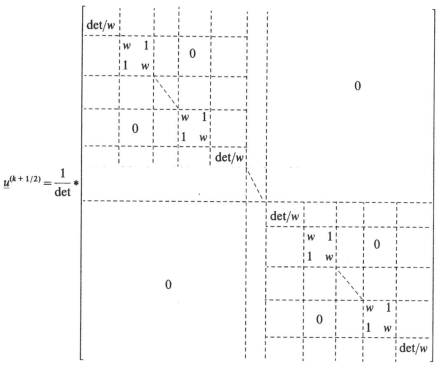

$$\underline{u}^{(k+1/2)} = \frac{1}{\det} *$$

$$* \begin{bmatrix} u_1^{(k)} + ru_1^{(k+1/4)} \\ u_2^{(k)} + u_3^{(k)} + ru_2^{(k+1/4)} \\ u_2^{(k)} + u_3^{(k)} + ru_3^{(k+1/4)} \\ \vdots \\ u_n^{(k)} + ru_n^{(k+1/4)} \\ \vdots \\ u_{n^2-n+1}^{(k)} + ru_{n^2-n+1}^{(k+1/4)} \\ u_{n^2-n+2}^{(k)} + u_{n^2-n+3}^{(k)} + ru_{n^2-n+2}^{(k+1/4)} \\ u_{n^2-n+2}^{(k)} + u_{n^2-n+3}^{(k)} + ru_{n^2-n+3}^{(k+1/4)} \\ \vdots \\ u_{n^2}^{(k)} + ru_{n^2}^{(k+1/4)} \end{bmatrix}. \qquad (8.6.13b)$$

The above matrix–vector multiplication will give the values of $\underline{u}^{(k+1/2)}$.

(iii) $(k+3/4)^{th}$ *time step*   Now the third intermediate level (the $(k+3/4)^{th}$ iterate). If we reorder the mesh points column-wise to the y-axis, equation (8.6.10c) is transformed to

$$\underline{u}_c^{(k+3/4)} = (G_1 + rI)^{-1}[G_1\underline{u}_c^{(k)} + r\underline{u}_c^{(k+1/2)}].$$

Then we have

$$\underline{u}_c^{(k+3/4)} = \frac{1}{\det} *$$

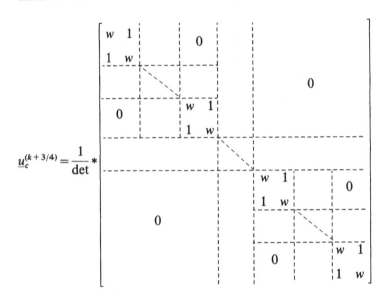

$$* \begin{bmatrix} u_1^{(k)} + u_{n+1}^{(k)} + ru_1^{(k+1/2)} \\ u_1^{(k)} + u_{n+1}^{(k)} + ru_{n+1}^{(k+1/2)} \\ \vdots \\ u_{n^2-2n+1}^{(k)} + u_{n^2-n+1}^{(k)} + ru_{n^2-2n+1}^{(k+1/2)} \\ u_{n^2-2n+1}^{(k)} + u_{n^2-n+1}^{(k)} + ru_{n^2-n+1}^{(k+1/2)} \\ \vdots \\ u_n^{(k)} + u_{2n}^{(k)} + ru_n^{(k+1/2)} \\ u_n^{(k)} + u_{2n}^{(k)} + ru_{2n}^{(k+1/2)} \\ \vdots \\ u_{n^2-n}^{(k)} + u_{n^2}^{(k)} + ru_{n^2-n}^{(k+1/2)} \\ u_{n^2-n}^{(k)} + u_{n^2}^{(k)} + ru_{n^2}^{(k+1/2)} \end{bmatrix}, \tag{8.6.13c}$$

where c stands for column-wise ordering.

Hence the values of $\underline{u}^{(k+3/4)}$ will be obtained from the above matrix–vector multiplication.

(iv) $(k+1)^{th}$ *time step* At the fourth and final level (the $(k+1)^{th}$ iterate), in a similar manner, equation (8.6.11d) is transformed to

$$\underline{u}_c^{(k+1)} = (G_2 + rI)^{-1}[G_2\underline{u}_c^{(k)} + r\underline{u}_c^{(k+3/4)}].$$

Then we have

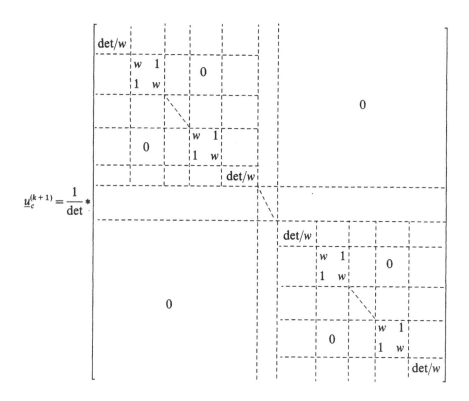

$$
\underline{u}_c^{(k+1)} = \frac{1}{\det} * \begin{bmatrix} u_1^{(k)} + ru_1^{(k+3/4)} \\ u_{n+1}^{(k)} + u_{2n+1}^{(k)} + ru_{n+1}^{(k+3/4)} \\ u_{n+1}^{(k)} + u_{2n+1}^{(k)} + ru_{2n+1}^{(k+3/4)} \\ \vdots \\ u_{n^2-n+1}^{(k)} + ru_{n^2-n+1}^{(k+3/4)} \\ \vdots \\ u_n^{(k)} + ru_n^{(k+3/4)} \\ u_{2n}^{(k)} + u_{3n}^{(k)} + ru_{2n}^{(k+3/4)} \\ u_{2n}^{(k)} + u_{3n}^{(k)} + ru_{3n}^{(k+3/4)} \\ \vdots \\ u_{n^2}^{(k)} + ru_{n^2}^{(k+3/4)} \end{bmatrix},
$$

(8.6.13d)

and the final values of $\underline{u}^{(k+1)}$ will be obtained from the above matrix–vector multiplication (Figure 8.6.4).

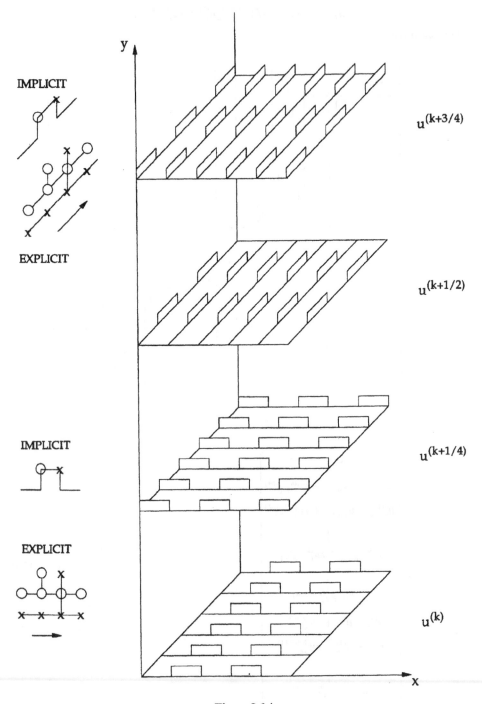

Figure 8.6.4

References

Abdullah, A. R., The Study of Some Numerical Methods for Solving Parabolic Partial Differential Equations, *Ph.D. Thesis*, Loughborough University of Technology, 1983.

Albasiny, E. H., The Solution of Non-Linear Heat-Conduction Problems on the Pilot ACE, *Proc. Inst. Elect. Engrs.*, B103, Suppl., No. 1, pp. 158–162, 1956.

Albrecht, J., Zum Differenzenverfahren bei Parabolischen Differentialgleichungen, *Z.A.M.M.*, 37, pp. 202–212, 1957.

Ames, W. F., *Nonlinear Partial Differential Equations in Engineering*, Academic Press, 1965.

Ames, W. F., *Numerical Methods for Partial Differential Equations*, 2nd Ed., New York, Academic Press, 1977.

Arms, R. J., Gates, L. D. and Zondek, B., A Method of Block Iteration, *J. SIAM*, 4, pp. 220–229, 1956.

Benson, A., The Numerical Solution of Partial Differential Equations by Finite Difference Methods, *Ph.D. Thesis*, Sheffield University.

Biringen, S., A Note on the Numerical Stability of Convection–Diffusion Equation, *J. Comp. and Appl. Maths.*, Vol. 7, No. 1, 1981.

Birtwistle, G. M., The Explicit Solution of the Equation of Heat Conduction, *Computer Journal*, Vol. 11, 1968.

Caldwell, J. and Smith, P., Solution of Burger's Equation with a Large Reynolds Number, *Appl. Math. Modelling*, Vol. 6, Oct. 1982, pp. 381–385.

Carasso, A. and Parter, S. V., An Analysis of Boundary Value Techniques for Parabolic Problems, *Mathematics of Computation*, 24, No. 110, 1970, pp. 315–340.

Carré, B. A., The Determination of the Optimum Accelerating Factor for Successive Overrelaxation, *Comp. J.*, 4, pp. 73–78, 1961.

Clark, J. and Barakhat, H. Z., On the Solution of the Diffusion Equation by Numerical Methods, *Jour. of Heat Transfer*, Trans., A.S.M.E., Series C, pp. 421–427, 1966.

Collatz, L., *Numerical Treatment of Differential Equations*, Springer, Berlin, 1960.

Conte, S. D., and Royster, W. C., Convergence of Finite-Difference Solutions to a Solution of the Equation of the Vibrating Rod, *Proc. Amer. Math. Soc.*, 7, No. 4, pp. 742–749, 1956.

Conte, S. D., A Stable Implicit Finite-Difference Approximation to a Fourth-Order Parabolic Equation, *J. Assoc. Computing Machinery*, Vol. 1, p. 111, 1954.

Crandall, S. H., An Optimum Implicit Recurrence Formula for the Heat Conduction Equation, *Quart. Appl. Math.*, 13, No. 3, pp. 318–320, 1955.

Crank, J. and Nicolson, P., A Practical Method for Numerical Integration of Solutions of Partial Differential Equations of Heat-Conduction Type, *Proc. Cambridge Philos. Soc.*, Vol. 43, p. 50, 1947.

Cuthill, E. H. and Varga, R. S., A Method of Normalised Block Iteration, *J. ACM.*, 6, pp. 236–244, 1959.

Dahlquist, G. G. and Bjorck, A., *Numerical Methods*, Prentice Hall Inc., 1974.

Danaee, A., A Study of Hopscotch Methods for Solving Parabolic Partial Differential Equations, *Ph.D. Thesis*, University of Technology, Loughborough, Leics., 1980.

Danaee, A. and Evans, D. J., A New Block Hopscotch Method for Solving Partial Differential Equations, *Conf. Num. Meth. for Coupled Problems*, Edit, E. Hinton, P. Bettess and R. W. Lewis, pp. 21–39, 1981, Pineridge Press.

Douglas, J., On the Numerical Solution of $\partial^2 u/\partial x^2 + \partial^2 u/\partial y^2 = \partial u/\partial t$ by Implicit Methods, *J. Soc. Industr. and Appl. Math.*, 3, No. 1, pp. 42–65, 1955.

Douglas, J., The Solution of the Diffusion Equation by a High Order Correct Difference Equation, *J. Math. Phy.*, 35, pp. 20–37, 1956.

Douglas, J., A *Survey of Numerical Methods for Parabolic Differential Equations*, Advances in Computers 2, pp. 1–54, New York, Academic Press, 1961.

Douglas, J., Alternating Direction Methods for Three Space Variables, *Numerische Mathematik*, 4, pp. 41–63, 1962.

Douglas, J. and Gunn, J. E., A General Formulation of Alternating Direction Method Part I, *Numerische Mathematik*, 6, pp. 428–453, 1964.

Douglas, J. and Jones, B. F., On Predictor-Corrector Methods for Non-linear Parabolic Differential Equations, *J. Soc. Indust. Appl. Math.*, 11, pp. 195–204, 1963.

Douglas, J. and Rachford, H. H., On the Numerical Solution of Heat-Conduction Problems in Two and Three Variables, *Trans. Amer. Math. Soc.*, Vol. 82, p. 421, 1956.

Dufort, E. C. and Frankel, S. P., Stability Condition in the Numerical Treatment of Parabolic Differential Equations, *Math. Tables and Other Aids to Computation*, Vol. 7, p. 135, 1953.

D'Yakonov, E. G., On the Application of Disintegrating Difference Operators, *Z. Vycisl. Mat. iMat, Fiz.*, 3, pp. 385–388, 1963.

Evans, D. J., A Stable Explicit Method for the Finite-Difference Solution of a Fourth Order Parabolic Partial Differential Equation, *Computer Journal* 8, No. 3, pp. 280–287, 1965.

Evans, D. J., Simplified Implicit Methods for the Finite Difference Solution of Parabolic and Hyperbolic Partial Differential Equations, *J. IMA*, 2, pp. 1–13, 1966.

Evans, D. J., On the Solution of Certain Skew-Symmetric Quindiagonal Linear Systems, *Int. J. Comp. Maths.*, Vol. 8, pp. 271–284, 1980.

Evans, D. J., Group Explicit Iterative Methods for Solving Large Linear Systems, *Int. J. Comp. Maths.*, 17, pp. 81–108, 1985.

Evans, D. J. and Abdullah, A. R. B., *A New Explicit Method for the Solution of* $\partial u/\partial t = \partial^2 u/\partial x^2 + \partial^2 u/\partial y^2$, *Int. J. Comp. Maths.*, 14, pp. 325–353, 1983.

Evans, D. J. and Abdullah, A. R. B., Group Explicit Methods for Parabolic Equations, *Int. J. Comp. Maths.*, 14, pp. 73–105, 1983.

Evans, D. J. and Abdullah, A. R. B., The Group Explicit Method for the Solution of Burger's Equation, *Computing*, 32, pp. 239–253, 1984.

Evans, D. J. and Benson, A., The Successive Peripheral Block Overrelaxation Method, *J. Inst. Math. Applic.*, 9, pp. 68–69, 1986.

Evans, D. J. and Danaee, A., A New Group Hopscotch Method for the Numerical Solution of Partial Differential Equations, *SIAM J. Numer. Anal.*, Vol. 19, No. 3, pp. 588–598, 1982.

Evans, D. J. and Sahimi, M. S., The Alternating Group Explicit (AGE) Iterative Method for Solving Parabolic Equations, 1-2 Dimensional Problems, *Int. J. Comp. Maths.*, 24, pp. 250–281, 1988.

Evans, D. J. and Yousif, W. S., Explicit Group Overrelaxation Methods for Partial Differential Equations, *Maths. and Comp. in Sim.*, 28, pp. 453–466, 1986.

Fairweather, G. and Gourlay, A. R., Some Stable Difference Approximations to a Fourth-Order Parabolic Partial Differential Equation, *J. Mathematics of Computation*, Vol. 21, pp. 1–11, 1966.

Fairweather, G. and Mitchell, A. R., A High Accuracy Alternating Direction Method for the Wave Equation, *J. Inst. Math. Applics.*, 1, pp. 309–316, 1965.

Forsythe, G. E. and Wasow, W. R., *Finite Difference Methods for Partial Differential Equations*, New York, John Wiley, 1960.

Fox, L., *Numerical Solution of Ordinary and Partial Differential Equations*, Pergamon Press, 1962.

Frank, W., Solution of Linear Systems by Richardson's Method, *J. ACM.*, 7, pp. 274–286, 1960.

Frankel, S. P., Convergence Rates of Iterative Treatments of Partial Differential Equations, *M.T.A.C.*, 4, pp. 65–75, 1950.

Friedman, A., *Partial Differential Equations of Parabolic Type*, Prentice-Hall, 1964.

Fröberg, C. E., *Introduction to Numerical Analysis*, 2nd Ed., Addison–Wesley Pub. Co., Reading, Mass., 1969.

Gerald, C. F., *Applied Numerical Analysis*, 2nd Ed., Addison–Wesley, p. 435, 1978.

Gladwell, I. and Wait, R., *A Survey of Numerical Methods for Partial Differential Equations*, Clarendon Press, 1979.

Gault, R. J., Hoskins, R. F., Milner, J. A. and Pratt, M. J., *Computational Methods in Linear Algebra*, London, Stanley Thornes Pubs., 1974.

Gourlay, A. R., Hopscotch: A Fast Second-Order Partial Differential Equation Solver, *J. Inst. Maths. Applics.*, 6, pp. 375–390, 1970.

Gourlay, A. R., Splitting Methods for Time Dependent Partial Differential Equations, Paper presented in the *State of the Art in Numerical Analysis*, edit. D.E.H. Jacobs, p. 761, 1977.

Gourlay, A. R. and McGuire, G. R., General Hopscotch Algorithm for the Numerical Solution of Partial Differential Equations, *J. Inst. Maths. Applics.*, Vol. 7, pp. 216–227, 1971.

Gourlay, A. R. and McKee, S., The Construction of Hopscotch Methods for Parabolic and Elliptic Equations in Two Space Dimensions with a Mixed Derivative, *J. of Computational and Appplied Maths.*, 3, No. 3, pp. 201–206, 1977.

Greenspan, D., *Discrete Numerical Methods in Physics and Engineering*, Academic Press, 1974.

Greig, I. S. and Morris, J. L., A Hopscotch Method for the Korteweg-de-Vries Equation, *J. Comp. Phys.*, 20, pp. 64–80, 1976.

Guittet, J., Une Novelle Methode de Directions Alterneés à q Variables, *Jour. Math. Anal. and Appl.*, 17, pp. 199–213, 1967.

Hockney, R. W., A Fast Direct Solution of Poisson's Equation using Fourier Analysis, *J.A.C.M.*, 12 pp. 95–113, 1965.

Isaacson, E. and Keller, H. B., *Analysis of Numerical Methods*, John Wiley and Son, 1966.

Jain, M. K., *Numerical Solution of Differential Equations*, Wiley Eastern Ltd., 1979.

Kahan, W., Gauss-Seidel Methods for Solving Large Systems of Linear Equations, *Ph.D.*, Univ. of Toronto, Canada, 1958.

Kellog, R. B., An Alternating Direction Method for Operator Equations, *J. Soc. Indus. Appl. Math.*, Vol. 12, No. 4, pp. 848–854, 1964.

Larkin, B. K., Some Stable Explicit Difference Approximations to the Diffusion Equation, *Math. Comp.*, Vol. 18, p. 196, 1964.

Lax, P. and Richtmyer R. D., Survey of the Stability of Linear Finite Difference Equations, *Comm. Pure Appl. Math.*, Vol. 9, p. 267, 1956.

Lees, M., Alternating Direction Method for Hyperbolic Differential Equations, *J. Soc. Indust. Appl. Math.*, 10, pp. 516–522, 1962.

Lees, M., A Linear Three Level Difference Scheme for Quasilinear Parabolic Equations, *Maths. of Computation*, 20, pp. 516–522, 1966.

Lieberstein, H. M., *A Course in Numerical Analysis*, Harper and Row, New York, 1968.

Liu, S. L., Stable Explicit Difference Approximations to Parabolic Partial Differential Equations, *AICh. E. Journal*, Vol. 15, No. 3, pp. 334–338, May 1969.

Lowan, A. N., The Operator Approach to Problems of Stability and Convergence of Solutions of Difference Equations and the Convergence of Various Iteration Procedures, *Scripta Mathematica Studies*, New York, 1957.

Madsen, N. K. and Sincovec, R. F., General Software for Partial Differential Equations in *Numerical Methods for Differential System*, Ed., L. Lapidus and W. E. Schiesser, Academic Press, Inc., pp. 229–242, 1976.

McGuire, G. R., Hopscotch Methods for the Solution of Linear Second Order Parabolic Partial Differential Equations, *M.Sc. Thesis*, Dundee University, 1970.

Meis, T. and Marcowitz, U., *Numerical Solution of Partial Differential Equations*, Springer–Verlag, 1981.

Mikhlin, S. G., *Linear Equations of Mathematical Physics*, Holt, Rinehart and Winston, Inc., 1967.

Milne, W. E., *Numerical Solution of Differential Equations*, J. Wiley, New York, 1953.

Mitchell, A. R., *Computational Methods in Partial Differential Equations*, John Wiley & Sons., 1969.

Mitchell, A. R. and Griffiths, D. F., *The Finite Difference Method in Partial Differential Equations*, John Wiley and Sons, 1980.

Morton, K. W., Stability of Finite Difference Approximations to a Diffusion Convection Equation, *Int. J. for Num. Meth. in Engn.*, Vol. 15, pp. 677–683, 1980.

Noye, J., *Numerical Simulation of Fluid Motion*, Ed. J. Noye, North-Holland Pub. Co., 1978.

O'Brian, G. G., Hyman, M. A. and Kaplan, S., Numerical Solution of Partial Differential Equations, *J. Math. Phys.*, 29, pp. 223–251, 1951.

Papamichael, N. and Whiteman, J. R., A Cubic Spline Technique for the One-Dimensional Heat Conduction Equation, *JIMA*, 11, pp. 111–113, 1973.

Parter, S. V., Multiline Iterative Methods for Elliptic Difference Equations and Fundamental Frequencies, *Num. Math.*, 3, pp. 305–319, 1961.

Peaceman, D. W. and Rachford, H. H., The Numerical Solution of Parabolic and Elliptic Differential Equations, *J. SIAM*, Vol. 3, p. 28, 1955.

Peaceman, D. W. and Rachford, H. H., Numerical Calculation of Multi-Dimensional Miscible Displacement, *Soc. Petroleum Eng. Journal*, 24, pp. 327–338, 1962.

Peaceman, D. W., Fundamentals of Numerical Reservoir Simulation, *Elsevier Scientific Pub. Co.*, Amsterdam, New York, 1977.

Price, H. S., Varga, R. S. and Warren, J. E., Application of Oscillation Matrices to Diffusion-Convection Equations, *J. Math. and Physics*, 45, p. 301, 1966.

Ralston, A., *A First Course in Numerical Analysis*, McGraw Hill, New York, 1965.

Ramos, J. I., Numerical Solution of Reactive–Diffusive Systems, Part 1: Explicit Methods, *Int. J. Comp. Maths.*, 18, pp. 43–66, 1985. Part 2: Methods of Lines and Implicit Algorithms, *Int. J. Comp. Maths.*, 18, pp. 141–162, 1985. Part 3: Time Linearisation and Operator–Splitting Techniques, *Int. J. Comp. Maths.*, 18, pp. 289–310, 1985.

Richardson, L. F., The Approximate Arithmetical Solution by Finite Differences of Physical Problems Involving Differential Equations, with an Application to the Stresses in Masonry Dam, *Philos. Trans. Roy. Soc.*, London, Ser. A210, pp. 307–357, 1910.

Richtmyer, R. D. and Morton, K. W., *Difference Methods for Initial-Value Problems*, New York, Interscience, 1967.

Samarskii, A. A., Local One Dimensional Difference Scheme for Multi-Dimensional Hyperbolic Equations in an Arbitrary Region, *Z. Vycisl. Mat. Imat. Fiz.*, 4, pp. 21–35, 1964.

Saul'yev, V. K., *Integration of Equations of Parabolic Type by the Method of Nets*, G.J. Tee, Transl., Pergamon, New York, 1964.

Smith, G. D., *Numerical Solution of Partial Differential Equation*, 2nd Ed., Oxford University Press, 1978.

Stewart, G. W., *Introduction to Matrix Computations*, Academic Press, 1973.

Stone, H. L. and Brian, P. L. T., Numerical Solution of Convective Transport Problems, *A.I.Ch.E. Journal*, Vol. 9, No. 5, 1963.

Thomas, L. H., Elliptic Problems in Linear Difference Equations Over a Network, *Watson Scientific Computing Laboratory*, Columbia University, New York, 1949.

Varga, R. S., *Matrix Iterative Analysis*, Prentice-Hall Inc., Englewood Cliffs, New Jersey, 1962.

Von Rosenberg, D. V., *Methods for the Numerical Solution of Partial Differential Equations*, American Elsevier, New York, 1969.

Wachspress, E. L., *Iterative Solution of Elliptic Systems and Applications to the Neutron Diffusion Equation of Reactor Physics*, Prentice Hall, New Jersey, 1966.

Yanenko, N. N., *The Method of Fractional Steps for Multidimensional Problems in Mathematical Physics*, 1971.

Young, D., On the Richardson's Method for Solving Linear Systems with Positive Definite Matrices, *J. Math. Phys.*, 32, pp. 243–255, 1954.

Young, D. M., *Iterative Solution of Large Linear Systems*, Academic Press, New York, 1971.

Zienkiewics, O. C., *The Finite Element Method in Engineering Sciences*, McGraw-Hill, New York, 1971.

Index

Printed and bound by CPI Group (UK) Ltd, Croydon, CR0 4YY

23/10/2024

01777679-0002